W9-BVH-448

Small Gasoline Engines

Service and Repair

by Rex Miller and Mark Richard Miller

An Audel® Book

Macmillan Publishing Company
New York

Collier Macmillan Canada
Toronto

Maxwell Macmillan International
New York Oxford Singapore Sydney

THIRD EDITION

Macmillan Publishing Company
866 Third Avenue
New York, NY 10022

Maxwell Macmillan Canada, Inc.
1200 Eglinton Avenue East, Suite 200
Don Mills, Ontario M3C 3N1

Macmillan Publishing Company is part of the Maxwell Communication Group of Companies.

Library of Congress Cataloging-in-Publication Data

Miller, Rex, 1929–
Small gasoline engines: service and repair / by Rex Miller and Mark Richard Miller.
p. cm.
Includes index.
ISBN 0-02-584991-3
1. Internal combustion engines, Spark ignition—Maintenance and repair. I. Miller, Mark Richard. II. Title.
TJ790.M552 1993
621.43'4'0288—dc20 92-8901
CIP

10 9 8 7 6 5 4 3 2 1

Printed in the United States of America

Contents

Preface

The purpose of this book is to supply the practical information needed to repair and service small gasoline engines. This information will enable you to understand and work with small gasoline engines in the many places they are used.

The book was designed for students in vocational classes as well as for people who would like to know more about repairing their own small engines. Although it contains some theory of operation, its main purpose and emphasis lie in the practical, everyday application of repair principles, practices, and procedures.

Many illustrations are included showing a variety of parts and techniques found in present-day practice in the field. Obviously, not all related problems can be presented here since there is a great deal of ingenuity required by the person on the job. We have followed standard procedures as recommended by the manufacturers of the engines.

In order to familiarize the reader with the metric measurement system (although it may be some years before it becomes a standard in the United States), symbols approved by the International Organization for Standardization are used where applicable throughout the book. They belong to the SI—the International System of Units—and are given in parentheses, following data given in customary units.

It is the hope of the authors that the book will serve as an appetizer to those really interested in going into the field of small-gasolineengine service and repair.

REX MILLER
MARK R. MILLER

Acknowledgments

As in the case of any book, many people have given generously of their time in assisting the authors of *Small Gasoline Engines.* Their efforts, time, and suggestions have been of great value. We would like to take this opportunity to thank all of them for their contributions.

Many manufacturers of parts, engines, and equipment have contributed illustrations, written material, and suggestions for inclusion in this book, and they have done so with a great deal of professionalism.

Manufacturers have been identified with each of the illustrations furnished. The list below gives their complete names. These manufacturers and agencies are but a sampling of the many who make the small gasoline engine the handy worksaver it is. In addition, the technical data furnished by these companies and agencies are much appreciated. Without their assistance this project could not have been objective or technically accurate.

We would like especially to thank Briggs & Stratton and Tecumseh for their wholehearted cooperation and contributions. When reading the book, you will see that without them the book would be very weak on technical information.

American Honda Motor Co., Inc. Power Equipment Division
Briggs & Stratton Corporation
Chrysler Corporation, Marine Product Group

Frederick Manufacturing Company (Silver Streak)
Johnson Motors, Outboard Marine Division
Kioritz Corporation
Lisle Corporation
Onan Corporation
Snap-on Tools Corporation
Tecumseh Products Co. (Lauson Power Products Engine Division)

SECTION ONE

Basic Principles

CHAPTER 1

Safety

In this chapter you will be introduced to the proper way of handling tools, gasoline, carbon monoxide, and test equipment. Each of these topics is important for your personal safety and well-being. The attitude you have toward the work you are doing makes a great deal of difference in whether or not you will be able to do this work safely and without harm to yourself, the environment, or others. Some aspects of this work with small engines demand your full attention. You can be burned, poisoned by fumes, and injured by moving or rotating equipment.

Your attitude and your ability to keep your mind on your work are important in this type of instruction. Pay close attention to the work rules and follow them religiously to prevent damage to the equipment and, most important, yourself.

Attitudes

Your attitude is important when you are working in any type of shop activity. There are tools and equipment that must be respected for what they can do and what they will do if improperly used. You are the key to a safe working area. You must have a place to put your tools and equipment. A tool must be placed out of the way so it is not tripped over or knocked down at the wrong time. It must also be located so it can be reached when needed. To keep your work area clean and neat, you must take time before and after a job for these housekeeping duties. Those who ignore this type of work will be spending more time looking for a wrench or a part than it takes to do the job. Carelessness is a very expensive proposition. The way you view the situation is part of your attitude. If you start with good work habits and maintain them even when doing a rush job, you and the person for whom you're doing the job will benefit.

Working with small engines is not dangerous, if you think about safety. "Prevent accidents before they happen" is more than just a slogan. It can be done. Just ask yourself if what you're doing is safe or if it is just a quick way to get the job done.

Just remember that your attitude makes the difference. It makes the difference in whether the job gets done. It makes the difference in whether the job is done correctly. It makes the difference in whether you work safely or skin your knuckles, cut your fingers, burn your arms on a muffler, or any of a dozen things that can hamper your doing the job well and quickly, and without incident. You are the key to a safe working environment. Keep a positive outlook and take a little time to do what must be done to make your job a safe job.

Handling Substances Safely

Gasoline

Make sure gasoline is stored in a metal safety can (Fig. 1–1). The can should have a spring-closed lid. Fumes are easily ignited, and any opening in the can will emit fumes. One spark and an explosion or fire will result. Do not use plastic bottles or glass jars for gasoline storage.

Do not use gasoline-soaked rags to clean an engine. Refueling hot engines is not advised. Make sure the engine is cool before you refuel. Use a funnel to ensure that gasoline does not drip onto any part of the engine.

Place rags wet with oil, gasoline, paint, or solvents in a safety container with a lid (Fig. 1–2). The container should be labeled appropriately—"Gasoline," "Rags," or "Solvent." The color of the can should be red to draw attention to it. Make sure it is emptied each day.

Fig. 1–1. A proper size and safety-enclosed can should be used to store gasoline. Can should also be labeled "Gasoline."

All combustible liquids should be stored in a fireproof room or cabinet (Fig. 1–3). Label the cabinet or room "FLAMMABLE—Keep Fire Away." A fire extinguisher should be placed near the flammable area. It should be the proper type of extinguisher for the type of fire expected at that location (Fig. 1–4).

Fig. 1–2. Old rags and oily materials should be kept in a metal container with the proper label. The container should be emptied each day.

Fig. 1–3. Gasoline and flammable materials should be stored in a safe room or a properly-marked storage cabinet.

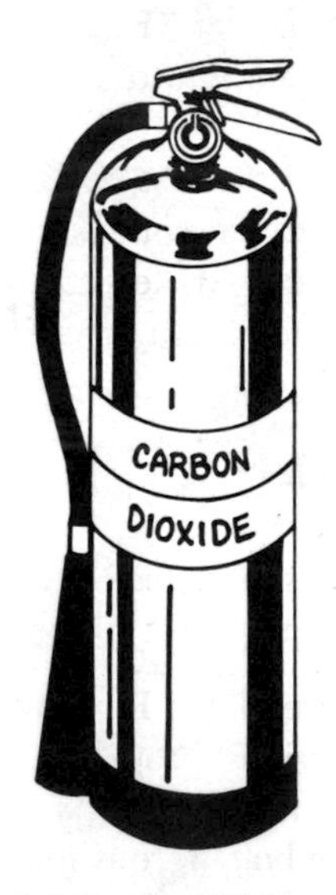

Fig. 1–4. The fire extinguisher should be painted red and marked with its contents and the type of fire it is to be used on.

Compressed Air

Compressed air can be dangerous when used improperly. However, it is a useful tool when properly put to the task. There are a few safety precautions that need reviewing before you take the hose and use it.

1. Check the hose connections before you turn on the air.
2. Hold the air hose nozzle when you turn it on or off. Don't allow it to slip while you are turning it on.
3. Wear safety glasses or goggles when using compressed air.
4. Do not use compressed air to clean your clothes, the floor, or the workbench. If compressed air is directed toward an open cut in the skin, it can remove large sections of skin, causing very dangerous infections. Be careful when using the hose.
5. Don't place the hose on the tabletop or on the floor while it still has the air on. It is possible for the hose to whip around and strike someone or damage other tools and equipment.

Carbon Monoxide

Carbon monoxide is a by-product of internal combustion, produced by incomplete combustion (Fig. 1–5). Incomplete combustion takes place with a limited amount of air—adding more air produces carbon dioxide.

Carbon monoxide is a colorless and odorless gas, so you do not know when it is present. Carbon monoxide is very dangerous to human beings as it can suffocate you. Make sure you work with running engines only in a well-ventilated room or, preferably, *outdoors*.

Using Hand Tools Safely

When using hand tools it is best to follow a few simple rules to protect both you and the equipment on which you are working.

1. Be sure your hands are clean. Remove the grease and oil before using a tool so that it will fit properly and not slip.
2. Make sure you have the proper size tool for the job with the proper fit on the nut or bolt. Screwdrivers should fit the slot on the screwhead. Leverage should be obtained by the proper tool, not by adding an extension of pipe or some other device.

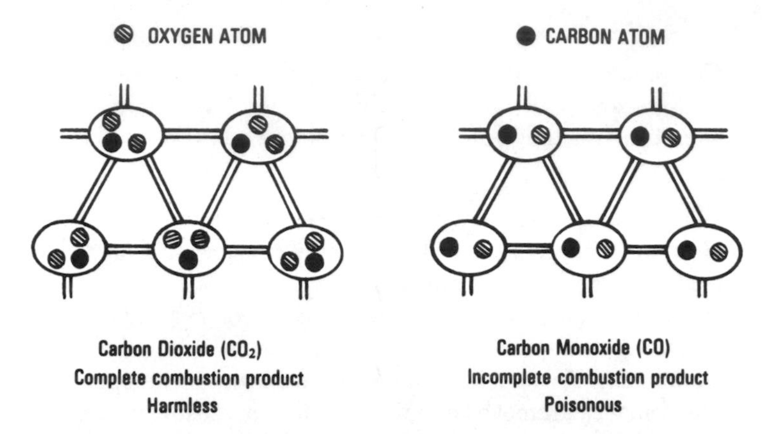

Fig. 1–5. Carbon dioxide is a normal by-product of combustion when it occurs in an atmosphere where plenty of air is present. Carbon monoxide is generated wherever combustion takes place with limited amounts of oxygen. Note how the carbon atoms combine with oxygen to produce each of the gases. Both are odorless and colorless. Carbon dioxide is harmless unless a complete room is filled with it. Carbon monoxide is very poisonous to humans.

3. If the tools are sharp-edged or pointed, make sure they are respected and the edges and points are used properly.
4. When you use a pointed or sharp-edged tool, make sure it is pointed away from your body. Don't allow it to slip and strike someone nearby.
5. Wear a face shield, goggles, or safety glasses when filing or cutting metal (Figs. 1–6 and 1–7). Don't allow flying chips to strike others standing or working nearby.
6. When passing hand tools to someone, make sure you pass them with the handles first.
7. Clamp small work in a vise. Don't work on a piece of equipment unless it is properly secured.

Using Test Equipment Safely

In order for test equipment to give you reliable readings, it has to be functioning properly. For it to function properly, it has to be treated

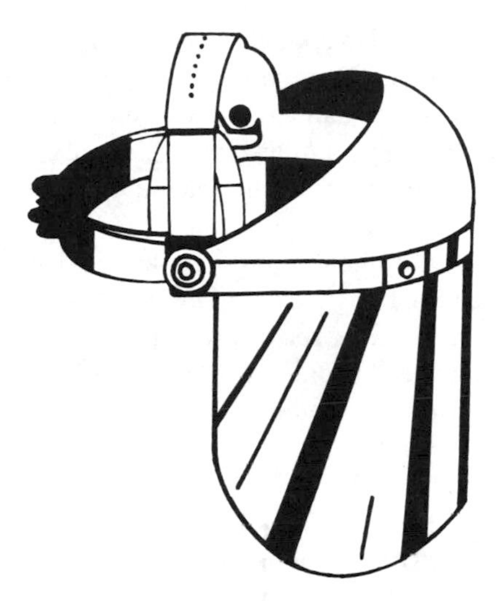

Fig. 1–6. A face shield is very good protection when working around flying chips and other small articles.

with care. Check the instruction manual that comes with the equipment to determine proper maintenance and use.

Test equipment can include a number of devices. One of the most useful is the ohmmeter. Make sure the ohmmeter is used with the circuit power *off*. It has its own power supply.

The ammeter used to test generator output should be chosen to fit the output of the generator. Make sure it is placed in series with the circuit being tested.

Fig. 1–7. Safety glasses are made of tempered glass and protect your eyes from flying chips and other objects.

The voltmeter is used to check the output of the generator. Do not use the voltmeter to check the voltage output of the magneto. The magneto's high voltage can damage the voltmeter. If the voltmeter has been properly protected by a *high-resistance probe,* it can be used to check the high-voltage output of the magneto. If the special probe is not available, do not test the spark voltage.

Electrical test equipment should be handled with care and not thrown into a toolbox. Rough treatment can damage the equipment and affect the accuracy of the readings.

Compression testers should also be handled with care. They should not be thrown into a toolbox, but replaced gently into their own protective boxes.

Working Safely with Small Engines

In order to operate the small gasoline engine, it is necessary to be aware of some of its limitations and possible dangers. Some good rules to follow are:

1. Check the engine for any loose connections or loose nuts and bolts.
2. Check with the owner to make sure all parts of the engine are in proper operating condition.
3. Check the fuel line for leaks.
4. If the engine is run indoors, make sure there is proper ventilation to prevent carbon monoxide poisoning.
5. Keep your head and your hands away from the running engine. Moving parts may injure you or catch on your hair or clothing.
6. Wear a face shield, goggles, or safety glasses. Don't forget to protect your hearing with ear protectors or headsets (Fig. 1–8).
7. Don't run the engine at high speeds for any period of time.
8. Keep in mind that mufflers get very hot while running. Keep all fuel and your hands away from the mufflers until they cool. The engine also becomes hot, so approach it cautiously.

By using common sense you should be able to work with the small gasoline engine with safety. The amount of damage you cause yourself

Fig. 1–8. A gasoline engine makes a lot of noise when running. In order to protect your hearing it is best to wear ear protectors. They resemble headsets but keep out the extra-loud noises.

or others will depend upon your following these simple rules. Keep the shop in order. Be alert. And don't allow yourself to become so involved that you are not aware of dangerous situations.

The small gasoline engine can be a source of pleasure and satisfying work. It can be maintained easily and inexpensively. Don't make it too dangerous and too costly by relaxing your safety consciousness. Always be alert!

The Safety Brake

Since June 30, 1982, manufacturers of lawnmowers have been required to provide some way of stopping the engine, especially the blade of the lawnmower, within 3 seconds of the time the control handle is released. This is to protect the operator, who may inadvertently place a foot or hand under the lawnmower while the blade is turning. A member of our family lost five toes under a lawnmower in the days before the automatic stopping requirement. The safety brake is sometimes referred to as the *dead man's clutch,* because if a person had a stroke or a heart attack and lost control of the mower or snow blower, it could do a great deal of damage.

Tecumseh engines use two flywheel braking methods to comply with the law of June 30, 1982. The flywheel brake systems have one braking system located on the inside edge of the flywheel and the other on the outside edge of the flywheel. Both locations can be used to provide safety by shutting down the engine and lawnmower blade within seconds after the operator releases an engine/blade control at the handle of the lawnmower. Tecumseh's methods will be discussed in detail in a later chapter in this book.

Another method used for braking the blade only utilizes a clutch in conjunction with either a top- or side-mounted recoil starter or 12-volt electric starter. The blade stops within 3 seconds after the operator lets go of the blade control bail at the operator position. The engine continues to run. The starter rope handle is on the engine.

The other method uses a recoil starter, top- or side-mounted, with the rope handle on the engine instead of within 24 inches of the operator position. This method is acceptable if the mower deck passes the 360-degree foot probe test. A specified foot probe must not contact the blade when applied completely around the entire blade housing. This alternative can be used with engine-mounted brake systems and typical bail controls. The blade stops within three seconds after the operator releases the blade control bail at the operator position and the engine is stopped.

Chapter Highlights

1. Attitude is important when working with engines.
2. Housekeeping duties are necessary in repair facilities.
3. Gasoline must be properly stored.
4. Old rags and inflammable materials must be stored in metal containers.
5. Compressed air can be dangerous when improperly used.
6. Wear safety glasses at all times in the shop.
7. Carbon monoxide can kill.
8. Proper ventilation is needed for working with engines.
9. Simple rules must be followed when using hand tools.
10. Test instruments must be handled with care.
11. Magnetos put out a high voltage.
12. Mufflers get very hot when the engine is running.

13. Protect your hearing around running engines.
14. Keep your hands away from the running engine.
15. The dead man's clutch is there for a reason, but allow time for the blade to stop before looking under the mower.
16. Tecumseh engines use two methods to stop rotating blades when the engine is turned off.
17. A clutch method is used to stop the blade while the engine continues to run on some mowers.
18. The starting rope handle has to be located within 24 inches of the operator position.

Vocabulary Checklist

attitude
carbon dioxide
combustible
consciousness
dead man's clutch
face shield
flywheel brake methods
fumes
goggles
leverage
magneto
muffler
ohmmeter
resistance
safety brake
solvent
ventilation

Review Questions

1. Why do you need a metal box or cabinet to store flammable materials?
2. What are the important things to remember when handling gasoline?
3. What should be worn when working with compressed air?
4. Why is carbon monoxide dangerous?
5. How should hand tools be kept?
6. How should a hand tool be passed from one person to another?
7. How must an ohmmeter be used?
8. What is a voltmeter? How can it be damaged?
9. How should test equipment be handled?

10. Why is it necessary to be careful around a running engine?
11. What is the purpose of the safety brake?
12. How does the safety brake work?
13. What is a dead man's clutch?
14. When did manufacturers of lawnmowers start providing a method of stopping the rotating blade?

CHAPTER 2

Hand Tools

The small-engine repairperson has to work with a number of tools. Some of them are of the everyday variety, but others may be special-purpose types. Special-purpose tools have a specific function in regard to a certain machine or device. Some electrical work is also included in small-engine repair. Special tools for electrical troubleshooting and repair are therefore necessary.

What follows is a brief discussion of the more important tools used by the small-engine repairperson.

General Tools for the Shop

Pliers and Clippers

Pliers come in a number of sizes and shapes designed for special applications. They are available with either insulated or uninsulated handles. Although pliers with insulated handles are always used when working on or near "hot" wires, they must not be considered sufficient protection alone. Other precautions must be taken.

Long-nose pliers are used for close work in panels or when working with small enclosures. *Slip-joint,* or *gas, pliers* are used to tighten locknuts or small nuts (Fig. 2–1).

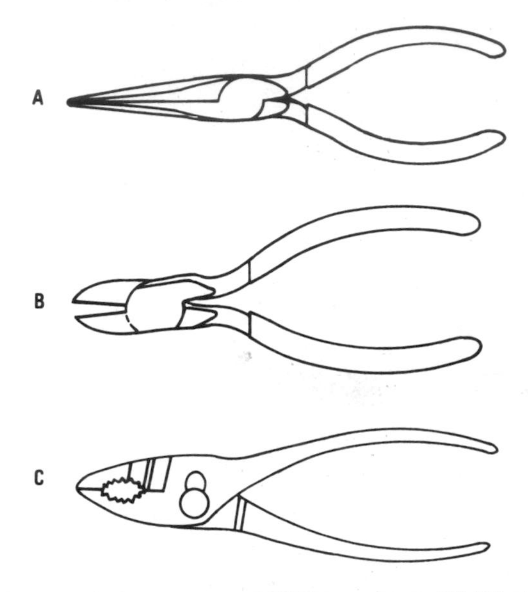

Fig. 2–1. Pliers. (A) Long-nose. (B) Wire cutters. (C) Slip-joint or gas pliers.

Wire cutters are used to cut wire to size. They come in handy for working with the electrical system of the small engine.

Screwdrivers

Screwdrivers are made in many sizes and tip shapes. Those used by electricians and small-engine repairpersons should have insulated handles. One variation of the screwdriver is the *screwdriver bit.* It is held in a brace and used for heavy-duty work. For safe and efficient use, screwdriver tips should be kept square and sharp and should be selected to match the screw slot (Fig. 2–2).

The *Phillips-head screwdriver* has a tip pointed like a star. It is designed to be used with a Phillips screw. These screws are commonly found in production equipment. The presence of four slots, rather than one, ensures that the screwdriver will not slip in the head of the screw. There are a number of sizes of Phillips-head screwdrivers. They are designated as No. 1, No. 2, and so on. The proper point size must be used to prevent damage to the slot in the head of the screw (Fig. 2–3).

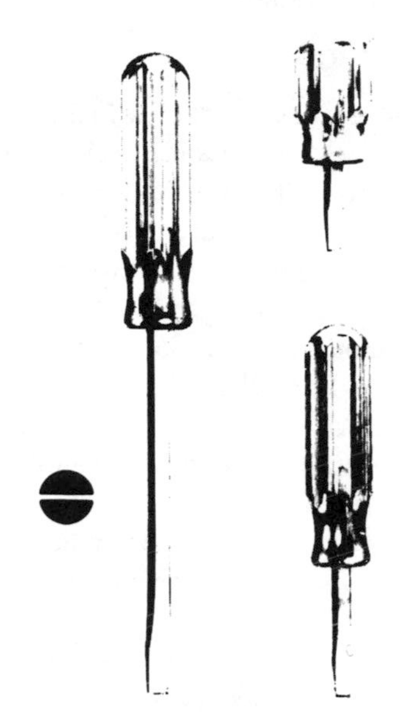

Fig. 2–2. Square-blade screwdrivers.

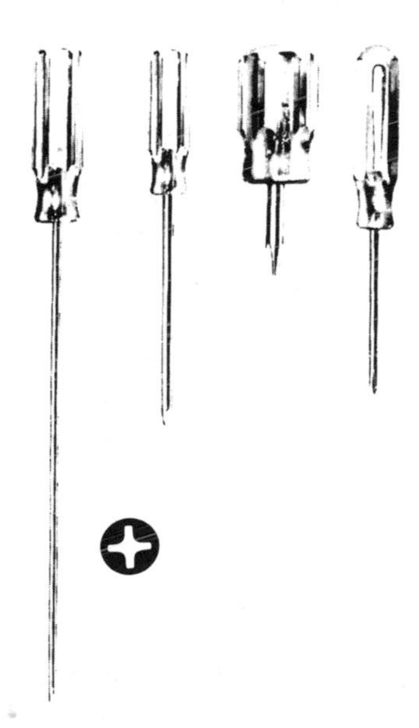

Fig. 2–3. Phillips-head screwdrivers.

Wrenches

Three types of wrenches are shown in Fig. 2–4. The adjustable open-end wrenches are *crescent wrenches. Monkey wrenches* are used on hexagonal and square fittings such as machine bolts, hexagon nuts, or larger fasteners. *Pipe wrenches* are used for pipe and conduit work. They should not be used where crescent or monkey wrenches can be used. Their construction will not permit the application of heavy pressure on square or hexagonal objects. Continued misuse of these tools in this manner will deform the teeth on the jaw faces and mar the surfaces of the material being worked.

Never use an extension on a wrench handle. If possible, always pull on a wrench handle and adjust your stance to prevent a fall if something lets go. Never use a hammer on any wrench other than a striking face wrench. Discard any wrench with broken or battered points.

Never use hand sockets on power or impact wrenches. Select the right size socket for the job. Select only heavy-duty impact sockets for use with air or electric impact wrenches. Replace sockets showing cracks or wear. Keep sockets clean (Fig. 2–5).

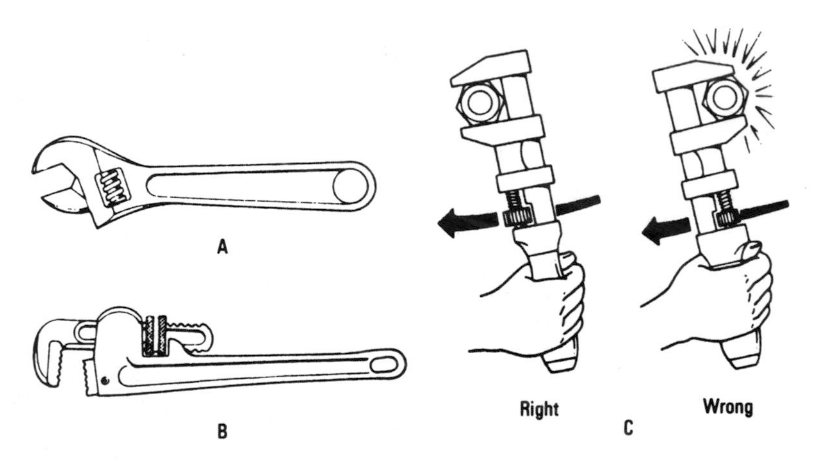

Fig. 2–4. Wrenches. (A) Crescent wrench. (B) Pipe wrench. (C) Monkey wrench.

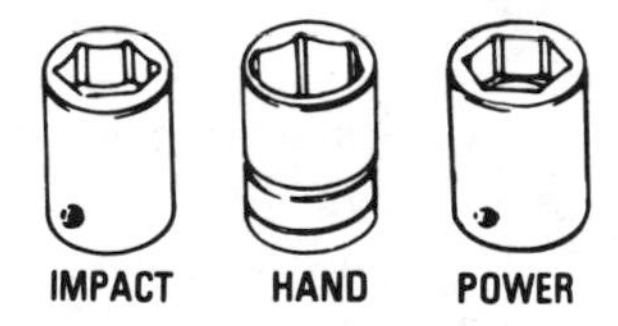

Fig. 2–5. Sockets for a hand wrench, an impact wrench, and a power wrench. *(Courtesy Snap-on Tools)*

Hammers

Hammers are used either in combination with other tools such as chisels, or in nailing equipment to building supports. Fig. 2–6 shows a carpenter's *claw hammer* and a machinist's *ball-peen hammer*.

The ball-peen hammer is not used to hammer nails or other similar jobs. The claw hammer is used on nails. The ball-peen is used for small-engine repair.

Never strike one hammer against a hardened object such as another hammer (Fig. 2–7). Always grasp a hammer handle close to the end. Strike the object with the full force of the hammer. Never work with a hammer having a loose head. Discard any hammer when the face chips or mushrooms.

Never use a *punch* or *chisel* which is chipped or mushroomed with a hammer (Fig. 2–8). If possible, hold a chisel or punch with a tool holder. When using a chisel on a small piece, clamp the piece firmly in a vise and chip toward the stationary jaw. Wear approved eye protection when using percussion tools.

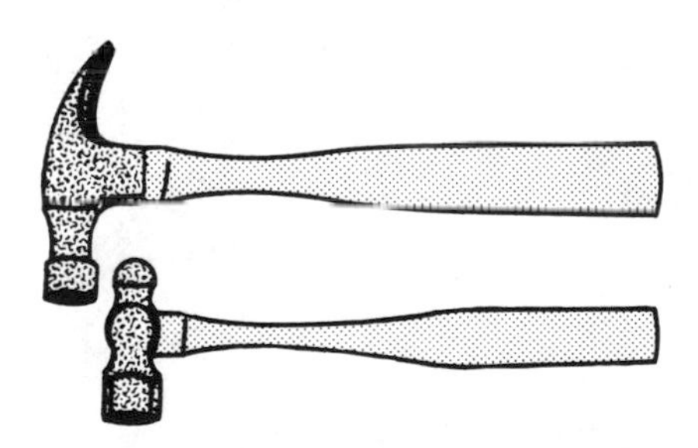

Fig. 2–6. Claw and ball-peen hammers.

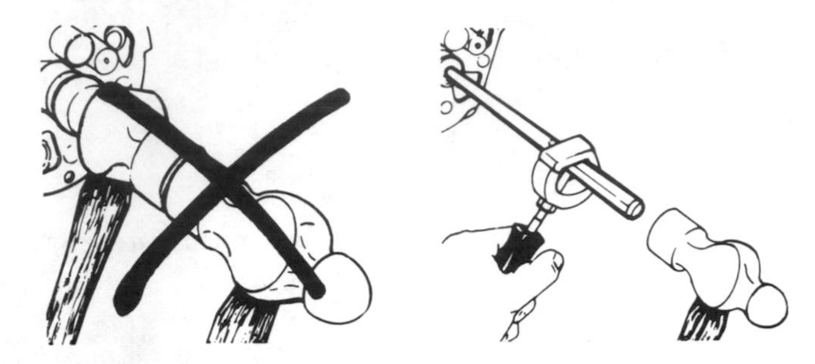

Fig. 2–7. Never strike one hammer against the face of another hammer. *(Courtesy Snap-on Tools)*

Extension-Cord Lamp

The *extension-cord lamp* shown in Fig. 2–9 normally is supplied with a long extension cord. It is used by the repairperson when there is insufficient light for the area being worked on. Heavy-duty or service-duty light bulbs are made for the rough handling these lamps get in daily use.

Special Tools for Small-Engine Repair

There are many special-purpose tools designed for specific jobs in the small-engine-repair trade. The intent of tool design is to make jobs easier and faster. Those who are in the business or trade are very much

Fig. 2–8. Never use a punch or chisel with a mushroomed end. File down the mushroomed pieces. *(Courtesy Snap-on Tools)*

Fig. 2–9. Extension-cord lamp.

aware of time and cost factors. These tools often make a difference between a profit and a loss for the small business.

Commonly Used Tools

Briggs & Stratton makes the most common small engines. The company suggests the following tools to make the job of repair and maintenance easier and quicker.

Spark Tester This small unit prevents your getting shocked when testing the quality of the spark generated by the magneto. Some models should not have the spark tester used on them (Fig. 2–10).

Remove the spark plug. Spin the flywheel rapidly, with one end of the ignition cable clipped to the spark tester. With the other end of the tester grounded on the cylinder head you have a completed setup. If the spark jumps the 0.166″ (4.2 mm) tester gap, you may assume the ignition system is functioning satisfactorily.

Another use of the tester is when the engine runs but misses during operation. A quick check to determine if ignition is or is not at fault can be made by inserting the spark tester between the ignition cable and the spark plug. A spark miss will be readily apparent. While conducting this test on Magna-Matic-equipped engines—models 9, 14, 19, and 23—set the tester gap at 0.060″ (1.5 mm).

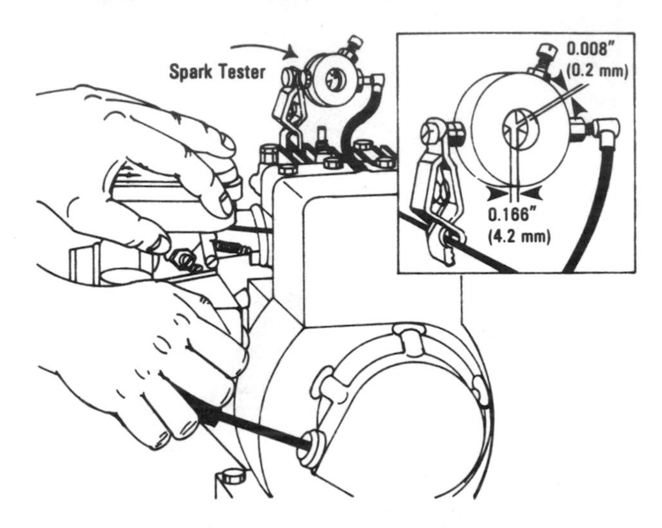

Fig. 2–10. Using the spark tester. *(Courtesy B & S)*

Spark Plug Gapper and Wire Gage—Clean spark plugs with a pen knife or a wire brush and solvent. Set the tester gap at 0.030" (0.75 mm) for all models. If the electrodes on the spark plug are burned away or the porcelain cracked, replace with a new plug. Do not use abrasive cleaning machines on small-engine spark plugs.

Fig. 2–11 shows the adjustment of a spark plug gap with a *wire gage.* Fig. 2–12 shows an efficient, low-cost tool that makes it easy to set spark plug electrodes. The spring steel gage wheel makes sure an accurate gap is set.

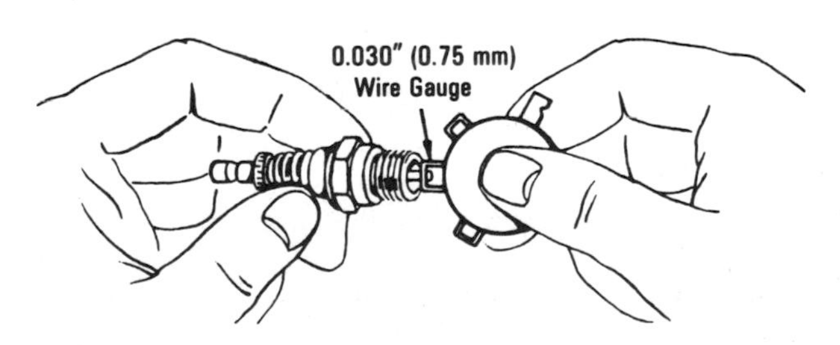

Fig. 2–11. Adjusting spark plug gap. *(Courtesy B & S)*

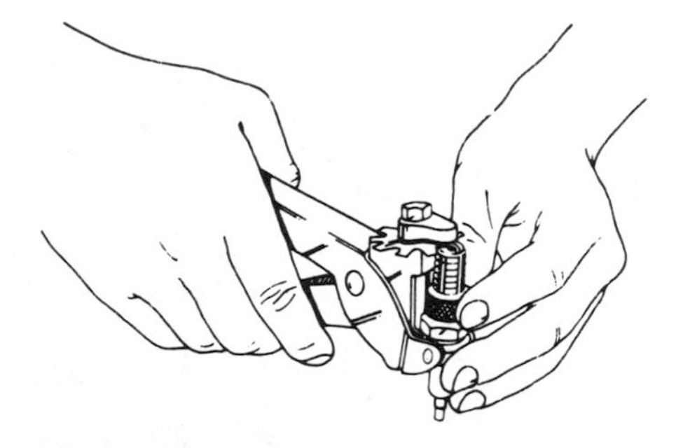

Fig. 2–12. **Spark plug gapper.** *(Courtesy Lisle)*

Feeler Gage—Fig. 2–13 shows a feeler gage. This tool is used to set points and can be used to check spark plug gap. The gage shown contains twelve blades each ¼″ (6.35 mm) wide by 1¾″ (44.45 mm) long with the following thicknesses:

0.011″ = 0.2794 mm	0.015″ = 0.3810 mm	0.019″ = 0.4826 mm
0.012″ = 0.3048 mm	0.016″ = 0.4064 mm	0.020″ = 0.5080 mm
0.013″ = 0.3302 mm	0.017″ = 0.4318 mm	0.022″ = 0.5588 mm
0.014″ = 0.3556 mm	0.018″ = 0.4572 mm	0.025″ = 0.6350 mm

Another type of feeler gage is shown in Fig. 2–14. Here the feeler gage is used to adjust the breaker point gap. This is done by turning the crankshaft until the points open to the widest gap. When adjusting breaker point assemblies, move the condenser forward or backward with a screwdriver until a gap of 0.020″ (0.5 mm) is obtained. Breaker point assemblies as shown in Fig. 2–15 are adjusted by loosening the lockscrew and moving the contact point bracket up or down.

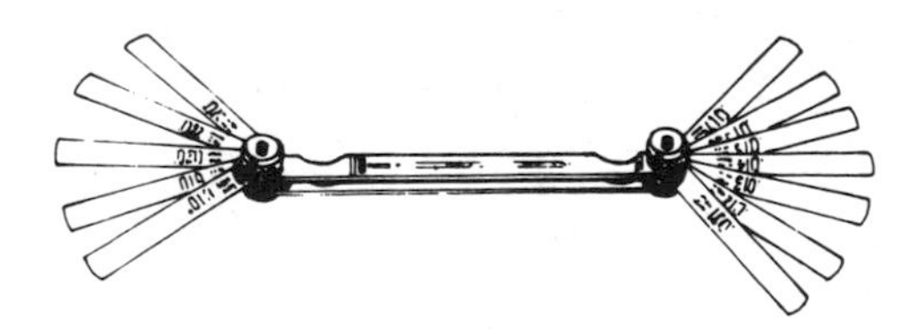

Fig. 2–13. **Feeler gage.** *(Courtesy Silver Streak)*

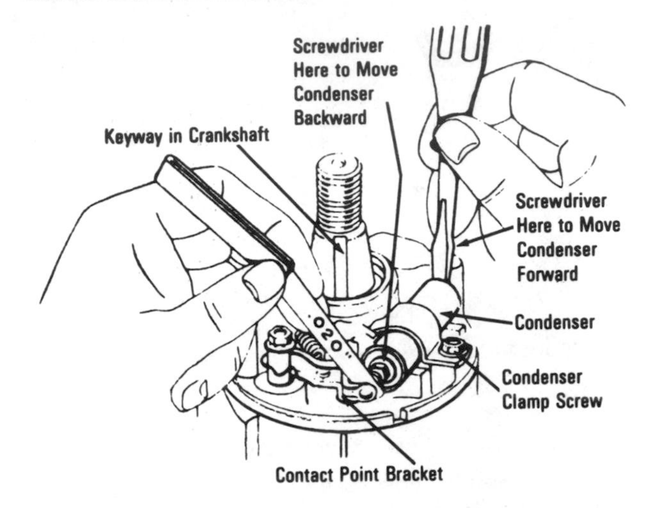

Fig. 2–14. Adjusting breaker point gap using a feeler gage. *(Courtesy B & S)*

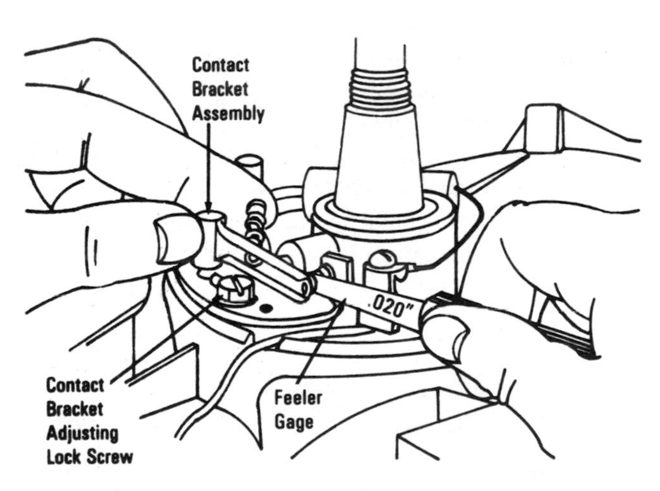

Fig. 2–15. Adjusting breaker point gap. *(Courtesy B & S)*

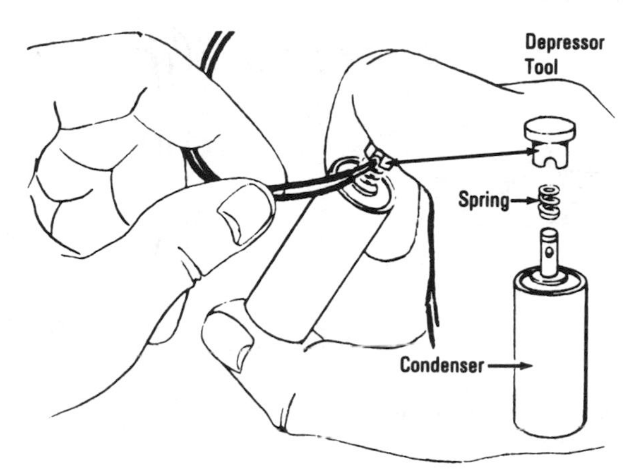

Fig. 2–16. **Assembling condenser and ignition wires.** *(Courtesy B & S)*

In some cases, the condenser has the primary wire fastened to it by a lockwasher and nut. One way to release the wire is to press the top of the condenser with your hand to release the spring tension that holds the wire in the slot (Fig. 2–16). There is a tool designed to make this job easier—the condenser spring compressor tool (Fig. 2–17). This tool holds the spring in a compressed position and allows the easy removal or installation of the primary wire in the condenser post.

Valve Spring Compressor—There are three methods used to hold the valve spring retainers (Fig. 2–18). To remove the types shown in Fig 2–19A and B, use the compressor shown in Fig. 2–18(A). Adjust the jaws until they just touch the top and bottom of the valve chamber. This will keep the upper jaw from slipping into the coils of the spring. Push the compressor in until the upper jaw slips over the upper end of

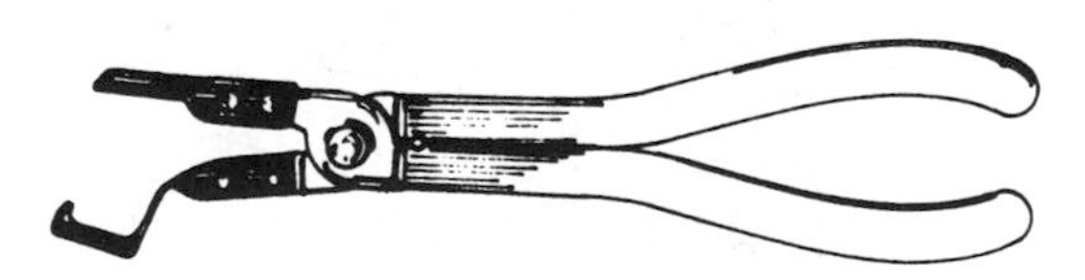

Fig. 2–17. **Condenser spring compressor.** *(Courtesy B & S)*

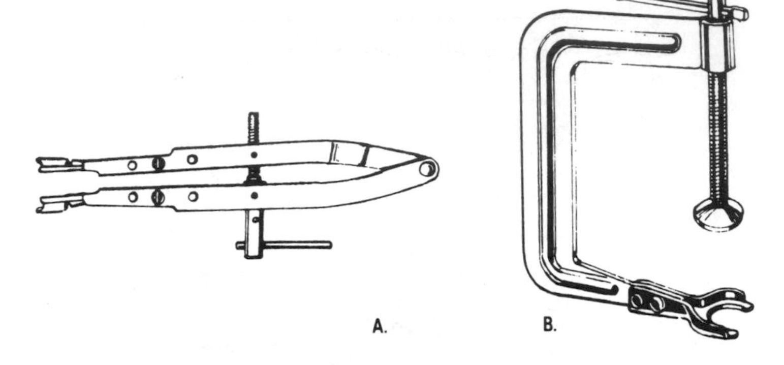

Fig. 2–18. (A) Valve spring compressor. (B) Small-engine valve lifter. *(Courtesy Silver Streak)*

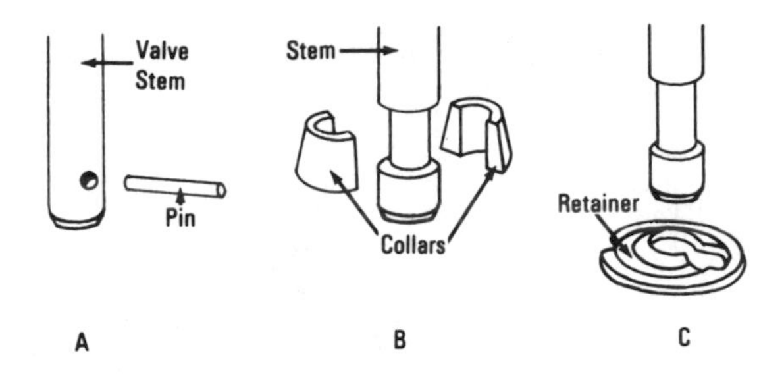

Fig. 2–19. Valve spring retainers. *(Courtesy B & S)*

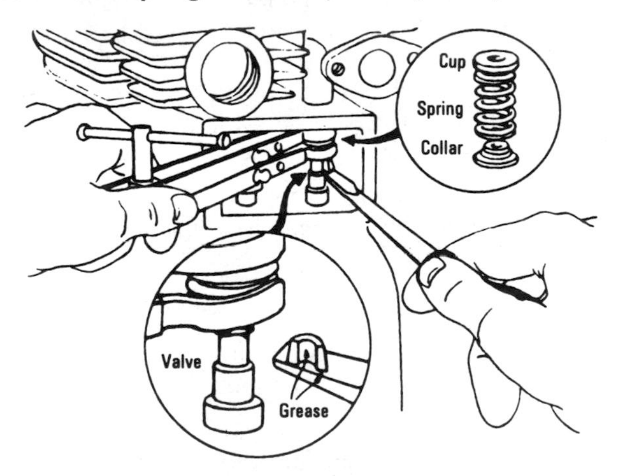

Fig. 2–20. Removing a valve spring. *(Courtesy B & S)*

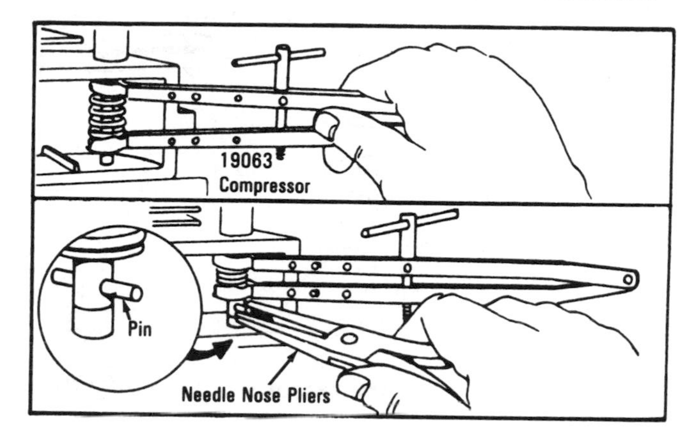

Fig. 2–21. Removing a valve spring. *(Courtesy B & S)*

the spring. Tighten the jaws to compress the spring (Fig. 2–20). Remove the collars or pin and lift out the valve. Pull out the compressor and spring (Figs. 2–21 and 2–22).

Fig. 2–23 shows how older-model compressor tools can be modified by grinding.

Starter Clutch Wrench—This tool is used with a ½" (12.7-mm) drive torque wrench (Fig. 2–24). On rope starter engines, the ½" (12.7-mm) diameter thread flywheel nut is *left-handed* and the ⅝" (15.9-mm) diameter thread is *right-handed* (Fig. 2–25). The starter clutch used on rewind and windup models has a right-hand thread (Fig. 2–26). Remove the clutch using the starter clutch wrench shown in Fig. 2–26 or the one shown in Fig. 2–24 with a ½" (12.7-mm) drive.

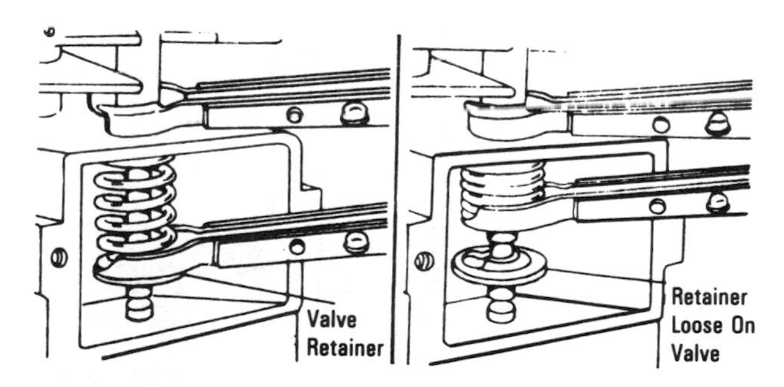

Fig. 2–22. Removing the retainer and spring. *(Courtesy B & S)*

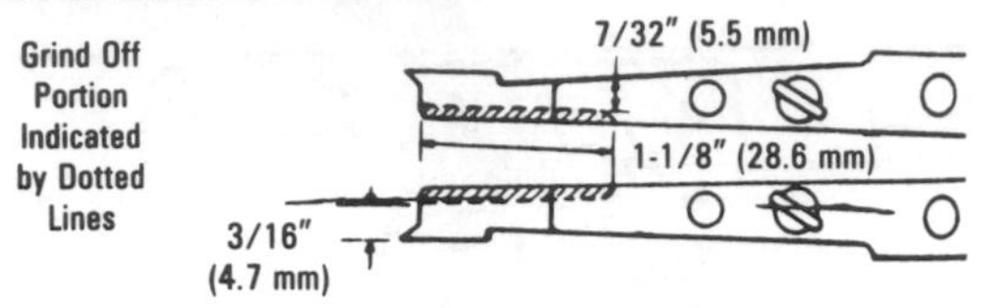

Fig. 2–23. Modifying the spring compressor. *(Courtesy B & S)*

Fig. 2–24. Starter clutch wrench used with a square ½" (12.7-mm) drive.

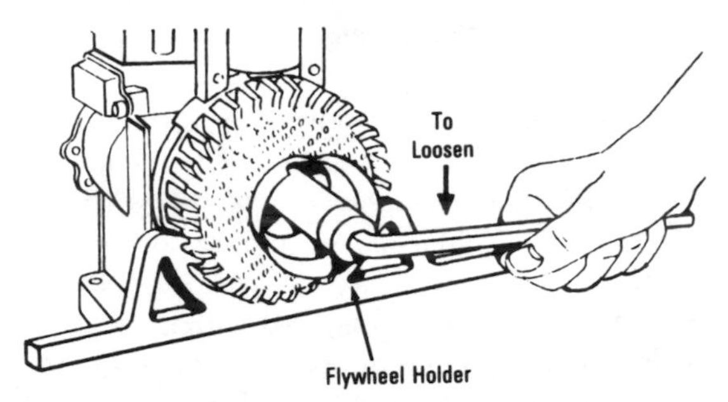

Fig. 2–25. Loosen flywheel rope starter with the wrench. Note the location of the wheel holder. *(Courtesy B & S)*

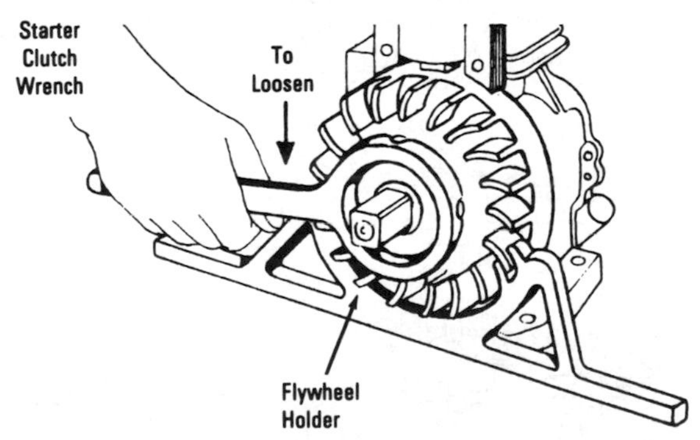

Fig. 2–26. Loosening the flywheel rewind starter and windup starter engines. *(Courtesy B & S)*

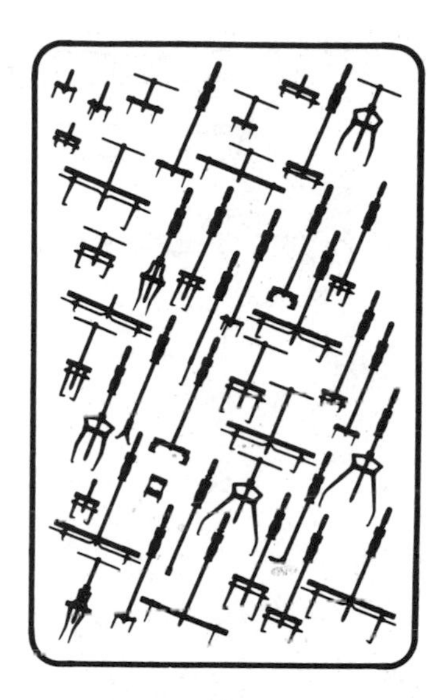

Fig. 2–27. **Forty types of flywheel pullers.** *(Courtesy Snap-on Tools)*

Flywheel Puller—There are many types of pullers. Fig. 2–27 shows forty types. Each has a special purpose or can be used for a number of purposes. The flywheel puller used on small engines is very small in comparison to others with long screws and adapters.

Some flywheels have two holes provided for use of a flywheel puller (Fig. 2–28). Leave the nut loose on the threads of the crankshaft

Fig. 2–28. (A) Removing the flywheel with a puller. (B) Universal-type flywheel puller. *(Courtesy B & S and Silver Streak)*

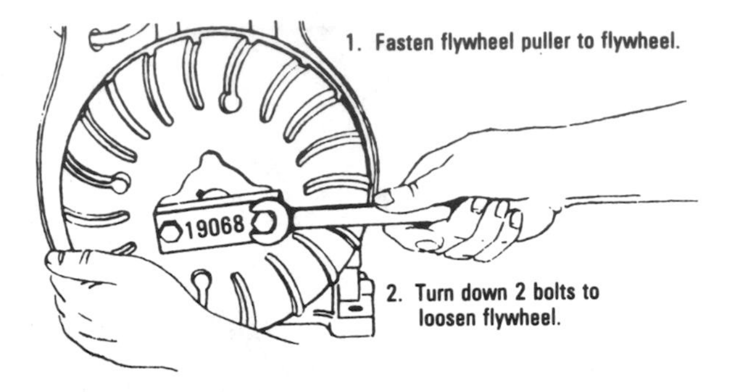

Fig. 2–29. Removing the flywheel. *(Courtesy B & S)*

for the puller to bear against. Small cast-iron flywheels do not require a flywheel puller.

To remove small cast-iron flywheels without puller holes, support the flywheel with a gloved hand. Exert an upward pull. With a rawhide hammer strike the outside rim of the flywheel a sharp blow. Several blows may be required on an extremely tight flywheel. Take care not to damage the flywheel fins, magnets, or ring gear (Fig. 2–29).

Flywheel Wrench—The flywheel wrench shown in Fig. 2–30 is made to put leverage where it is needed in order to remove the nut holding the flywheel in place. To hold the flywheel before it is removed, it is a good idea to use the flywheel holder. Fig. 2–31 shows three sizes for the 4- and 5-horsepower engines, 3- and 3½-horsepower engines, and the 6-, 7- and 8-horsepower engines. The flywheel holder is shown in use in Fig. 2–25.

Piston-Ring Compressor—It is next to impossible to install the piston with rings inserted in the cylinder without a piston-ring compressor (Fig. 2–32). The proper tool should be used. Note the difference between the compressor used for cast-iron pistons and that used for alu-

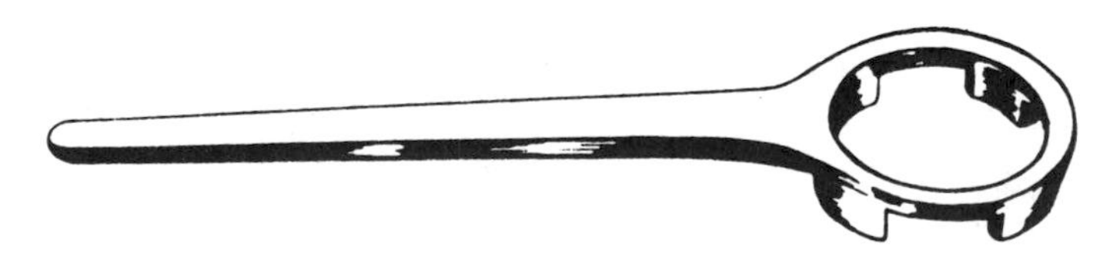

Fig. 2–30. Flywheel wrench. *(Courtesy Silver Streak)*

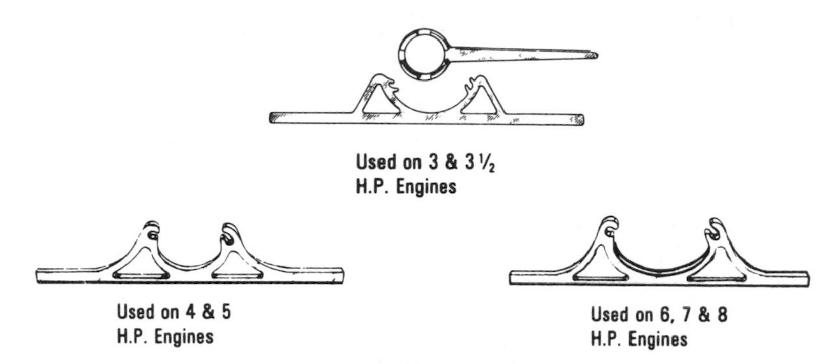

Fig. 2–31. Flywheel holders. *(Courtesy Silver Streak)*

minum pistons. On all aluminum pistons, use the compressor shown in Fig. 2–32B. On all cast-iron models, use the compressor shown in Fig. 2–32A. Turn the piston and compressor upside down on the workbench and push downward so the piston head and the edge of the compressor band are even while you tighten the compressor. Draw the compressor up tight to fully compress the rings. Then you can loosen the compressor very slightly (Fig. 2–33).

DO NOT ATTEMPT TO INSTALL PISTON AND RING ASSEMBLY WITHOUT A RING COMPRESSOR.

Ring-Groove Cleaner—This tool is used for the removal of heavy carbon deposits in the piston-ring grooves. A cutter blade fits into the tool and ring groove. The turning of the cleaner by hand will remove the

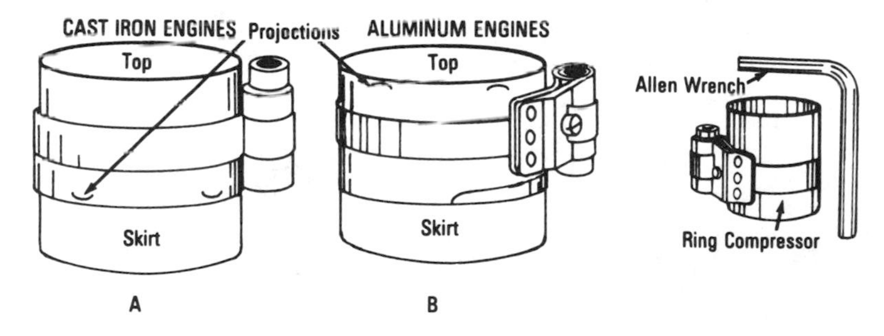

Fig. 2–32. (A) Piston ring compressor. (B) Compression rings. (C) Ring compressor with wrench. *(Courtesy Silver Streak and B & S)*

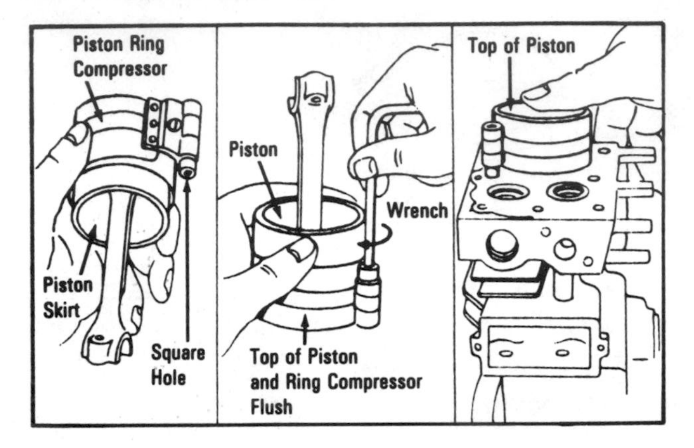

Fig. 2–33. Installing a piston assembly. *(Courtesy B & S)*

excess carbon buildup. This will allow a better fit for new rings (Fig. 2–34).

Piston-Ring Installer—A tool has been designed to reduce ring breakage in the process of putting rings into the grooves (Fig. 2–35). Insert the ring ends into the jaws of the tool. Then expand gently and slip over the piston and into the groove.

Piston-Ring Expander—Fig. 2–36 shows several piston-ring expanders. They also help to remove or install rings without breaking or distorting them.

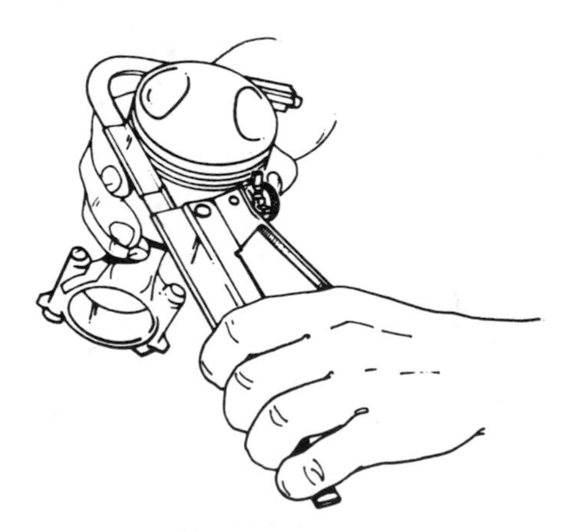

Fig. 2–34. Piston-ring groove cleaner. *(Courtesy Lisle)*

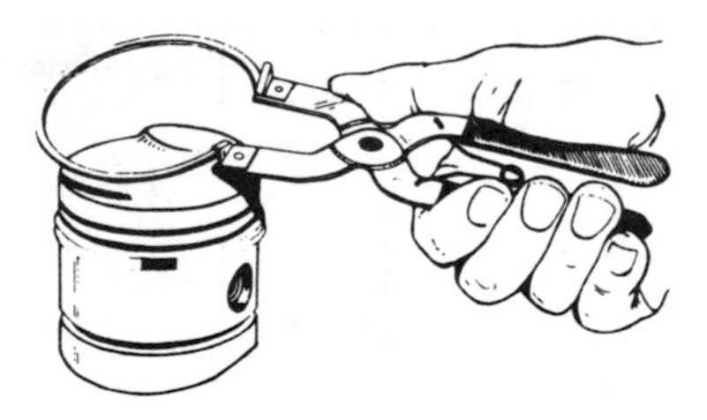

Fig. 2–35. Piston-ring installer. *(Courtesy Lisle)*

Miscellaneous Tools

There are any number of tools designed for doing a job and making it easier.

Carburetor Adjustment Tool—It fits into the idling adjustment screw slot and can turn up to a 20° angle. It is designed not to slip out of the screw slot (Fig. 2–37).

Carburetor Pin Tool—This tool aids in removing the carburetor linkage pins (Fig. 2–38). It removes a pin, retains it, and can be used to reinstall the pin. "E" clips can be handled easily with this type of tool.

Shroud Holder and Spring Rewinder—One of the tricky jobs in small-engine repair is the rewinding of the spring used in the starter. A de-

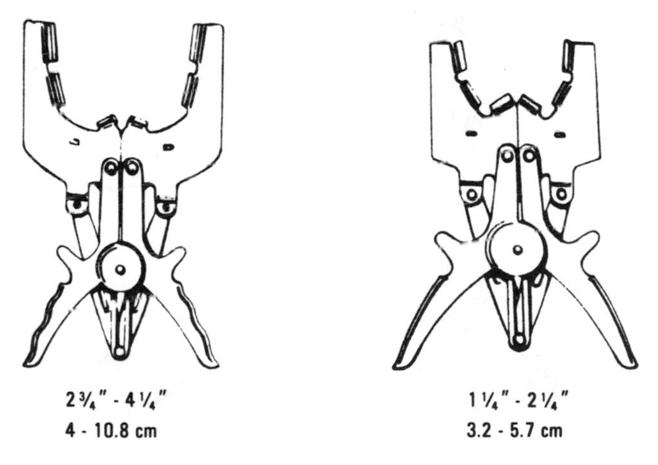

Fig. 2–36. Piston-ring expanders. *(Courtesy Silver Streak)*

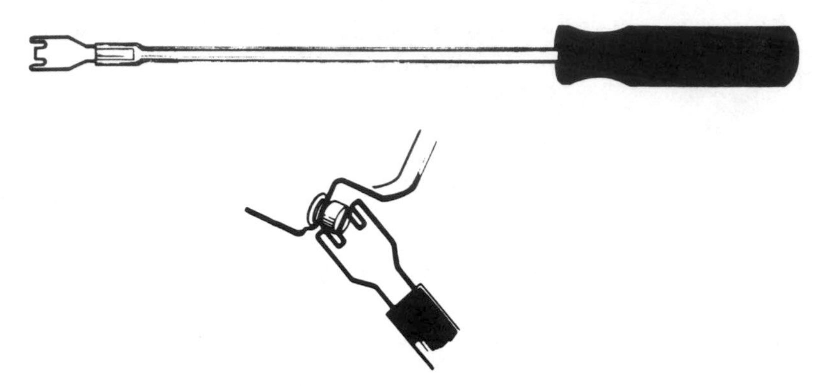

Fig. 2–37. **Carburetor adjustment tool.** *(Courtesy Lisle)*

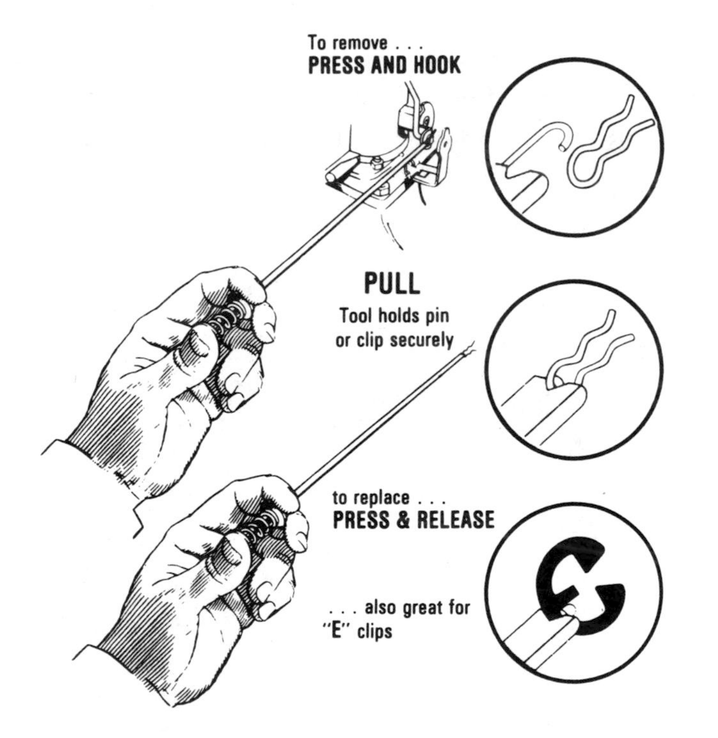

Fig. 2–38. **Carburetor pin tool.** *(Courtesy Lisle)*

vice has been designed to hold the shroud while the spring is rewound (Fig. 2–39). The spring is inserted at 3 (Fig. 2–39B) and the rope is inserted at 8. The tool handle swings around (see 6) and rewinds the rope and starter spring into the shroud.

Wire Bender—Throttle wire is a problem when it breaks sharply at the end of the control wire. A handy, inexpensive tool that makes wire bending simple is shown in Fig. 2–40. A perfect "Z" bend can be made by following steps A, B, C, and D in Fig. 2–40.

A. Hold the tool in your left hand. The notched part is in an upright position facing your right hand. Insert the wire in the hole. The hole is located directly below the notch.

B. Bend the wire down toward your right hand until it is at a 90° angle.

C. Next, rotate the tool until the wire is in the center of the notch.

D. Complete the job by bending the wire up 90°

Safety precautions should be followed whenever you bend wire. Be sure to protect your eyes.

There are many other tools used to grind valves, to hone the cylinders, and to do various specialized jobs. These tools will be described as you read the troubleshooting and repair procedures on the following pages. However, some special tools are shown here.

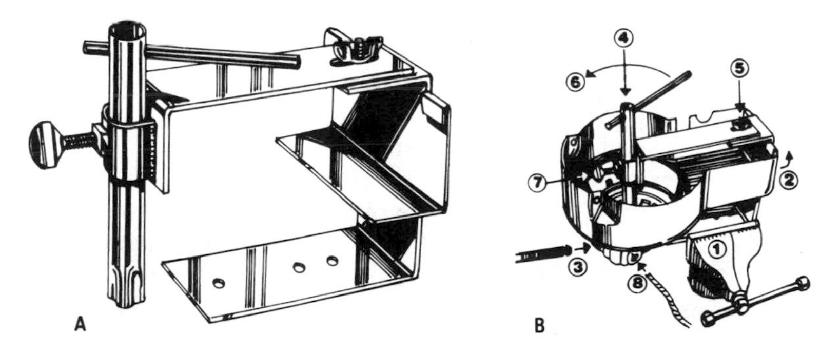

Fig. 2–39. (A) Shroud holder and spring rewinder. (B) Using the spring rewinder. *(Courtesy Silver Streak)*

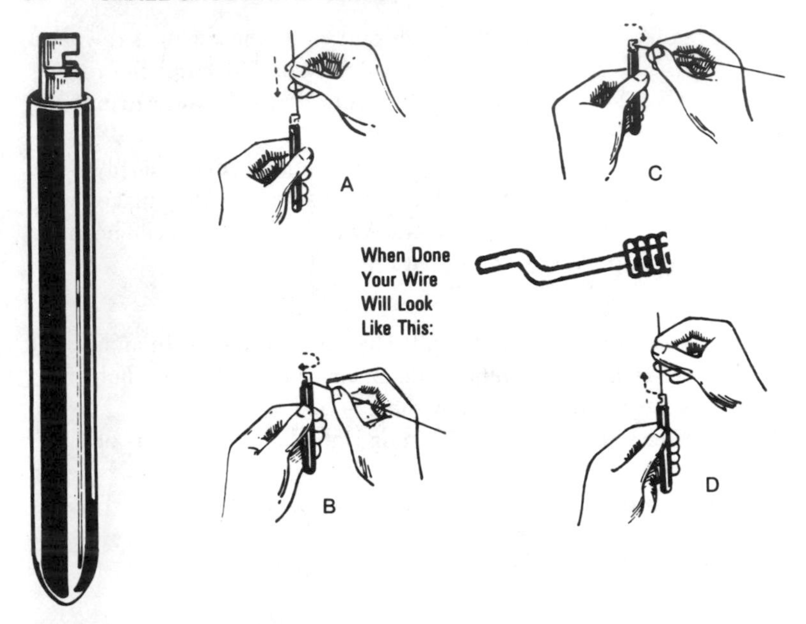

Fig. 2–40. Step-by-step method of using the wire bender. *(Courtesy Silver Streak)*

Flywheel Strap Wrench—This wrench (Fig. 2–41) can be used to hold the flywheel steady while the socket wrench loosens the nut holding the flywheel in place. The nut is turned in the opposite direction from the direction of crankshaft rotation.

Oil Seal Remover Tool—Oil seal work can be tedious and can cause problems when the repaired engine leaks oil all over a clean garage floor because seals have not been replaced. If the crankshaft is in place, remove the oil seal by using the proper oil seal puller (Fig. 2–42).

Oil Seal Driver—Replacing the seals is very important, as previously mentioned. Use the proper driver-protector and place it over the crankshaft as shown in Fig. 2–43. Drive it into position using the universal driver. The seal will automatically be set to the proper depth. Use a seal-protector sleeve every time the oil seal is put onto or pulled off the crankshaft (Fig. 2–44).

Float-Setting Tool—The float-setting tool (Fig. 2–45) is used to make sure the float is properly set to prevent flooding the engine. Use the

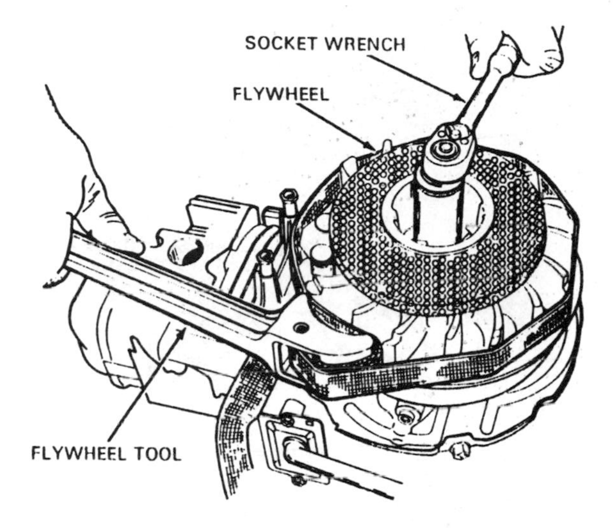

Fig. 2–41. **Flywheel strap wrench.** *(Courtesy Tecumseh)*

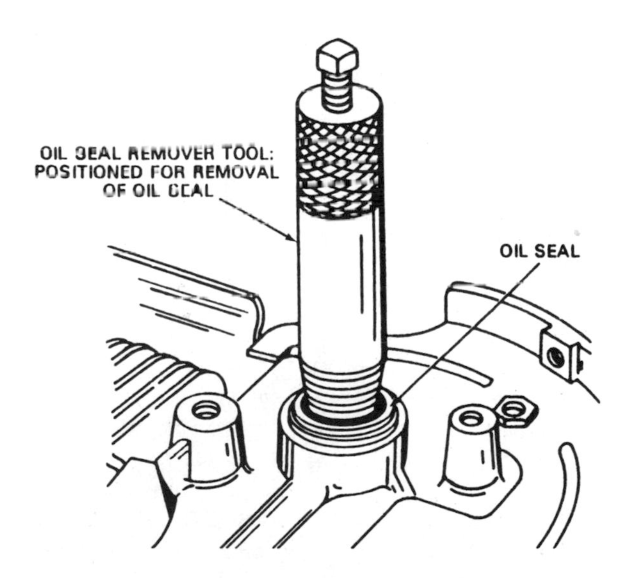

Fig. 2–42. **Oil seal remover tool.** *(Courtesy Tecumseh)*

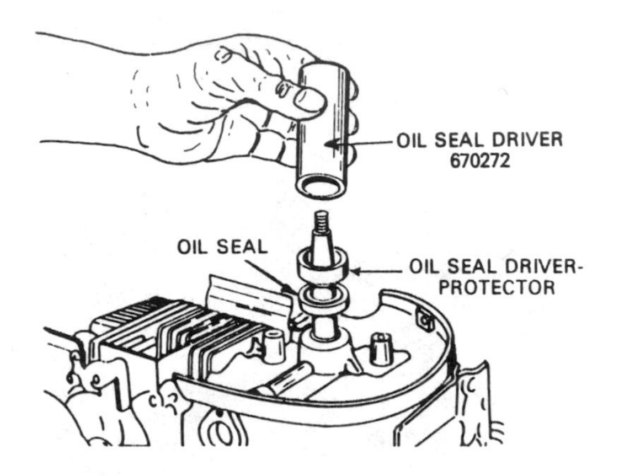

Fig. 2–43. Oil seal driver. *(Courtesy Tecumseh)*

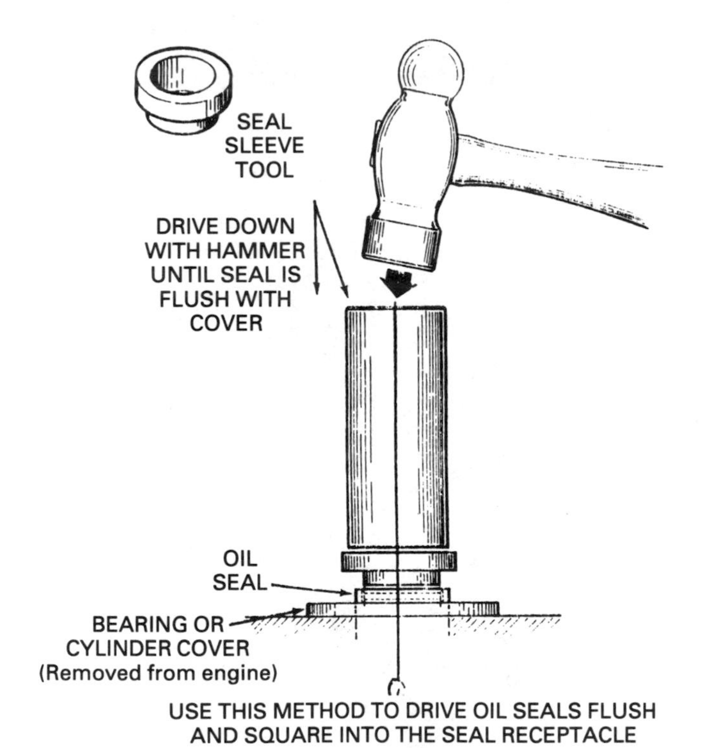

Fig. 2–44. Oil seal sleeve tool. *(Courtesy Tecumseh)*

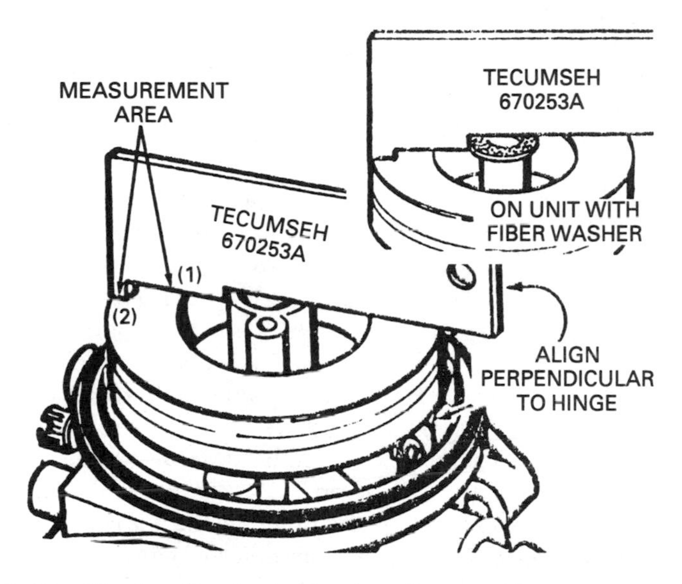

Fig. 2–45. Float-setting tool. *(Courtesy Tecumseh)*

tool as shown in Fig. 2–45. The toe of the float must be within the tolerances of the float-setting tool. The toe *must be under step (1)* and *can touch step (2)* without gap for the proper setting. If the float is too high or too low, it can be adjusted by removing the float and bending the tab accordingly.

If only a minor adjustment is called for, it may be made without removing the float. But, be careful inserting a small-bladed screwdriver in to bend the tab. Be sure not to affect other parts. NOTE: Some carburetors have a fiber washer between the bowl and casting. Use the fiber washer with the float-setting tool as shown in Fig. 2–45.

Electrical Meters

The most useful electrical meter is the volt-ohm-ammeter. It is sometimes referred to as a VOM, or volt-ohm-milliammeter. In most small-engine work the current part of the meter needs to be in amperes rather than milliamperes.

Voltmeter

The *voltmeter* is used to measure voltage. Voltage is the electrical pressure needed to cause current to flow. The voltmeter is used across the line or across a motor, generator, or whatever is being used as a consuming device.

Voltmeters are nothing more than ammeters calibrated to read volts. There is, however, an importance difference. The voltmeter has a very high internal resistance. Thus, very small amounts of current flow through its coil (Fig. 2–46). This high resistance is produced by multipliers. Each range on the voltmeter has a different resistor to increase the resistance so the line current will not be diverted through it (Fig. 2–47). The voltmeter is placed across the line, whereas the ammeter is placed in series. You do not have to break the line to use the voltmeter. The voltmeter has two leads. If you are measuring DC, you have to observe polarity. The red lead is positive (+) and the black lead is negative (–). However, when AC, or alternating current, is used, it does not matter which lead is placed on which terminal.

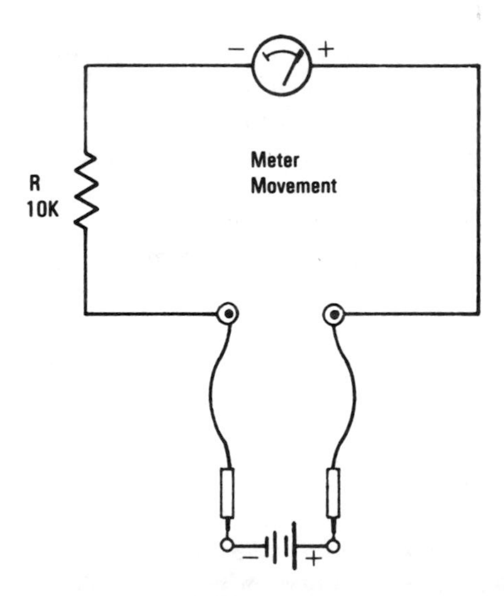

Fig. 2–46. A voltmeter circuit with high resistance in series with the meter movement allows it to measure voltage.

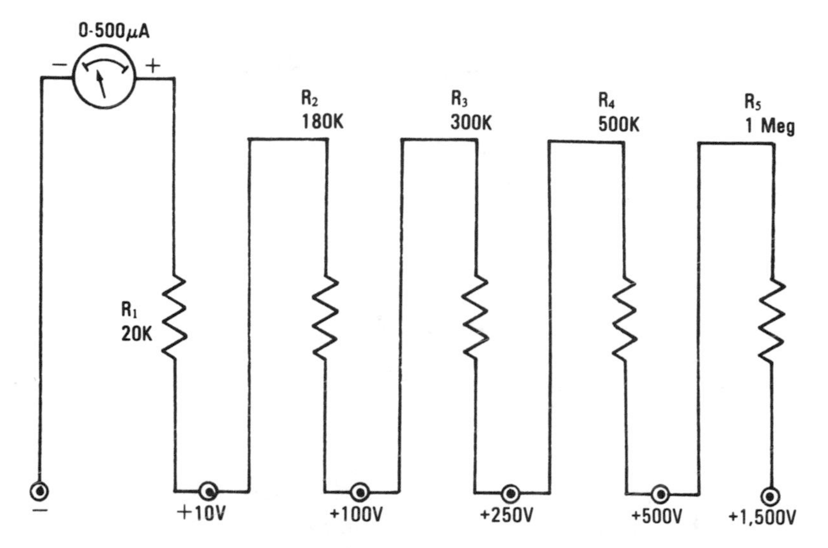

Fig. 2–47. Different types of multirange voltmeters. This view shows the interior of the meter box or unit.

If the voltage is not known, use the highest scale. Turn the range switch to a point where the reading is in the mid-range of the meter movement. Normal line voltage in most locations if 120 volts. When line voltage is lower than normal, it is possible for the equipment to draw excessive current. This will cause overheating and eventual failure due to burnout. The correct voltage is needed for the equipment to operate according to its designed specifications. The voltage range is usually stamped on the nameplate of the device. Low voltage is usually 10 percent lower than normal.

Ohmmeter

The *ohmmeter* measures resistance. The basic unit of resistance is the ohm (Ω). Every device has resistance. That is why it is necessary to know the proper resistance before trying to troubleshoot a device by using an ohmmeter. The ohmmeter has its own power source (Fig. 2–48). Do not use an ohmmeter on a line that is energized or connected to a power source.

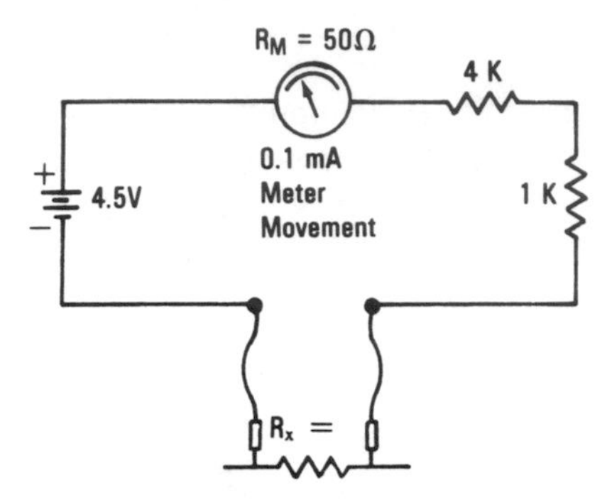

Fig. 2–48. Internal circuit of an ohmmeter.

An ohmmeter can read the resistance of the windings of a generator or motor. If the correct reading has been given by the manufacturer, it is then possible to see if the reading has changed. If the reading is much lower, it may indicate a shorted winding. If the reading is infinite (∞), it may indicate that there is a loose connection or an open circuit.

Ohmmeters have ranges (Fig. 2–49). The R × 1 range means the scale is read as is. R stands for resistance. If the R × 10 range is used, the scale reading must be multiplied by 10. If the R × 1000 range is selected, the scale reading must be multiplied by 1000. If the meter has R × 1 Meg range, the scale reading must be multiplied by 1,000,000 (a meg is one million).

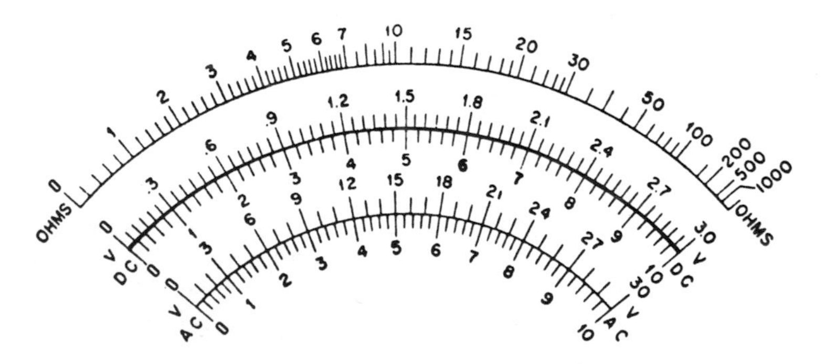

Fig. 2–49. A multimeter scale. Note the ohm and volt scales.

Multimeter

The *multimeter* is a combination of meters (Fig. 2–50). It may have a voltmeter, ammeter, and ohmmeter in one case. This is the usual arrangement for fieldwork. This way it is possible to have all three meters in one portable combination. It should be checked for each of its functions.

Nuts

A number of nuts of different shape and size are used in the small-gasoline-engine-repair trade. Fig. 2–51 shows a selection of the nuts with their proper names. Also note the bolts, washers, and head styles in Fig. 2–52.

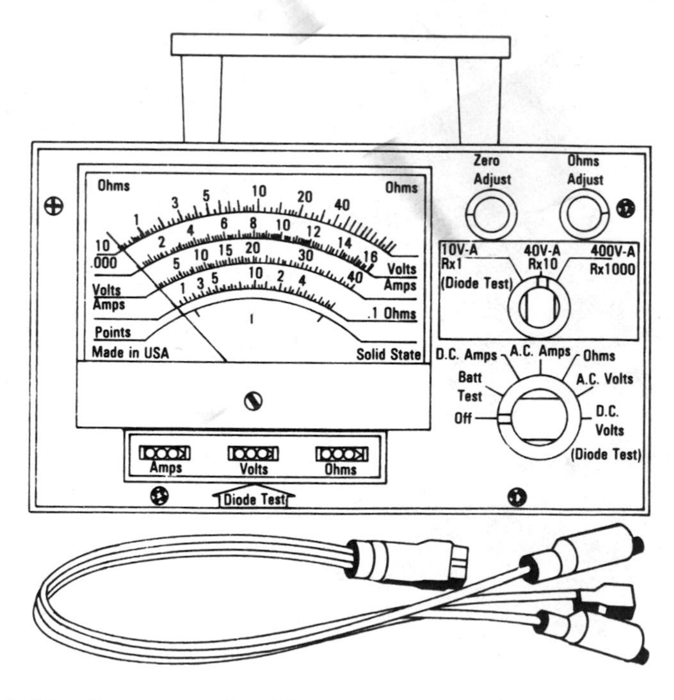

Fig. 2–50. Two types of multimeters.

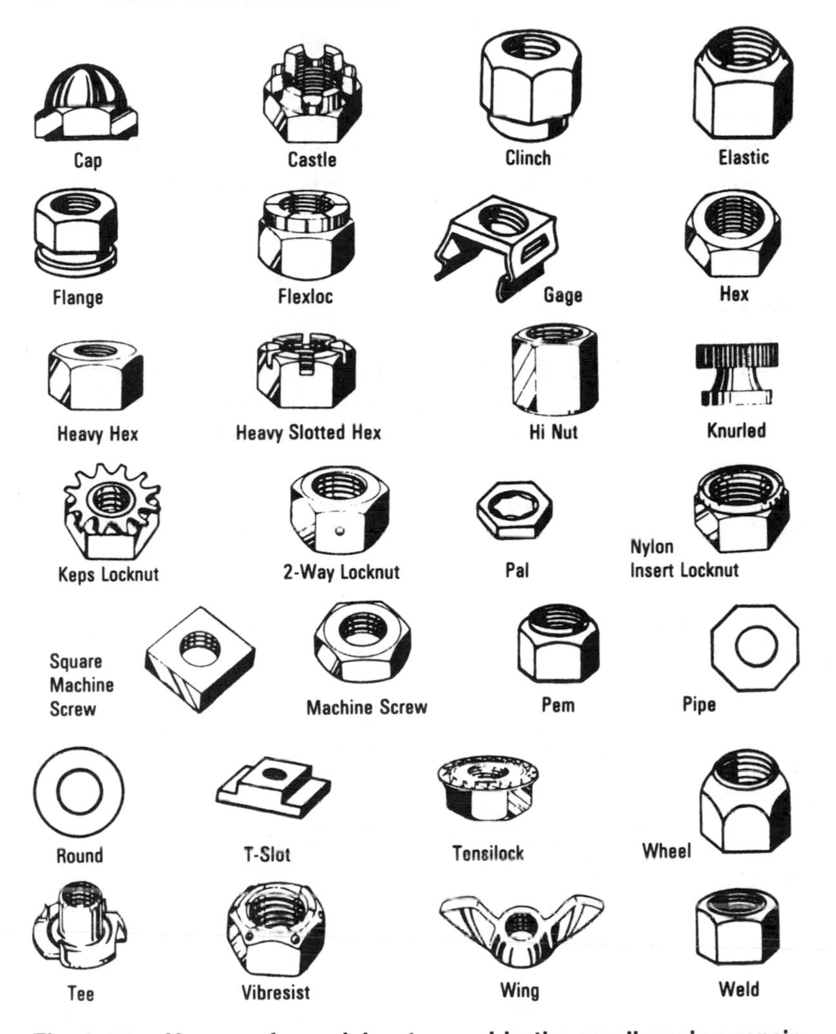

Fig. 2–51. Names of special nuts used in the small-engine repair trade.

HEAD STYLES
GUIDE CHART

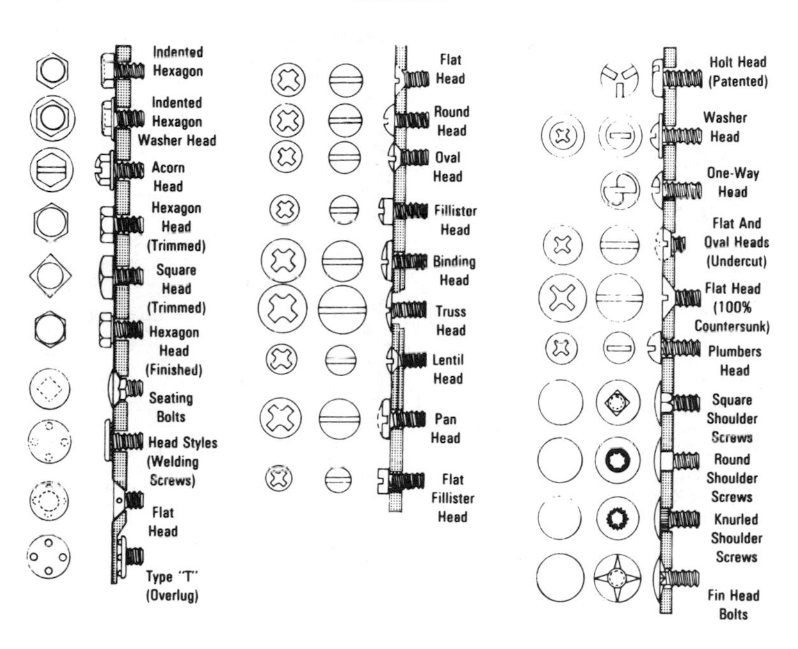

HEAD STYLES

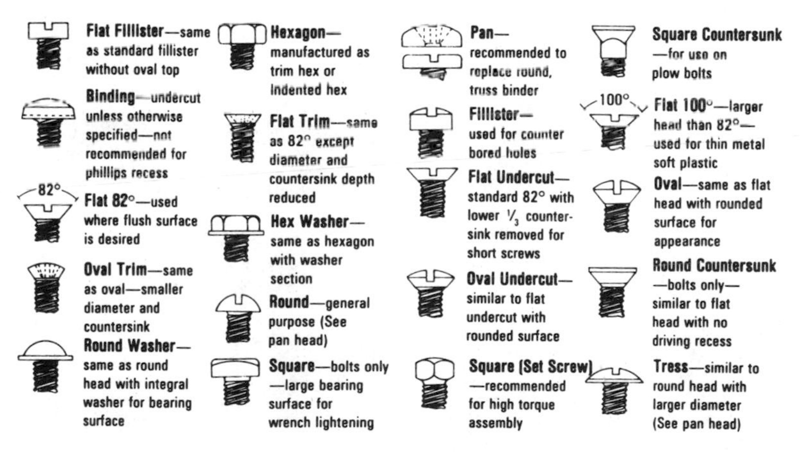

Fig. 2–52. Washers, bolts, and head styles.

WASHERS

ASTM AND SAE GRADE MARKINGS FOR STEEL BOLTS AND SCREWS

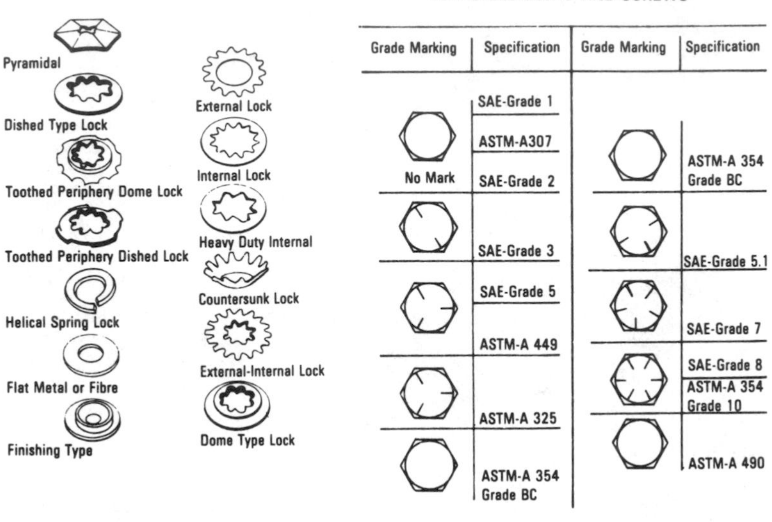

Grade Marking	Specification	Grade Marking	Specification
No Mark	SAE-Grade 1 ASTM-A307 SAE-Grade 2		ASTM-A 354 Grade BC
	SAE-Grade 3		SAE-Grade 5.1
	SAE-Grade 5 ASTM-A 449		SAE-Grade 7
	ASTM-A 325		SAE-Grade 8 ASTM-A 354 Grade 10
	ASTM-A 354 Grade BC		ASTM-A 490

DRIVING RECESSES

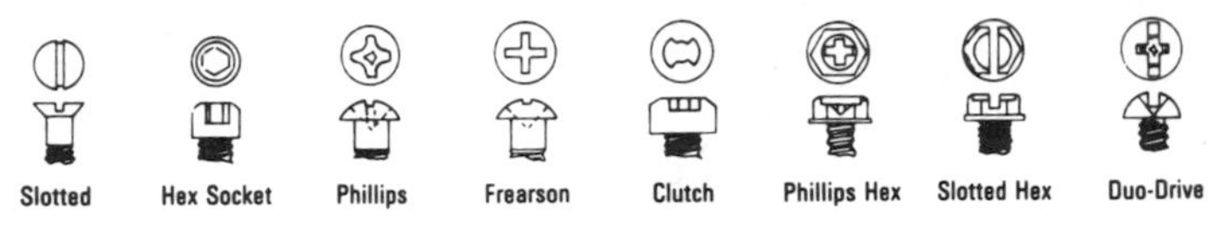

STANDARD BOLT STYLES

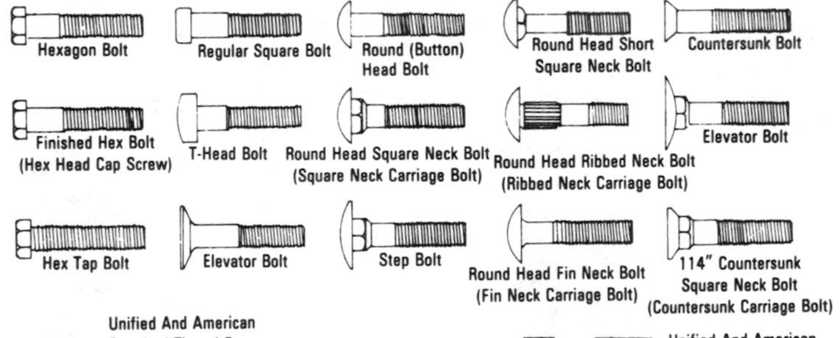

THREADS

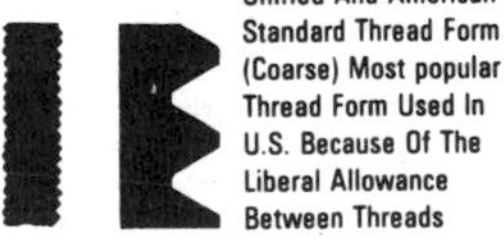

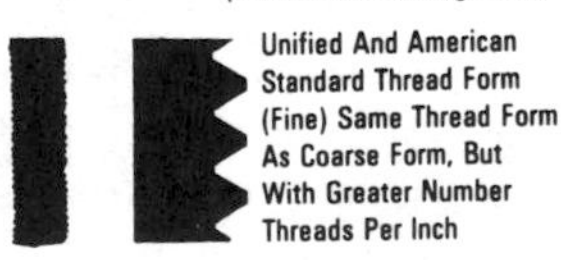

Fig. 2–52. (Cont'd)

Chapter Highlights

1. Pliers can be used for a number of repair functions.
2. Screwdrivers are an absolute necessity for engine repair.
3. The crescent and monkey wrenches are similar, yet different.
4. Never use hand sockets on power or impact wrenches.
5. Don't use a claw hammer where a ball-peen hammer is called for.
6. Never use a punch or chisel that is chipped or mushroomed with a hammer.
7. Heavy-duty or service-duty light bulbs are made for the rough duty given to extension-cord lamps.
8. The spark tester is used to check if the ignition is working.
9. The spark plug gap is very important.
10. The feeler gage is used to check the spark plug gap.
11. The wire gage can be used to check spark plug gaps.
12. Breaker points are set using the feeler gage.
13. The valve spring compressor is needed to remove the valve springs.
14. There are many types of flywheel pullers.
15. The flywheel wrench is handy for removing the flywheel of engines.
16. The piston-ring compressor is needed to replace the piston into the cylinder.
17. The ring-groove cleaner is needed to clean the grooves before new rings are installed.
18. A piston-ring expander is needed to remove the old rings.
19. Throttle wire can be bent to a perfect "Z" if a special tool is used for bending.
20. The voltmeter can be used to check electric motors and starters.
21. The multimeter can be used to check a number of electrical functions.
22. A flywheel strap wrench can be used to hold the flywheel steady while the socket wrench loosens the nut holding it in place.
23. Oil seal removal is best done with a tool designed for the job.
24. An oil seal driver aids in the proper placement of the oil seal and insures no oil leakage later.
25. A float-setting tool is very much needed to prevent a flooding engine. Float specs are available for each engine.

Vocabulary Checklist

ball-peen hammer
breaker points
carburetor adjustment tool
carburetor pin tool
chisel
claw hammer
crescent wrench
feeler gage
float-setting tool
flywheel puller
flywheel strap wrench
flywheel wrench
ignition
monkey wrench
multimeter
oil seal driver
oil seal remover tool
Phillips-head screwdriver
piston-ring compressor
pliers
punch
ring-groove cleaner
slip-joint pliers
spark plug
spark tester
starter clutch wrench
throttle wire
valve spring compressor
voltmeter
wire gage

Review Questions

1. How are long-nose pliers different from slip-joint pliers?
2. Why should screwdriver tips be kept square and sharp?
3. Describe the Phillips-head screwdriver.
4. How is a crescent wrench different from a monkey wrench?
5. How does a carpenter's hammer differ from a ball-peen hammer?
6. Why should you wear safety glasses or goggles when using a chisel or punch?
7. What does a spark tester do?
8. Why would you not use an abrasive machine to clean spark plugs?
9. What is a feeler gage?
10. How does a valve spring compressor work?
11. What is a starter clutch wrench used for?
12. Why would you need a flywheel puller?
13. What does a piston-ring compressor do?
14. Why would you need a ring-groove cleaner?
15. Why wouldn't you use an ohmmeter on a circuit with the power applied?

16. What is a multimeter? How is it used?
17. What purpose does the flywheel strap wrench serve?
18. Why is an oil seal remover tool necessary?
19. Why do you need an oil seal protector when removing or installing an oil seal?
20. How do you make sure the engine is not flooded with gasoline?

CHAPTER 3

Power Generation

Mechanical and electrical devices are used to generate power. Many electrical devices are mechanically driven. The driving force for an electrical generator is usually a small gasoline engine. These engines make very small, compact units for portable operation in the field where emergency power is required (Fig. 3–1).

In this chapter we will be taking a look at some of the things to consider in the generation of power. The emphasis will be on the mechanical aspects of power generation. Factors such as friction, energy, and power will be covered. Pressure, density, force, speed, and torque are also important considerations in a study of small gasoline engines. Motion, momentum, and hydraulics play a part as well in the generation of power. Heat transfer is important since it is a heat engine we are considering for a power source. The transfer of the heat to the surrounding air is very much related to the design of the engine. Heat transfer is designed into an engine from the first. It may be a water-cooled or an air-cooled engine (Fig. 3–2). Each type has its particular design requirements. Some of these design requirements will be studied here.

A. Four-Cycle TVS 10S Rotary Mower Engine
(Courtesy Tecumseh)

B. Two-Cycle TVS 600 Vertical Shaft Rotary Mower Engine
(Courtesy Tecumseh)

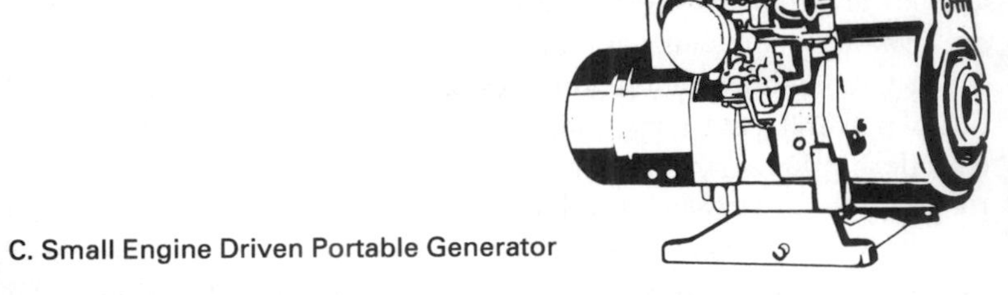

C. Small Engine Driven Portable Generator

Fig. 3–1. (A) A four-cycle rotary lawn-mower engine. (B) A two-cycle rotary lawn-mower engine with vertical shaft. (C) Small gasoline engine drives a portable electrical generator for emergency power.

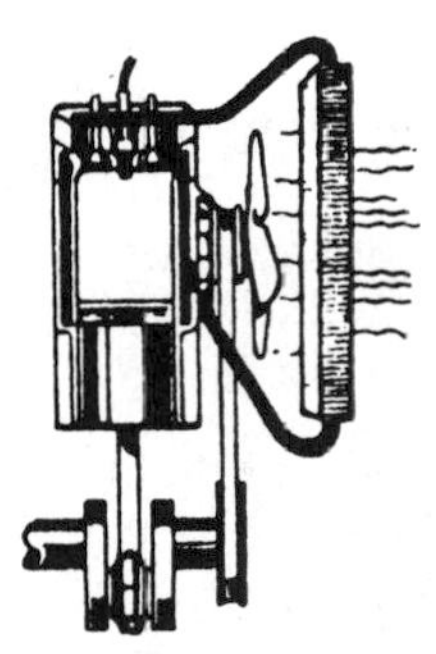

Fig. 3–2. Air-cooled (left) and water-cooled (right) engines. *(Courtesy Onan)*

Friction, Energy, and Power

When we talk of engines, we think of machines. An engine is a machine. It produces power. This machine transforms potential energy into mechanical power. What is energy? Well, it takes a definition to make it clear. Keep in mind that all the universe is made up of matter and energy.

Matter is the label applied to everything that makes up our universe and has weight and occupies space. There are any number of examples of matter. Your hair, your body, your clothes, and even dust. All these and everything else around you is considered matter. Even the air you breathe is matter. Matter has no motion of its own.

Energy

Energy is the other term that needs a definition in order to make our study meaningful. Energy has been defined as that which puts life into matter. We can't measure energy; we can't weigh it or measure it in terms of the space it occupies.

You may think of energy as the light from the sun or the heat from the bonfire that warms you. You may think of energy as the spark of electricity that causes lightning or the spark between the electrodes of a spark plug.

Energy is the ability to do work. It has to be thought of as having no measurable weight or dimensions of its own. Energy occupies space and acts on matter. There are natural actions of energy that shape the world. Following are examples of what energy does all around you.

Energy Forms—There are five active forms of energy. You may think of them as being:

Heat energy produces warmth.
Light energy produces light and color.
Kinetic energy is the movement of a body of matter.
Chemical energy is the change brought about by one substance acting on another.
Electrical energy produces shocks and magnetism.

All these forms of energy are active forms; they are going on all the time. Any one form of energy is, in most cases, accompanied by another form. They cause changes in everything we know in the universe. The world changes constantly since the wind is blowing and water is flowing. Each of these activities creates energy. They can also be used to generate electrical energy. The transformation of energy from one form to another is very important today (Fig. 3–3).

Potential Energy—Potential energy is energy at rest. It can be represented by anything that possesses the ability to be converted. Oil can be changed or converted to heat energy when it burns. This heat can cause water to become steam. The steam can be used to turn turbines and produce electrical energy.

Kinetic Energy—Kinetic energy is energy at work or in action. It can be defined as the motion of a body of matter. Keep in mind that a moving body possesses mass and volume. It also has dimensions and velocity.

Friction

Friction is present in all machines. Mechanical friction presents itself in the resistance to movement created by two surfaces touching one another.

Friction is used to advantage in stopping cars and other moving objects. The energy is changed from kinetic to heat energy and is dissipated by the movement of air over the rough surfaces.

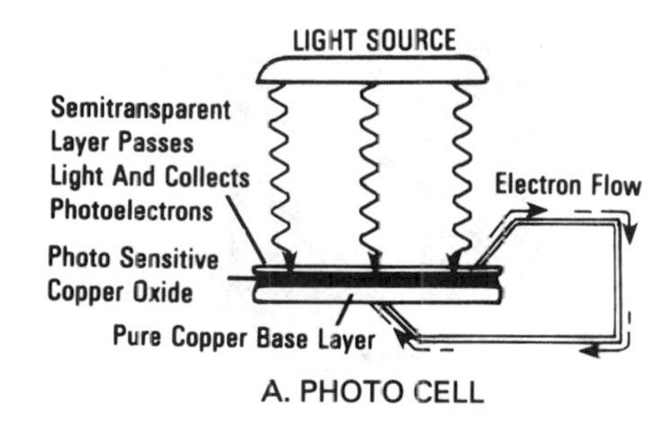

A. PHOTO CELL

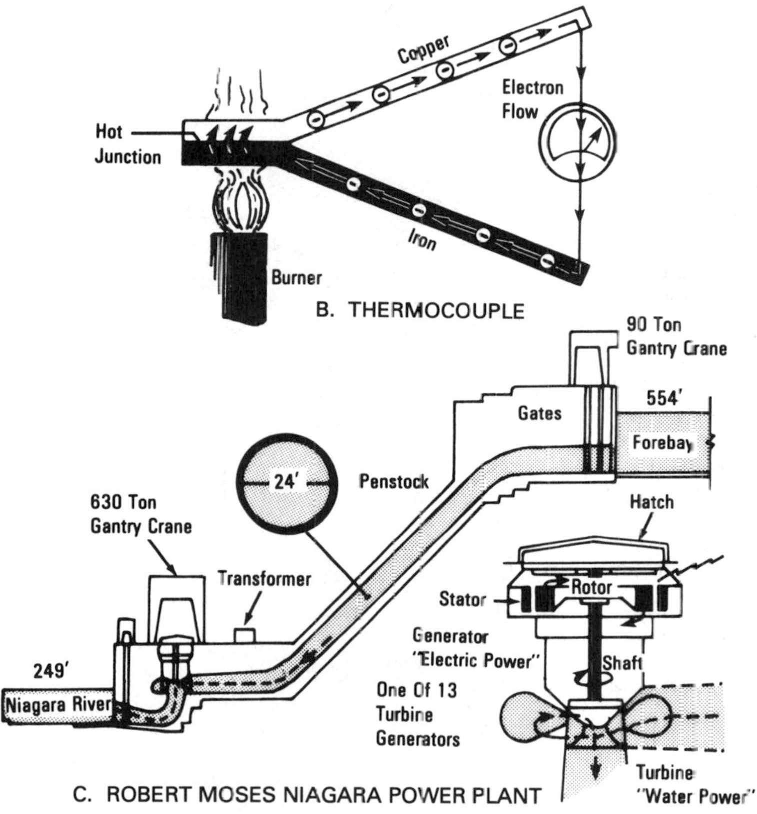

B. THERMOCOUPLE

C. ROBERT MOSES NIAGARA POWER PLANT

Fig. 3–3. Transformation of energy by: (A) Light. (B) Heat. (C) Falling water. (D) Battery (chemical means). (E) Atomic energy. (F) Wind.

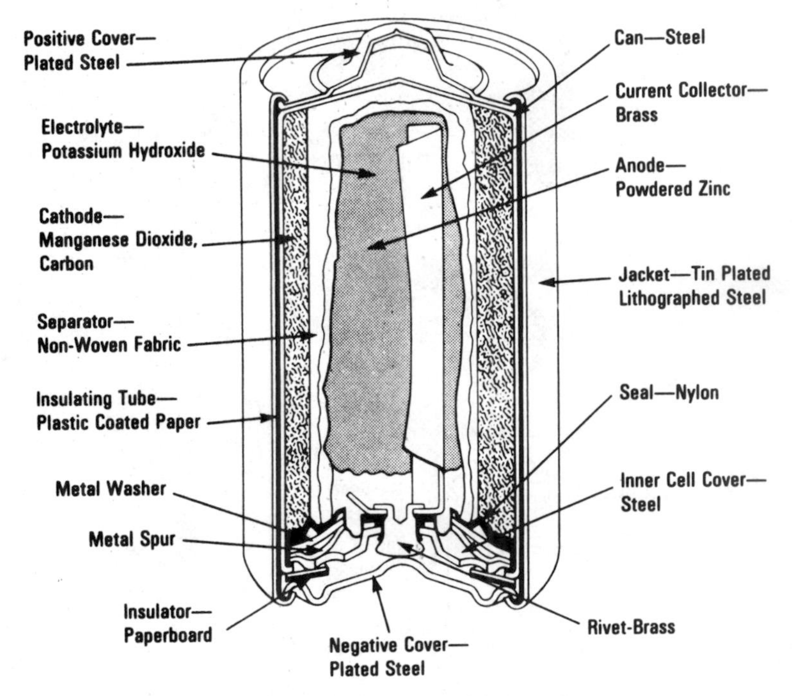

D. CUTAWAY OF CYLINDRICAL ALKALINE CELL

Fig. 3–3. (Cont'd)

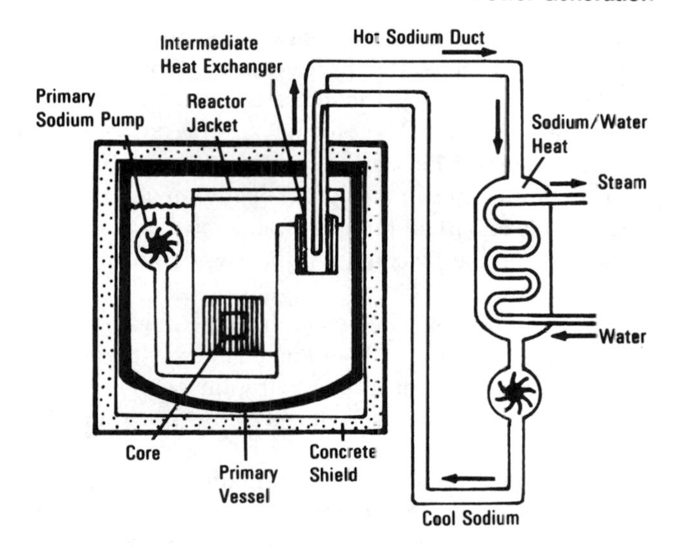

E. LIQUID-METAL, FAST BREEDER REACTOR (EDISON)

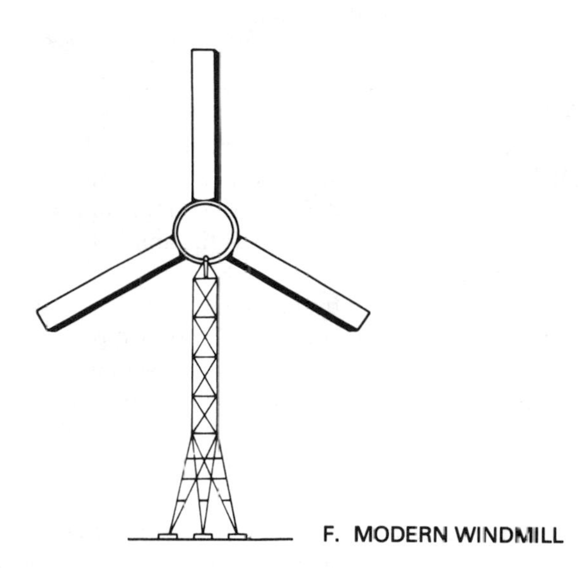

F. MODERN WINDMILL

Fig. 3–3. (Cont'd)

All engines have friction. As you will learn later in this book, only enough horsepower is generated by an engine idling to overcome its internal friction. Overcoming friction becomes of great importance when designing an engine. The efficiency of the energy conversion from gasoline, or potential energy, to the engine in motion, or kinetic energy, makes a difference in the cost of operation of the engine. The cost of operation is important to the overall appeal of a machine. Cost determines whether the machine will be successful in the marketplace.

You experience friction when you rub your hands together. If you do it rapidly, you generate heat. Your kinetic energy (the movement of your hands) is converted to heat energy by the friction of the hands rubbing together. The friction is produced by the rough surface of your hands.

In engines you use oil to produce a thin film coating between two moving, or one moving and one stationary, surfaces (Fig. 3–4). The thin film of oil reduces the friction since the oil is very slippery. The less friction in an engine, the more power can be generated and made useful. It also aids in keeping the heat produced by the engine down to a level where it can be dissipated.

Power

Power is defined as the rate of doing work. It is measured by the watt (W) in electrical terms and in metric measurement. Power is defined as horsepower in the English system of measurement.

One horsepower is defined as the ability to lift 33,000 pounds one foot in one minute.

Work can be done by any number of means. It is defined as moving an object through a distance. When we put work, energy, power, and friction together, we can define the action of an engine.

An engine is nothing more than an energy-converting machine. It changes potential energy into kinetic energy (Fig. 3–5). Electrical energy can be converted into mechanical energy; light energy can be converted into electrical energy; electrical energy can also be converted into mechanical energy; and chemical energy (battery) can be converted into electrical energy. These energy converters make the study of the generation of power interesting.

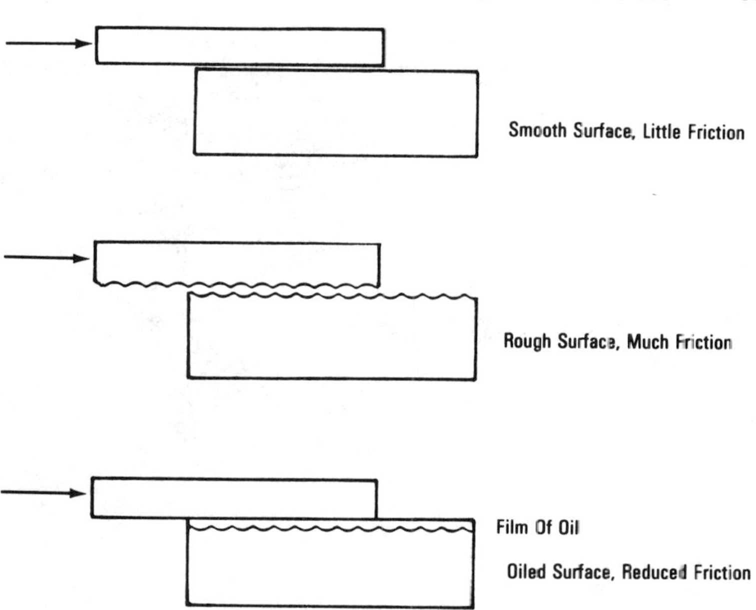

Fig. 3–4. Smooth surfaces have little friction. However, they still have some roughness, even though they are polished. Rough surfaces have much friction. It takes much energy to move the surfaces past one another. A thin film of oil on the surfaces makes most of the friction disappear.

Pressure, Density, Force, Speed, and Torque

Pressure, density, force, speed, and *torque* are words used to describe the operation of a gasoline engine. Each term has a meaning, and is defined below. Next, these terms will be used in reference to the gasoline engine and its operation.

Pressure is defined as the force applied to a unit area (Fig. 3–6).
Density is how much of something you have in a given area (Fig. 3–7).
Force is the cause of the acceleration or movement of material bodies (Fig. 3–8).

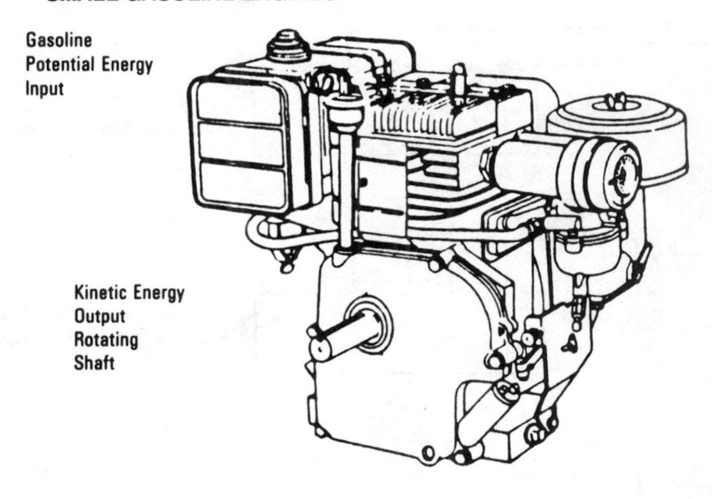

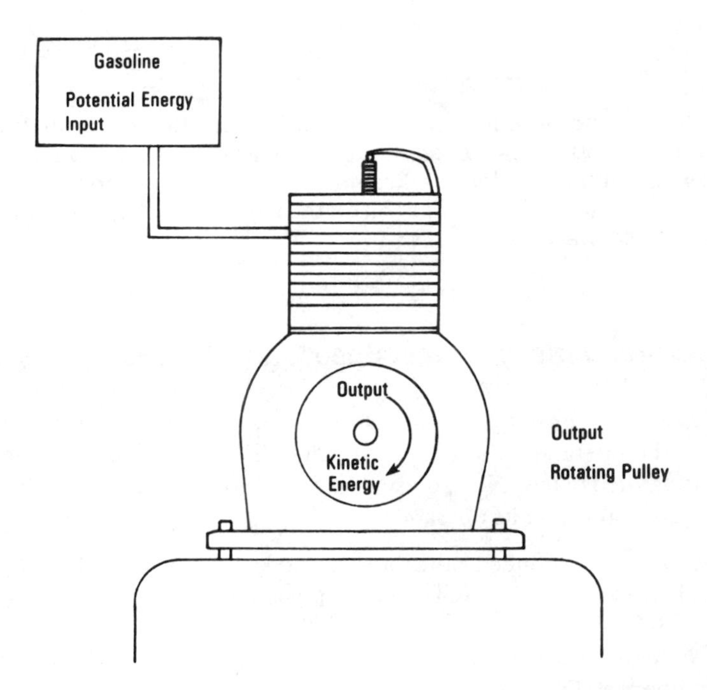

Fig. 3–5. A gasoline engine converts potential energy to kinetic energy.

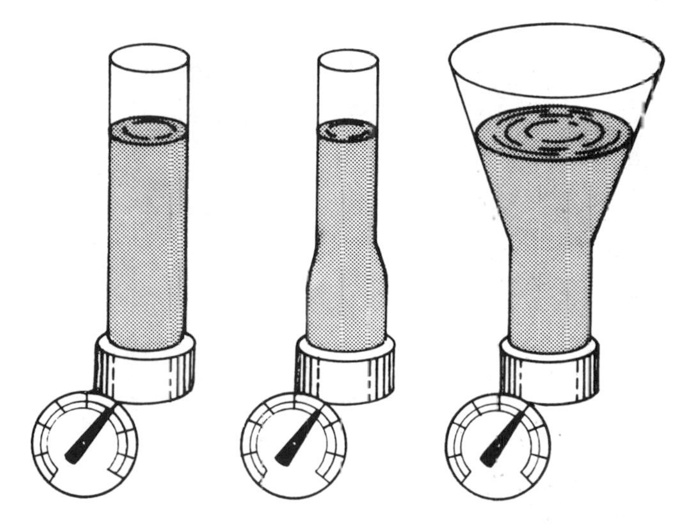

Fig. 3–6. Each of these containers has the same area at the bottom. Each has the same *pressure* exerted on that area. The volumes are different, but the level is the same in each container.

Speed is how fast something travels or moves (Fig. 3–9).

Torque is a tendency to produce a twisting action (Fig. 3–10).

These terms are used to define a gasoline engine's performance. They need to be understood before you become more deeply involved with the operation of a mechanical device such as an engine.

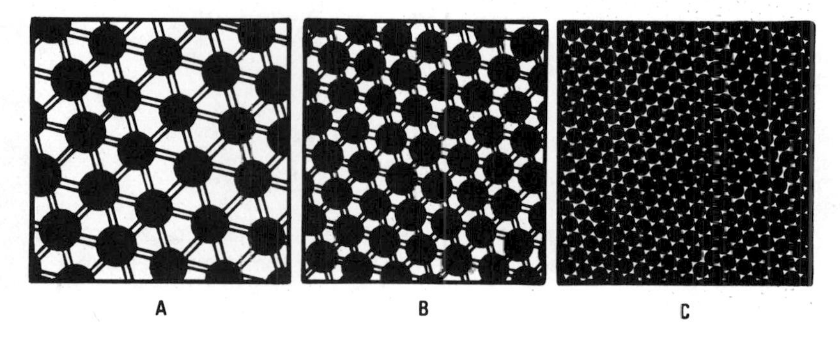

Fig. 3–7. Density is how much of something you have in a given area.

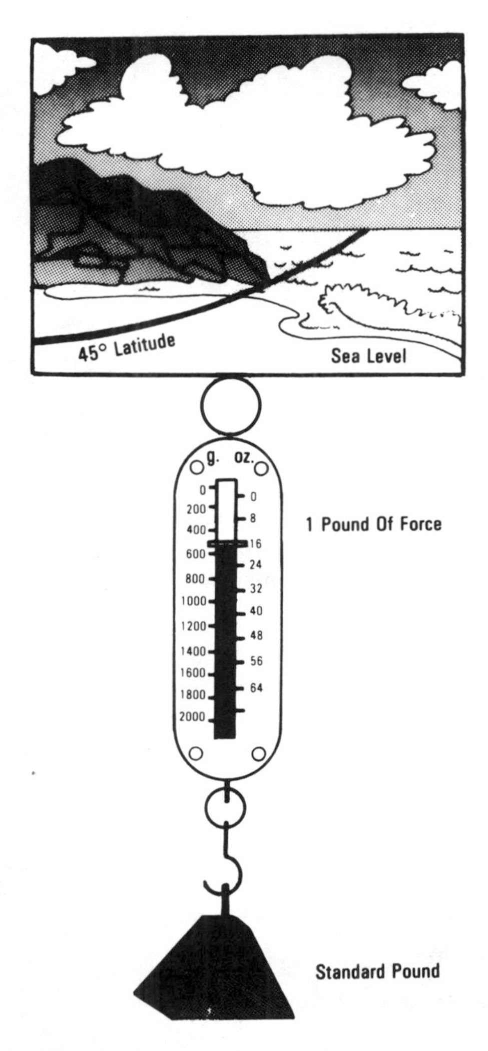

Fig. 3–8. At 45° latitude and at sea level, one pound of force is needed to overcome the earth's attraction for the standard pound.

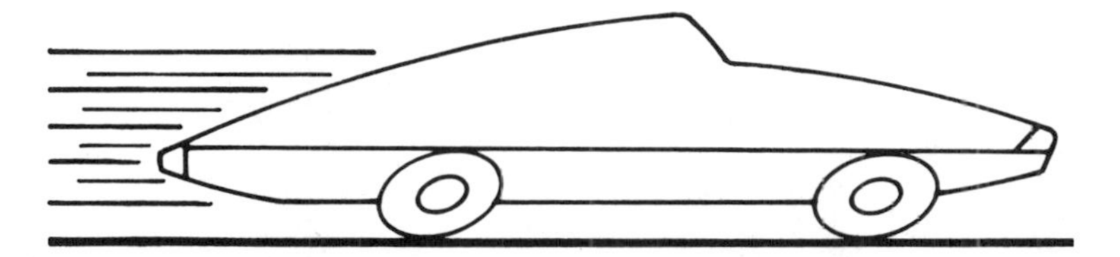

Fig. 3–9. Speed is how fast something travels in a given time, for instance, miles per hour or feet per second.

Pressure Produced by Force

Force can be transferred through the use of a fluid or a gas. Mechanical force exerted on an engine's piston is converted to pressure. The pressure causes the piston to move. This is the object in getting an internal combustion engine to run or operate.

The air has pressure. The atmospheric pressure of air at sea level is 14.7 pounds per square inch (Fig. 3–11). Gravity holds the air close to the earth. The earth is surrounded by an atmosphere of air. The air is made up of molecules, or small parts of matter. Without gravity the air would expand outward and move away from the earth. The entire atmosphere of air is pushing down on the earth with a pressure of 14.7 pounds per square inch. This is referred to as *atmospheric pressure,* the weight of a column of air one inch square resting on the earth or ocean.

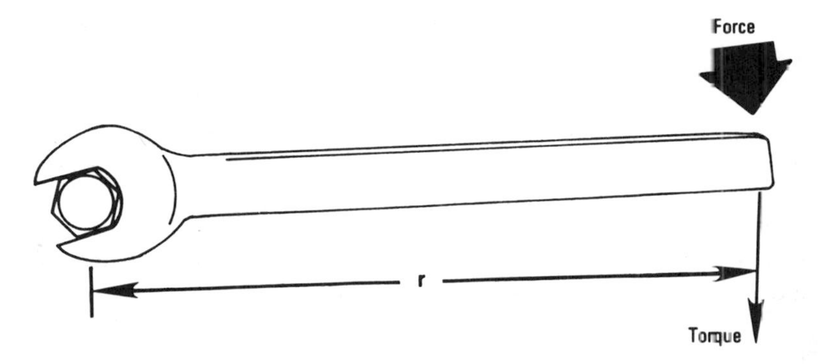

Fig. 3–10. The wrench places torque or twisting action on the bolt to cause it to rotate. Torque is another way to measure force. Torque equals force times radius (*r*).

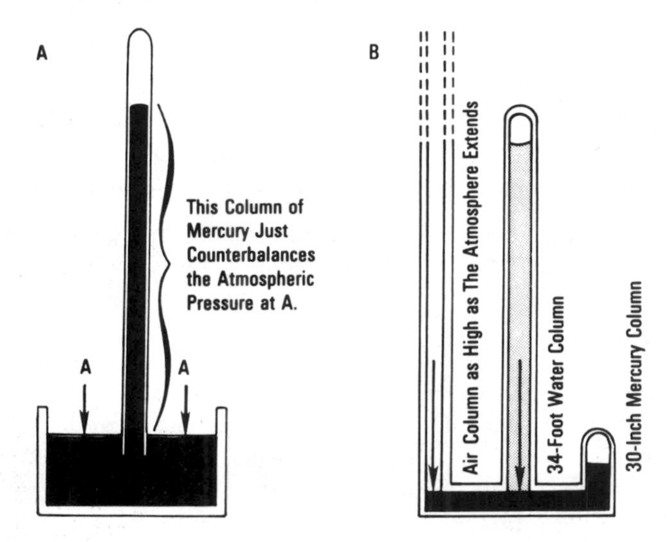

Fig. 3–11. (A) At sea level air is counterbalanced by a column of mercury 30″ (76 cm) high. (B) Pressure exerted by a column of air as high as the atmosphere is the same as that exerted by a 34′ (10.36-m) water column or a 30″ (76-cm) column of mercury.

Air pressure will become very important when we discuss the operation of the carburetor on a gasoline engine.

Density

Molecules contain smaller particles known as atoms (Fig. 3–12). Each element (such as iron) contains a certain number of atoms. These atoms are so small they cannot be seen, even with high-powered microscopes. Certain combinations of atoms make up molecules. Molecules are larger than atoms. Water is a molecule of hydrogen and oxygen. To be precise, water consists of two atoms of hydrogen and one atom of oxygen. It is written as H_2O (Fig. 3–13).

Molecules differ in size and weight. The spacing between the molecules and atoms differs according to the substance. If one molecule or atom is heavier or contains more atoms or molecules per cubic inch, it is said to be denser. The more molecules you have in a square inch, the denser the area. Air becomes more dense as you go below sea level, since there are more molecules of air pressing down on you. The oppo-

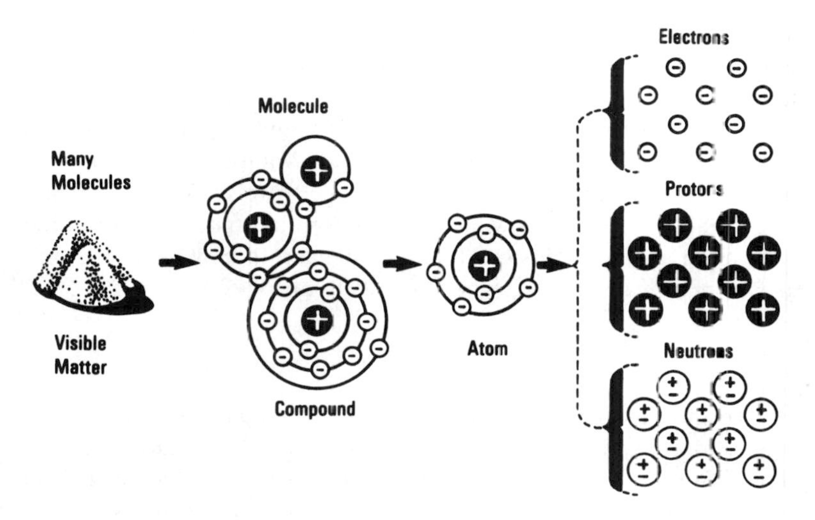

Fig. 3–12. Molecular structure.

site is true as you go above sea level. There are fewer molecules of air as you go higher in altitude, meaning that the air pressure is less. This also lowers the temperature at which water will boil.

Speed

Speed is the rate of motion. How fast you do something is the speed of getting it done. You may be able to travel from home to school

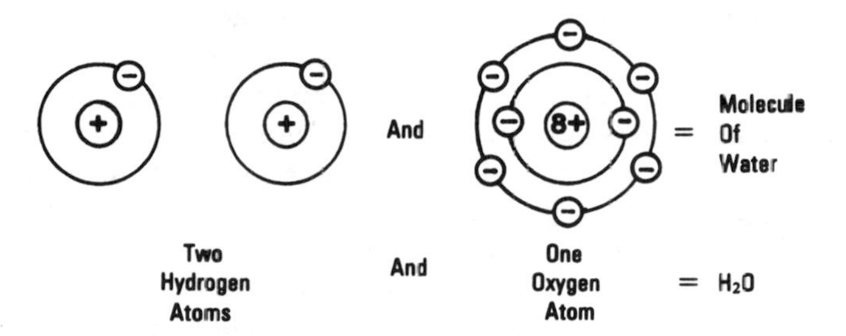

Fig. 3–13. Two or more atoms linked are called a molecule. Here are two hydrogen atoms and one oxygen atom forming a molecule of the compound water (H_2O).

at a very fast rate. Your rate of motion, or how fast you traveled from home to school, is your speed. Speed is measured in miles per hour, feet per second, or meters per hour.

An engine has so many revolutions per minute (rpm—also written as r/min). This is the speed of the engine. Revolutions per second refers to how many times an engine flywheel revolves in one second. Speed can also refer to how many revolutions per minute occur in the rotation of a wheel. This can then be changed to how many feet per second or miles per hour a wheel or engine turns.

Torque

Torque has to do with twisting action. When torque is used in reference to an engine, it means the rotating force delivered to the crankshaft. Torque is measured in foot-pounds (abbreviated as ft-lbs). Torque is also measured in inch-pounds (abbreviated as in-lbs).

In most of the books dealing with small gas engines you will find torque measured in ounce-inches or foot-pounds. Actually, torque is measured in most scientific applications in such units as *pound-feet* and *inch-ounce*. Physicists and engineers use the pound-feet, inch-ounce, and Newton-meter as the unit of measurement for torque.

In most common usage, torque is measured in inch-pounds (in/lb) or Newton meters (N-m). In some instances, it is measured in inch-ounces (in/oz). It can also be expressed in terms of angular velocity. Angular velocity is the expression used by mechanical engineers and robotics technicians when the torque is taken from a rotating engine shaft.

Torque specifications for cylinder head bolts, a flywheel nut, a spark plug, an oil drain plug, or other parts in the engine are usually given in inch-pounds or foot-pounds. This is so they correspond with torque wrench markings and old accepted practices.

This is similar to using the term *condenser* when referring to a *capacitor*. The only people who still use condenser are those who work with gasoline engines. It is hard to get the term out of the books and classrooms, even though it is wrong and no longer proper usage.

A torque wrench is used to place 150 pounds of torque on a bolt. This is more accurately described as 150 foot-pounds of torque. It will take the same amount of torque, 150 foot-pounds, to loosen the bolt. However, it would have to be applied in the opposite direction. When we talk about engines and the torque generated, it is in terms of how much force is required to bring the crankshaft to a dead stop.

Once motion is added to torque, the velocity of the motion is measured in terms of rpm. An engine develops torque continuously at a given number of rpm. This torque causes the engine to produce power that can do work. Torque in an engine will increase with an increase in rpm. There are conditions under which this will not happen. For instance, if more fuel is added to the engine to burn, there is an increase in the torque produced. However, there is a point where the conversion of burned fuel to torque starts to decrease. This takes place when the additional fuel is totally used or expended within the engine to overcome increased frictional losses. Friction losses increase rapidly with the increase in combustion heat after a certain temperature is reached. Increased compression load and other factors are to be considered in the decrease in torque production in an engine.

This can be summed up by saying that in any engine there is a point of diminishing torque, up to which increased rpm will increase the power output. Beyond this point, increasing the rpm will decrease the power output. Engine efficiency will start to decrease even before this point of diminishing torque is reached.

Motion, Momentum, and Hydraulics

Motion

Motion is defined as the act, process, or instance of changing place or position. We associate motion with movement. Movement and motion are almost the same in meaning. However, for our purpose some differences should be noted. Movement means to change places, or to move from here to there. Motion is the act or process of changing from one place or position to another. Some object has to be moved in order to create motion. Engines are motion machines. They generate power by the motion of several parts in sequence and under proper timing.

Energy is found in common things such as sunshine, falling water, and electricity. Whenever we have motion, heat, or light, we have energy being used. Energy is the ability to do work or the ability to produce motion, heat, or light. We use energy to produce motion or start an engine.

Keep in mind that kinetic energy is energy in motion (Fig. 3–14).

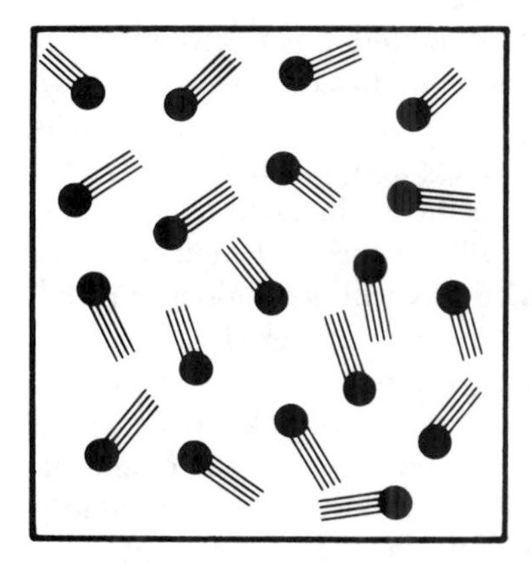

Fig. 3–14. Kinetic energy is energy in motion. Molecules are in motion.

Heat energy is the number of atoms or molecules moving and the speed at which they move. Light can be seen as a series of waves. The waves move from the source of light and carry energy with them.

The energy produced on earth originally came from the sun. Its form has been changed to do work or to serve people. The conversion of energy in an efficient manner is really the product of years of study and experimentation. Today's internal combustion engine is the product of years of research and development in energy conversion.

Momentum

Momentum is the product of the mass of an object and its velocity. The faster an object is moving, the greater is its momentum. For instance, if a ball is handed to you, you hardly feel it. However, if a ball is thrown to you, the hand that catches it will feel a great deal of pressure and even pain in some cases. The momentum of the ball had to be destroyed for you to catch and hold it. The greater the distance traveled by the ball, the more momentum it acquired.

The important characteristics of motion are inertia, momentum, and acceleration. Inertia is the tendency of an object at rest to remain at rest until acted upon by some object or external force to place it in motion. An object has a tendency to remain in motion once it has been placed in motion until acted upon by some external force. The effects of inertia must be overcome to make a car move. The effects of inertia must be overcome to stop a car once it is moving.

Keep in mind that momentum is the measured force produced by a moving object. The faster a car travels, the more momentum it has. Also, a heavier object has more momentum than a lighter object. In order to stop a car, its momentum must be destroyed by its brakes. When you catch a ball, its momentum must be dissipated by your hand. This shows that the momentum of any object, the car or the ball, in motion is a measurement of its inertia (Fig. 3–15).

Acceleration is defined as an increase in an object's speed. You have to overcome the inertia of a stationary object in order tc place it in motion. However, the force needed to cause an object to increase its

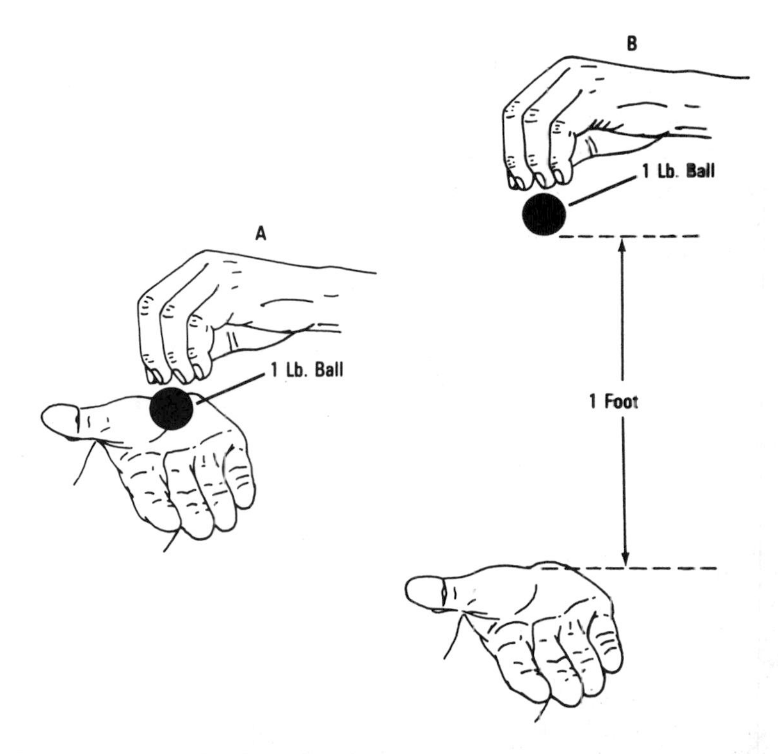

Fig. 3–15. (A) Weight is lowered onto your hand. Very little momentum is present. Impact is hardly felt. (B) If ball is dropped 1′ (2.54 cm), the impulse and change of momentum are much greater and can be felt. The impact may even injure your hand. Momentum is equal to mass times velocity.

speed depends on the weight of the object and the rate of increase desired. Therefore, the rate of increase is the measurement of acceleration (Fig. 3–16). Acceleration is measured in feet per second per second. The first "per second" indicates the rate of increase. The second "per second" indicates the time it takes the rate increase to take place. Deceleration is the opposite of acceleration. It is the rate at which the speed of a moving object like a ball or car is reduced. It is measured the same as acceleration.

Hydraulics

The term *hydraulics* comes from the Greek word *hydro,* meaning water, and *aulis,* meaning tube or pipe. The study of hydraulics started with the study of water at rest and in motion. It was expanded to include the study of any liquid.

Hydraulics is the science that deals with the manner in which liquids act in tanks, pipes, and enclosures. It deals with the use of the properties of liquids and their behavior on submerged surfaces.

Hydraulics is used to operate many tools and mechanisms. The brake system for the automobile is hydraulically operated. Door stops use hydraulics to ease the closing of doors. Shock absorbers also use hydraulics to dampen vibrations in cars and other vehicles.

In a hydraulic press, the piston fits tightly in the cylinder (Fig. 3–17). The liquid cannot pass between the piston and the cylinder wall.

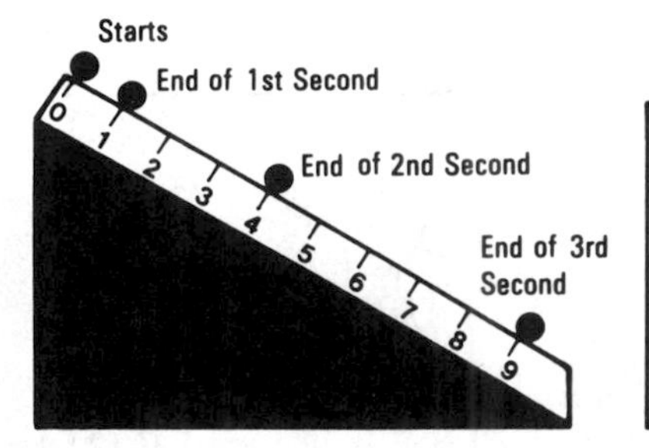

Elapsed Time (sec)	Total Distance (ft)	Distance Each Second (ft)	Final Velocity (ft/sec)
1	1	1	2
2	4	3	4
3	9	5	6
4	16	7	8
5	25	9	10

Fig. 3–16. A ball rolling down an inclined plane represents uniform acceleration. Distance rolled each second is 2′ (61 cm) greater than that rolled the second before. Its gain in velocity, or acceleration, is 2′ (61 cm) per second per second.

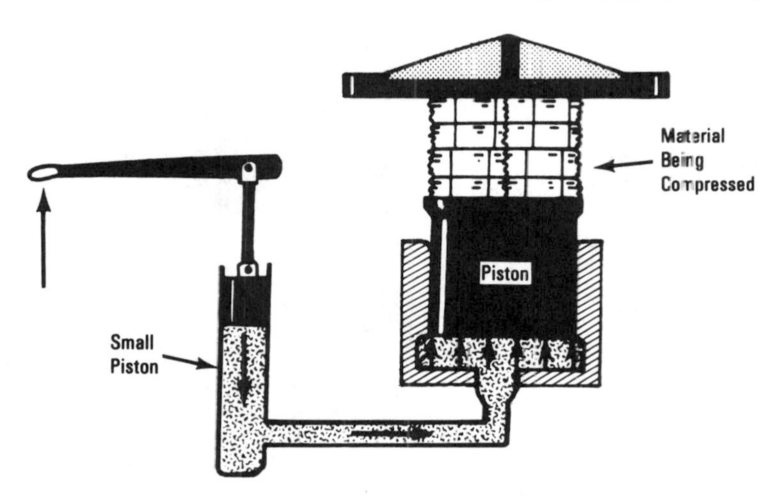

Fig. 3–17. In a hydraulic press the piston fits tightly against the cylinder. A small force on the piston transfers the energy to the larger piston.

A 1-pound weight placed on the small piston shown in Fig. 3–18 can balance a 100-pound weight on the large piston because the large piston has an area 100 times that of the small piston.

It is possible to multiply force by using a hydraulic-press-type arrangement. A mechanical advantage is obtained and efficiency is improved by this type of arrangement. Hydraulics has become one of the leading methods used to transmit and control power. Transmissions use these principles to shift gears automatically. Many applications of hydraulics are found in today's world (Fig. 3–19). Everything from elevators to power brakes and power steering uses the principles of hydraulics. Small gasoline engines are used to operate hydraulic pumps that can push a large sewer pipe under a road without disturbing the ground or having to cut through the road surface.

Heat Transfer

All heat engines convert heat energy into mechanical energy. The small gasoline engine is a heat engine. Efficiency is determined by the proportion of heat supplied as compared to the mechanical output of

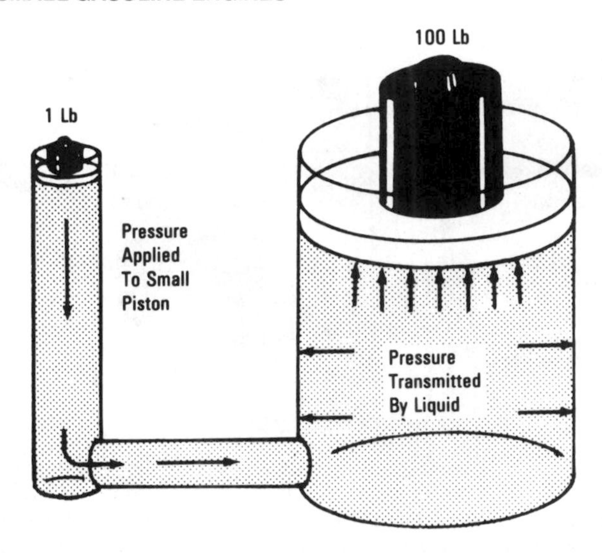

Fig. 3–18. A 1-pound weight placed on the small piston can balance a 100-pound weight on the large piston.

Fig. 3–19. A pressurized washer-sprayer is powered by a small gasoline engine.

the useful work done. Heat is generated by the use of a fuel. In this case, the fuel is gasoline.

It is important to keep the temperature of an engine within its design limits. If it gets too hot, the heat will cause the thin film of oil covering the moving parts to evaporate or migrate. This allows the surfaces to produce more friction and more heat. If too much heat is produced and too much friction is present, the engine can seize or freeze (stick). The parts will no longer move. That means that the heat generated by the engine must be transferred to some other medium. In most cases, this medium is the air that surrounds the combustion chamber (Fig. 3–20).

Heat Movement

Matter is made of atoms and molecules. These atoms and molecules are constantly moving. As they move, they tend to collide with

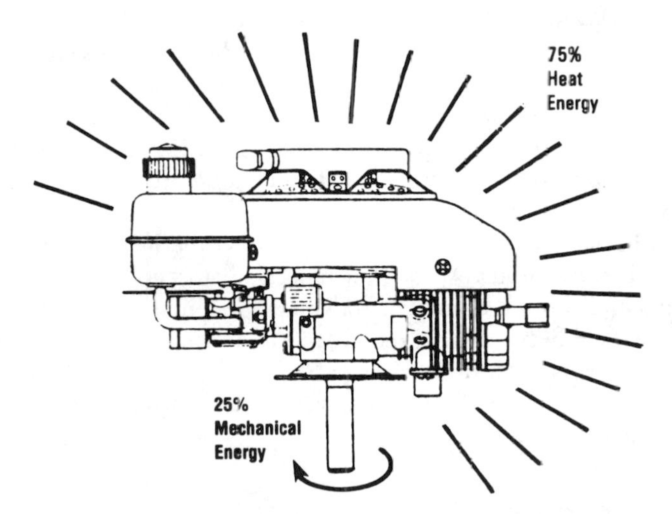

Fig. 3–20. Air has to move over the engine surfaces to carry away the excess heat generated by the operation of the engine. About 75 percent of the energy generated from the burning of the gasoline is spent in heat energy. Only 25 percent is actually available for mechanical energy.

one another. That means the heat energy is transferred by momentum or impact. Molecular action increases as the heat increases. This causes the molecules to strike with greater force and frequency. Kinetic energy that causes increased molecular movement is transferred from one molecule to another.

Keep in mind that heat will always leave a hotter surface to move to a colder one. This will happen until both surfaces are the same temperature and everything around them is also the same temperature.

Two things prevent everything on earth from having the same temperature. First, it takes time for heat to travel and, second, chemical changes store up heat as latent energy. This is done within the molecules of matter. Latent heat is stationary until a chemical change again changes it or converts it to kinetic energy and puts it in motion.

There are three ways in which heat travels: convection, conduction, and radiation.

Conduction—Heat transferred by conduction is by direct contact. It depends on the ability of the materials through which heat must pass. You can feel a hot surface. This means you make contact with the surface and your hand reacts to the transfer of heat (Fig. 3–21).

Convection—Heat is carried from a hot body to a cold one by its movement through a medium such as air or liquid. When you are warmed by hot air from a furnace, it is through convection. The hot air is a gas that transfers the energy of the hot gas to you. The gas is hotter than your body, so you feel the transfer of energy from a hot body to a cold body by way of the hot gas (Fig. 3–22).

Radiation—Heat by radiation means that the heat is transferred by rays directly through space from a hot body to a cold body. Radiant heat depends on the color of a body and the temperature difference between the radiator and the receiver. Darker colors absorb more heat than lighter colors. Absorption increases with the difference in temperature between two bodies. Reflectors can be used to change the direction of the radiant-heat travel (Fig. 3–23).

Importance of Heat Transfer

Heat transfer is very important in engines. If there is too much heat, the engine block can be damaged. All internal combustion engines produce heat. The revolutions per minute and the number of cylinders will make a difference in how much heat must be dissipated.

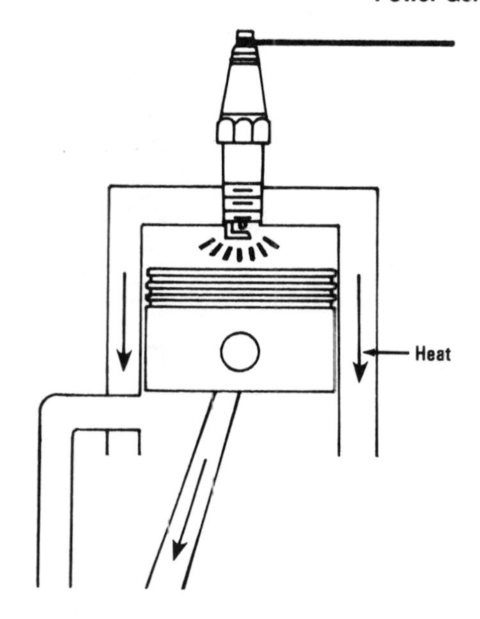

Fig. 3–21. Heat transfer by conduction.

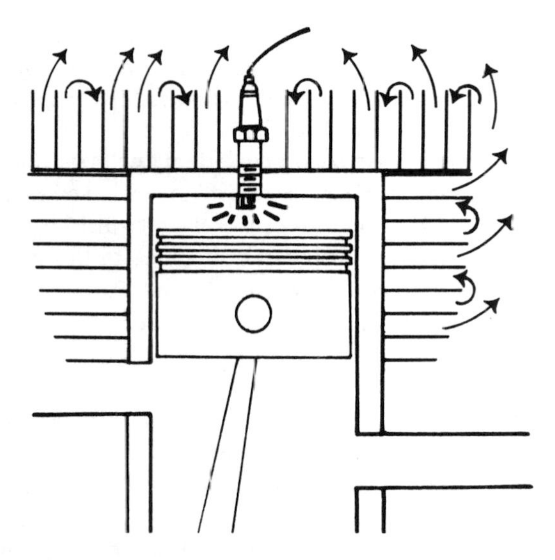

Fig. 3–22. Heat transfer by convection.

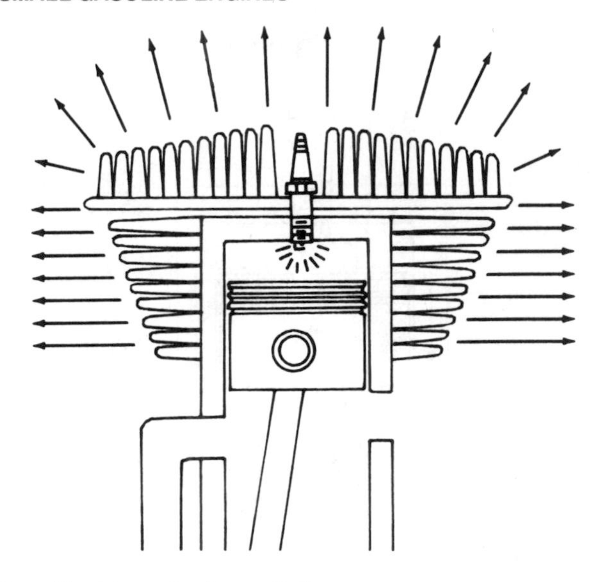

Fig. 3–23. Heat transfer by radiation.

There must be a system for getting rid of this heat. There are two ways to do this. One is the air-cooled engine, which uses large surface areas to dissipate the heat (Fig. 3–24). It can be modified by having a fan blow air over the surface area (Fig. 3–25).

The other method is the liquid-cooled engine. The liquid used is usually water. However, in instances where there will be freezing temperatures, the liquid is a mixture of water and antifreeze. Liquid is used to transfer the heat from the engine block to the radiator. The radiator usually has a fan that moves air through it. The radiator radiates the heat and the air picks it up and dissipates it over a wide area (Fig. 3–26).

As you can see, the climate in which the engine will be operated is important in the design of the heat-transfer method. Manufacturers must be aware of the temperature extremes under which their engines will operate. They must design the engines to operate properly under these conditions.

The lubricating system of an engine is subjected to great heat differences. The oil must be selected in terms of where it will operate.

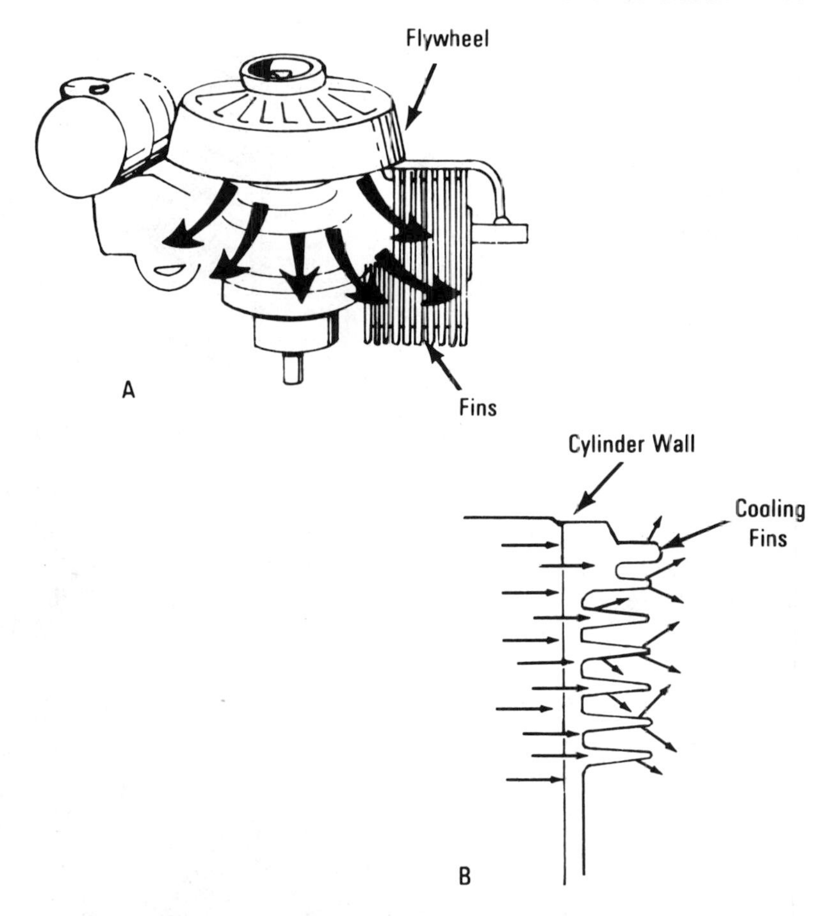

Fig. 3–24. Open cooling. The flywheel produces air movement over the cooling fins and the rest of the engine.

More about this later. Heat transfer does take place through the oil used for lubricating purposes.

Metal surfaces expanding due to heat must be taken into consideration. You don't want the piston, the cylinder block, or the cylinder head to expand at different rates and thereby produce problems. The expansion of an aluminum piston has also to be taken into consideration. Lubrication films are a consideration in the design of an engine.

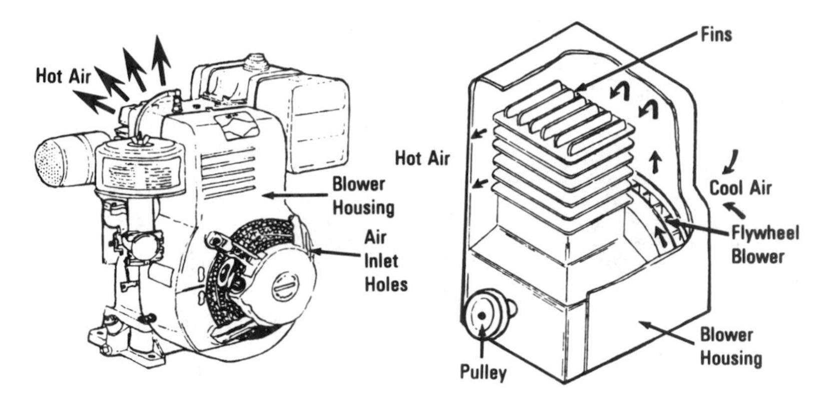

Fig. 3–25. The enclosed blower forces air over the metal surfaces of the engine.

This means that getting rid of excess heat is very important. Using the proper means of heat transfer then becomes a problem for the engine designer. It also becomes a concern of the operator and owner of the engine. That is why the manufacturer's operating specifications should be followed.

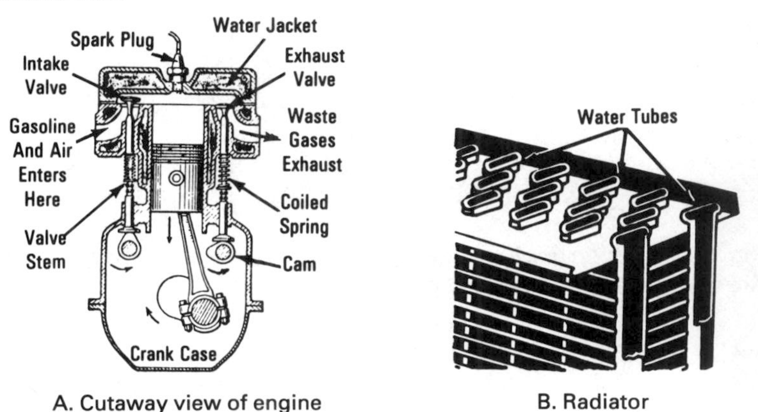

A. Cutaway view of engine

B. Radiator

Fig. 3–26. (A) Note the water-cooling jackets that surround the combustion chamber of the engine. (B) A radiator is used to transfer the heat from the cooling fluid to the surrounding air. Usually a fan is placed so it will draw air through the radiator fins and over the surfaces heated by the hot water flowing through the tubes in the radiator. Now, how do you think a radiator gets its name?

Chapter Highlights

1. Energy comes from the sun.
2. Engines and machines are thought of together.
3. An engine is a machine or energy converter.
4. Matter is anything that makes up our universe.
5. Energy is that which puts life into matter.
6. Heat energy produces warmth.
7. Light energy produces light and color.
8. Kinetic energy is the movement of a body of matter.
9. Chemical energy is change that is brought about by one substance acting on another.
10. Potential energy is energy at rest.
11. Kinetic energy is energy at work or in action.
12. Friction is present in all machines.
13. Power is the rate of doing work.
14. Work can be done by any number of means.
15. An engine is nothing more than an energy converter.
16. Pressure is the force applied to a unit area.
17. Density is how much of something you have in a given area.
18. Force is the cause of the acceleration of moving bodies.
19. Speed is how fast something travels or moves.
20. Torque is something that tends to produce a twisting action.
21. Air has pressure.
22. Molecules contain smaller particles known as atoms.
23. Force is transferred through the use of a fluid or gas.
24. Speed is the rate of motion.
25. Torque is measured in foot-pounds.
26. Motion is the act, process, or instance of changing place or position.
27. Momentum is the product of mass times velocity.
28. Acceleration is the increase on an object's speed.
29. Hydraulics deals with liquids.
30. Heat engines convert heat energy to mechanical energy.
31. Heat will always move from a hot surface to a cold surface.
32. There are three ways in which heat travels: conduction, convection, and radiation.
33. Heat transfer is very important in engines.
34. Lubricating systems of an engine are subjected to great heat differences.

35. Heat transfer dictates whether an air-cooled or a water-cooled engine is used for a particular job.

Vocabulary Checklist

absorption
atmospheric pressure
conduction
convection
density
energy
foot-pounds
force
friction
heat transfer
horsepower
hydraulics
kinetic energy
lubrication
molecule
momentum
motion
potential energy
power
pressure
radiation
revolutions per minute
speed
torque
work

Review Questions

1. What is friction? How do you use it to walk?
2. What is kinetic energy?
3. What is power? How is it defined?
4. What is an engine?
5. Define the following words:
   a. pressure
   b. density
   c. force
   d. speed
   e. torque
6. What is atmospheric pressure?
7. What is the abbreviation for revolutions per minute?
8. How is torque measured?
9. What is momentum?
10. What is hydraulics?
11. Where is energy found?

12. What are the important characteristics of motion?
13. What type of engine is the small gasoline engine?
14. What is matter made of?
15. What are the three ways heat travels?

CHAPTER 4

Engine Operation

The operation of gasoline engines is covered in this chapter. Different types of internal combustion engines are detailed. The four-stroke-cycle, the two-stroke-cycle, and the rotary combustion types are examined according to their operational principles and advantages and disadvantages. Practical applications for each type of engine are given to aid you in selection and identification later.

Classification of Engines

The term *engine* is used instead of motor because of the method of obtaining energy used to produce power. An engine uses its own fuel source to produce power; a motor is a device that uses an externally generated source of energy to produce mechanical energy. A motor converts electrical energy into mechanical energy, and also converts a liquid, gas, or solid material into mechanical energy. Engines are classified according to the type of combustion they employ to generate mechanical energy.

The *internal combustion* engine uses the principle of burning fuel inside a cylinder or container to produce power. The combustion or burning of fuel takes place inside the engine. This internal combustion takes place in three ways, which identify engines. There is the four-stroke-cycle engine, which needs four separate strokes to produce a power cycle (Fig. 4–1). Fig. 4–2 shows an assembly line putting the engines together.

The two-stroke-cycle engine completes the power cycle in only two strokes of the piston (Fig. 4–3). The rotary engine is entirely different because it does not use a piston and reciprocating motion to produce power (Fig. 4–4).

So far we have taken a look at only one of the three types of internal combustion engines. The *turbine* type of internal combustion engine is used on some aircraft. The principle involved is that of kinetic energy being directed to a rotating wheel. The wheel, in turn, is connected to a rotary shaft. This rotating motion is then harnessed to drive the means of propulsion. See Fig. 4–5 for an example of a turbine engine.

The third type of internal combustion engine is the *jet.* The jet engine compresses air and fuel and explodes them into a stream of hot, expanding gases. The expanding gases are directed out the back of the engine, forcing the jet engine forward by exerting pressure on the air behind the engine. Rockets operate on the same principle but carry their own oxygen and fuel so they can operate in the near vacuum conditions of outer space (Fig. 4–6).

The piston engine is the most commonly used and will be studied

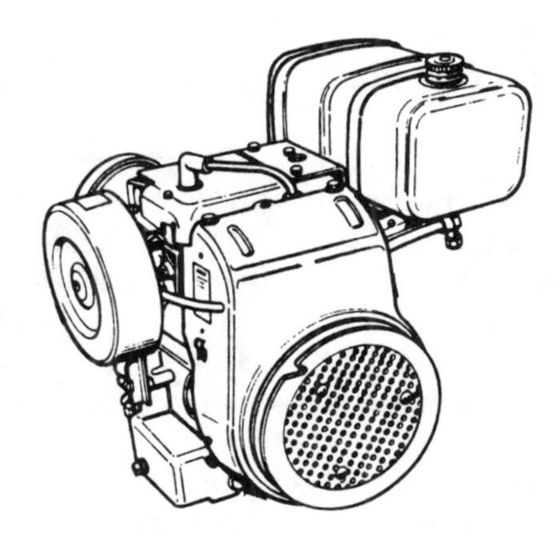

Fig. 4–1. Four-stroke-cycle engine. *(Courtesy B & S)*

Fig. 4–2. Engine assembly for 10-hp, four-cycle engine. *(Courtesy Tecumseh)*

here. It uses the exploded gases to drive a piston downward to cause a crankshaft to rotate. The rotating shaft is harnessed to drive any number of machines.

Principles of Engine Operation

In the reciprocating internal combustion engine, we make use of the straight-line piston movement. The piston is connected to the flywheel (Fig. 4–7). The downward motion of the piston caused by the exploding gases above the piston causes the shaft to rotate 180°. The momentum picked up by the shaft causes it to rotate the additional 180° to produce a complete cycle. This means the piston returns to its starting point and is ready to move again. The up-and-down motion of the piston is the basis for this type of engine operation.

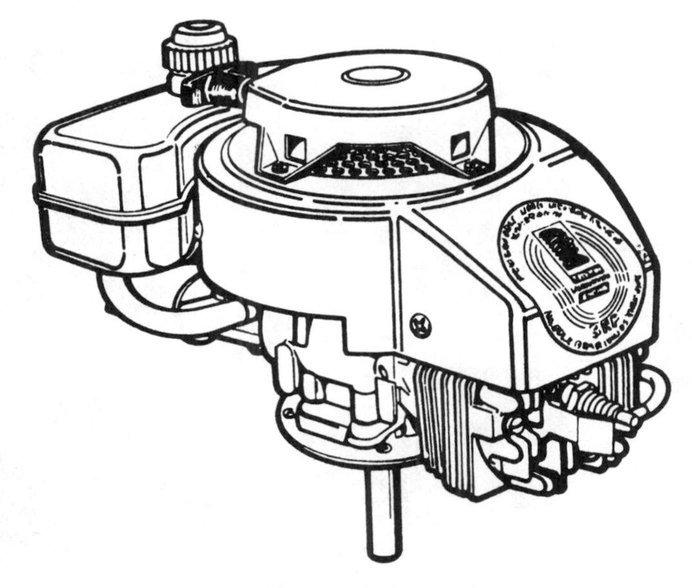

Fig. 4–3. Two-stroke-cycle engine. *(Courtesy Tecumseh)*

Let's take a closer look at what parts of the engine provide the desired mechanical energy. The cylinder consists of a cylindrical bore inside a cylinder block, with the piston fitted into this cylinder bore (Fig. 4–8). The piston must be able to move up and down freely. In order to make the piston fit tightly enough to compress the air and fuel

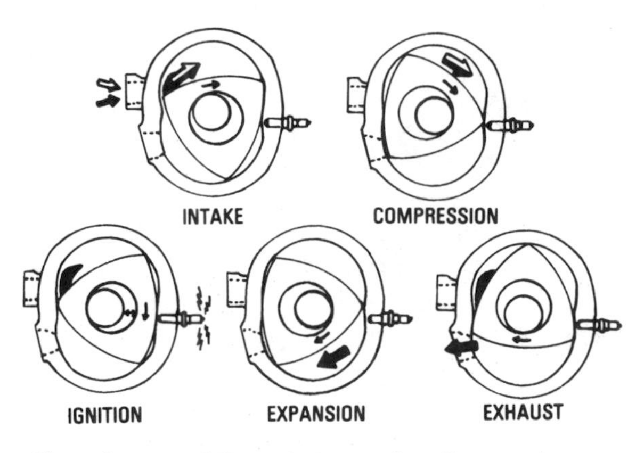

Fig. 4–4. Five phases of the rotary combustion engine.

Fig. 4–5. Turbine engine.

mixture in the combustion chamber, it is necessary to fit the piston with rings. These rings provide a seal so that the expanding gases cannot escape around the piston. The rings also support the piston when it moves up and down inside the cylinder bore.

One end of the cylinder bore is closed by a cylinder head. This piece causes the engine to have a combustion chamber between the top of the piston and the cylinder head. The combustion chamber contains the mixture of fuel and air that is exploded by a spark plug at the proper time.

A head gasket is placed between the cylinder head and the cylinder block. This gasket seals the chamber tightly and prevents leakage of hot gases when the engine is operating.

On the other end of the cylinder block is an area enclosed by a pan called the crankcase. The crankshaft, its bearings, plus the oil for lubrication are located in the crankcase.

The crankshaft is held in place with main bearings. These bearings allow the crankshaft to rotate freely. The piston is connected to the

Fig. 4–6. Jet engine.

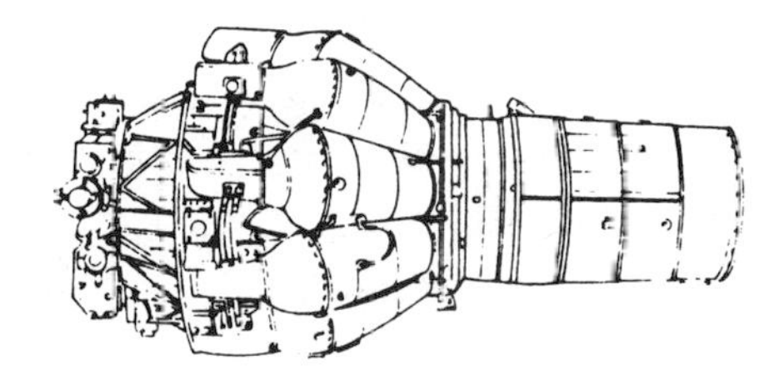

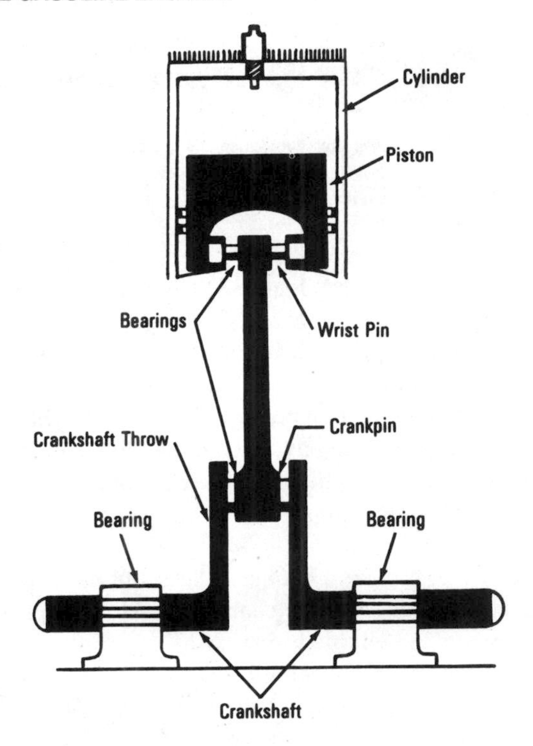

Fig. 4–7. Power-generating parts of a piston engine.

shaft by a connecting rod. A wrist pin connects the piston to the rod. Study Fig. 4–7 again to identify all of these parts.

Generation of Continuous Power

The internal combustion engine with reciprocating motion needs five distinct events to cause continuous power. These are (Fig. 4–9):

1. *Intake:* Filling the combustion chamber with a fresh charge of fuel and air, properly mixed.
2. *Compression:* Compressing the charge of air and fuel as much as possible in the combustion chamber.
3. *Ignition:* Firing the compressed charge in the combustion chamber.

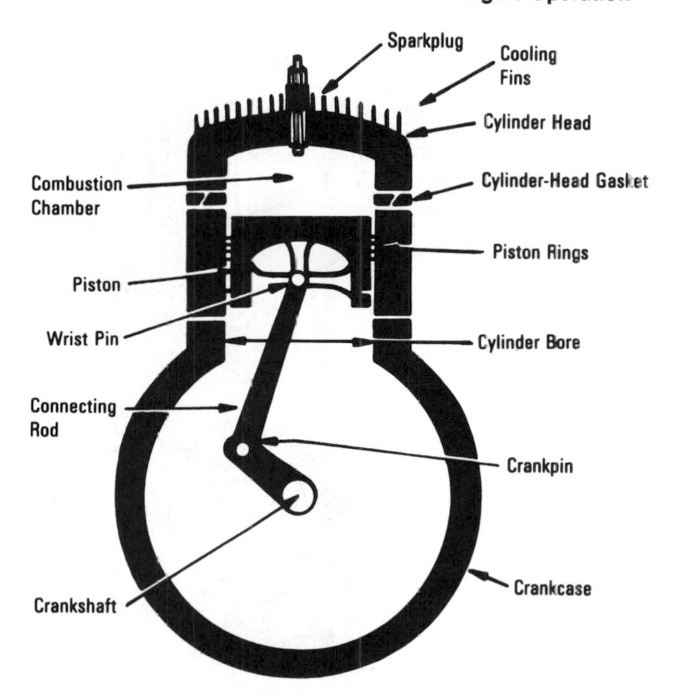

Fig. 4–8. Parts of the piston engine.

4. *Power:* Piston pushed outward as a result of the exploding charge of fuel and air in the combustion chamber.
5. *Exhaust:* Discharging the burned gases from the combustion chamber.

Note how the completion of these five events causes the crankshaft to rotate (Fig. 4–9) 360°, or to make one complete revolution.

By repeating this sequence of events over and over it is possible to produce continuous power through internal combustion.

Timing

As mentioned previously, the timing of the ignition spark is very critical to the operation of an internal combustion engine. The spark must be present at the exact moment when the maximum power will

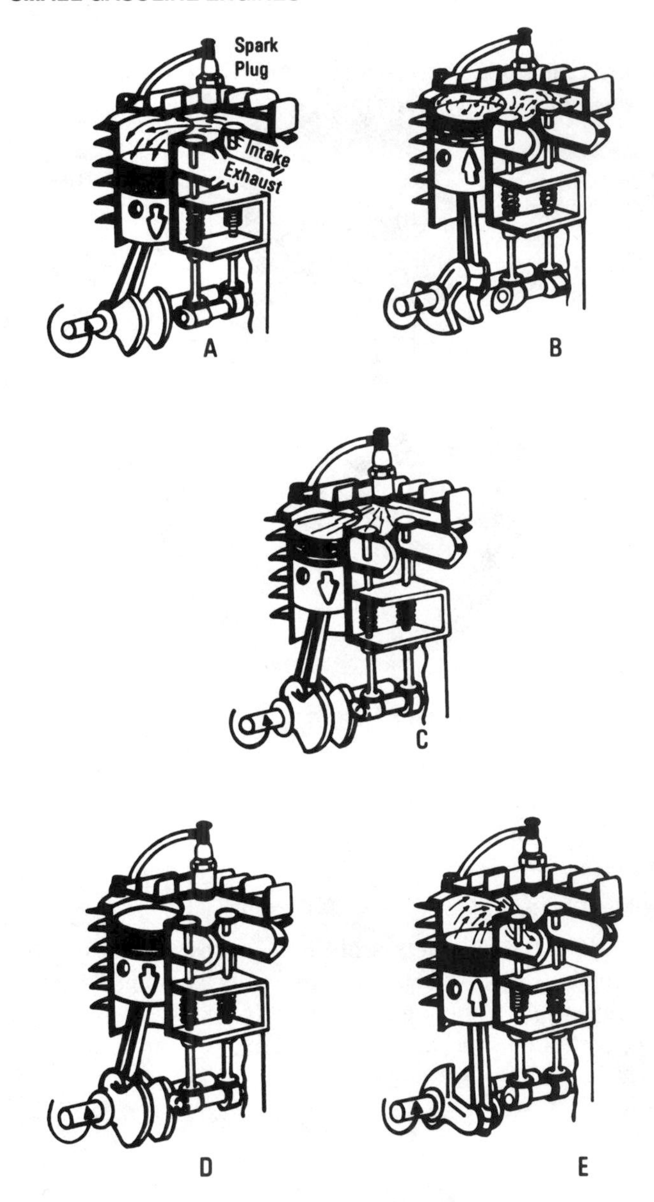

Fig. 4–9. The five events of a four-stroke-cycle engine.

be generated by the exploding fuel mixture. This timing is done by a number of methods. They will be discussed later in the book.

Carburetion

The basic purpose of a carburetor is to produce a mixture of fuel and air that will operate an engine. Atmospheric pressure must be used to make it economical enough to use on a small engine.

Atmospheric pressure may vary slightly due to altitude and temperature. However, it is a constant potent force that tends to equalize itself in any given area. It is the weight of the air in the atmosphere pushing in all directions and is commonly figured between 13 and 15 pounds per square inch. Actually, the atmospheric pressure at sea level is 14.7 pounds per square inch. We know that air moves from a high-pressure area to a low-pressure area. The greater the difference in pressure between the two areas, the greater the velocity of the air. We artificially create low-pressure areas in the carburetor and thus obtain movement of either air or fuel.

Venturi—A venturi is a restricted or constricted area in a carburetor. It is used to develop a vacuum that will draw in fuel (Fig. 4–10). To understand the operation of a venturi, think about an everyday occurrence. Have you ever noticed that the wind blowing through a narrow space between two buildings always seems to be much stronger in

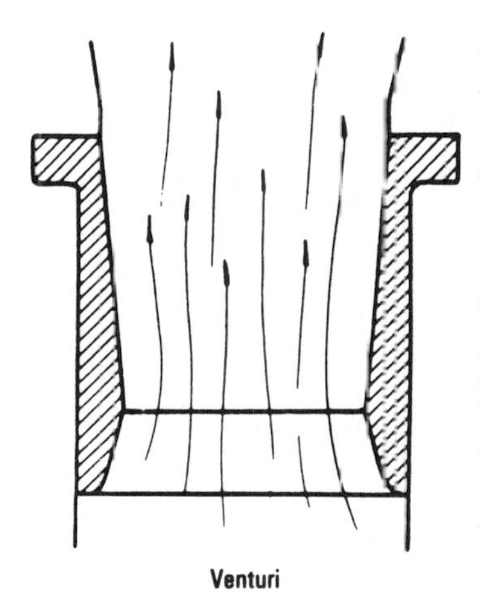

Fig. 4–10. Venturi effect caused by a restriction.

the open? In other words, its velocity is greater. The same thing can be seen in a river. The current is always faster in a narrow, shallow place than in the deep, wide pools.

These narrow spaces can be called venturis. The great amount of air or water suddenly forced through a constricted space has to accelerate in order to maintain the volume of flow. This is the way a venturi is placed in a carburetor. The shape is carefully designed to produce certain air flow patterns.

An *airfoil* is similar to the tube placed in Fig. 4–11. The tube is standing still and has the same pressure on all sides. Once the air is moving around it, or the tube itself is moving, there is a difference of pressure on the two sides. This becomes important when we place a nozzle across a wind stream. The venturi and the airfoil play important roles in the making of a carburetor where you want to mix air and fuel. The idea is to burn the gasoline in vapor form instead of liquid form. The carburetor's job is to mix the air and fuel so that a vapor is drawn into the combustion chamber.

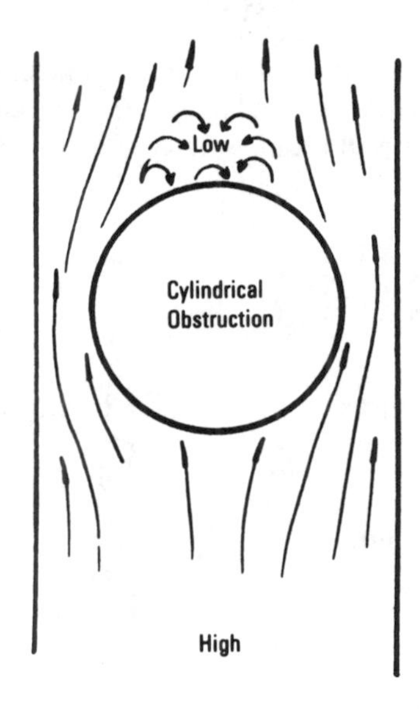

Fig. 4–11. An airfoil creates a difference of air pressure.

Fig. 4–12 shows a gravity-feed carburetor. In this type, the fuel flows down from a tank mounted above the carburetor. Here is shown the position of the nozzle and the fuel level. The fuel in the bowl seeks its own level, which is well below the discharge holes. Notice that the discharge holes are in the venturi, the place of greatest air velocity. As the piston in the cylinder moves down with the intake valve open, it creates a low-pressure area that extends down into the carburetor throat and venturi.

Two things start to happen. The air pressure above the fuel in the bowl pushes the fuel down in the bowl and up in the nozzle to the discharge holes. At the same time the air rushes into the carburetor air horn and through the venturi, where its velocity is greatly increased.

The nozzle extending through this air stream acts as an airfoil, creating an area of still lower pressure on the upper side. This allows the fuel to stream out of the nozzle through the discharge holes into the venturi, where it mixes with the air and becomes a combustible mixture ready for firing in the cylinder.

A small amount of air is allowed to enter the nozzle through the vent. This air compensates for the difference in engine speed and prevents too rich a mixture at high speed.

Other things must also be considered for making a practical carburetor. They will be examined in detail in Chapter 7.

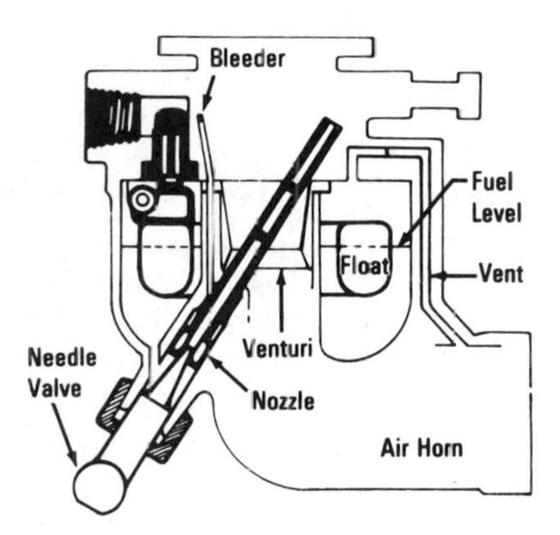

Fig. 4–12. Small-gas-engine carburetor with the venturi and nozzle located.

Construction of Reciprocating Engines

The four-stroke-cycle engine is the type most often used for small engines. The two-stroke-cycle engine is also used for smaller devices and outboard motors for boats.

Four-Stroke-Cycle Engines

The automobile uses the four-stroke-cycle engine for its power plant. A limited number of rotary combustion types were produced in the 1970s. The two-stroke-cycle engine has been used primarily for snowmobiles, lawn mowers, chainsaws, weed cutters, model airplanes, and outboard motors for boats. There are advantages and disadvantages for each type of engine.

Fig. 4–13 illustrates the parts in a four-stroke-cycle engine used in a lawnmower. This type has a vertical crankshaft so it can be attached to a lawnmower blade. The same type of engine is shown in Fig. 4–14, but with a horizontal crankshaft for devices driven from the side of the engine. This type is used to drive generators for alternating-current power plants. Examine the illustrations and become familiar with the parts and their names. You will be using them as you study more about the engines.

Four strokes of the engine, shown in Fig. 4–15, are needed to make a complete revolution of the crankshaft. In *A* the downward stroke of the piston creates a partial vacuum. The intake valve opens and atmospheric pressure forces the gas-and-air mixture into the combustion chamber. In *B*, the intake and exhaust valves are closed as the piston starts its upward movement. During this upward movement the piston compresses the mixture from 70 to 125 pounds per square inch, or about 1⁄10 of the space it would ordinarily occupy. Next, in *C* the compressed fuel-and-air mixture is ignited by an electric spark from the spark plug. The heat of the combustion causes an expansion of gases within the combustion chamber and the cylinder. This forces the piston downward. The exhaust valve opens on the exhaust stroke (*D*) and allows the burned gases to escape as the piston moves upward. Once the chamber is exhausted of burned gases, the cylinder is ready for another cycle.

Note the steps in the four-stroke-cycle engine: intake, compression, power, and exhaust. Each stroke plays an important part in mak-

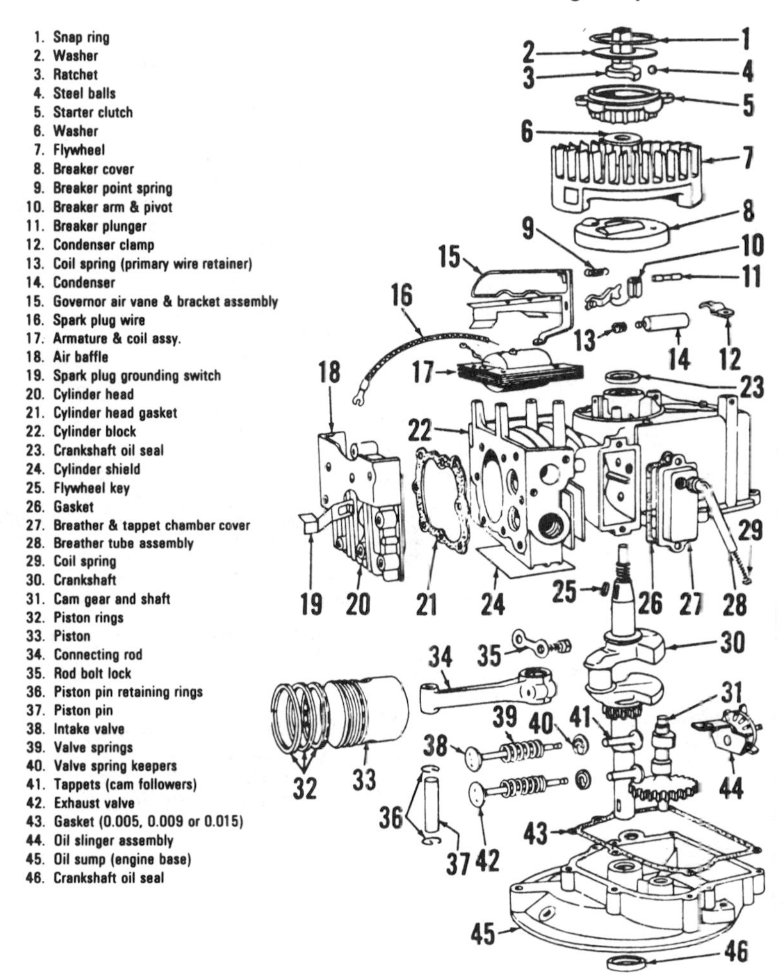

Fig. 4–13. Exploded view of a four-stroke-cycle engine with vertical crankshaft. *(Courtesy B & S)*

ing the engine operate. In the four-stroke-cycle engine the crankshaft has to rotate 360° twice in order to complete the operation.

There is also a four-stroke-cycle diesel engine. This engine is known as the compression-ignition engine. It uses the pressure and heat of the compression stroke to cause the fuel to ignite. Fig. 4–16

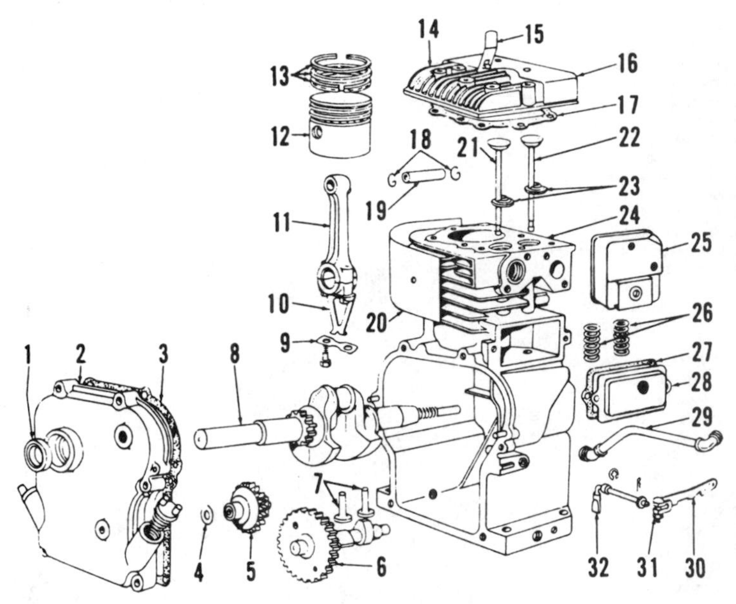

1. Crankshaft oil seal
2. Crankcase cover
3. Gasket (0.005, 0.009 or 0.015)
4. Thrust washer
5. Governor assembly
6. Cam gear and shaft
7. Tappets (cam followers)
8. Crankshaft
9. Rod bolt lock
10. Oil dipper
11. Connecting rod
12. Piston
13. Piston rings
14. Cylinder head
15. Spark plug ground switch
16. Air baffle
17. Cylinder head gasket
18. Piston pin retaining rings
19. Piston pin
20. Air baffle
21. Exhaust valve
22. Intake valve
23. Valve spring retainers
24. Cylinder block
25. Muffler
26. Valve springs
27. Gasket
28. Breather & tappet chamber cover
29. Breather pipe
30. Governor lever
31. Clamping bolt
32. Governor crank

Fig. 4–14. (A) Exploded view of a four-stroke-cycle engine with horizontal crankshaft. *(Courtesy B & S)*

Fig. 4–14. (B) Final inspection and assembly of four-cycle engines. *(Courtesy Tecumseh)*

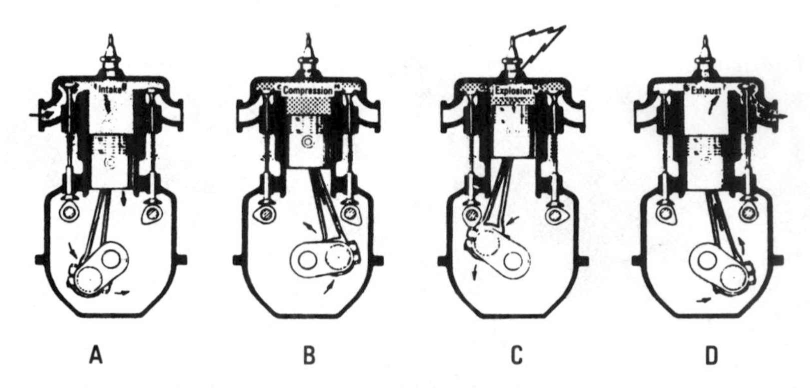

Fig. 4–15. Four strokes of a gasoline engine. *(Courtesy Lawn Boy)*

shows how this type of engine operates. In *A* the fresh air is drawn into the combustion chamber. This is just like the first step in the spark-ignition engine above. On the second stroke (*B*) the intake and exhaust valves are closed as the piston moves upward. This compresses the air to about 625 pounds per square inch, or 1⁄15 of its original space. At the proper moment (*C*), fuel is injected into the chamber where the very hot air is compressed. High compression creates enough heat to ignite the injected fuel. This causes the piston to be forced downward as the fuel burns or explodes. The next step or stroke (*D*) sees the exhaust valve open to permit the burned gases to escape. The piston moves upward to expel the burned gases. Once the four strokes are com-

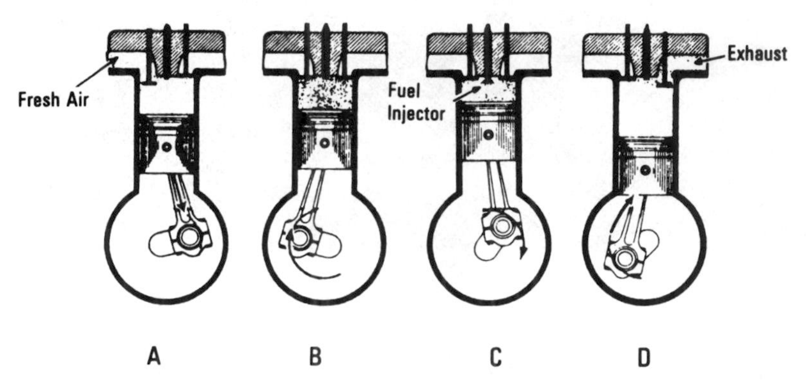

Fig. 4–16. Four strokes of a diesel engine.

pleted, the cylinder is ready to start over again. The operation of an internal combustion engine relies on a series of rapid strokes to create power.

Two-Stroke-Cycle Engines

The two-stroke-cycle engine completes its operation in two strokes instead of four. It does require some different considerations, however. The two-cycle engine needs an oil-and-gasoline mixture for fuel because the crankcase is used to bring in the fuel.

Fig. 4–17 shows how the two-cycle engine operates. In *A*, the fuel-and-air mixture enters the combustion chamber on one side of the cylinder. The domed piston head directs the mixture to the top of the combustion chamber. It also assists in sliding the exhaust gases out through the exhaust ports. In *B*, the loop scavenge design uses the vacuum-activated reed valve. Here the ports are located on three sides of the cylinder. The intake ports are on two sides opposite each other and the exhaust ports are shown by the three holes above the head of the piston.

The flat piston is used in this design. As the mixture shoots into the combustion chamber through the two sets of intake ports, it collides with the piston and is directed to the top of the combustion chamber, looping when it strikes the cylinder head. This forces all spent gases out through the open exhaust ports.

In *C*, the loop scavenge design is again used, but the reed valve is eliminated. The carburetor has been moved from the lower crankcase end to the cylinder. Along with the intake and exhaust ports, a third port has been added. The third port forms the passageway from the carburetor to the crankcase and is opened and closed by the piston skirt as the piston moves back and forth in the cylinder.

The carburetor appears to block the path from the crankcase to the intake ports. However, these gases pass around the carburetor and the third port.

In *D*, essentially the same loop scavenge, third-port design is shown as in *C*. The only difference is the two small reed valves located on the adapter cover from the carburetor to the third port. They are opened early by the vacuum created as the piston starts to rise. An extra amount of fuel-air mixture passes through these reed valves before the piston skirt opens the third port. This extra charge of fuel-air mixture increases the engine horsepower.

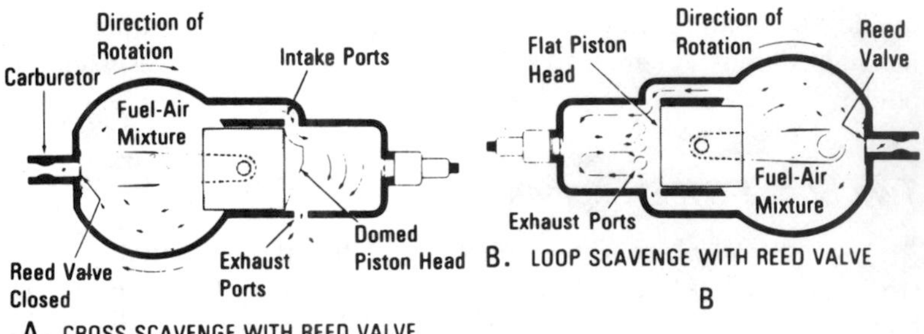

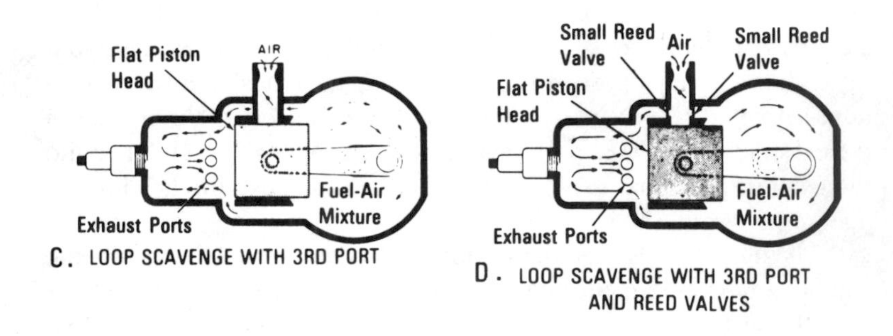

Fig. 4–17. (A) Two-stroke-cycle engine using a cross scavenge with reed valve. The reed valve is a flap or flutter valve that is activated by crankcase pressure. A reduction in crankcase pressure opens the valve, allowing the fuel-air mixture and oil mixture to enter the crankcase. Increased crankcase pressure closes the valve, prohibiting the escape of the fuel-air and oil mixture back through the carburetor. (B) Two-stroke-cycle engine with loop scavenge with a reed valve. (C) Loop scavenge with a third port. The reed has been eliminated. (D) Loop scavenge with a third port and reed valves. *(Courtesy Tecumseh)*

Fig. 4–18 shows the moving parts of the four-cycle and two-cycle engines so you can compare them. The two-cycle runs at a higher rpm in order to produce power. Four-cycle engines require an oil sump and a circulation system for lubrication of moving parts. The two-cycle engine does not need a lubrication system. It is able to obtain lubrication

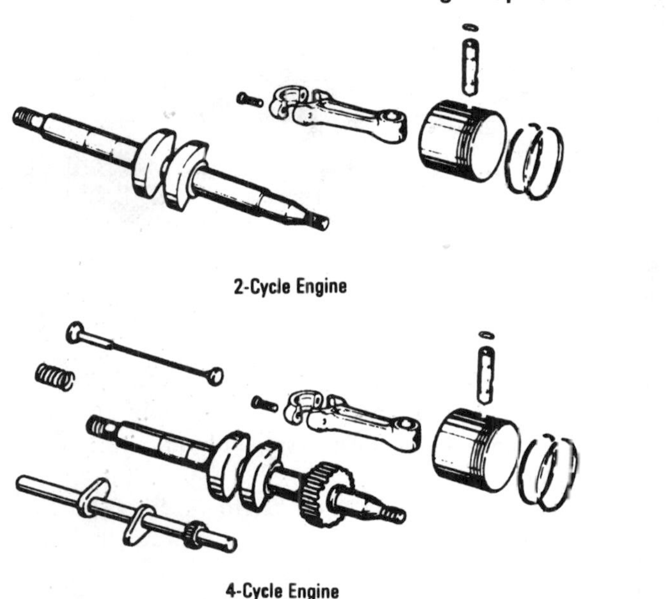

Fig. 4–18. The moving parts of a four-cycle and a two-cycle engine. Note the smaller number of moving parts in the two-cycle.

by using an oil-fuel mixture. These engines are lightweight and portable. They do run at a higher speed than four-cycle engines so they can produce the required power and still remain relatively small and portable. This makes them ideal for devices requiring small gasoline engines, such as model airplane engines, chainsaws, and weed cutters.

A quick comparison of the two- and four-stroke-cycle engines is made in Fig. 4–19. Note how two power strokes are made in the two-cycle operation in the same time as one is made in the four-cycle engine.

Rotary Combustion Engines

The rotary combustion engine operates on the same five events that the four-stroke-cycle reciprocating engine does: intake, compression, ignition, power, and exhaust. The rotor housing serves as the cylinder block. The rotor is the piston assembly. However, it does have fewer parts than the piston-type engine. The main advantage of the

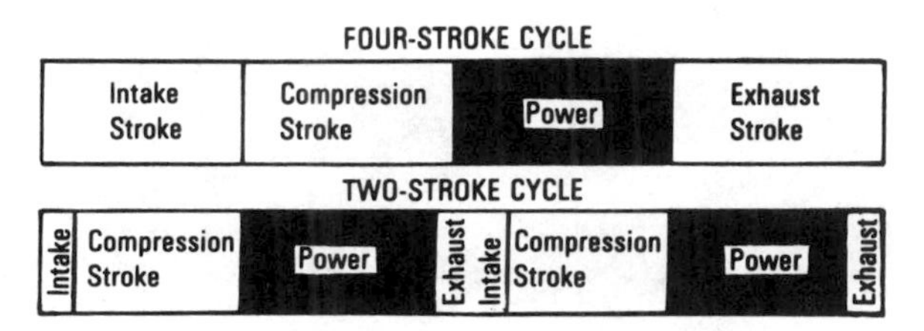

Fig. 4–19. Four-stroke-cycle and two-stroke-cycle engine cycles compared.

rotary combustion engine (called a Wankel engine) is its weight and size. Though it does not exceed the fuel consumption rate of the four-stroke-cycle engine, it runs more smoothly and quietly, with a hum rather than a roar.

Examine the illustration of the engine in Fig. 4–4. Note that the three chambers are formed by the sides of the rotor and the wall of the housing. The shape, size, and position are constantly being changed by the rotor's clockwise rotation. The rotor opens the intake port. The port has no speed-restricting valve mechanism. The fuel-and-air mixture enters through the port as it does in the four-cycle engine. As the rotor passes the port, it closes it and the compression phase begins. The next phase is ignition, the firing of the charge of air and fuel. The burning fuel creates the expansion phase, or power stroke. The expanding gases push the rotor until the exhaust port opens. The exhaust then escapes as the rotor seal passes the port hole. This completes the cycle. The cycle is repeated to keep the rotor moving and the power production constant. Some small engines were made with this design. Mazda made the engine for its whole line of cars for a few years. General Motors and Ford had rights to develop the design, but found no particular fuel consumption advantage in the Wankel. Furthermore, it would not meet pollution requirements.

Valve Operations

The intake and exhaust valves have already been mentioned in this chapter. They do what their names imply. One allows the fuel-air mixture into the combustion chamber; the other opens to allow the very hot exhaust gases to escape from the combustion chamber and cylinder. As can be seen, this means that one operates at a much higher temperature than the other. Each four-stroke-cycle engine must have two valves for each cylinder. These valves have to be timed to open at the right time and to close when their job is done. That means some type of cam arrangement in most instances. Therefore, we will take a closer look at the valves and cams in this part of the chapter.

Fig. 4–20 illustrates the location of the valves and their operation. In *A* the intake stroke takes place. This means the exhaust valve is

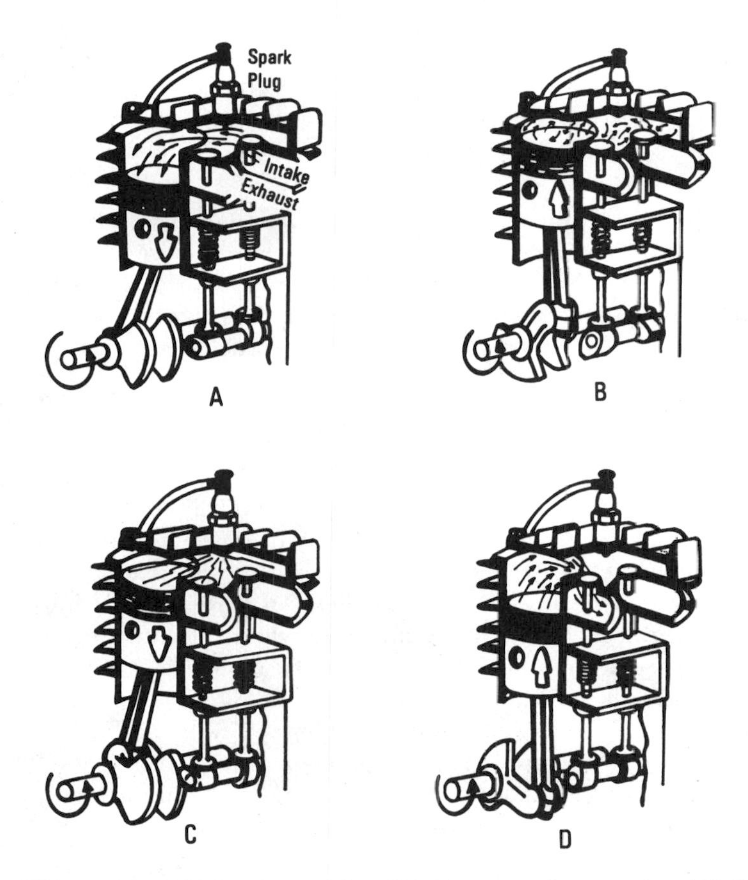

Fig. 4–20. (A) Intake stroke. (B) Compression stroke. (C) Power stroke. (D) Exhaust stroke. *(Courtesy B & S)*

closed and the intake valve is open. The piston is moving downward and the air-fuel mixture is drawn into the cylinder. Then, in *B*, the intake valve closes and the piston moves upward on the compression stroke. The air-fuel mixture becomes greatly compressed in the small space between the top of the piston and the cylinder head.

C shows the spark firing inside the combustion chamber. The ignition of the fuel-air mixture forces the piston down as the gases (hot from ignition) expand. This is the power stroke. Note how both valves are closed at this point. Next, in *D*, the exhaust valve opens and the upward movement of the piston pushes the burned gases out of the open exhaust valve. Then the exhaust valve closes. The intake valve opens. The engine is ready to repeat the cycle.

Timing

It is generally conceded that the valves are the most important factor in good compression. They operate under more severe conditions than any other parts of the engine. This is particularly true of the exhaust valve. The valves open and close in little less than one revolution. When the engine is operating at 3000 rpm, each valve opens and closes in about 0.02 second.

Not only do valves have to open and close in 0.02 second, they have to seal well enough to stand pressures up to 500 pounds per square inch. When the engine is under full load, the exhaust valve is red-hot. The temperature of the valve under full load may be 1200°F or more (Fig. 4–21). The intake valve is cooled by the incoming fuel mixture. The exhaust valve is subjected to high-temperature exhaust gases passing over it on the way out of the cylinder. It is therefore very difficult to cool the head of the exhaust valve. The cylinder head, the cylinder, and the top of the piston are exposed to this same heat, but these parts are cooled by air from the flywheel fan and oil from the crankcase. Very special steel is required in the exhaust valve to enable it to withstand the corrosive action of the high-temperature exhaust gases.

The small gasoline engine is a single-cylinder engine with only two valves, as compared to twelve or sixteen in an automobile engine. The fewer the valves, the more important each one becomes.

In a one-cylinder engine, one bad valve can cause a great drop in horsepower, or cause the engine to stop entirely. In a multicylinder

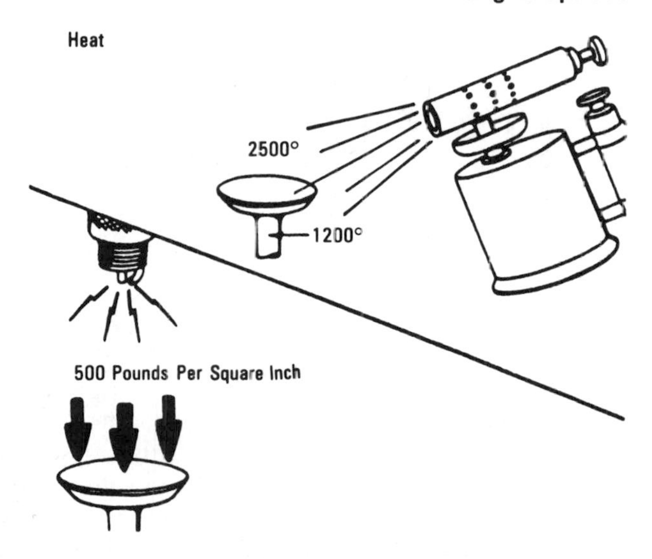

Fig. 4–21. Heat and pressure on valves. *(Courtesy B & S)*

engine, one valve may fail and only 1/6 or 1/8 of the power is affected as the bad cylinder may be motorized by the other good cylinders. Hence, good valve condition is even more important in a one-cylinder engine than it is in multicylinder engines.

Fig. 4–22 shows some problems with valves. When valves are removed, they should be cleaned with a wire brush wheel to remove the carbon deposits. The carbon also has to be removed from the valve guides. When the valves are clean, they should be checked for smoothness and fit. While valves become burned to some extent, it is very seldom that a valve seat or face will be very badly burned. Dished or necked valves are almost never found in repair work on small engines.

Valve-seat burning is usually caused by an accumulation of carbon or fuel lead either on the valve stem or on the valve face, or it may be from insufficient tappet clearance. These deposits on the valve stem or on the face will hold the valve open, allowing the hot flames of the burning fuel to eat away the valve face and seat. A dished valve is one that has a sunken head. This is caused by operating at too high a temperature with too strong a spring. Highly leaded fuels can also erode the head. A necked valve is one in which the stem directly beneath the head has been badly eaten away by heat or corrosion.

Fig. 4–22. Types of valve failures. *(Courtesy B & S)*

Valve Sticking

Valve sticking is caused by fuel lead, gum, or varnish forming on the valve stem and in the valve guide. Most of the deposits are formed by fuel lead. Since the amount of lead in different fuels varies, the rate of deposit buildup naturally varies. When an exhaust valve no longer closes properly due to excess lead deposits, the hot gases escaping from the combustion chamber heat up the valve stem and guide excessively. This causes the oil on the valve stem to oxidize into varnish, which holds the valve partially open and causes burning. Intake valves sticking may be caused by the use of fuels having an excessively high gum content. Fuels stored for too long may contain high amounts of gum.

Fig. 4–23 illustrates the parts of the valve and the location of the valve in the engine. If burning occurs in a rather limited area on the valve face, it indicates that something may have caused the valve to tip. This could be due to a bent valve stem or a deposit on one side of the valve seat or stem. Such a condition would leave an opening for the passage of hot exhaust gases, which could burn the valve so badly that it could not be refaced. These valves must be discarded.

The important parts of a valve are the head, margin, face, and stem. They make contact with the seat and the valve guide in the cylinder. The margin is the edge of the valve head. As a general rule, a valve should be discarded when the margin is reduced to less than half of its original thickness (Fig. 4–24).

The margin on a Briggs & Stratton valve is 1⁄32″ (0.794 mm). Therefore, when it becomes less than 1⁄64″ (0.397 mm), the valve should be discarded. Remember, this is after all pit and burn marks have been removed from the valve face. If the valve is bent, the face will be ground unevenly; and if the margin becomes too thin on one

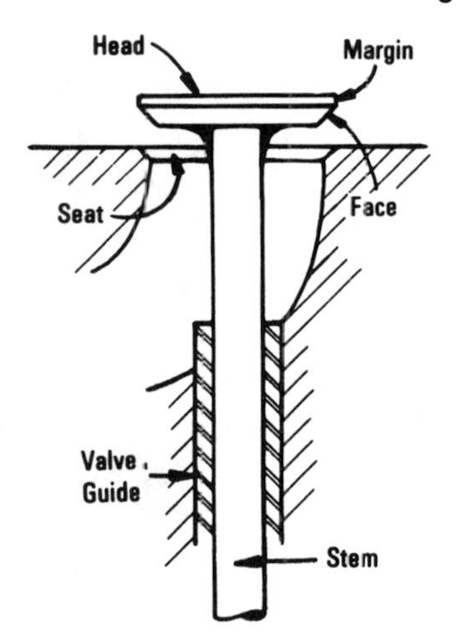

Fig. 4–23. Valve parts. *(Courtesy B & S)*

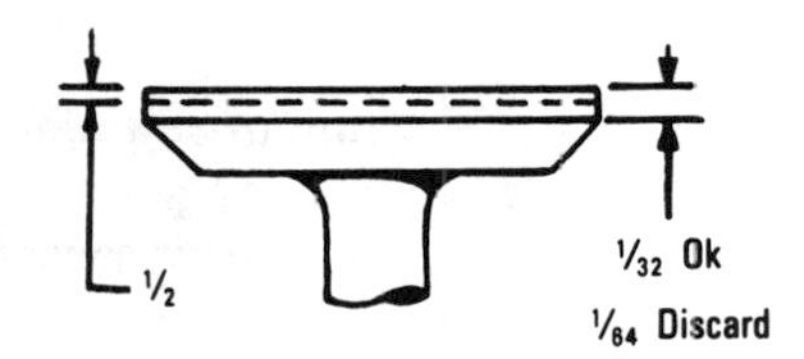

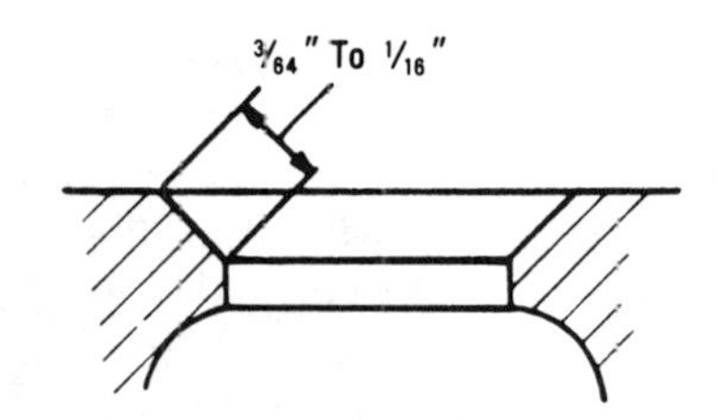

Fig. 4–24. Dimensions of valves. *(Courtesy B & S)*

side, the valve should be discarded. A valve with too thin a margin will not be able to withstand the heat and will quickly crack and burn. After facing the valves and the valve seats to 45°, place a little grinding compound on the valve face and very lightly lap the valve to the seat. Use fine-grinding compound to remove any grinding marks and give a clear picture of the valve seat width.

The valve seat width is usable to 5⁄64″ (1.984 mm). But a new seat should be between 3⁄64″ (1.191 mm) and 1⁄16″ (1.587 mm). It should be in the center of the valve face. After the valve seat and faces are ground, the valve should be installed in the guide, the cam gear turned to the proper position, and the tappet clearance checked. Refer to Chapter 15 for the engine tappet clearance. Usually the clearance will be too small and the end of the valve stem will have to be ground off to obtain the proper valve clearance. Care should be taken not to overheat the end of the valve stem while this grinding is taking place. Be sure the end is square with the stem. It is recommended that the valve springs and retainers be assembled immediately after setting the tappet clearance, to prevent dirt getting under the valve seat.

Power-Producing Components

So far we have looked at the carburetion, the ignition, and the basic construction of the small one-cylinder gasoline engine. There are some important parts which play a role in making the machine a power-producing device. These parts are the piston, piston rings, connecting rod, crankshaft, and flywheel.

Piston Displacement

Piston displacement is measured in cubic inches or cubic centimeters. This is the amount of volume the piston will occupy and then vacate in its downward movement. Top dead center (TDC) means the piston is at its topmost position of upward travel. Bottom dead center (BDC) is the lowest the piston will go in its downward travel (Fig. 4–25). The piston's length of travel up and down the cylinder is called the stroke length. If you take the stroke length and multiply it by the square of the radius of the piston and multiply that by pi (π), a constant of 3.14159, you can find the cubic inch displacement of the piston:

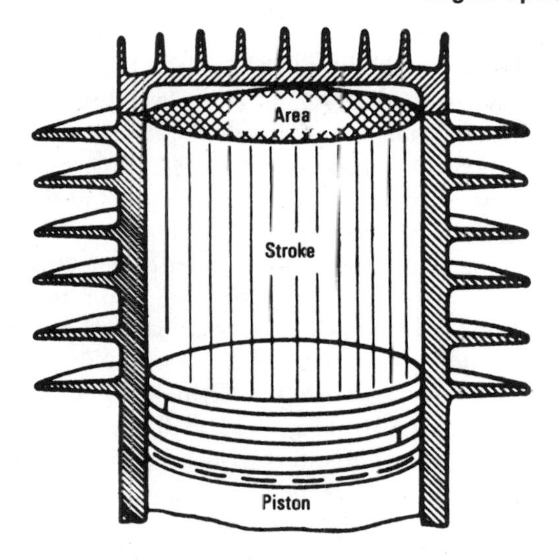

Fig. 4–25. Piston displacement measurements.

Displacement = stroke (in inches) × radius² × π

An example would be to take a piston of 2⅜″ (6 mm) bore or diameter. The radius is 1.1875″ which is the diameter divided by 2, or 1³⁄₁₆″ (3 mm). The stroke is 1⅞″ (4.75 mm). Substitute the information in the formula and you have (1.1875″ = 1³⁄₁₆″; 1.875″ = 1⅞″) 1.1875″² = 1.41015625″:

1.41015625″ × 1.875″ × 3.14159 = 8.3065 cubic inches

So the displacement of the engine is 8.3 cubic inches. Manufacturers tell you the displacement so you know the engine's size. The displacement determines the maximum power the engine can produce. It is the amount of charge the engine pulls into the cylinder each time the piston moves downward.

Compression Ratio

Compression ratio is another term used in regard to the piston and cylinder operation in the production of power. It is the ratio of the maximum volume inside the cylinder at the beginning of the compression stroke divided by the minimum volume remaining once the piston has reached the end of its compression stroke (Fig. 4–26). That means the measurement of the engine at BDC and TDC. In the case of the engine we used for our piston displacement calculation, if the piston displacement is 8.3 cubic inches, this means it has the 8.3 cubic inches at BDC. The air-fuel mixture is then compressed into a space at the top of the piston and between TDC and the cylinder head of about 1.3 cubic inches. That would give the engine or cylinder a compression ratio of 8.3 to 1.3, or an 8 to 1 ratio, written as 8:1. That would be a very high compression ratio for a one-cylinder engine, so it would be very hard to turn over or start manually. Therefore, most of these small engines have a 4:1 ratio, meaning that the 8.3-cubic-inch engine would have a TDC volume of 2.075 cubic inches. Most small gasoline engines will have compression ratios of around 6:1. Automobile engines (where electric starters are used) have even higher compression ratios. This indicates the engine efficiency. The greater the compression ratio, usually the greater percentage of the fuel's energy is released on ignition.

Diesel engines have very high compression ratios to ignite the fuel without a spark. They also have very heavy-duty moving parts to withstand the extra forces placed on the entire engine.

Do not change the compression ratio of an engine unless you know that the engine parts will withstand the added stress.

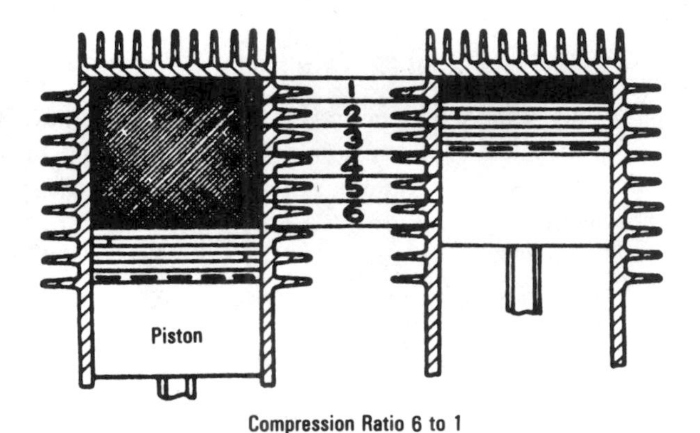

Fig. 4–26. Compression ratios.

Pistons

Pistons may be made of cast iron or aluminum. Each type has distinct advantages and disadvantages. The aluminum type is lighter. However, it will expand at a rapid rate when heated. This has to be taken into consideration when designing the piston. The greatest disadvantage of the cast-iron piston is its weight. See Fig. 4–27 for typical pistons used in small engines.

Piston Rings

Pistons have rings or grooves where rings are placed to improve their ability to compress air and fuel without too much "blow-by" (Fig. 4–28). There are two types of piston rings. The compression rings are placed in the top groove on the piston. In some cases there are two compression rings to improve fit between the piston and the cylinder wall. They stop gas leakage past the piston and into the oil sump below. Oil rings are placed below the compression rings. The oil ring is also shaped a little differently. The purpose of this ring is to allow the crankcase oil in a four-cycle engine to lubricate the piston when it is moving up and down within the cylinder bore.

Keep in mind that two-cycle engines do not need oil rings since the oil and gas are mixed and the oil is available when the fuel charge is pulled into the combustion chamber. The bottom ring in a two-cycle engine is called a scraper, or carbon control ring.

Note that gaps are staggered when the rings are placed on the piston. The gaps are open when the engine is cold, and loss of compression may occur if the gaps are aligned.

A number of ring types are available. Just make sure you have the right type and size for the engine on which you are working.

Piston Rods

The piston rod is called a connecting rod since it connects the piston to the crankshaft (Fig. 4–29). The wrist-pin end connects to the piston. Inside the circular part is a bushing. This has to be lubricated as described in Chapter 9. The end of the rod connected to the crankshaft has a larger hole. It also has a bushing or bearing. Different manufacturers use different types of bearings or bushings. Three types of bear-

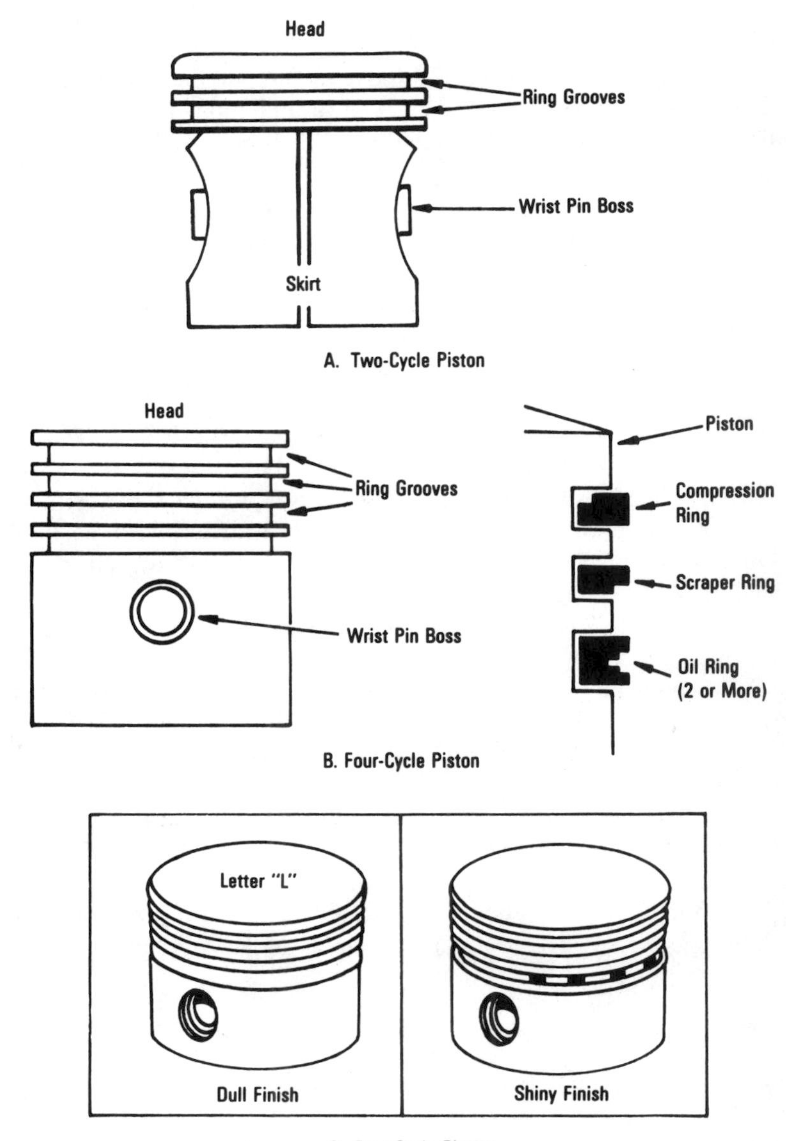

Fig. 4–27. Pistons of two- and four-cycle engines. *(Courtesy B & S)*

Note especially the center compression ring. The scraper groove should always be down toward the piston skirt. Be sure the oil return holes are clean and carbon is removed from all grooves. NOTE: Install expander under oil ring, when required.

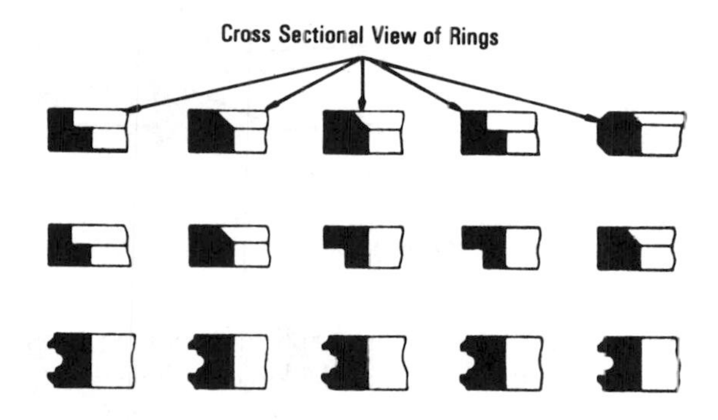

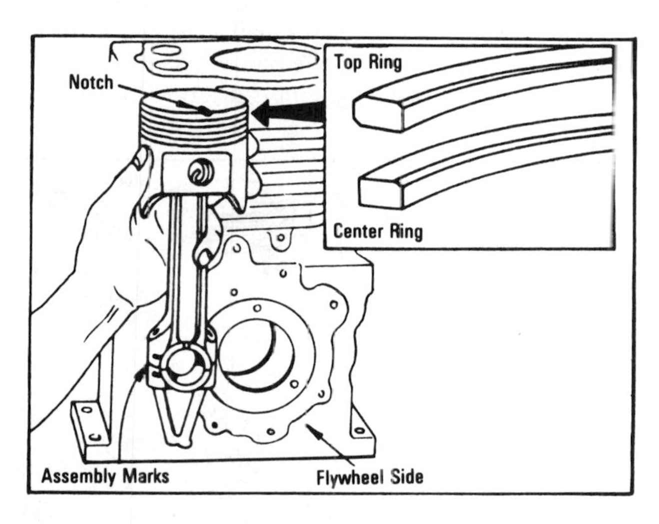

Fig. 4–28. Piston rings and their fits. *(Courtesy B & S)*

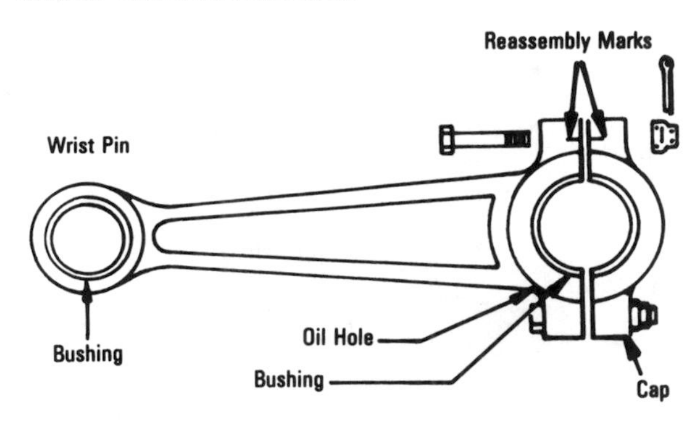

Fig. 4–29. A piston connecting rod.

ings will be discussed later in this chapter. The wrist pin slips through the piston and the rod. It may have locking pin holes or it may have a locking cap screw to keep it in place and attached. Some of the rods have a sleeve-type bearing. The wrist pin floats in this type. In such cases the wrist pin is fastened into the piston to keep it from rotating or moving endways (Fig. 4–30).

The lower end of the connecting rod has a cap that is bolted to the rod. This secures it around the crankshaft crankpin (Fig. 4–31). In some instances the crankshaft has the power takeoff end and the flywheel end in two pieces. The crankpin is inserted to hold the crank-

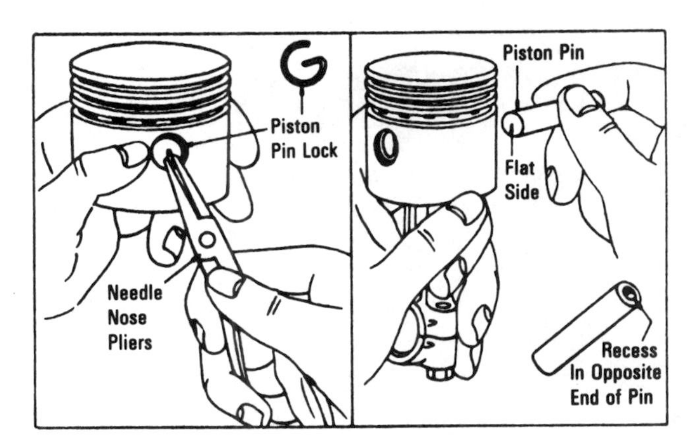

Fig. 4–30. Wrist pins.

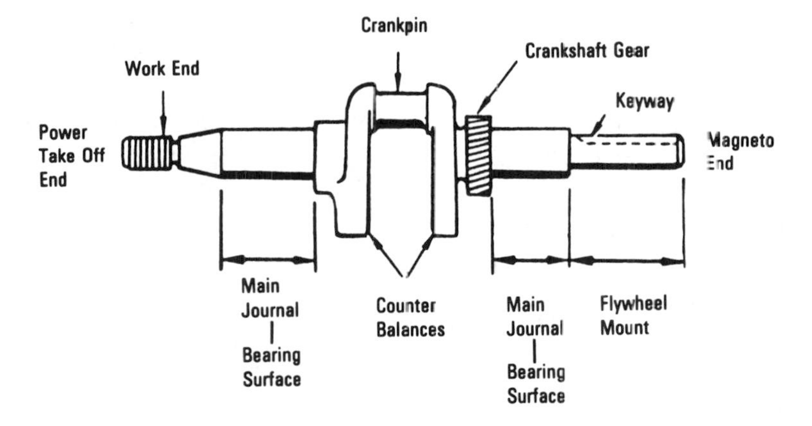

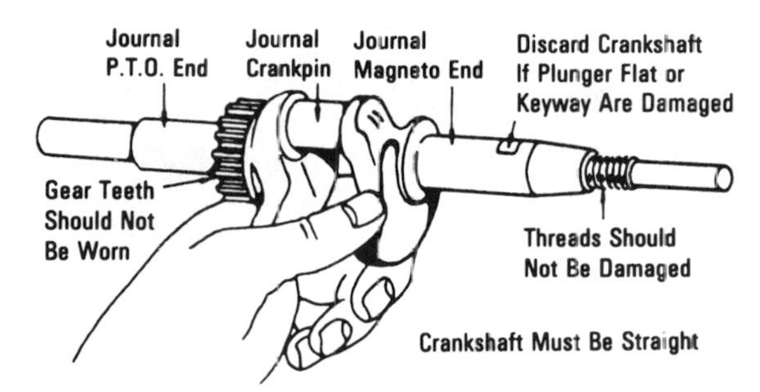

Fig. 4–31. Crankshaft.

shaft together. This way it is possible to use roller bearings here instead of the usual split sleeve-type bearing.

Keep in mind that the total weight of the piston, rings, cap, wrist pin, bearings, and bolts determines the mass of the moving portion of the engine. The momentum of this assembly determines the engine performance. It is very important to make sure that all the parts used are those recommended by the manufacturer. The inertia of the assembly can be altered by using the wrong weight bolt or pin. Engine performance is directly affected. The proper rod and the proper connecting parts must always be selected for replacement.

These parts will wear in and mate to the cylinder and surrounding surfaces. It is not necessary to replace parts just because you took the engine apart. For parts that are matched, meaning that if one part is replaced, the other must also be replaced, it is best to keep them matched. That is, if you have to replace parts, replace all matched parts; if there is no sign of wear beyond what that manufacturer recommends, do not replace any of the matched parts.

Crankshaft

The crankshaft is an interesting means of changing the reciprocating motion of the engine to rotary motion. The crankshaft has a work end, or power takeoff end (PTO), and a flywheel end. In between are the main journal, the crankpin, the counterbalances, and a keyway and crankshaft gear (Fig. 4–31).

The crankshaft is the most important part of the engine. It takes the force of the combustion explosion and rotates with it to produce rotary motion. The crankshaft rotates with it to produce rotary motion. The crankshaft rotates the flywheel, the camshaft, and the load attached to it. The camshaft provides the proper timing for the ignition. The cam also causes the intake and exhaust valves to open and close at the right time (Fig. 4–32). The crankshaft has to be balanced to run smoothly. That is why the counterbalances are shaped to meet the

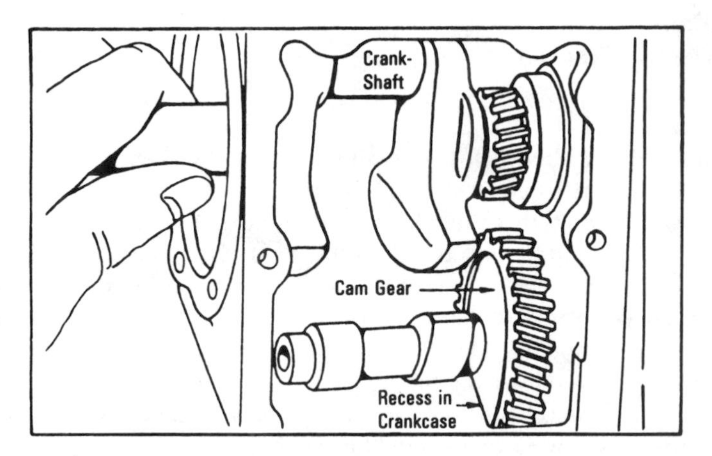

Fig. 4–32. Locations of cam gears and crankshaft. *(Courtesy B & S)*

weight distribution of the piston, rod, and connecting components. Each crankshaft is designed to fit a particular type of engine. There are hundreds of designs.

The main journals are the surfaces that ride in a bearing to support the forces being applied to the crankshaft. They have to have special arrangements for lubrication to make sure the main bearings are not excessively worn.

Bearings and Bushings

Counterbalances are placed on the crankshaft to make sure the engine runs smoothly. They are the heavy, lopsided parts of the crankshaft. They have been carefully designed to take the downward force of the piston and distribute it so that the shaft spins smoothly. The overall weight of the piston assembly is equalized by the counterbalances. A counterbalance can develop an inertia at the opposite shaft side which relieves the strain on the shaft.

If the counterbalances are heavy enough, they can offset the flywheel function. In fact, in some engines the flywheel function is taken over by the counterbalances.

These counterbalances and the shaft need to be supported. This is where bearings and bushings come into use (Fig. 4–33). The rotating members of the engine are supported by at least one, and in most instances two or more, bearing surfaces. If metal is allowed to rub against metal, it won't take long before one of the pieces wears excessively and alignment is ruined. This would produce a very lopsided engine with much noise and vibration. A bearing or bushing is much cheaper to replace than repairing the damage caused by excessive wear.

Sleeve bearings are used to lessen friction of a rotating shaft. Bearings provide support and allow free rotation of shafts or other moving parts. Their selection is usually made after considering the load, the operating conditions, temperature, shock, vibration, contamination, and shaft rigidity.

Journal bearings are plain or sleeve bearings used for shaft rotation and support. They are most often machined or cast of a relatively soft material and contained in a suitable housing. Antifriction bearings have rolling rather than sliding contact points. They have less surface contact area and require less lubrication. They may be ball, roller, or needle bearings (Fig. 4–33). The success of any bearing surface de-

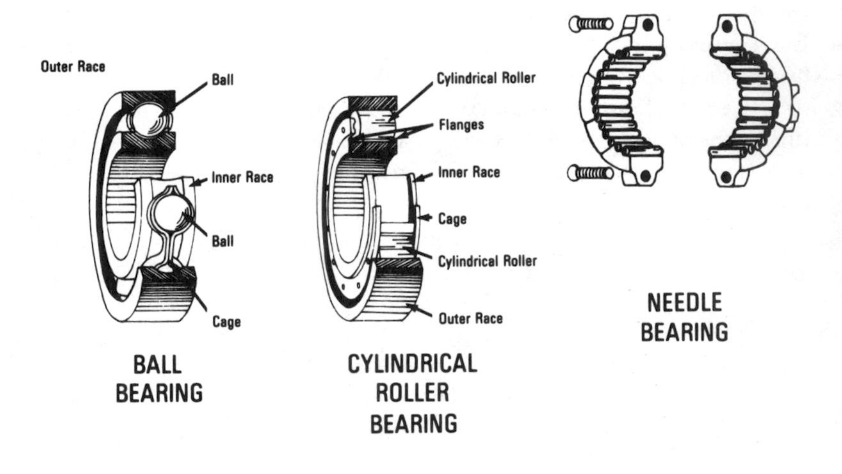

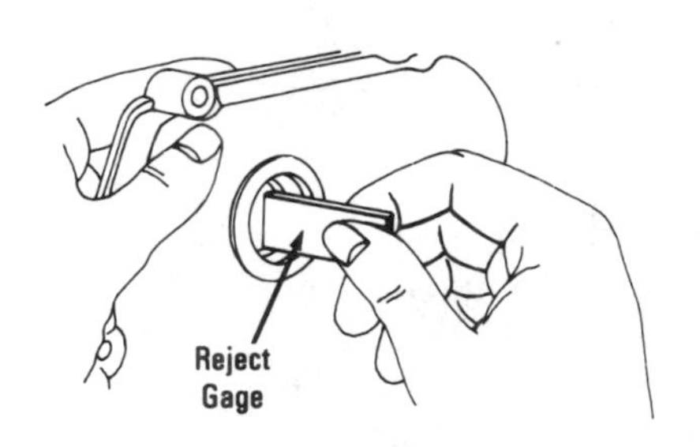

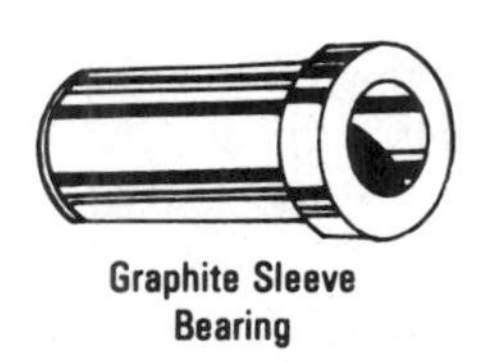

Fig. 4–33. Bearings can be in roller, needle, or tapered roller design. *(Courtesy B & S)*

pends on the light film of oil lubricating and separating the two operating surfaces. Ball bearings may be packed with grease and sealed. They then require little or no lubrication during the life of the bearing. Ball bearings, needle bearings, or taper rollers are more expensive than the sleeve bearing. Damaged bearings are always replaced with new ones when signs of wear are apparent.

A *bushing* is a removable metal liner that does not need a separate container. It is similar to a sleeve bearing. Bushings are usually made in one piece, although some are made in two pieces. They are usually pressed into the hole they are to fit.

Bushings are made of metal that is softer than the shaft metal. Cast-iron and bronze bushings are standard in small gasoline engines. Some bushings are made of nylon or graphite and do not require lubrication. Other types usually have a method of lubrication incorporated into their design. Bushings should be replaced if found worn. Replace with the same size specified by the manufacturer (Fig. 4–34).

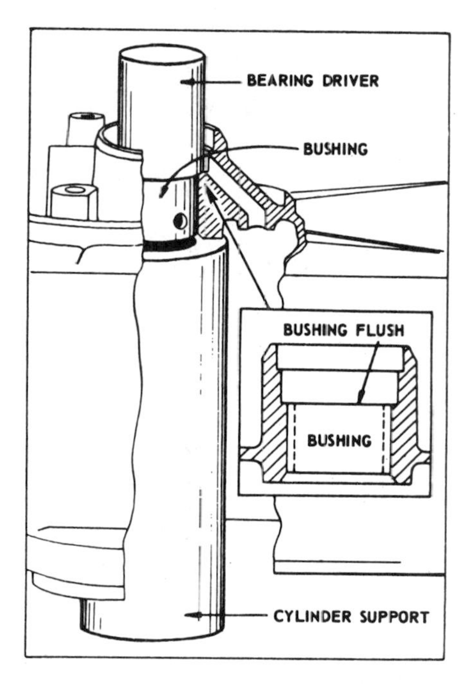

Fig. 4–34. Pressing in a new bushing. *(Courtesy B & S)*

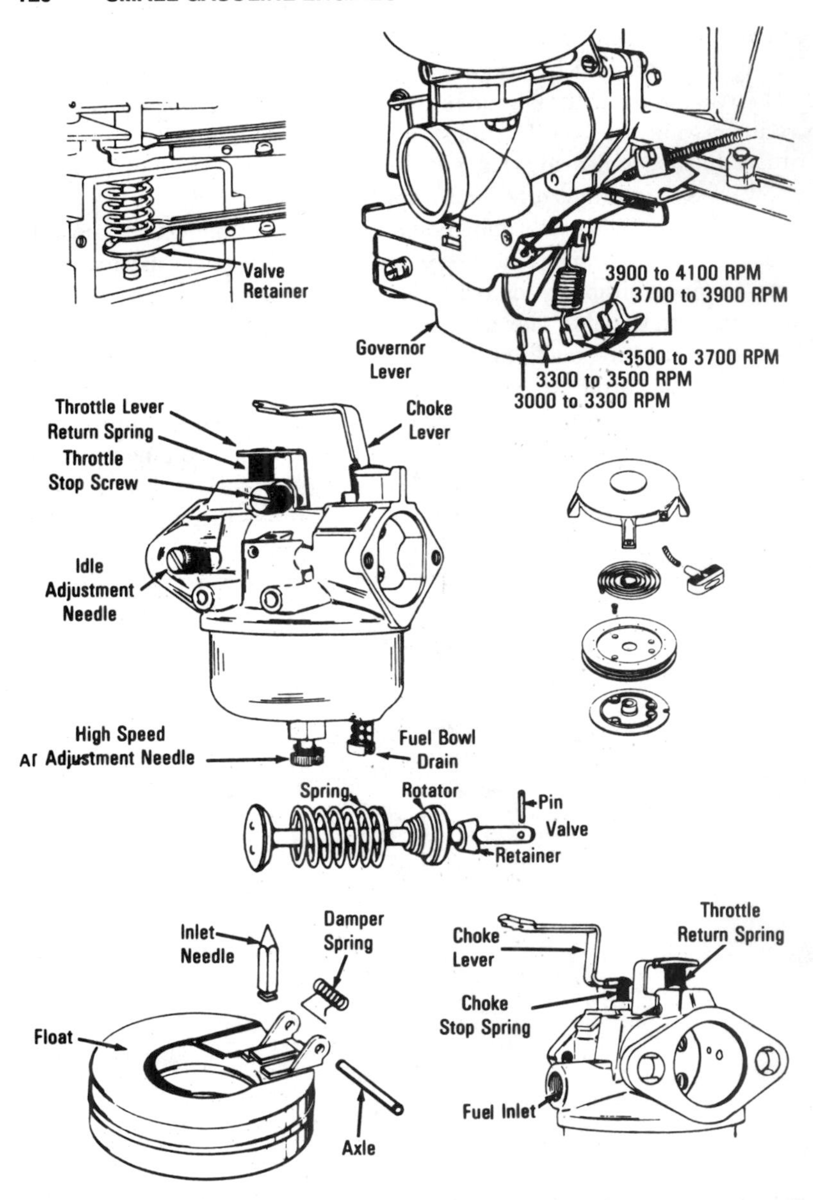

Fig. 4–35. Springs play an important part in the operation of a small gasoline engine. *(Courtesy B & S)*

Gears and Springs

Gears and springs are an essential part of any engine. Gears are used to make sure the timing of the valves and ignition is proper. Springs are used to make sure the adjustments stay where they belong or are allowed to vary only a certain amount. Carburetors use springs on the idle adjustment, the governor, and the needle valve. Springs are located in the engine to make sure the exhaust and intake valves close tightly. Fig. 4–35 shows the placement of the valve springs.

Gears mesh with one another so they don't slip or work loose. Belts have a tendency to slip and cause timing to become either delayed or slightly accelerated—enough to cause misfiring.

Gears are used also in the transmission of power from the engine to the device being driven. The gear is a simple lever. Gears are usually mounted on shafts to actually move something in a different direction or to rotate it at a different speed. Note that gear teeth do not mesh at the tips or ends. They meet at the midpoint between the gear teeth. This means you should inspect this area for wear when repairing an engine.

The main purpose of gears in a small engine is for the timing of the

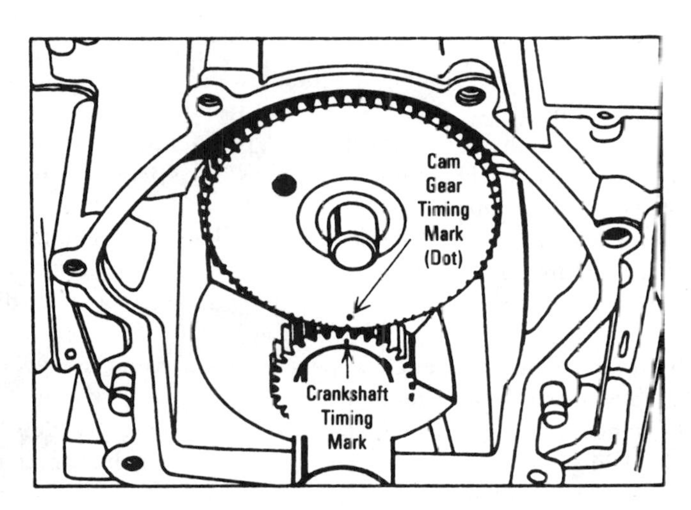

Fig. 4–36. Gears play an important role in timing the valve action and the production of ignition. *(Courtesy B & S)*

ignition and the intake and exhaust valves. They are aligned with a dot on the crankshaft and the gear. This assures that the timing is correct (Fig. 4–36).

Chapter Highlights

1. Engines are different from motors.
2. Internal combustion engines are very useful.
3. Two-stroke engines run faster than four-stroke engines.
4. The turbine engine is also an internal combustion type.
5. Reciprocating internal combustion engines make use of the straight-line piston movement to produce rotary power.
6. A head gasket is used to seal the cylinder head between the top and the cylinder block.
7. The internal combustion engine needs five distinct events to cause continuous power to be generated.
8. Carburetion is needed to mix the air and fuel.
9. Timing is critical to the proper operation of the engine.
10. The venturi is a principle used in carburetors.
11. The airfoil is needed to generate a difference of pressure between two sides of the tube.
12. A gravity-feed carburetor has the fuel flow down from the tank to the carburetor.
13. The diesel engine does not need a spark plug for ignition.
14. The two-stroke-cycle engine completes its operation in two strokes instead of four.
15. The two-stroke-cycle engine does not need an extra tank for oil. Oil and gas are mixed and used as fuel.
16. Four-cycle engines require an oil sump and a circulation system for oil.
17. The rotary combustion engine operates on the same five events that the four-stroke-cycle engine does.
18. The temperature of a valve under load may be as high as 1200°F.
19. Valve sticking is caused by fuel lead, gum, or varnish forming on the valve stem and in the valve guide.
20. The valve should be discarded when the margin or edge of the valve is reduced to less than half the original thickness.

21. Piston displacement is measured in cubic inches or cubic centimeters.
22. Compression ratio refers to the piston and cylinder operation. It is the ratio of the maximum volume inside the cylinder at the beginning of the compression stroke divided by the minimum remaining once the piston has reached the end of its compression stroke.
23. Diesel engines have very high compression ratios.
24. Pistons may be made of cast iron or aluminum.
25. Piston rings require a compressor to allow them to be placed inside the cylinder bore.
26. The piston rod is also called the connecting rod. It connects the piston to the crankshaft.

Vocabulary Checklist

airfoil
air-fuel mixture
atmospheric pressure
bearing
bottom dead center
bushing
combustion chamber
compression ratio
connecting rod
counterbalance
crankcase
crankpin
crankshaft
cylinder bore
cylinder head
energy
four-stroke cycle
gears
gum
head gasket
internal combustion
jet engine
momentum
piston displacement
piston engine
piston ring
piston rod
port
reciprocating engine
rotary combustion engine
rpm
sequence
sleeve bearing
spark-ignition chamber
tappet clearance
timing
top dead center
turbine engine
two-stroke cycle
valve seat
venturi
wrist pin

27. The crankshaft is used to change the reciprocating motion of the engine to rotary motion.
28. Crankshafts are the most important part of the engine.
29. Counterbalances are used to make the engine run smoothly.
30. Sleeve bearings are used to lessen the friction of a rotating shaft.
31. Journal bearings are plain or sleeved bearings used for shaft rotation and support.
32. A bushing is a removable metal liner that does not need a separate container.
33. The main purpose of gears is for timing of the ignition.

Review Questions

1. What is the difference between a motor and an engine?
2. How does an internal combustion engine operate?
3. What are the two strokes that a two-stroke-cycle engine uses?
4. Explain briefly how a turbine engine operates.
5. How does a jet engine operate?
6. What's the difference between a cylinder and a cylinder bore?
7. What is a crankcase? Where is it located?
8. What is a crankshaft? What does it do?
9. Describe these five events:
   a. intake
   b. compression
   c. power
   d. exhaust
   e. ignition
10. What does a carburetor do?
11. What is a venturi?
12. What is an airfoil?
13. What's the difference between a four-stroke-cycle engine and a two-stroke-cycle engine?
14. What is a reed valve? What type of engine uses reed valves?
15. Describe briefly how a rotary combustion engine works.
16. How long does it take a valve to open at 3000 rpm?
17. What is the temperature of an exhaust valve when it is operating at full speed?
18. Why is valve sticking a problem in a one-cylinder engine?

19. What is a tappet?
20. What is meant by compression ratio?
21. What is a piston?
22. What is the purpose of piston rings?
23. What is a bushing?
24. What is a sleeve bearing? A ball bearing?

SECTION TWO—

Operations of the Small Gasoline Engine

CHAPTER 5

Electricity and Magnetism

In this chapter you will be shown how magnetism and electricity work together. Magnetism is used to produce electricity for the spark plug in small gasoline engines. Electricity can be used to produce magnetism and cause solenoids and relays to energize.

This chapter deals with the six most often used methods of producing electricity. Current, voltage, and resistance are the three most important factors in electricity or electrical work. You will be able to see what each actually is and how each is measured. You will be able to see how electrical power is generated by the use of small gasoline engines driving electrical generators.

Magnetism and Magnets

People have known about magnetism for centuries. However, it was not until a few hundred years ago that people discovered how it affected electricity. Magnetism is needed for the power we generate to run motors, lights, and everything else electrical. Magnets or magnetic effects can be found in almost everything modern and electrical. The simple doorbell uses it, as does the memory of a computer.

The Chinese are said to have been aware of some of the effects of magnetism as early as 2600 B.C. They discovered that certain stones, when suspended freely, had a tendency to line up in a north-south direction. Because of the directional quality of these stones, they were later called *lodestones*—leading stones. This name was given to them because they were used in primitive compasses.

The word *magnet* originated with the ancient Greeks. Magnets are mentioned in Greek writings as early as 800 B.C. The Greek word *magnetite* was used to describe these natural magnets, or lodestones. When certain metals are held close to the lodestone, they are attracted to its surface.

Magnetic materials are those materials attracted by magnets. Iron, nickel, and cobalt are the three magnetic materials. These materials can be magnetized.

Nonmagnetic materials cannot be magnetized. They are not attracted by magnets. Paper, wood, glass, and tin are nonmagnetic materials.

Artificial Magnets

Natural magnets can be found in the United States, Norway, and Sweden. They do not have any practical use today. We can make stronger magnets with electric coils. These stronger magnets are known as *artificial magnets.*

Artificial magnets can be produced by magnetic materials and can be made in various sizes and shapes. They are used extensively in electrical devices. Basically, there are two ways of making artificial magnets. These two ways are by using a coil and a battery and by stroking (Fig. 5–1).

If a coil and battery are used, the material to be magnetized is inserted into a coil of insulated wire. A heavy flow of electrons is passed through the wire from the battery. Magnets can also be produced by stroking a magnetic material with magnetite (a natural magnet) or with an artificial magnet. The forces causing magnetization are represented by lines of force (Fig. 5–2).

Some materials hold their magnetism better than others. Because of this, artificial magnets are usually classified as *permanent* or *temporary.* This classification depends on their capability to retain their magnetism after the magnetizing force is removed. Permanent magnets are made from substances such as hardened steel and certain alloys. Har-

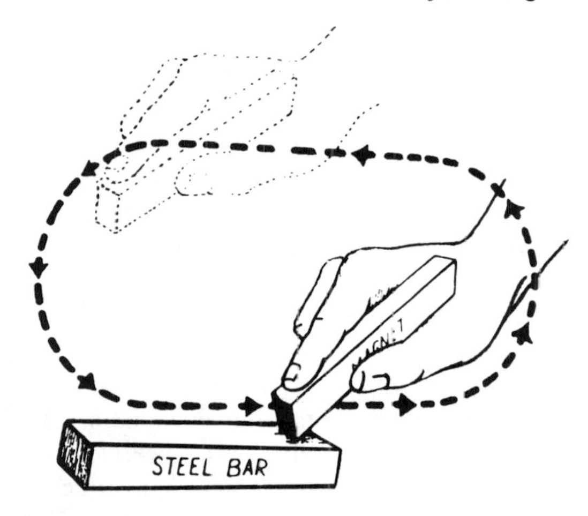

Fig. 5–1. Stroking a piece of steel with a permanent magnet can magnetize the steel bar.

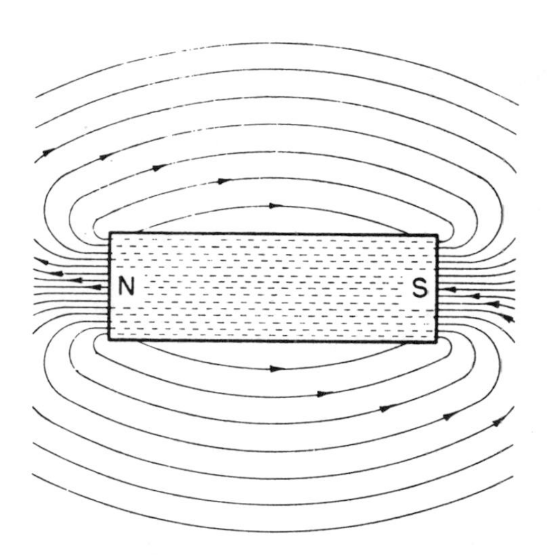

Fig. 5–2. Lines of force around a magnet come from the north pole and enter the south pole.

dened steel, however, is rather hard to magnetize because of the opposition it offers to the magnetic lines of force. This opposition is known as *reluctance.* In other words, because it is hard to magnetize, hardened steel is said to have a high reluctance.

A material with low reluctance is fairly easy to magnetize. Soft iron and annealed silicon steel have low reluctance. However, they keep only a small part of their magnetism once the force is removed. Materials that lose their magnetic strength easily are called temporary magnets. The magnetism that remains in the material is called *residual magnetism.* The ability of a material to retain some residual magnetism is called *retentivity.*

The difference between a permanent and a temporary magnet has been indicated in terms of reluctance. A permanent magnet has a high reluctance while a temporary magnet has a low reluctance. Magnets are also described in terms of their *permeability.* This refers to the ease with which magnetic lines of force distribute themselves throughout the material. A permanent magnet that is produced from a material with high reluctance has low permeability. A temporary magnet produced from a material with a low reluctance has high permeability.

Magnetic Poles

The magnetic force surrounding a bar magnet is not uniform. There is a great concentration of force at the ends of a magnet, and there is a very weak concentration at the middle of the magnet. You can see this if you dip a magnet into iron filings. Only a few particles will cling to the middle while many will cling to the ends. The two ends of a magnet are called the *poles.* Magnets usually have two poles. Both poles have equal strength.

Law of Magnetic Poles—If a bar magnet is suspended freely on a string, it will align itself in a north-south direction. If the experiment is repeated, the same thing will happen. The same pole will always swing toward the North Pole. This pole is called the north-seeking pole or simply the north pole. The other pole of the magnet is the south-seeking pole.

The compass is a practical application of this knowledge. A compass is a device that has a freely rotating, magnetized pointer. The needle always points toward the North Pole. Remember, the north pole of one magnet is always attracted toward the south pole of another magnet. *Like poles repel; unlike poles attract.*

The magnetic North Pole of the earth actually attracts the south pole of a magnet. Therefore it must be assumed that the early users of magnetism for navigation did not fully understand magnetism. They called the end of the compass that pointed toward the geographic North Pole the north pole.

The fact that a compass needle always aligns itself in a north-south direction indicates that the earth is a huge natural magnet. The magnetic North Pole is about 15° from its geographic axis. This locates the magnetic poles some distance from the axis.

Magnetic Fields

Meters and motors use magnetism, as do a number of other electrical devices. The use and control of magnetic fields is important in the study of electricity.

Lines of force have been identified because of what happens to iron filings poured over a magnet (Fig. 5–3). These lines do not actually exist. They are imaginary lines used to illustrate the pattern of the magnetic field. These magnetic lines are assumed to come from the north pole of a magnet. They then pass through the surrounding space and enter the south pole. The lines of force then travel inside the magnet from the south pole to the north pole. This completes the loop.

When two magnetic poles are brought closer together, the mutual attraction or repulsion of the poles produces a more complicated pattern than a single magnet. These magnetic lines of force can be plotted by placing a compass at various points throughout the magnetic field (Fig. 5–4).

Magnetic lines of force are imaginary. However, they make it easier to explain certain properties of a magnet. The following are characteristics of magnetic lines of force.

- Magnetic lines of force will never cross one another.
- Magnetic lines of force are continuous and will always form closed loops.
- Parallel magnetic lines of force traveling in the same direction repel one another. Parallel magnetic lines of force traveling in

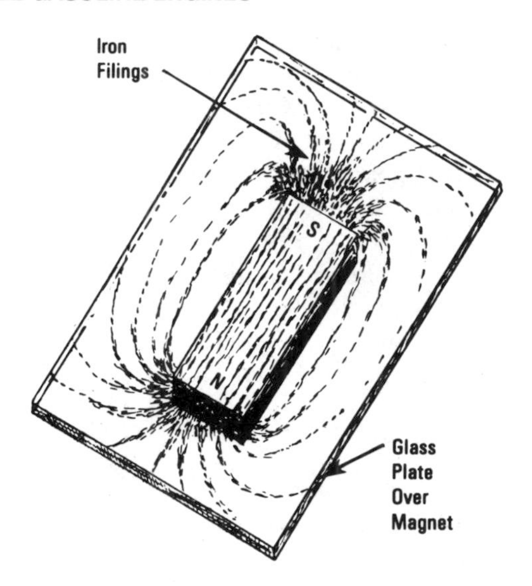

Fig. 5–3. Lines of force about a magnet can be shown by placing powdered iron fillings on a piece of glass placed over a magnet.

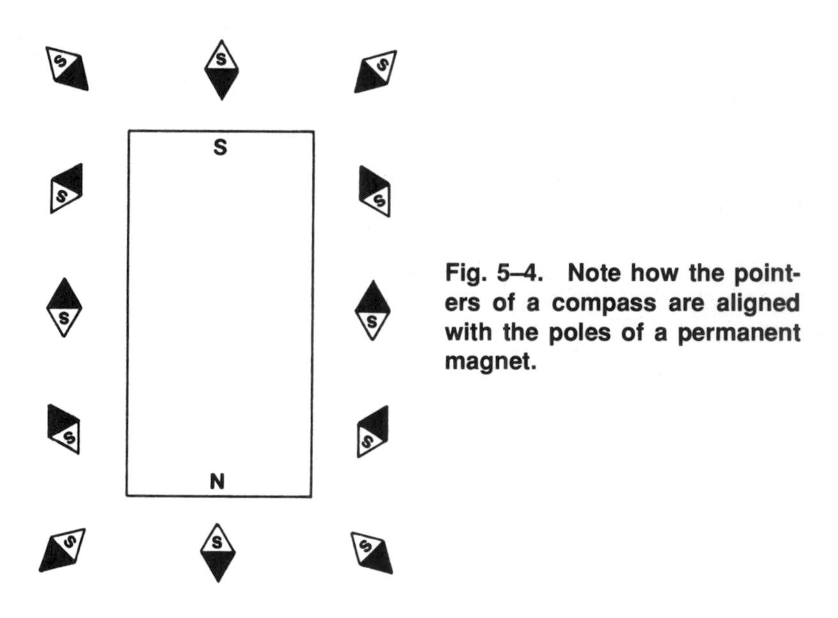

Fig. 5–4. Note how the pointers of a compass are aligned with the poles of a permanent magnet.

opposite directions unite with each other. They form into single lines traveling in a direction determined by the magnetic poles creating the lines of force.

- Magnetic lines of force tend to shorten themselves. Therefore, the magnetic lines of force between two unlike poles pull the poles together.
- Magnetic lines of force pass through all materials, both magnetic and nonmagnetic.
- Magnetic lines of force always enter and leave magnetic material at right angles to the surface.

This information will be helpful in the study of meters, motors, generators, and the magneto type of ignition system of small gasoline engines.

Magnetic Flux—The total number of lines of force leaving or entering a pole of a magnet is called *magnetic flux*. The number of flux lines per unit area is called *flux density*.

Care of Magnets

Metal that has been magnetized can lose its magnetism if handled improperly. If it is jarred or heated, the alignment of molecules will be disturbed. This will result in the loss of some of its effective magnetism. High temperatures and jarring or rough treatment of devices using magnets must be avoided. If a meter using a horseshoe magnet is mishandled, the meter may give an inaccurate reading. Do not heat the magnets in the flywheel of a small gasoline engine. This can cause it to lose its magnetism and be unable to produce electricity.

Generating an Electromotive Force (EMF)

Ways of producing an electromotive force, or an emf, are many. There are, however, six ways used today that are commercially acceptable. There are some exotic or unusual methods also. The six standard methods of producing electricity, or an emf, are:

- Friction
- Chemical reactions
- Heat

- Light
- Pressure
- Magnetism

Friction

Friction is produced by the movement of one object against another. Friction produces electricity. For example, walking across a thick carpet in a dry room can cause your body to take on an electrical charge. What happens if you touch metal? You receive a shock, don't you? Here is how this works: Electrons are present on a surface (in this instance, the carpet). When this surface comes in contact with a surface that has fewer electrons (your body), this second surface obtains a static charge. That's why you receive a shock when you touch metal. You are discharging the excess electrons picked up from the carpet.

Static electricity is used in air purifiers to attract dust particles from the air. It is used also in some photocopying processes.

Charged Bodies—Walking across a carpet in a dry room illustrates one of the basic laws of electricity: Like charges repel each other and unlike charges attract each other. The law of charged bodies may be demonstrated by a simple experiment. Two pith balls (paper or pulp) are suspended near one another. Use thread to support them (Fig. 5–5). Pith balls are light balls used in physics experiments.

A hard rubber rod is rubbed to give it a negative charge (Fig. 5–6). The rod is then held against the left ball. The rod (since it is negative) will give a negative charge to the left ball (Fig. 5–7A). When released,

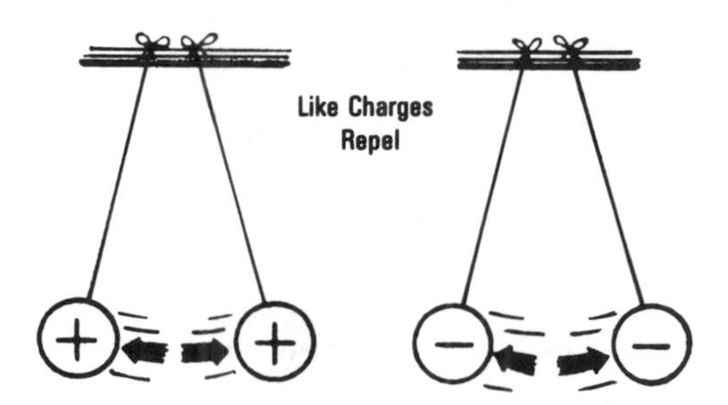

Fig. 5–5. Like charges repel when suspended at the end of a string so as to hang free and interact.

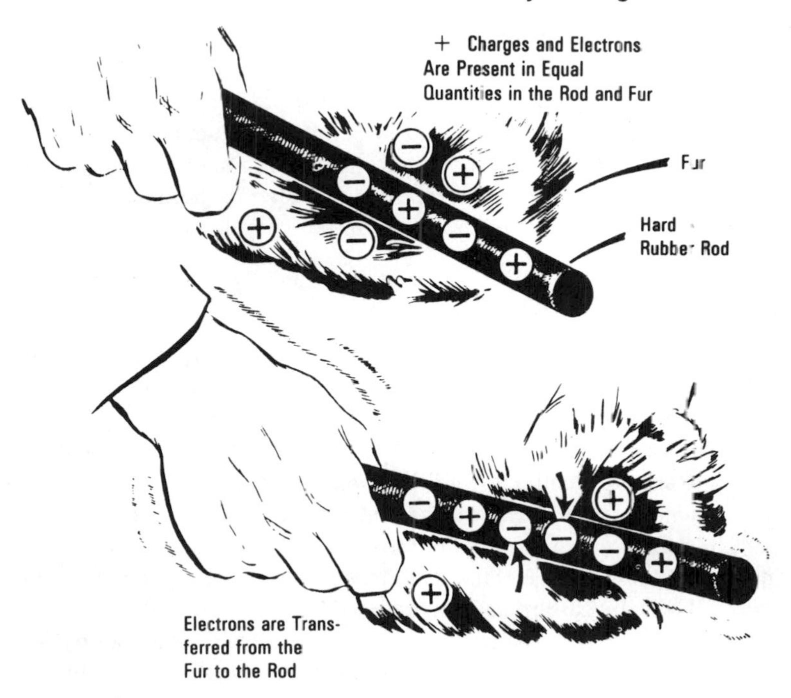

Fig. 5–6. Friction can cause a transfer of charges.

the two balls will touch (Fig. 5–7B). They will remain in contact until the lefthand ball takes on some of the negative charge from the right-hand ball. When they both have the same charge, they will swing

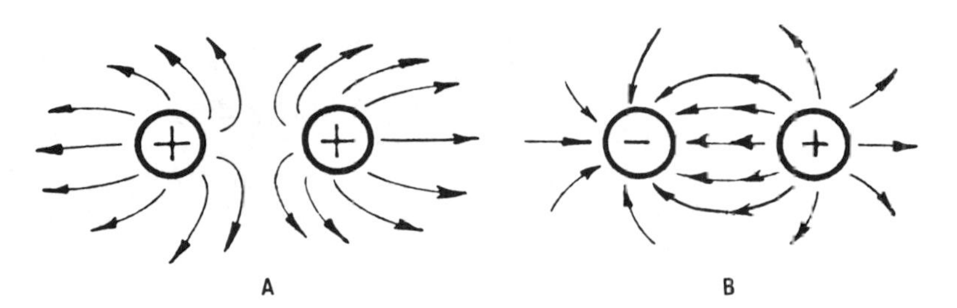

Fig. 5–7. Electrostatic charges also attract or repel according to their charge polarity.

apart. If positive charges are placed on both balls, the balls will also be repelled.

This experiment demonstrates the *electrical field of force.* This is the space around charged bodies in which their influence is felt. It extends between positive and negative charges. It can exist in air, glass, or paper. It can even exist in a vacuum. The electric field of force may also be called an *electrostatic field.*

Chemical Reactions

Chemical reactions can be used to produce electricity. This is direct current (DC). A cell or battery produces DC by chemical action. A battery is made up of more than one cell.

A *cell* is a device that changes chemical energy into electrical energy. The simplest cell, known as *galvanic* or *voltaic,* is shown in Fig. 5–8. It is made of a piece of carbon and a piece of zinc.

The solid parts of the cell are the *electrodes.* An electrode is a conductor that establishes electrical contact with a nonmetallic part of the circuit. Water and sulfuric acid make up the liquid part of the cell. This liquid part is called the *electrolyte.* It may be a salt, an acid, or an alkaline solution. An electrolyte is a nonmetallic electrical conductor. Since this electrolyte is liquid, this particular cell is a *wet cell.*

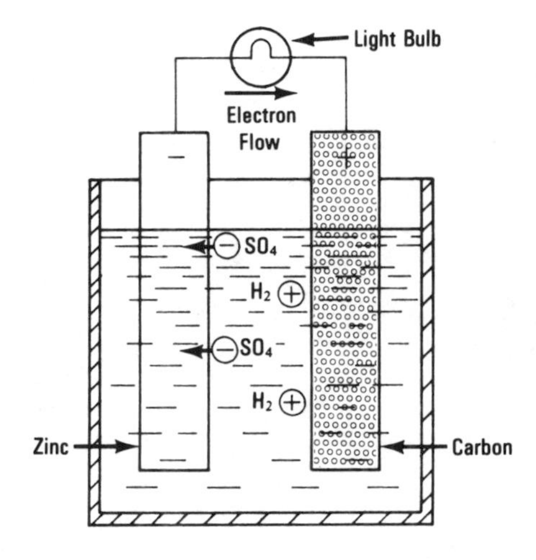

Fig. 5–8. A piece of zinc and a piece of carbon suspended in a solution of sulfuric acid will make a battery.

The electrodes carry the current from the electrolyte to the circuit. A dry cell with a paste electrolyte of sal ammoniac is shown in Fig. 5–9.

In the simple galvanic cell and the automobile battery the electrolyte is liquid. In the *dry cell* it is in paste form. Common flashlight batteries are dry cells.

Primary Cell—A cell that cannot be recharged is called a *primary cell.* Such a cell cannot be recharged because the chemical action eats away one of the electrodes. This is usually the negative electrode. When this happens, the electrode must be replaced. In the case of the dry cell—for example, a flashlight battery—it is cheaper to throw away the cell and buy a new one.

Secondary Cell—A *secondary cell* is a cell that can be recharged. The electrode and the electrolyte are altered by chemical action when the cell produces electricity. To continue to produce electricity, these cells must be restored to their original condition. This is done by forcing an electric current through them. This charging current runs in the opposite direction from that which the cell produces. A meter hooked into the circuit will show the direction of current flow both charging and discharging. This is obvious when you check the ammeter in a car's electrical system.

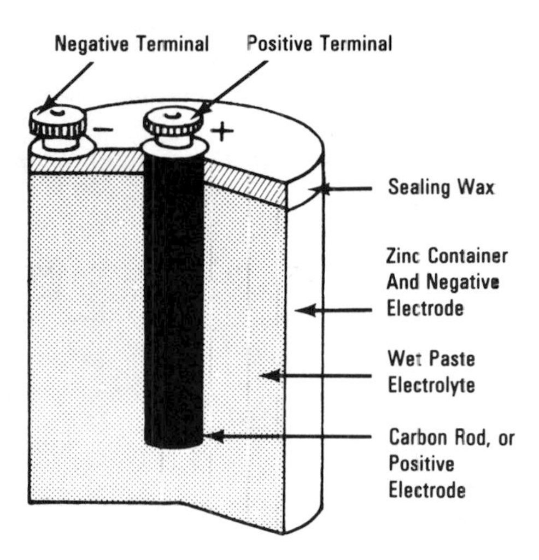

Fig. 5–9. A piece of zinc and a piece of carbon placed in a wet paste (sal ammoniac) electrolyte will produce a battery.

Heat

In 1821, Thomas Johann Seebeck discovered that electricity could be produced by methods that used heat. Seebeck worked with dissimilar materials to get results. When two dissimilar conductive materials are brought together, there is a movement of electrons. Movement is always from the more dense material to the less dense material. This results in too few electrons in one material. The other material, however, has an excess of electrons. The addition of heat causes an increase in the transfer of free electrons. At a fixed temperature the transfer of electrons stops.

Seebeck's discovery led to the development of the thermocouple. A *thermocouple* is made up of two dissimilar metals. Each metal has a different number of electrons. When heat is applied, the electrons move (Fig. 5–10).

Thermocouples can work as thermostats. A *thermostat* is an automatic device for regulating temperature. The furnace in your home or apartment building is probably regulated by a thermostat. For a thermocouple to work as a thermostat it must generate enough electricity to energize a coil of wire. The coil of wire is a *solenoid* that can cause a valve to be turned on or off.

Light

Albert Einstein developed the theory that a beam of light is made of small bundles of energy. Later experiments by J. J. Thompson proved that light energy can force electrons out of a metal. Their dis-

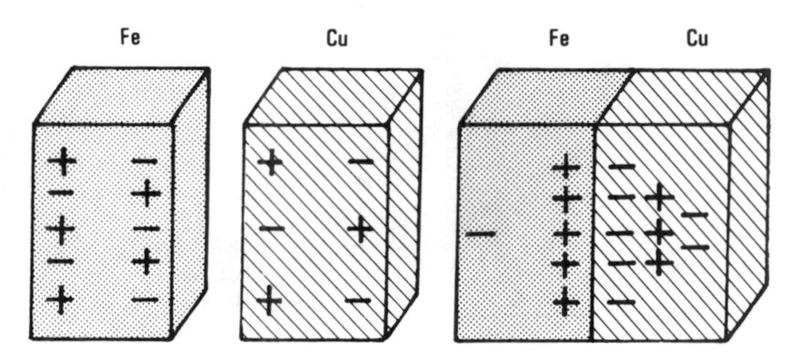

Fig. 5–10. Two dissimilar metals can produce electricity when heated.

coveries led to the development of the photoelectric cell. A *photoelectric cell* is a device that uses light energy to produce an electromotive force (emf). An electromotive force is any force that causes electrons to move, thereby producing electricity.

Fig. 5–11 shows how a photoelectric cell is made. The cell consists of an iron base plate. A coating of selenium is placed on the plate. Next, a layer of material such as plastic is placed over the selenium. This layer allows electrons to pass in only one direction. A thin layer of gold or silver is placed on top of the selenium. Finally, a copper ring is pressed against the gold or silver film. This is the electrical connection to the film.

When light energy falls on the cell, it travels through the film. It also goes through the barrier and collects on the metallic film. The film thus acquires a negative charge. The selenium and the back-plate become positively charged. The one-way nature of the barrier prevents the electrons from returning to the selenium. This way they cannot neutralize the charge. As a result the iron back-plate remains positive. The copper collector ring is negative.

A meter connected to this cell will indicate light intensity. Perhaps you have seen a photographic light meter. It works this way.

Silicon Solar Cell—Much research has been done with solar cells. A *solar cell* is a cell that is able to convert sunlight into electricity. The first experiments were not considered very valuable as a source of electrical power. Photoelectric cells and solar cells are one and the same and are today made of silicon.

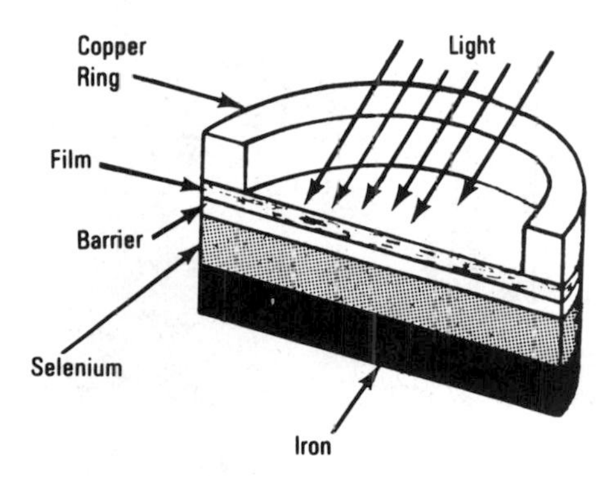

Fig. 5–11. The older types of photoelectric cells used selenium copper and iron to produce electricity.

Silicon is the most abundant element on earth. For example, sand contains silicon. A small amount of silicon is melted and a crystal is grown from this silicon. The crystal produced is cylindrical and made of 99.999999 percent pure silicon. The crystal is usually *doped* with a small concentration of another metal such as phosphorus, boron, or arsenic. Doped means to add an impurity. An impurity is anything other than the pure silicon, which is almost 100 percent pure. Pure silicon is a poor conductor. The small amount of impurity makes the crystal conduct negative charges of electrons. It will also conduct positive charges or holes if properly doped.

Phosphorus is used to make the negative type of crystal. Boron is used to make the positive type of crystal. Positive-type crystals are called p-type. Negative-type crystals are called n-type. The n- or p-type cylindrical crystal rod is then sliced into thin wafers.

The next step is to make the p-n junction. For example, if an n-type wafer is used, boron is spread freely at high temperature into its surface. This boron diffused into the n-type wafer makes the thin surface layer become n-type (Fig. 5–12).

When sunlight strikes the p-n junction, a negative-charge and a positive-charge pair are produced. The electron will tend toward the n-type silicon regions. This electricity can be made to flow in an external load to do useful work. Each wafer can develop about 0.4 to 0.5 volt in the sunlight for an indefinite period.

New designs and new materials will make the solar cell more useful and less expensive in the future. Power output has to be improved

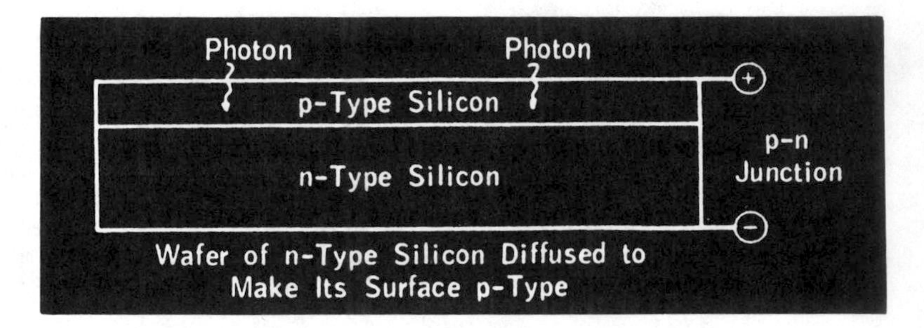

Fig. 5–12. Newer materials (silicon) can be used to produce a more efficient type of solar cell.

in order to make it economically possible to use this power source effectively.

Pressure

Nearly all known inorganic solids are crystalline in nature. *Inorganic* means nonliving; *crystalline* means that they have a crystal structure. However, you would need a microscope or strong magnifying glass to see this. For example, a close look at table salt under the magnifying glass shows that the small grains of salt are actually tiny cubes. Each of these cubes contains an equal number of positive and negative ions.

An *ion* is an atom with one electron more or one less than that atom would normally have. Since it contains an equal number of positive and negative ions, a cube of salt is electrically balanced.

A crystal (such as salt) made of ions is called an *ionic crystal.* One method of generating an emf uses ionic crystals such as quartz. These crystals will generate a voltage whenever stress is applied to their surfaces. Applying pressure causes the ions to move. Thus, if a crystal is squeezed, opposite charges will appear on two opposite surfaces of the crystal. If the force is reversed and the crystal is stretched, charges will again appear. However, they will be of the opposite polarity from those produced by squeezing. If a crystal of this type is vibrated, it will produce a voltage that alternates. Quartz or similar crystals can thus be used to convert mechanical energy into electrical energy. This is called the *piezoelectric effect.*

This method of generating electricity is not suitable for uses requiring high voltages. It is, however, useful in communications devices and systems. Crystals are used in microphones, phonograph pickups, and CB radios.

Magnetism

A horseshoe magnet has two poles, a north and a south pole. In Fig. 5–13 these are labeled *N* and *S*. Lines of force are thought of as leaving the north pole and entering the south pole. Note the direction of the arrows in Fig. 5–13. The area between these two poles is a magnetic force field. What would happen if you were to move a conductor (such as a copper wire) through the magnetic field? The conductor

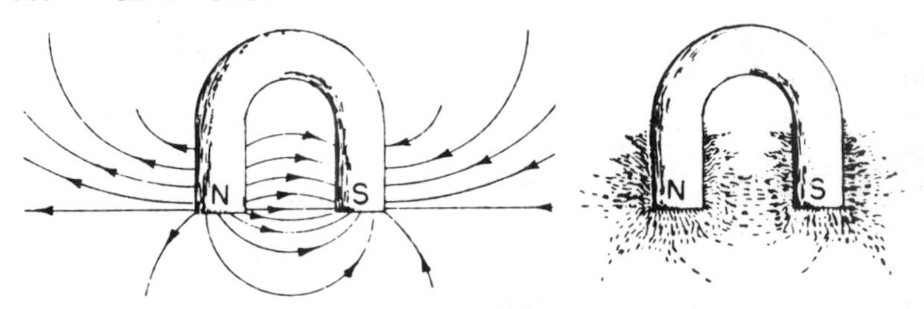

Fig. 5–13. Note the lines of force from the north to the south poles in a horseshoe-shaped magnet. Also note how the lines of force shape up with powered iron filings.

would cut the lines of force, creating a movement of electrons in the copper wire.

Whenever an electron moves, it generates a magnetic field. This magnetic field encircles the electron. Electrons moving along a conductor create a magnetic field about the conductor.

The amount of voltage induced in a conductor cutting a field of force depends on three things:

1. The strength of the magnetic field.
2. The speed at which the conductor cuts the field.
3. The length of the conductor within the field.

Electricity produced by magnetism is used to cause a spark in the small gasoline engine at the tip of the spark plug. The *magneto* is a device that uses a moving magnetic field to generate electricity.

Measuring Electricity

Current flow is measured in amperes (A). Amperes are closely related to other electrical measures you have heard of—volts, ohms, and watts.

Ampere and Coulomb

Scientists have discovered that electrons that flow past a point in a conductor can be measured. The number of electrons flowing past a given point in one second has to be 6.28×10^{18} in order to equal 1 ampere. That amount of electrons, more than six quintrillion, is called a coulomb when standing still. In simple terms, 1 coulomb on the move equals 1 ampere. One coulomb of electrons past a point in one second equals 1 ampere.

The ampere is broken down into smaller units for some uses, especially electronics. A microampere is one-millionth (0.000001) of an ampere. A milliampere is one-thousandth (0.001) of an ampere. The microampere is abbreviated as μA. Note how the Greek letter *mu* (μ) is used to indicate micro. The milliampere can be abbreviated as mA.

Even smaller amounts of current can be measured. The term *nano* is sometimes used. Nano means one-thousandth of a millionth. In transistor circuits, microamperes are common measurements. Milliamperes are used to measure electronic circuit currents.

Volt

The volt (V) is the unit of measure of electrical pressure. It is the electrical pressure exerted to make electrons move along a conductor. The volt was named for Count Alessandro Volta (1745–1827), who discovered the first source of continuous current.

In some cases the volt is too large a unit. In others it is too small. For instance, the signal strength of a radio station is measured in microvolts, one millionth of a volt.

However, in order to use electricity in terms of the small gasoline engine you need thousands of volts; k, or kilo, stands for thousand. Therefore, a 10-kV spark plug wire would have to be able to take 10,000 volts without arcing to the metal part of the engine. Long-distance transmission of electrical power is accomplished by stepping up the voltage. There is nothing unusual about long-distance lines having voltages of 238,000, 245,000, and 750,000 volts.

Electromotive Force (emf)

Three terms—*potential difference, electromotive force,* and *voltage*—are often used to mean the same thing. You will hear electricians refer to the "voltage of the generator" or the "emf of the generator." However, there are technical distinctions in the terms. To be absolutely correct, *emf* should be applied only to the force produced by a battery or generator. *Potential difference* applies to a total circuit or a part of a circuit—for example, "The potential difference or potential drop is 65 volts." The term *voltage* applies to the number of volts—for instance, "The lamp has a voltage rating of 120 volts."

Ohm

The directed movement of electrons makes up a current flow. Some materials offer little opposition to current flow. This opposition is known as resistance. The unit of measurement for the resistance is the ohm. The standard of measure for 1 ohm is the resistance provided at 0° Celsius (C) by a column of mercury having a cross-sectional area of 1 square millimeter and a length of 106.3 centimeters. A conductor has 1 ohm of resistance when 1 volt produces a current of 1 ampere. The symbol used to represent the ohm is the Greek letter omega (**Ω**).

Kilohm—In some cases there are thousands of ohms in a particular device. For example, a circuit may have 100,000 ohms. This may be written as 100k ohms. One thousand ohms is called a kilohm.

This is a common practice. The term 100k can be easily written on a schematic alongside the resistor. A schematic is a shorthand type of drawing with electrical symbols.

Megohm—Sometimes there may be a million ohms in a circuit or a resistor. In such a case, one million ohms is referred to as 1 megohm. Megohm is formed from mega and ohm. *Mega* means one million. You will see it written as, for example, a resistor of 2M. This means 2,000,000 ohms. Sometimes it may be marked as 2 megs instead of 2M. They both mean the same thing. M stands for million. In business dealings, however, the M stands for one thousand, being taken from the Roman numeral meaning thousand.

Power

Power is the rate at which work is done. Work is done whenever a force causes motion. If a mechanical force is used to move a weight, work is done. However, force exerted without causing motion is not work. An example of this is a compressed spring that exerts a force between two fixed objects.

When voltage exists between two points, no work is done. This is similar to the spring under tension. However, when voltage causes electrons to flow or move, work is done. The instantaneous rate at which this work is done is called the electric power rate. It is measured in watts (W). It takes 746 watts to equal 1 horsepower. That means an electric motor with 120 volts and wired to produce 1 horsepower will draw 746 watts of power from the line or source.

Sometimes small gasoline engines are used to turn generators. It takes about a 7- or 8-horsepower gasoline engine to produce 3000 watts. That is little more than 4 horsepower of electrical energy. So, you can see, the generator is not 100 percent efficient. It takes about twice the mechanical horsepower to produce the same electrical horsepower.

Wattage Rating

When a current is passed through a resistor, heat is generated. The resistor must be capable of dissipating (losing) this heat. If not, the carbon resistor will swell and break. The wire-wound resistor will smoke. It will also change resistance.

The capability of a resistor to lose heat is determined by its physical size. The diameter and length of a resistor determine its wattage rating.

The heat dissipated is measured in watts. Some of the more common wattage ratings for carbon resistors are ¼, ½, 1, and 2 watts. The wire-wound resistor usually starts at 3 watts and goes into the hundreds.

The watt is the unit of measure for electrical power. It is equal to the voltage times the current. Where power is represented by P, voltage by E, and current by I, the formula becomes $P = E \times I$. The power will change when either the voltage or current changes. For example:

Voltage across a resistor is 100 volts. Current through the resistor is 5 amperes. Power is equal to the voltage times the current.

$$P = E \times I$$

$$P = 100 \times 5$$

$$P = 500 \text{ watts}$$

There are three quantities in this formula. When any two of the electrical quantities are known, the third can be determined. A good way to find out how the formula works is to do a few sample problems.

Standard sizes of light bulbs are good examples of useful problems. In some cases you will want to know how many light bulbs of a certain wattage can be placed across a generator before it stalls the engine driving it.

Standard sizes of light bulbs available for home use are:

10 watts	15 watts	25 watts	40 watts
60 watts	75 watts	100 watts	150 watts
200 watts	250 watts	300 watts	

Most circuits for use at home are designed for no more than 15 amperes. It would be useful to know how many light bulbs you can safely plug into the circuit. You can use what you just learned to find out what the current would be for the light bulbs just mentioned.

Find the current drawn by a 75-watt light bulb. You already know that the power is 75 watts. E is 120, since the bulbs are designed to operate on that voltage. You do not know I (the current). Use the formula to find the answer.

$$P = E \times I$$

$$75 = 120 \times I$$

$$I = \frac{75}{120}$$

$$I = 0.62 \text{ amperes}$$

The resistance of the bulb can be found by the following formula: R stands for resistance. Again, P stands for power and E stands for voltage. Find the resistance of a 100-watt light bulb.

$$P = \frac{E^2}{R}$$

$$100 = \frac{120 \times 120}{R}$$

$$100 = \frac{14{,}400}{R}$$

$$R = \frac{14{,}400}{100}$$

$$R = 144 \text{ ohms}$$

Using the same formula, find the resistance of the rest of the bulbs. The power formula can also be stated as $P = I^2R$.

If an ohmmeter is used to measure the resistance of a light bulb, it will not be the same as what you have just figured. That is because the bulb has a hot and a cold resistance. You figured the hot resistance. The cold resistance is much lower. An ohmmeter cannot measure hot resistance because an ohmmeter cannot be used on a "live" circuit. The hot resistance has to be figured by the formula you just used.

Light bulbs have smaller filament resistance as the bulb wattage is increased. (The filament is the glowing part of the light bulb.) However, the size of the bulb (the glass part) increases as the wattage rating goes up. This is because more air is needed around the bulb to dissipate the heat generated by higher wattages.

How many 100-watt light bulbs can you safely operate on a 15-ampere house circuit? Wattage ratings are additive. That means each bulb's rating is added to give the total wattage on the line. An electric iron may use 10 amperes or be rated at 1200 watts. Hair dryers use 700, 1000, or 1200 watts. How much current do they draw? How many 100-watt light bulbs can you have on when you use a hair dryer that uses 1200 watts on the same circuit?

Magnetos

The magneto is a device used to generate the high voltage needed for the spark plug. This high voltage is generated by using magnetism. Magnetic lines of force cutting the coil of wire induce an emf or voltage in the coil. Most small engines use a permanent magnet embedded in the flywheel to generate the voltage needed to create the spark. The flywheel is usually turned with a rope or spring starter. This moving magnetic field cuts the stationary coil of wire and produces electricity.

The ignition coil, or magneto coil in this instance, is really a transformer. It has a primary and a secondary. On the Briggs & Stratton models, the turns in the primary are such that there are 60 turns in the secondary for every one in the primary. That means there is a 60 to 1 ratio of turns. If 170 volts are induced in the primary, there will be 10,200 volts in the secondary. This is sufficient voltage to jump the gap of the spark plug, usually 0.030″ (0.7620 mm). Fig. 5–14 shows the location of the magneto on an engine. This engine has an internal-breaker-point set, while Fig. 5–15 shows the location of the external-breaker-point model.

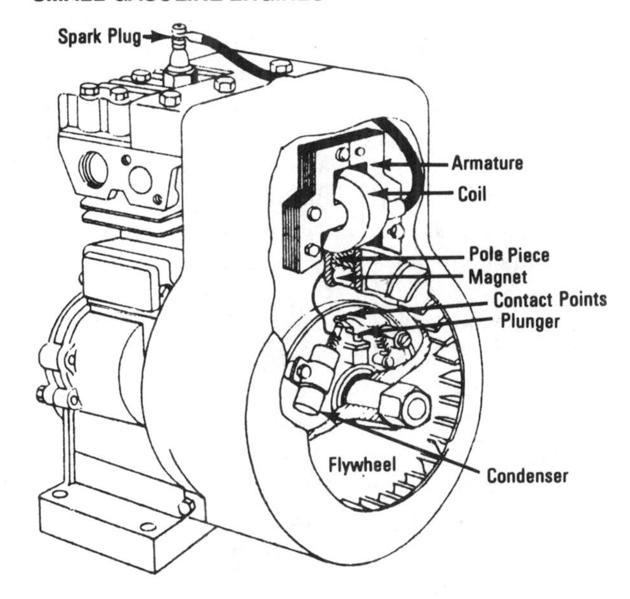

Fig. 5–14. Flywheel ignition with internally-mounted breaker points. *(Courtesy B & S)*

In Fig. 5–16 the armature is installed on the engine with the governor blade to allow sufficient space for easy movement. Fig. 5–17 shows the different types of armatures and how they vary slightly. The coils differ somewhat (Fig. 5–18). These are replacement coils. The Briggs & Stratton coils are the same with the exception of the length of the high-voltage wires.

The permanent magnets that make up the magneto are mounted inside the flywheel. Fig. 5–19 shows one type of arrangement. Ceramic magnets are used in some machines while others use an Alnico® magnet. Both have a very strong magnetic field. Magnets are not to be heated or dropped. They will lose their magnetism if subjected to too much abuse. Ceramic magnets are 3/8″ (9.525 mm) long while the Alnico magnet is 7/8″ (22.225 mm) long.

Fig. 5–20 shows the flow of magnetism through the iron core of the coil as the magnet in the flywheel approaches the armature. Arrows indicate the flow of magnetism. At this point there is very little magnetism at the top part of the core. The air gap at the top of the core pro-

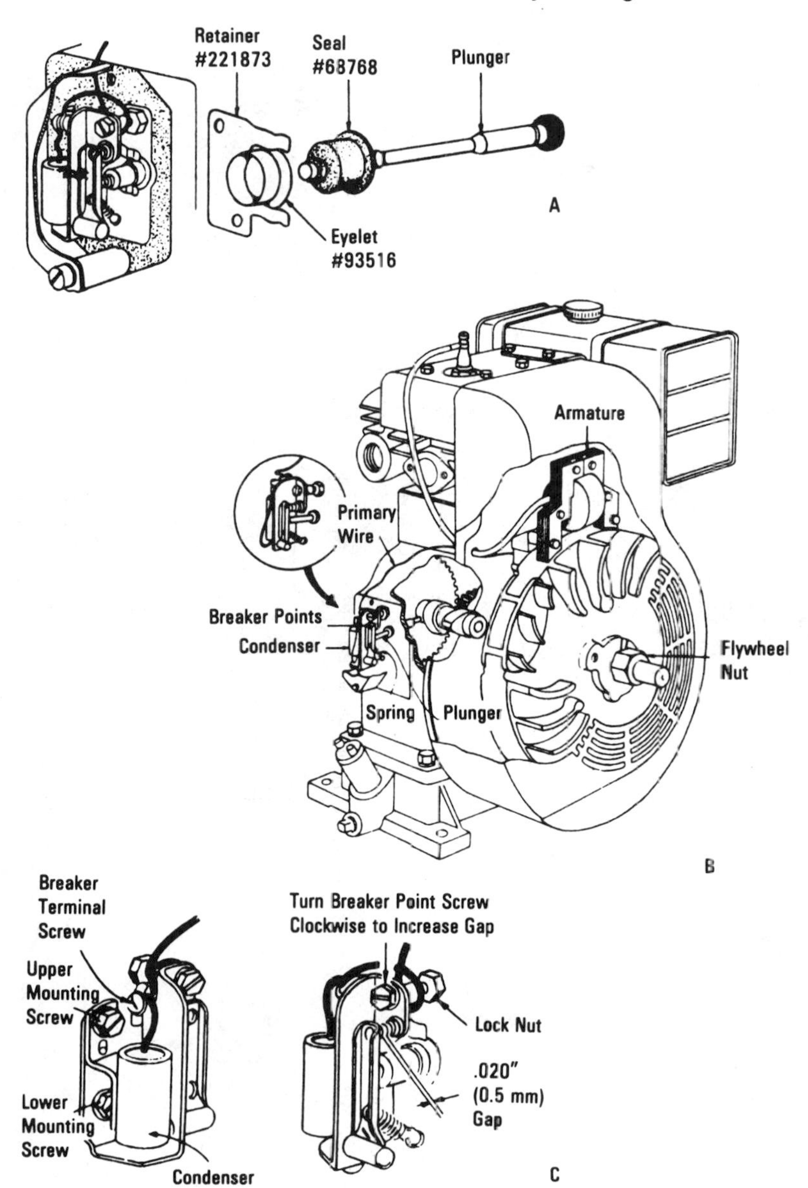

Fig. 5–15. Flywheel ignition with externally mounted breaker points. *(Courtesy B & S)*

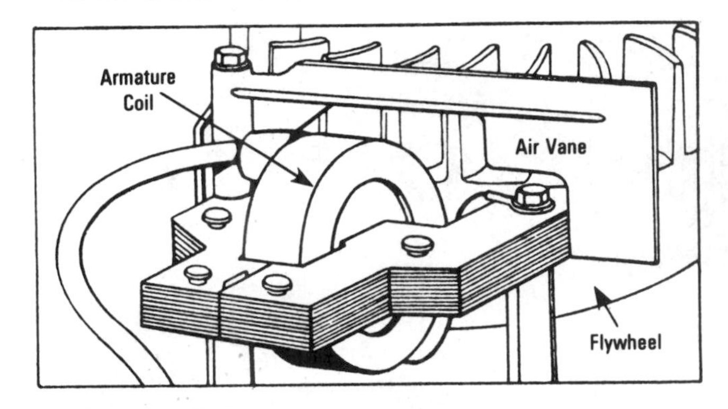

Fig. 5–16. Armature and governor blade. *(Courtesy B & S)*

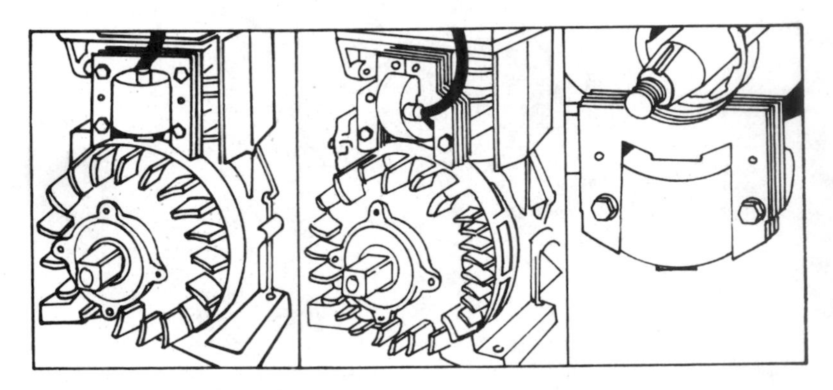

Fig. 5–17. Types of armatures. *(Courtesy B & S)*

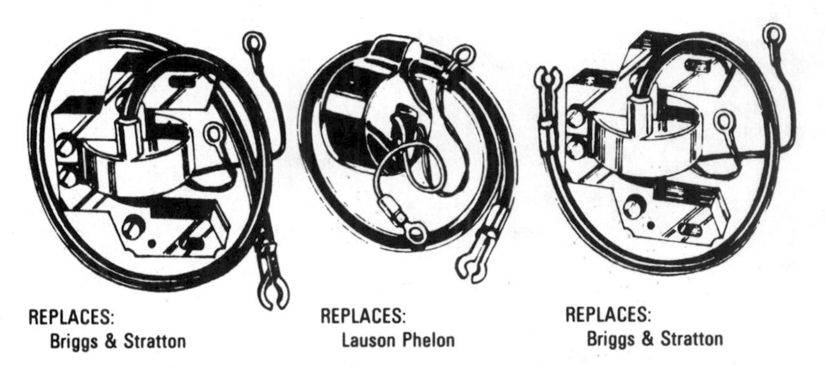

Fig. 5–18. Magneto coils. *(Courtesy Frederick Manufacturing)*

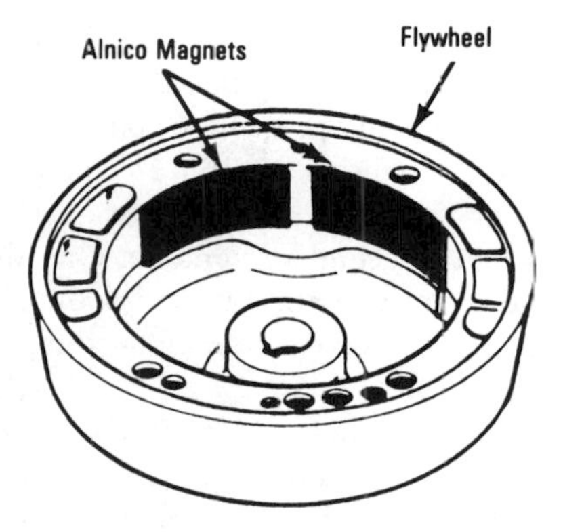

Fig. 5–19. **Flywheel with permanent magnets.** *(Courtesy B & S)*

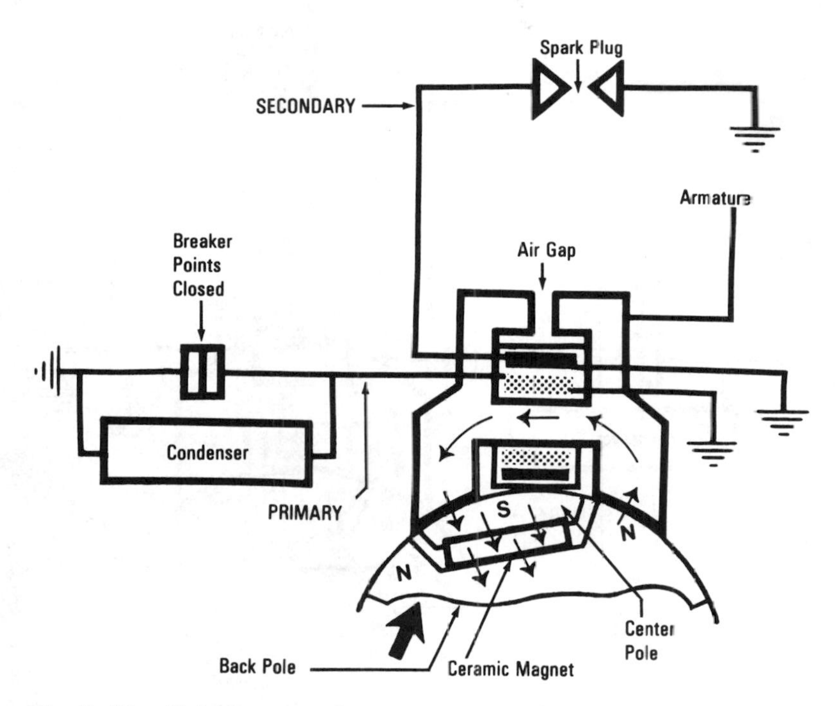

Fig. 5–20. **Using a ceramic magnet to produce a spark.** *(Courtesy B & S)*

duces reluctance to magnetism. Reluctance is to magnetism as resistance is to electric current flow. That means the magnetic field cuts the turns or windings of the primary part of the coil. It should be remembered at this point that the ignition coil is not a coil or inductor, as you may remember it from electrical courses. It is a transformer and, as such, has a primary and a secondary winding. One thing to keep in mind here, also, is the fact that the breaker points are closed. The breaker points complete the electrical circuit for the primary circuit of the transformer (ignition coil).

The flywheel with the magnet rotates (Fig. 5–21). As it rotates the position of the magnet changes. However, the magnetism continues to flow in the same direction. Note that it now flows through the center of the core and through the air gap at the top of the core. However, the

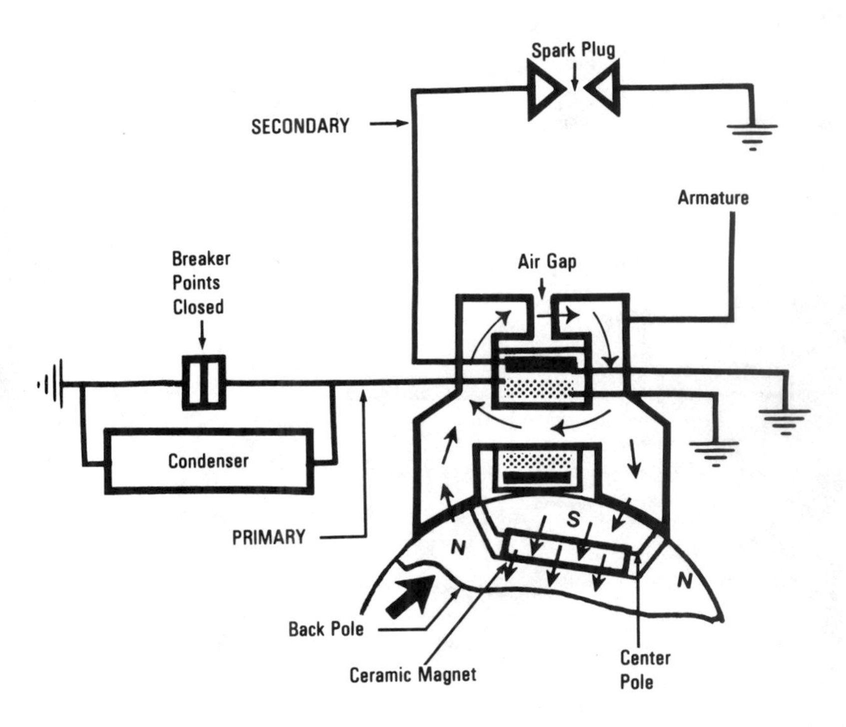

Fig. 5–21. Magnetic lines of force when the magnet is moved to this location. *(Courtesy B & S)*

flow of the magnetism is in the opposite direction through the outer portion of the core and through the air gap. This is because of the change in the flywheel direction and relative position of the magnet. The shunt air gap is important. It provides a path for the flux from the armature legs and the core. That means the current flow through the primary circuit is low. This in turn saves the breaker points from having too much arcing, which would destroy them. The small current through the breaker points makes for longer-lasting points.

The breaker points are opened by the cam arrangement built into the engine (Fig. 5–22). The plunger is activated by the camshaft lobe.

At the point where the breaker points open, the current stops flowing in the primary circuit. The electromagnetic circuit is also nonexistent at this point. Magnetism changes instantly from the flow shown in Fig. 5–21 to that shown in Fig. 5–23. The arrows indicate the complete reversal of the magnetism. The reversal occurred so fast that the flywheel magnet has not had a chance to move any noticeable distance.

This movement of the magnetism through the core of the ignition coil has induced an emf, or voltage, of 170 volts in the primary of the transformer (ignition coil). The ignition coil has a secondary that has 60 times as many turns as the primary. That means that the 170 volts are converted to 10,200 volts in the secondary. Note the simple circuit for the secondary. The high-voltage lead is connected from the ignition coil and over to the spark plug. The other end of the spark plug is connected to ground by virtue of its having screw threads on the body of the plug. Once the 10,200-volt current is applied to the top of the spark plug, its path through the spark plug means it must jump the 0.028″ to 0.30″ (0.7112- to 0.7620-mm) gap to the grounded side electrode. The collapsing magnetic field occurs in a short time interval. This means the spark plug must fire at the right time or the fuel-and-air mixture inside the cylinder will not be used to its maximum efficiency.

One item you should have noticed in the primary circuit is the condenser situated across the breaker points. This is a condenser with 0.16 to 0.24 microfarads of capacitance. The usual size marked on the condenser is 0.2 microfarads. The purpose of the capacitor across the points is to take the surge of current that occurs when the points are opening and keep it from creating an arc across the points. Arcing creates a temperature of around 2000°F. This can melt the metal of the contacts. Pitted contact points are caused by arcing. All of the arcing cannot be suppressed by the condenser, but it can be controlled for a

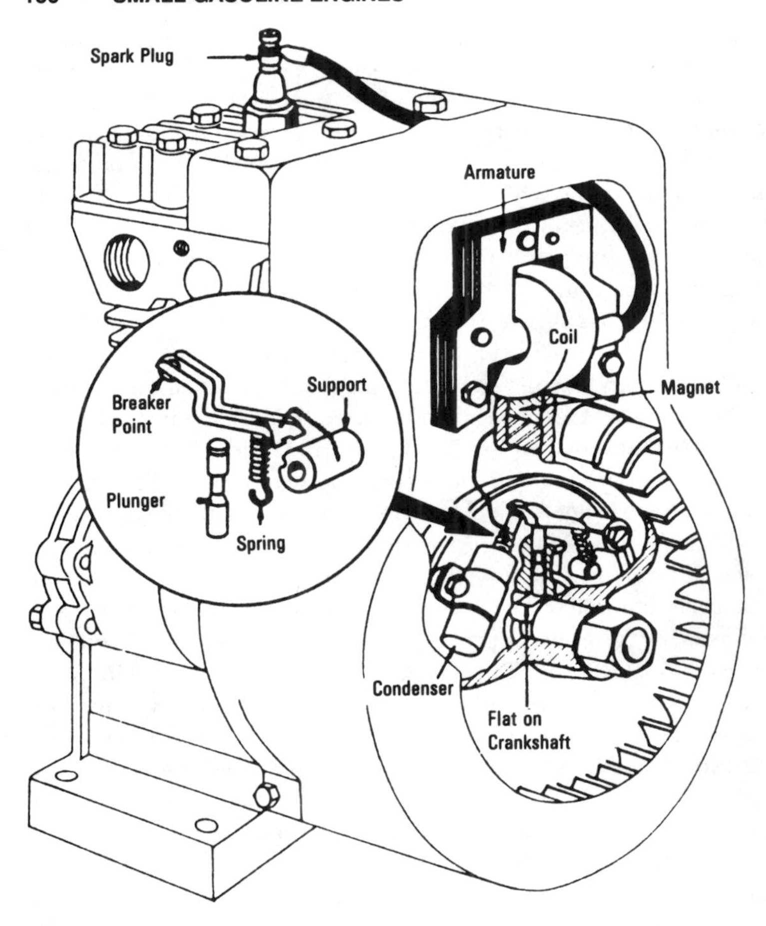

Fig. 5–22. Plunger-activated breaker points. *(Courtesy B & S)*

period of time to ensure that constant attention to or replacement of the breaker points is not necessary.

Testing the Magneto

The magneto can be tested by placing the spark tester (Fig. 5–24) in the secondary of the ignition coil. The flywheel is then spun vigorously. The spark should jump the 0.166″ (4.2164-mm) gap of the tester.

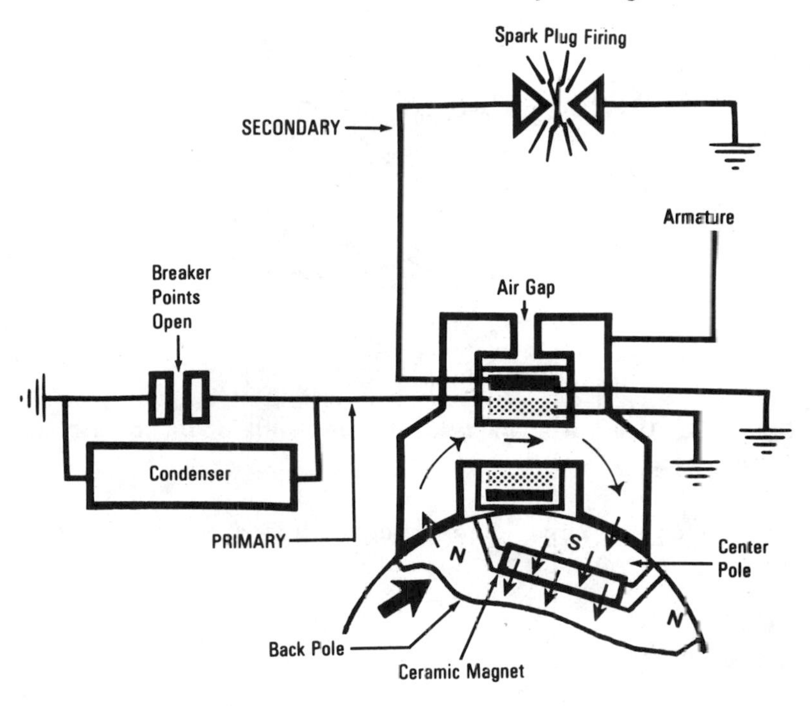

Fig. 5–23. Spark plug firing and relative position of magnet. *(Courtesy B & S)*

Damage to the coil can result if the engine spins more than just a few revolutions with the cable disconnected. Some models of the Briggs & Stratton engines should not have the spark test performed. These are models 9, 14, 19, and 23. They have the Magna-Matic ignition systems.

There are some variations in the magneto systems with different models. However, through the years the basic principles have remained the same. The magnet induces an emf into the primary coil and the opening of the primary circuit causes the magnetic field of the coil to collapse, inducing an emf in the secondary of the transformer.

On small engines, be sure that the flywheel *key* is not partially sheared (Fig. 5–25). This can cause the timing to be off enough to mean difficult starting. The soft metal key is used so that if the flywheel should become loose, the key will be sheared. This allows the flywheel to shift and stop the engine before any further damage can occur. The

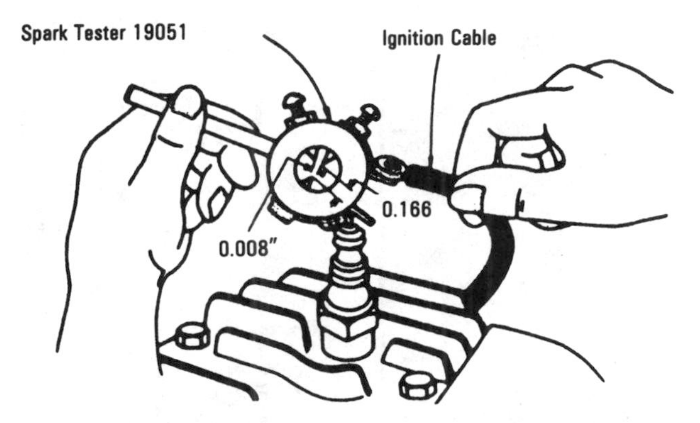

Fig. 5–24. Using a spark tester to check the spark for a plug. *(Courtesy B & S)*

flywheel key is a locater and not a driver. *A steel key should not be used.*

Condensers

The condenser is called a capacitor in an electronics circuit. Only auto mechanics and small-engine repairers still use the term *condenser*. The capacitor has two plates of metal foil. These plates store

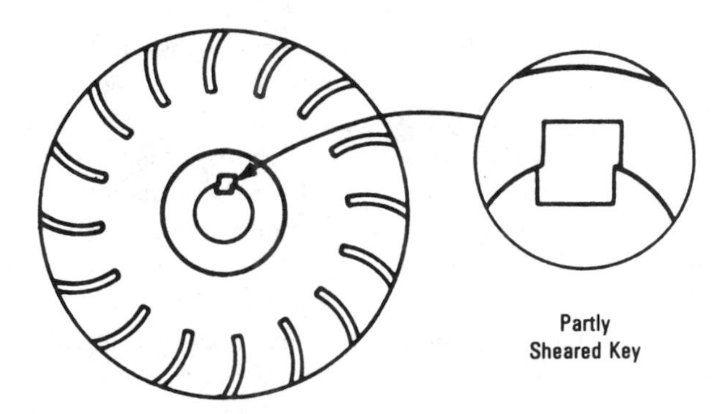

Fig. 5–25. Location of key in flywheel and the shearing action that can change timing. *(Courtesy B & S)*

the electrons and hold them until the circuit changes and needs them. The symbol for a capacitor (condenser) is ⊣⊢. The unit of measurement is the farad (F). However, the farad is too large and is broken down into smaller units. One of the smaller units is the microfarad, or one-millionth of a farad. It is abbreviated in a number of ways. The latest is μF. The Greek letter *mu* (μ) is used to designate the *micro*, or one-millionth (0.000001).

In Fig. 5–26 a number of different condensers are shown. They are made for different models of small engines. They are paper capacitors or condensers with aluminum foil on each side of the paper insulator. The roll of paper and foil is encased in a metal sleeve to protect it. Sometimes condensers will short after much use, and in some instances they will open. In either case it means the points will not work properly. If the points arc over and stick closed, then you know the condenser is open. If the condenser shorts, the points may be open but they are shorted out of the circuit. That means there will be no high voltage to the spark plug.

Breaker Points

Breaker points come in many sizes and shapes (Fig. 5–27). They are made of an insulated movable point that connects the coil primary lead and a stationary point that is grounded to the stator body. This provides the return path for the primary circuit. Replacement points are usually packed with a wick, the proper screws, and, in most instances, a new condenser.

The condition of the surface of the contact points and the distance between the points are the important parts of proper operation.

Spark Plugs

By examining the spark plug you can determine a number of engine problems. There are at least five conditions which indicate problems with plugs. They are described as normal, oily, sludgy, carboned, and burned or pitted electrodes. By inspecting the tips of the spark plug you can determine what is going on inside the engine under run-

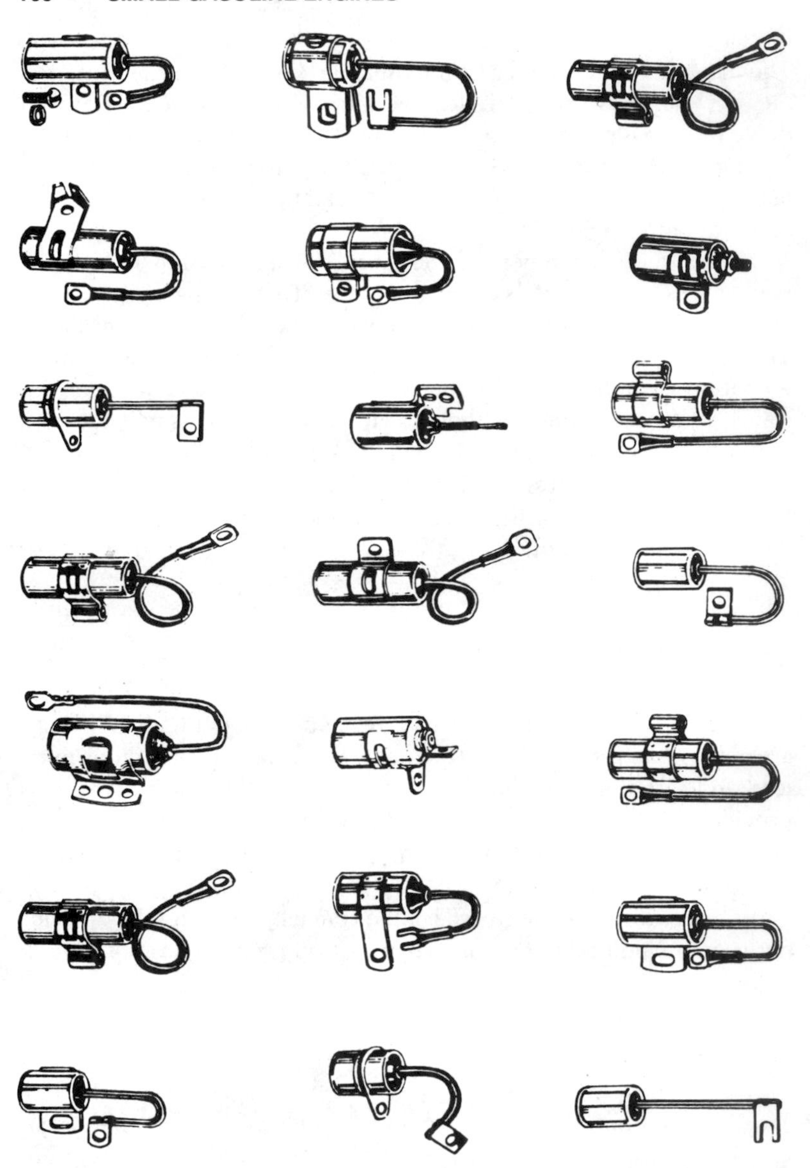

Fig. 5–26. Various types of capacitors (condensers) used in small gasoline engines. *(Courtesy Frederick Manufacturing)*

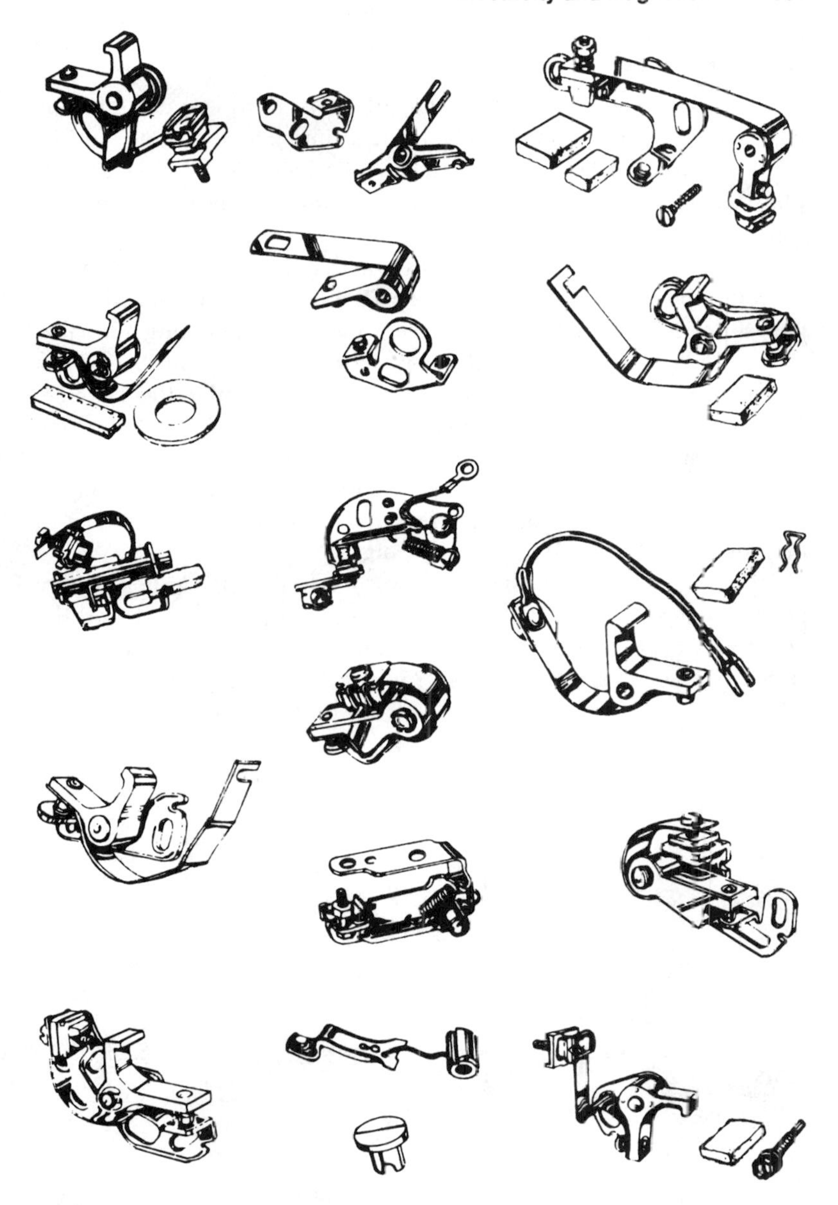

Fig. 5–27. Breaker points used in various types of small gasoline engines. *(Courtesy Frederick Manufacturing)*

ning conditions or why an engine will be hard to start or will not start at all.

Fig. 5–28 shows two types of spark plugs used in small gasoline engines. They can be removed by using a simple tool as shown in Fig. 5–29. The tool fits both the 13/16" (20.638-mm) and the 3/4" (19.050-mm) plug. Once the plug is removed, inspect it for an indication of engine trouble.

The *normal* plug (Fig. 5–30A) is one that does not have too many carbon deposits. The insulator tip may be brown or tan and there is very little pitting of the electrodes. The engine is functioning normally.

The *oily* plug (Fig. 5–30B) is one that has wet oil on it. If it is a two-stroke engine, it means you have put too much oil in the gasoline. The fuel mixture may be too rich. Some engines need an extra-rich mixture. In that case use a hotter plug.

In a four-stroke engine it means the engine is using more than the normal amount of oil. You may wish to decarbonize the engine. You may also want to substitute a hotter plug.

A *sludgy, wet* plug such as that shown in Fig. 5–30C indicates very poor oil control. There will be flaky, smudgy carbon deposits on the electrodes. This condition can indicate several problems in a four-stroke engine. There may be bad rings or the valve stem guides are

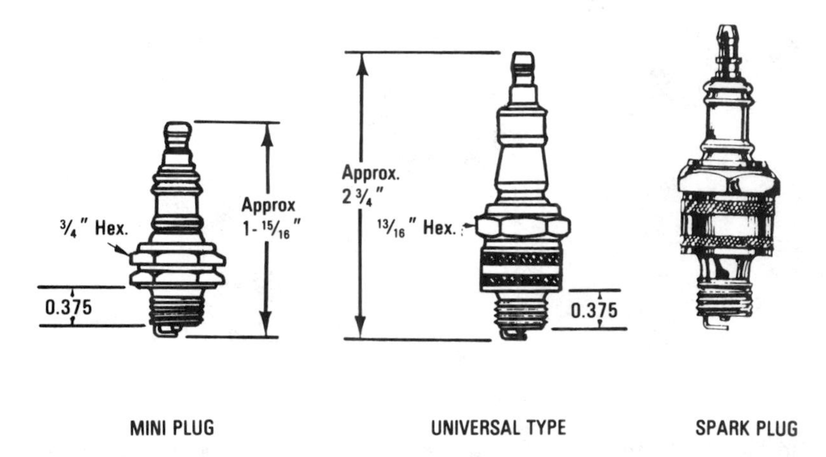

Fig. 5–28. The miniplug and the universal-type spark plugs.

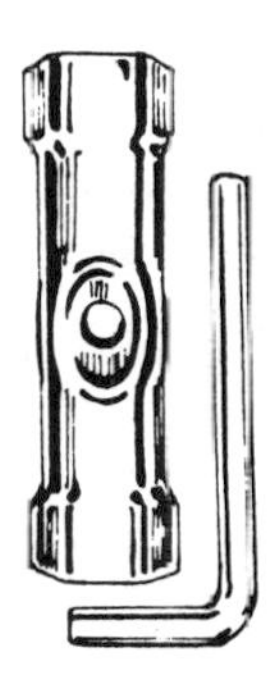

Fig. 5–29. Spark plug tool for installation and removal.

worn badly. More service is needed to correct it. In two-stroke engines there is very little possibility of this happening to a plug.

A *hard, dry carbon deposit* on the plug indicates one of two things: Either the fuel mixture is too rich at normal running speed or the plug is too cold for the engine. If the engine requires a rich mixture, switch to a hotter plug. Fig. 5–30D shows an example of the carboned plug.

Worn carburetor jets can cause the rich mixture. If they are okay, make a carburetor adjustment to "lean-out" the fuel supply to the engine. A carburetor bowl level can be too high and cause the same thing. Excessive fuel pump pressure can also cause the condition—that is, of course, if the engine has a fuel pump.

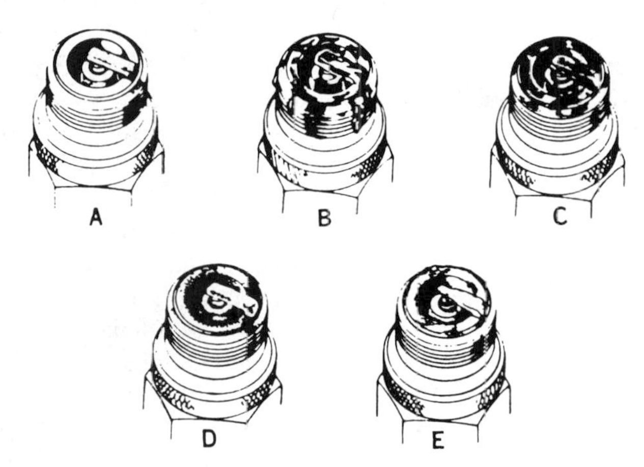

Fig. 5–30. Condition of the spark plug can tell you what is wrong with the engine.

Burned and/or pitted plugs with the insulator tip (Fig. 5–30E) gray in color means that the engine has been running too hot. This can be caused by a number of things. The wrong plug may have been chosen for this engine. If this is not the problem, look for the following causes:

- Compression too high; caused by internal carbon deposits.
- Wrong head gasket.
- Engine overloaded for too long.

Capacitor Discharge System

One type of discharge system uses an impulse coil, a zener diode, a diode, a capacitor, an SCR (silicon controlled rectifier), and an ignition coil. The ignition coil can be of the same type normally used on small engines. The flywheel has a magnet embedded in it. As the flywheel is rotated past the impulse coil, it causes an AC voltage to be generated. The first half of the AC current passes through the path shown in Fig. 5–31 (step 1). Electrons are pulled from the positive plate of the capacitor and placed on the negative plate. In order to do so, current must flow through the ignition coil primary. The diode allows current to flow in only one direction.

Once the capacitor has been charged to the voltage of the impulse coil, it holds its charge until something causes it to discharge. This discharge path is provided by the SCR shown in Fig. 5–31 (step 2). The trigger coil of the system is excited by the moving magnet in the flywheel. It generates a voltage that triggers the anode-cathode path of the SCR. The gate voltage applied to the SCR by the trigger coil reduces the resistance between the cathode and anode of the SCR.

This lowering of resistance makes a low-resistance path for the capacitor to discharge through the ignition coil primary and the SCR. By discharging rapidly, the capacitor dumps all its energy into the coil. This causes the ignition coil to generate the high voltage necessary for firing the spark plug in the secondary circuit of the ignition coil.

The diode places a high-resistance path in the way of the charged capacitor on the second half of the AC pulse generated by the impulse coil. The zener diode breaks down in either direction and allows current to flow whenever the voltage is great enough to cause it to conduct. It serves as a voltage-regulating device. The zener produces a

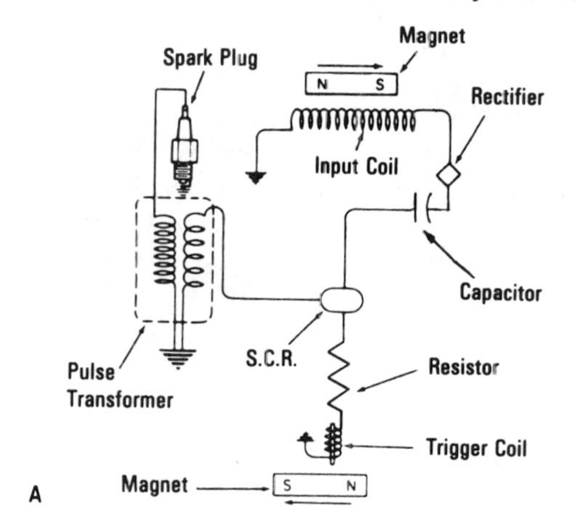

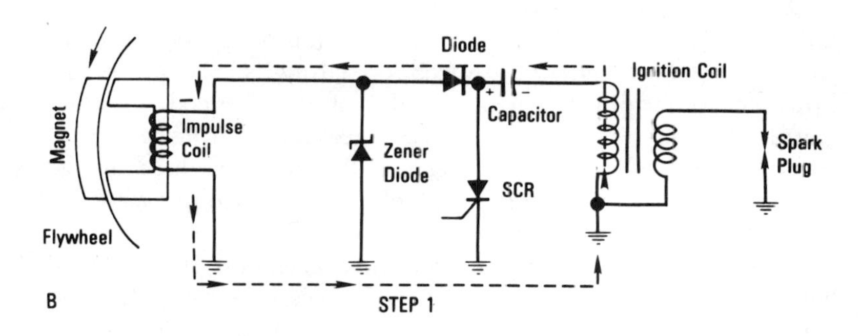

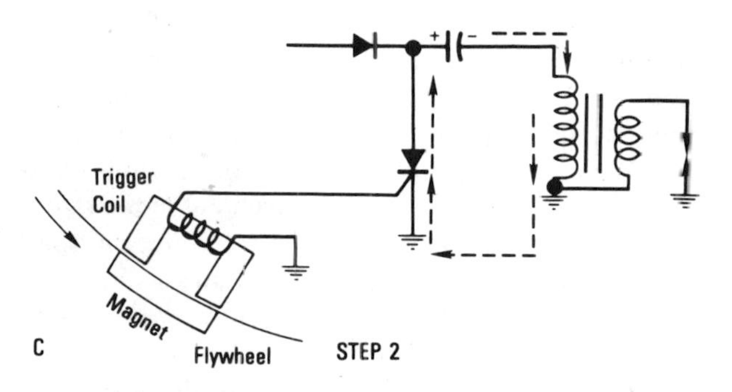

Fig. 5–31. Electronic ignition.

short circuit across the impulse coil whenever the voltage rises to high for the circuitry.

This type of system allows a large current to flow through the primary side of the ignition coil in a short period. There are no moving parts to the system, except for the magnets in the flywheel. No arcing or switching is noticed. The switching is done by the SCR. The advantage of the system is that there are no arcing points to change or clean.

In two-cylinder engines, there are two trigger coils to create a pulse through the ignition coil for each cylinder when needed.

Impulse Coupling

When an engine is first started, the voltage available from the hand cranking is relatively low. The impulse of electricity generated by the starting pull on the rope starter may cause it to backfire. This is due to the fact that the full engine speed requires a spark 20° to 35° ahead of top dead center. The speed for starting is very low, so the firing point is also different at low speeds than at high speeds, to obtain maximum power from the engine. In some cases backfiring occurs.

Impulse coupling is used to overcome the magneto's disadvantage at slow cranking speeds. Impulse coupling can change the spark advance at start and at run speeds.

This type of coupling is used with rotor magnetos. It is installed between the engine and the rotating magneto part. This way the driving force for the magneto rotation must pass through the coupling (Fig. 5–32).

The shell of the impulse coupling is the housing for the spring and hub. It is a rather strong spring. The hub at the magneto side is held stationary. The shell located on the engine side can be rotated until the spring is wound. Then the hub is released. The energy stored in the spring will kick the hub around at great speed. The hub rotation can first be retarded, then speeded up. The movement of the hub is transmitted to the magneto. This results in first retarding the spark, then speeding up the magneto to create a stronger spark.

Fig. 5–33 shows how the impulse coupling works. Note how a pawl is fastened to the hub side. This pawl, which has a stop pin on the outside, is used to hold the hub stationary until the spring is wound. Then, to release it you simply push the stop pin out of the way.

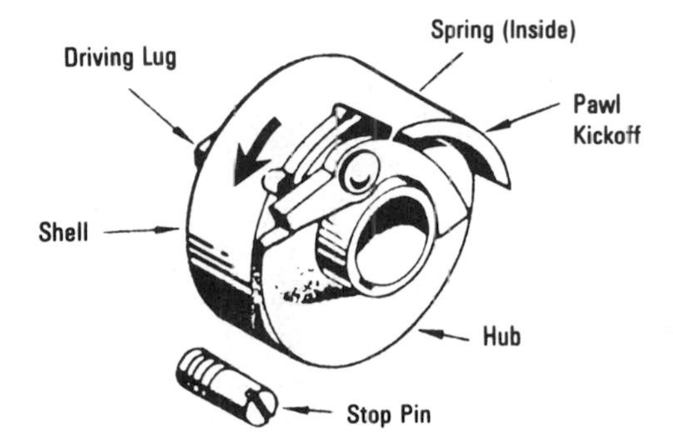

Fig. 5–32. Spark impulse coupling.

Manufacturer's specifications will indicate where this impulse coupling is referenced in regard to the crankshaft.

Solid State Ignition Systems

Today's small gas engine is using electronic circuitry to produce the spark that ignites the air-fuel mixture. There are, of course, a few different types of electronic systems utilized to produce the high voltage for the spark.

Tecumseh's solid state Capacitor Discharge Ignition (CDI) is an all-electronic ignition system which is encapsulated in epoxy for protection against dirt and moisture.

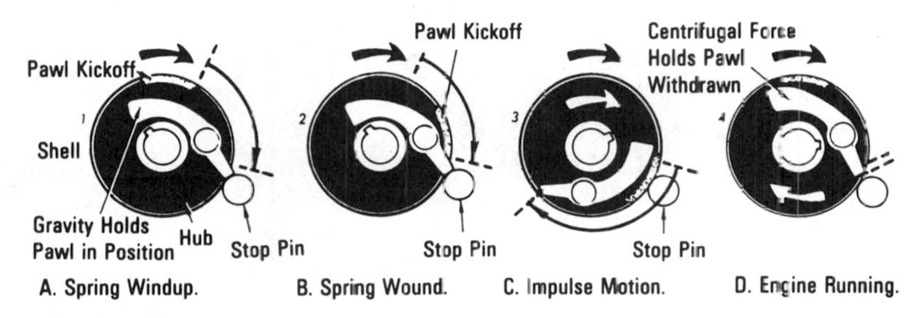

Fig. 5–33. One-pawl impulse coupling.

Operation

As the magnets in the flywheel (Fig. 5–34) rotate past the charge coil, an emf is generated sufficient to charge the capacitor. Diode D_1 is used to rectify this charge coil output so the capacitor is charged with DC or direct current. The charged capacitor holds its charge until the SCR is triggered and produces a low resistance path that puts the capacitor directly across the pulse transformer. The capacitor then dumps its charge into the transformer primary which in turn causes the secondary of the transformer (which is connected directly to the spark plug with a high tension wire) to generate the high voltage needed to produce the spark. Resistor R_1 is directly across the trigger coil which has a voltage induced in it by a passing magnet. Once the voltage across the resistor reaches a predetermined (set by manufacturer) point, the SCR fires or shorts allowing a low resistance path from its anode to cathode. The resistor is connected to the gate of the SCR.

The voltage presented to the spark plug is in the neighborhood of

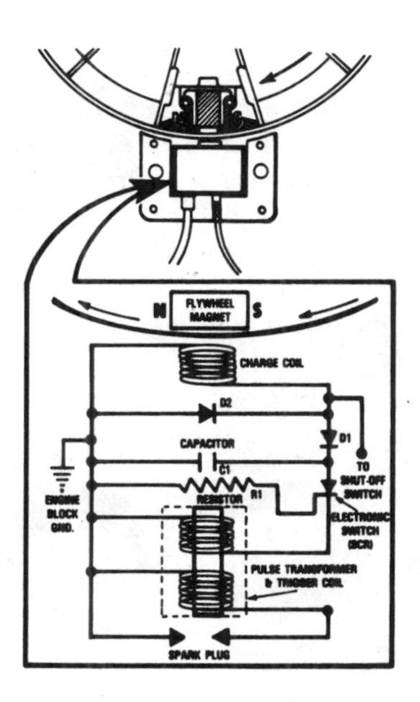

Fig. 5–34. Solid state ignition system. *(Courtesy Tecumseh)*

25,000 volts. It was stepped up from the 200 volts across the capacitor to 25,000 by the transformer. This voltage should cause the spark plug to fire, even if the spark plug gap is more than 0.035″.

Location

The solid state (CDI) electronic ignition system with all the components sealed in a module is located outside the flywheel. There are no components under the flywheel as is the case with the breaker-point magneto system.

The CDI system coil can be identified by the square configuration of the module and a stamping "Gold Key" to identify the proper flywheel key (Fig. 5–35).

The external solid state CDI module on the left in Fig. 5–36 has been replaced by the new module on the right. Both are interchangeable, have the same service part number, electrically test out the same, and are used on small frame 2- and 4-cycle engines as well as TVM 125, 140, V70, H and HH 50, 60, and 70 engines.

Both modules have two holes on the one leg of the lamination to secure the 350 milliampere charging system (Fig. 5–37). These two new external solid state (CDI) modules are used on different engines, are not interchangeable, and have different service part numbers. However, the center coil component is the same. It is permanently secured to two different types of lamination stacks. The laminations are

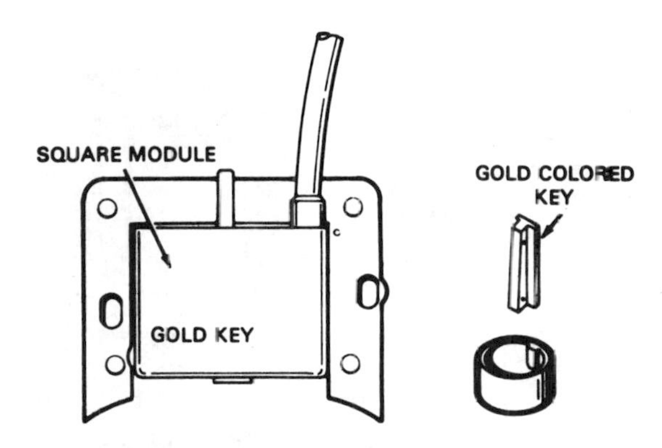

Fig. 5–35. Solid state module. *(Courtesy Tecumseh)*

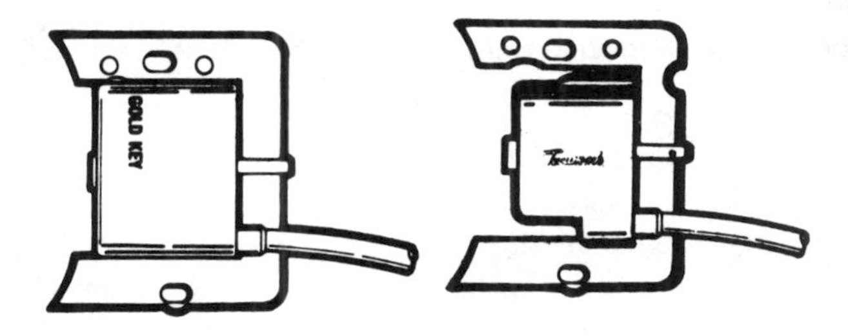

Fig. 5–36. External solid state CDI modules. *(Courtesy Tecumseh)*

different, so they have to be identified properly to function correctly when connected as part of the ignition system. As previously mentioned, the module on the left has two holes in the one leg of the lamination stack to secure the 350 mA charging system.

The module on the right has a different hole spacing and *will not* accept the 350 mA charging system. This module is used on TVM 170, 195, 220, HM 70, 80 and 100 engines.

The proper air gap setting between magnets and the laminations

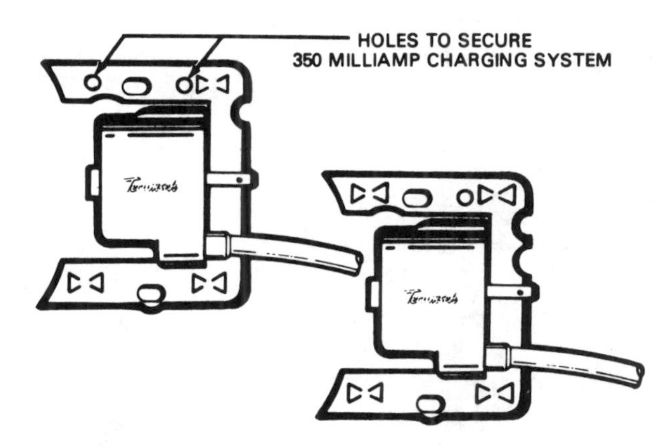

Fig. 5–37. New external solid state CDI modules. *(Courtesy Tecumseh)*

on both breaker point and CDI systems is 0.0125″. Place the 0.0125″ gage between the magnets and laminations and tighten mounting screws to a torque of 30 to 40 inch-lbs. Recheck gap setting to make certain there is proper clearance between the magnets and laminations (Fig. 5–38). Due to the variations between pole shoes, the air gap may vary from 0.005″ to 0.020″ when the flywheel is rotated. There is no further timing adjustment on external lamination systems.

Other Solid State Systems

Fig. 5–39 shows a transformer mounted outside the flywheel and all the other electronic components under the flywheel. To time an engine with this system, rotate the magneto fully counterclockwise and tighten the bolts.

The only on-engine check that can be made to determine whether the ignition system is working is to separate the high tension lead from the spark plug and check for a spark.

Visual checks can be made to made certain there are no leads or wires cracked or shorted.

All components can be checked using available test equipment. If testers are not available, use new replacement parts as a test for possible failed parts.

The system pictured in Fig. 5–40 has components located under the flywheel. To time an engine with this system, turn the magneto

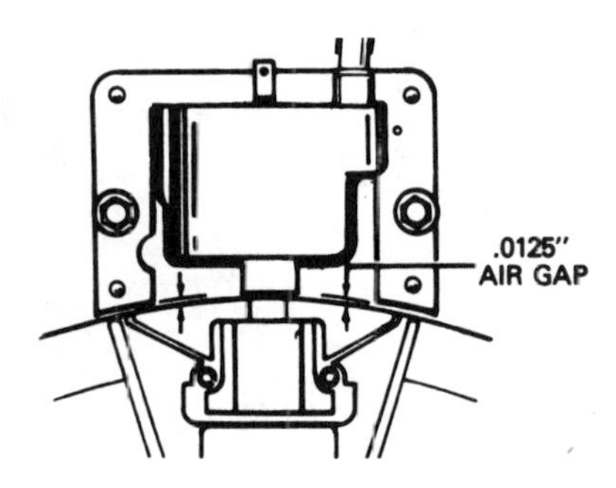

Fig. 5–38. Air gap between magnets and laminations. (*Courtesy Tecumseh*)

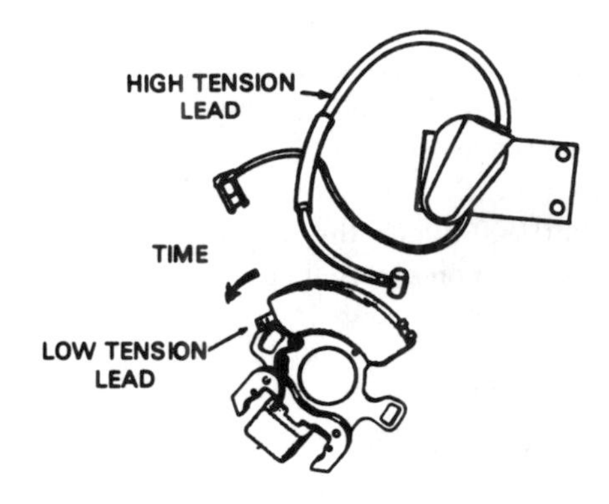

Fig. 5–39. Outside the flywheel transformer. *(Courtesy Tecumseh)*

fully clockwise and tighten the screws. Check the system by checking for a spark or using available test equipment.

The system shown in Fig. 5–41 also is located under the flywheel. All components are encapsulated into one module. No timing is necessary with this type. Check the system by checking for a spark or use available test equipment.

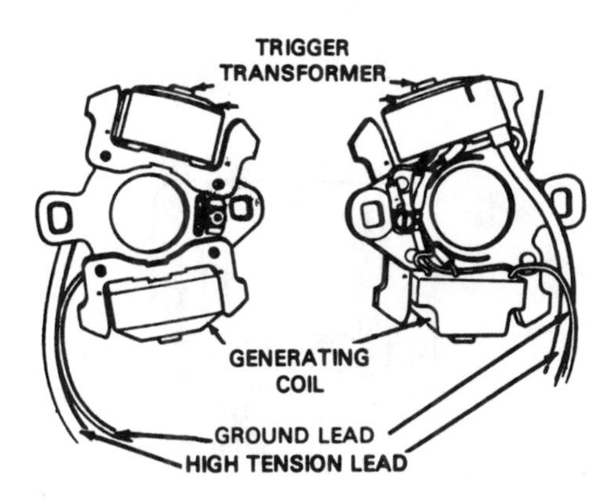

Fig. 5–40. These components are located under the flywheel. *(Courtesy Tecumseh)*

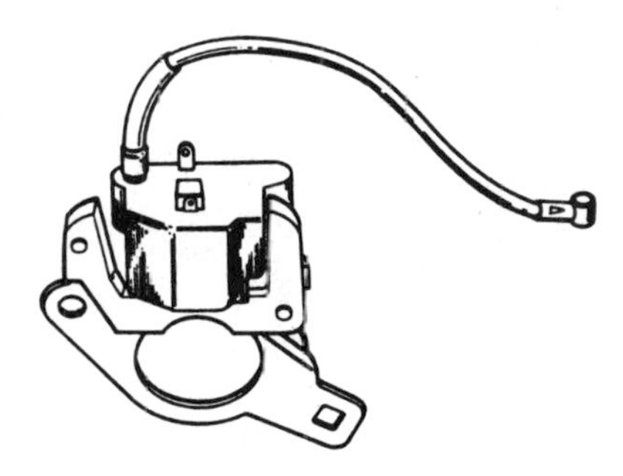

Fig. 5–41. An epoxy-encapsulated module that is mounted under the flywheel. *(Courtesy Tecumseh)*

Electrical Circuits on Small Engines

Oil Level Circuit

The oil alert system is an engine saver when it comes to running an engine for long periods of time without shutting it down for cooling or checking of liquid levels. If the engine oil level falls to a predetermined point, the *oil alert buzzer* will sound. This indicates that oil should be added.

The oil level switch (Fig. 5–42) is a reed switch. When the engine oil falls to a predetermined point, the float (with magnet) inside the switch unit in the oil pan will descend and approach the reed switch. As a result, electromagnetism in the reed switch will strengthen, the

Fig. 5–42. Oil level switch operation. *(Courtesy Honda)*

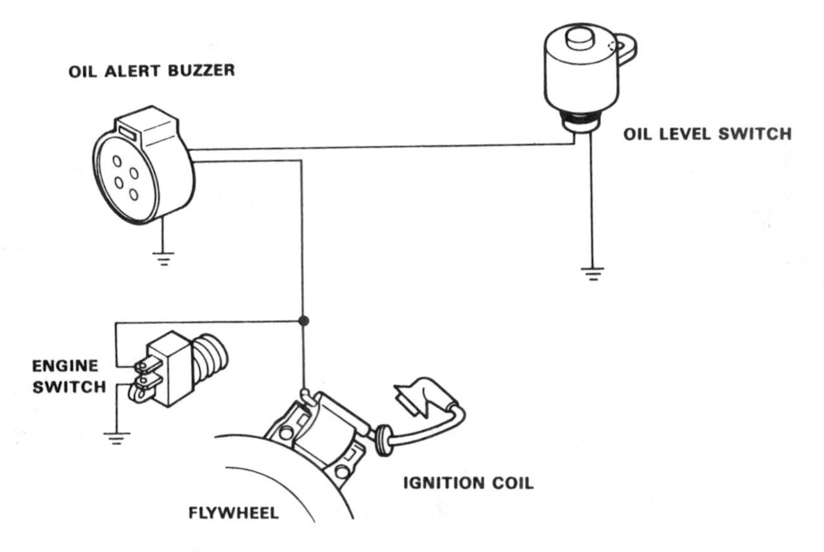

Fig. 5–43. Oil level alert buzzer controlled by oil level switch and engine switch. *(Courtesy Honda)*

contact points will be pulled together and the reed switch will make contact.

Fig. 5–43 shows the circuitry for the oil level switch and oil alert buzzer. Of course, the engine switch has to be closed in order to complete the electrical path. When the reed switch is closed due to a low oil level, electric current is sent to the buzzer unit from the primary side of the transistorized ignition coil, and the buzzer will sound, indicating that oil should be added.

Location of the oil level switch in the oil pan is shown in Fig. 5–44. Fig. 5–45 shows how to check the oil level switch. Check for continuity between the yellow and green switch leads with an ohmmeter. If you get a reading of infinity, the switch is off. If the meter reads zero ohms, then the switch is closed. Hold the switch in its normal position. The ohmmeter should read *zero* resistance. Hold the switch upside down. The ohmmeter should read *infinity.* You can inspect the float by dipping the switch into a container of oil. The ohmmeter reading should go from zero to infinity as the switch is lowered in the upright position.

To make sure the electrical warning circuit is operating properly, the buzzer should also be checked for proper operation. This is done by

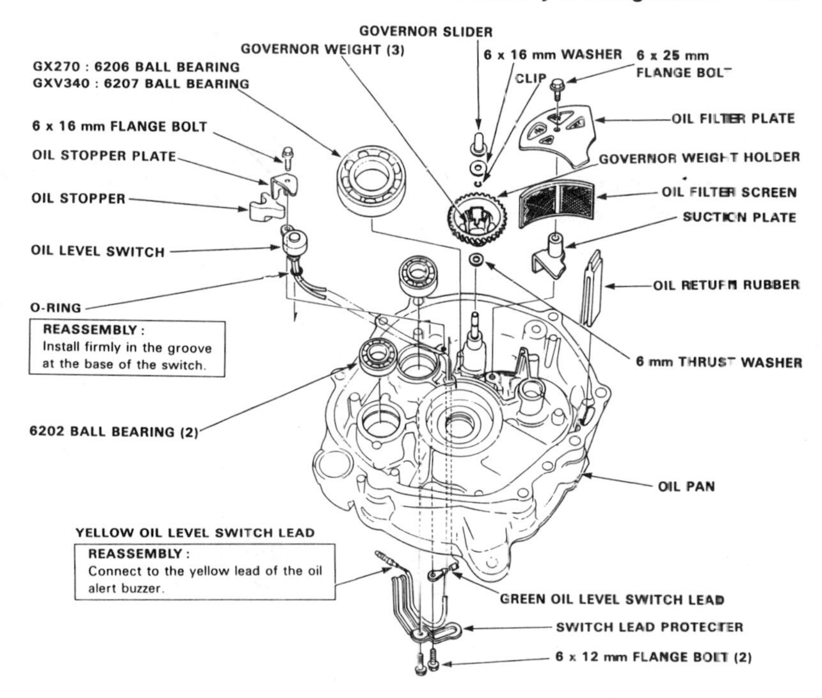

Fig. 5–44. Note how the oil level switch lead protector is installed. *(Courtesy Honda)*

first identifying the components and their respective locations (Fig. 5–46). The oil alert buzzer should have a black, a yellow, and a green lead (Fig. 5–47). Connect the buzzer's black lead to the engine's stop switch lead, and ground the green and yellow leads. Pull the recoil starter with the engine stop switch *off*. The buzzer should sound; if not, replace the buzzer.

The buzzer will not function properly and may be damaged if exposed to gasoline, physical shock, or ambient temperatures of more than 60°C (140°F). When installing the engine, allow plenty of clearance around the buzzer.

When replacing the oil level switch (Fig. 5–48), note how the leads are fed through the 10mm nut and how the o-ring is installed firmly in the groove. The two 6x12 screws will hold the switch in place in the oil sump.

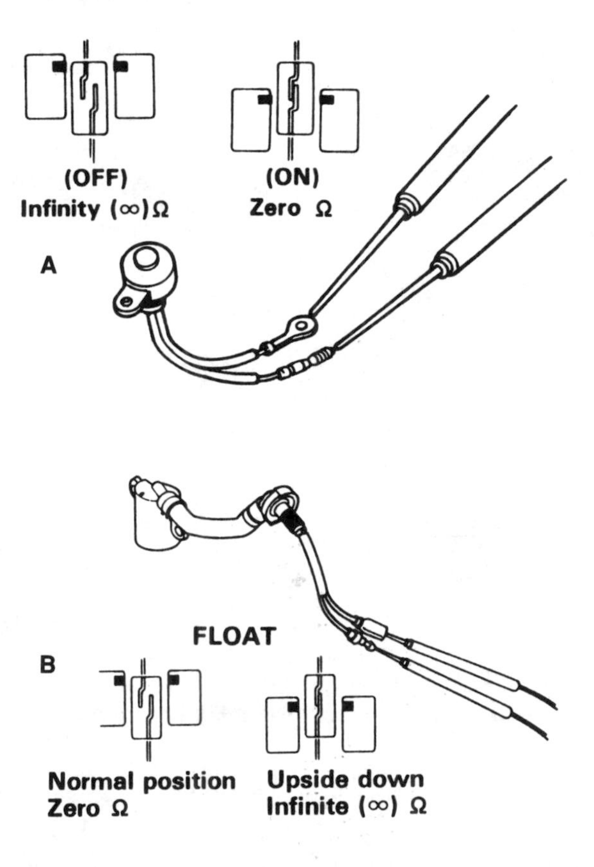

Fig. 5–45. (A) Checking out the oil level switch. (B) Oil level switch for GX120K1 and GX160K1 engines. *(Courtesy Honda)*

Fig. 5–49 shows where the engine stop switch is located on the GXV270 and GXV340 Honda engines. In order to check this switch for proper operation, see Fig. 5–50. Check for continuity between the terminals. There should be continuity when the switch is depressed and no continuity when the switch is released.

The engine switch is slightly different on the GX240 and GX340. The engine switch on an engine without oil alert is shown in Fig. 5–51. Turn the switch, and check for continuity between the wire and the fan cover with an ohmmeter. With the switch *on* it should have no continu-

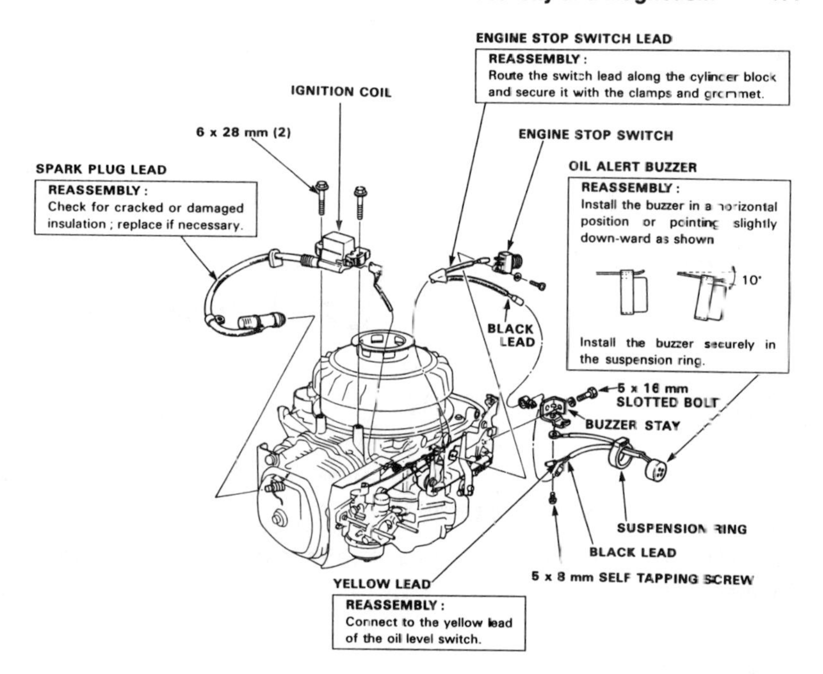

Fig. 5–46. Note location of the buzzer and its installation. *(Courtesy Honda)*

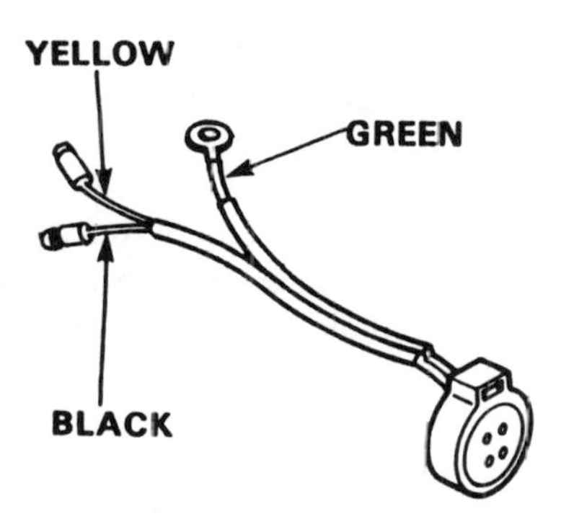

Fig. 5–47. Oil alert buzzer checkout. *(Courtesy Honda)*

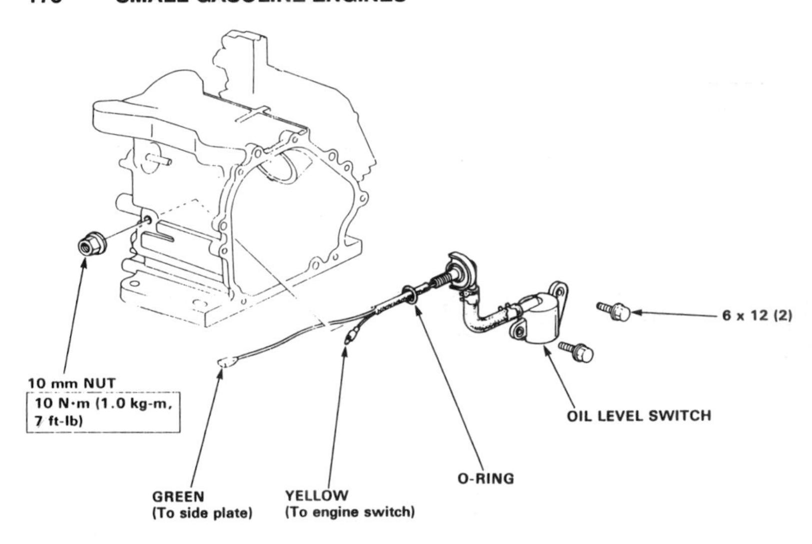

Fig. 5–48. Note how the oil level switch is installed in this model. *(Courtesy Honda)*

ity and with the switch in the *off* position it should have continuity. That means it grounds out the ignition when the switch is in the *off* position.

The engine with oil alert has to be checked for continuity between the engine switch black wire and the switch body with an ohmmeter (Fig. 5–52). A switch in the *on* position should have *NO* continuity; in other words, the switch is open. Switch in the *off* position should have continuity, which means it is shorted. In order to check the oil alert unit see Fig. 5–53. For the GX240 engine with oil alert, a battery with 1.5 to 6 volts is connected to the oil alert unit leads. The black lead gets the + terminal. The yellow lead is connected to the – terminal. When properly connected, the oil alert lamp should light. If the lamp does not light, replace the unit. Make sure a battery of *less than* 6 volts is used.

The GX340 engine with oil alert takes a little more time and effort to check out its unit (Fig. 5–54). DO NOT connect a battery to the GX340 oil alert unit. It may damage the diodes. Perform the following checks:

Table 5–1. Text Checks for GX340 Engine with Oil Alert (Ohmmeter Required)

Engine switch position / Terminals	OFF	ON
Black and Yellow	0.1 Ω max.	0.1 Ω max.
Black and Ground Yellow and Ground	0.1 Ω max.	1 Mv min.
Green and Ground	3.0–4.0 KΩ	3.0–4.0 KΩ
Black and Green Yellow and Green	3.0–4.0 KΩ	1 MΩ min.

(Courtesy Honda)

Lamp Coil Kits

Lamp coils are made for lighting 15W and 25W lamps at 6 volts, and with the proper arrangement, 15 watts and 25 watts at 12 volts. Fig. 5–55 shows how one coil is used to produce 6 volts at 15 watts or 6 volts at 25 watts. It also shows how two coils can be placed in series to produce 12 volts or 24 volts. By arranging the coils in series or parallel it is possible to obtain 6 volts, 12 volts, or 24 volts. Wattage rating for the lamps that the coils can light are 15-watt, 25-watt, 30-watt, and 50-watt. The GX 120K1 and GX 160K1 engines utilize these coils in various combinations depending on the purpose for which the engine is used (Fig. 5–56).

Note the difference between the standard flywheel with a magnet for a transistorized ignition and the flywheel for a lamp coil with two more magnets added (Fig. 5–57).

The GX240 and GX340 engines have lamp coils available for 6-, 12-, and 24-volt outputs for 25- and 50-watt lamps. They can be installed singly or in pairs (Fig. 5–58), to produce the output shown in Fig. 5–59. Note how the flywheels shown in Fig. 5–60 have the magnets arranged for various purposes.

The charging coil is located under the flywheel as shown in Fig. 5–61. Note how the charging coil is mounted under the flywheel with two 6x12mm screws to hold it in place.

Inspection—The charging coil (Fig. 5–62) can be checked without removing it. Using an ohmmeter, measure the resistance between the terminals. The charging coil resistance is usually 3.0 to 4.0 ohms. Now

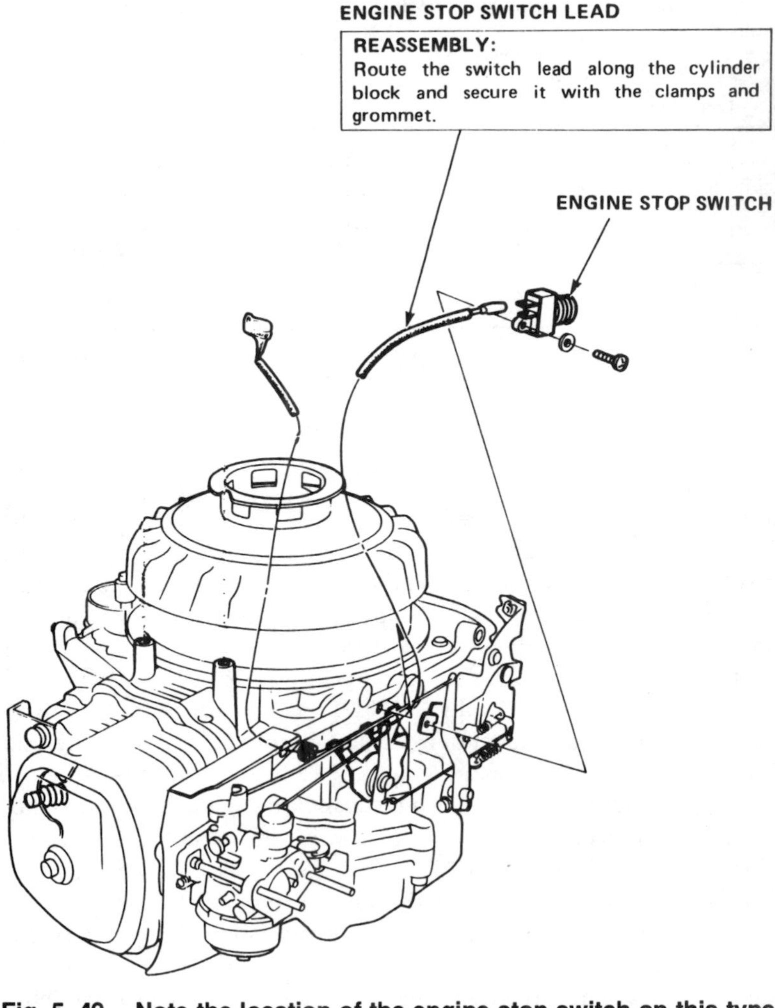

ENGINE STOP SWITCH LEAD

REASSEMBLY:
Route the switch lead along the cylinder block and secure it with the clamps and grommet.

ENGINE STOP SWITCH

Fig. 5–49. Note the location of the engine stop switch on this type of engine. *(Courtesy Honda)*

you can see the importance of a digital ohmmeter when it comes to checking very low resistances. An analog-type ohmmeter would probably just show something near zero without being able to differentiate between 0 and 4 ohms.

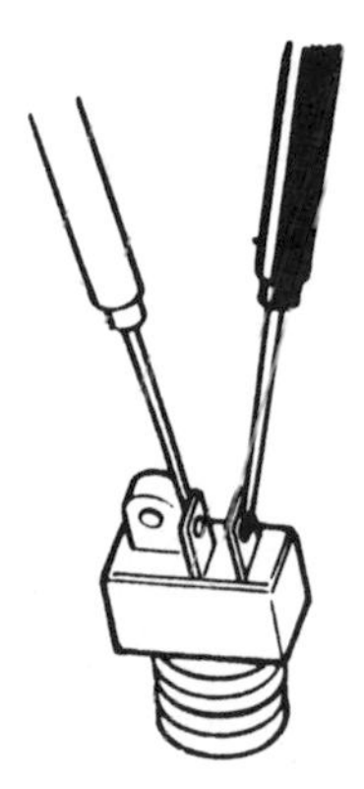

Fig. 5–50. Checking the engine stop switch on GXV270 and GXV340 engines. *(Courtesy Honda)*

Control Box

The electrical circuitry for the models of GX240 and GX340 with electric starters resembles that shown in Fig. 5–63. Note the control box and its components: the combination switch, the rectifier unit, the blades type fuse (same as used in most modern automobiles), and the circuit breaker.

The *combination switch* is checked for continuity between the wires shown in Table 5–2. Replace the switch if the correct continuity is not obtained (Fig. 5–64). Keep in mind that this switch has three positions: off, on, and start.

The *oil alert unit* in the GX240 is checked by connecting a battery with 1.5 volts to 6 volts to the unit leads as shown in Fig. 5–65. Replace the alert unit if the lamp does not light. The black/red lead goes to the battery + terminal and the yellow lead goes to the – terminal.

Table 5–2. Combination Switch Continuity Checks—GX240/GX340

Wire Color / Switch position	Black/ Red	Ground	Black/ White	White
OFF	○—	—○		
ON				
START			○—	—○

In the GX340 engine, measure the resistance between the terminals and do not connect a battery or you may damage the diodes (Fig. 5–66). Table 5–3 shows how the engine switch is either on or off and the terminals are checked with a digital ohmmeter so that the 0.1-ohm resistance can be detected, as well as the 1 megohm minimum in one of the positions.

The *rectifier* unit can be checked for continuity between its two terminals as shown in Fig. 5–67. There should be continuity in one direction, but not when the test leads are reversed. Replace the rectifier unit if there is continuity in both directions, as this means that the diodes are shorted. Also, if there is continuity in neither direction, it is open and should be replaced.

The *circuit breaker* is checked for continuity between its two terminals as shown in Fig. 5–68. There should be continuity in the *on* position (that is, with the button pushed in) and no continuity in the *off* position (that is, with the button pulled out). Replace the circuit breaker if the correct continuity is not obtained.

Electrical Diagrams

Fig. 5–69 shows the electrical diagram for Honda's GX120K1 and GX160K1 engines with electric starters, but without the oil alert. Here it is possible to see what is included in the control box. Note in Fig. 5–70 that the control box also contains the oil alert unit. Note how the oil alert unit is utilized to ground out the ignition coil when the oil level reaches the critical level.

As can be seen here in these electrical diagrams, it is a rather easy electrical circuit when there is no starter, battery, or alarm system to

Table 5–3. Checking out the Oil Alert Unit on GX340 Engine

Engine switch position / Terminals	OFF	ON
Black/Red and Yellow	0.1 Ω max.	0.1 Ω max.
Black/Red and Ground Yellow and Ground	0.1 Ω max.	1 MV min.
Green and Ground	3.0–4.0 KΩ	3.0–4.0 KΩ
Black/Red and Green Yellow and Green	3.0–4.0 KΩ	1 MΩ min.

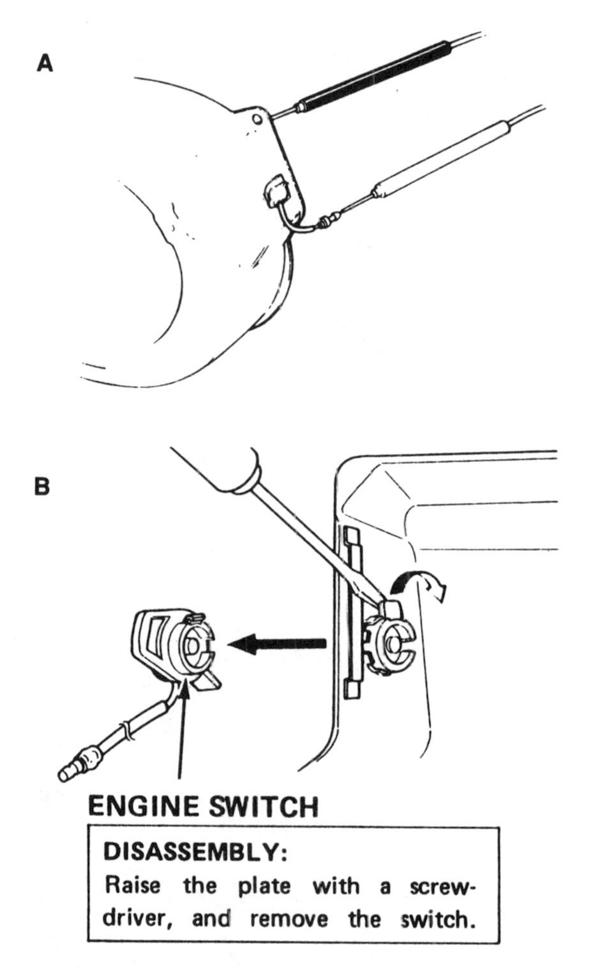

Fig. 5–51. (A) Engine switch without oil alert is simple to check out. (B) Checking out and removing the engine switch. *(Courtesy Honda)*

contend with. Whenever more devices are added, the electrical system becomes more complicated, meaning that a better understanding of the electrical circuity is needed by the person who services or repairs these engines.

More electrical circuits are shown in another chapter in this book.

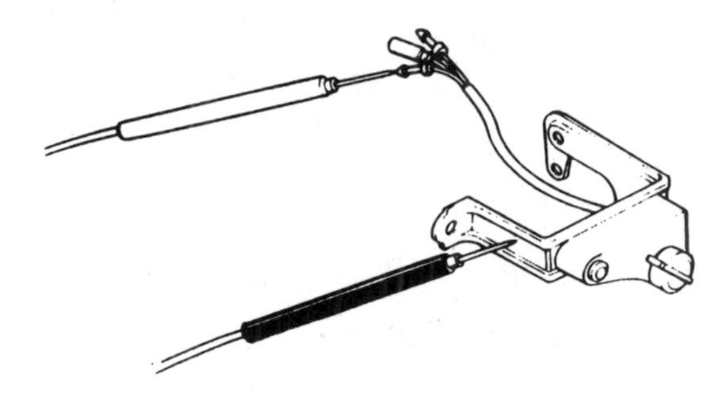

Fig. 5–52. Checking the engine switch with an oil alert on the engine. *(Courtesy Honda)*

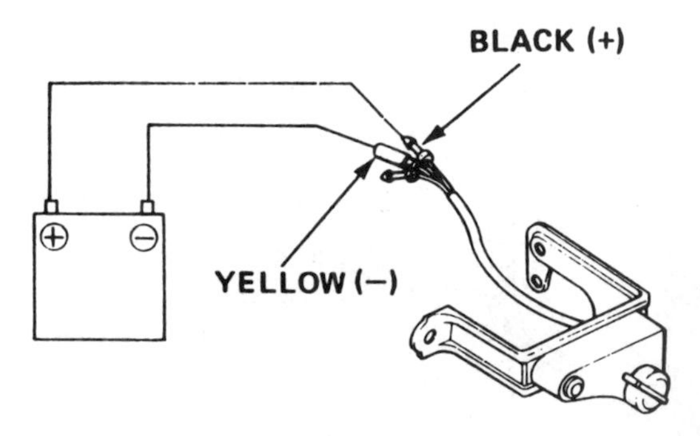

Fig. 5–53. Checking the oil alert circuit on the GX240 engine. *(Courtesy Honda)*

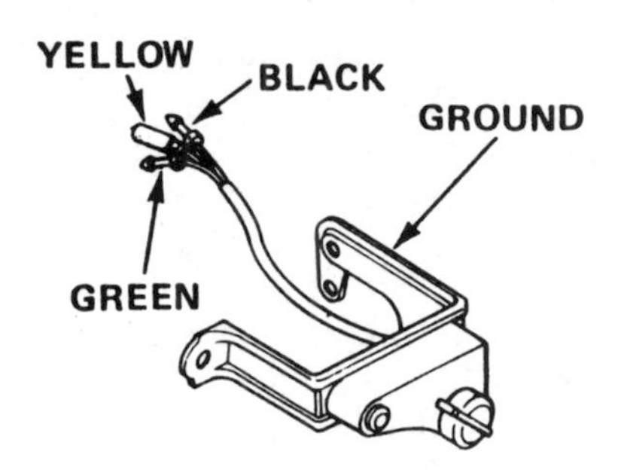

Fig. 5–54. Measuring the GX340 oil alert. *(Courtesy Honda)*

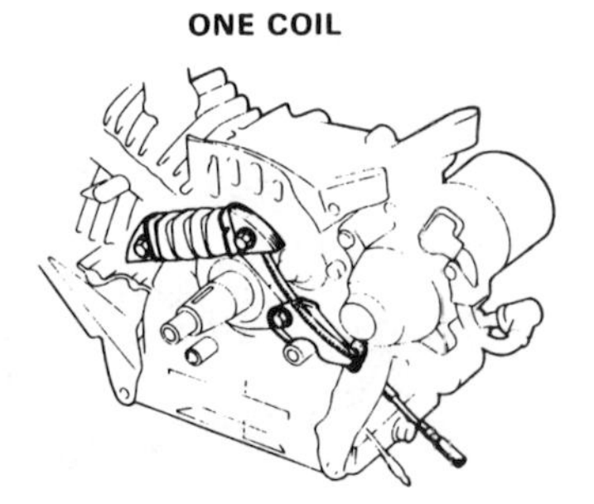

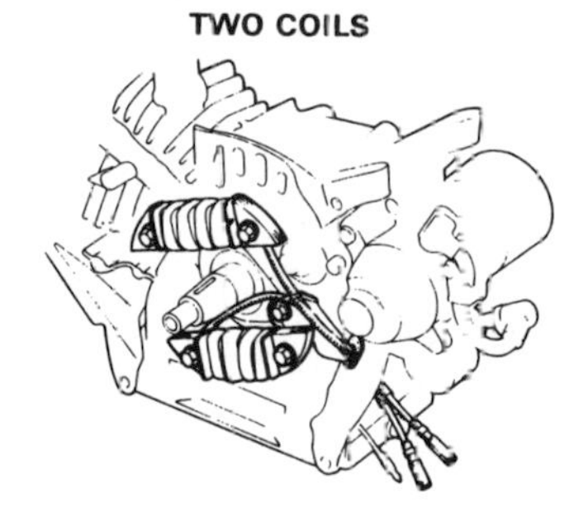

Fig. 5–55. Location of one coil and two coils on engines. *(Courtesy Honda)*

LAMP COIL TYPES	ONE COIL	TWO COILS (SERIES)	TWO COILS (PARALLEL)
6 V 15 W	6 V 15 W	12 V 15 W	6 V 30 W
6 V 25 W	6 V 25 W	12 V 25 W	6 V 50 W
12 V 15 W	12 V 15 W	24 V 15 W	12 V 30 W
12 V 25 W	12 V 25 W	24 V 25 W	12 V 50 W

※Use the parallel connector (No. 32105-ZEI-000) for the parallel two coils.

Fig. 5–56. Lamp coil types for GX120K1 and GX160K1 engines. *(Courtesy Honda)*

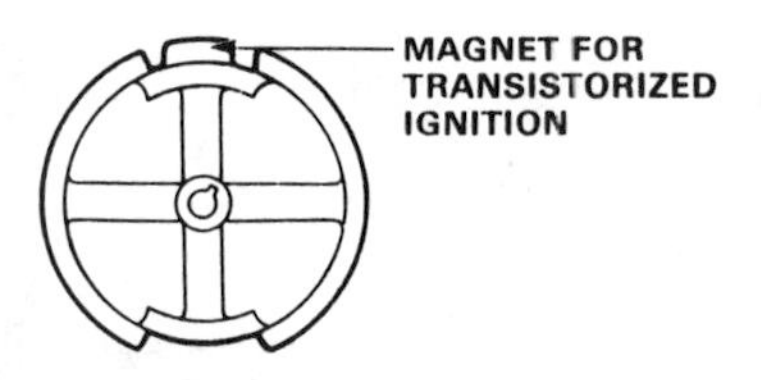

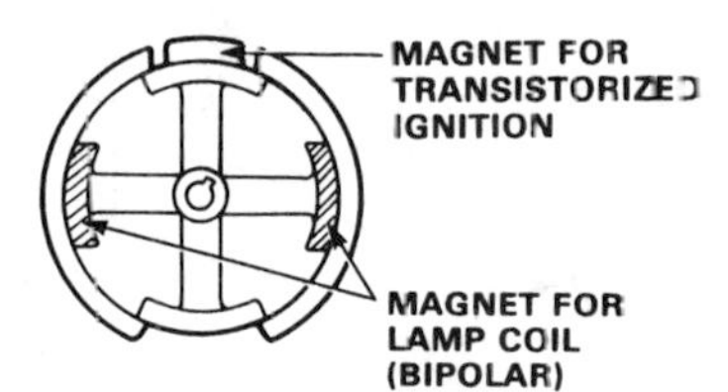

Fig. 5–57. Standard flywheel magnet and flywheel with magnets for lamp coil and transistorized ignition. *(Courtesy Honda)*

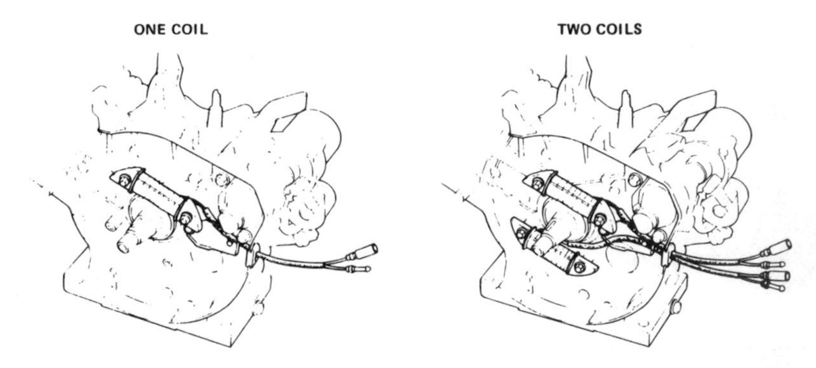

Fig. 5–58. Lamp coil kit—one coil and two coils. *(Courtesy Honda)*

LAMP COIL TYPES	ONE COIL	TWO COILS (SERIES)	TWO COILS (PARALLEL)
		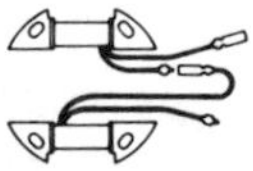	
6V–25W	6V–25W	12V–25W	6V–50W
12V–25W	12V–25W	24V–25W	12V–50W

Fig. 5–59. Lamp coil types. *(Courtesy Honda)*

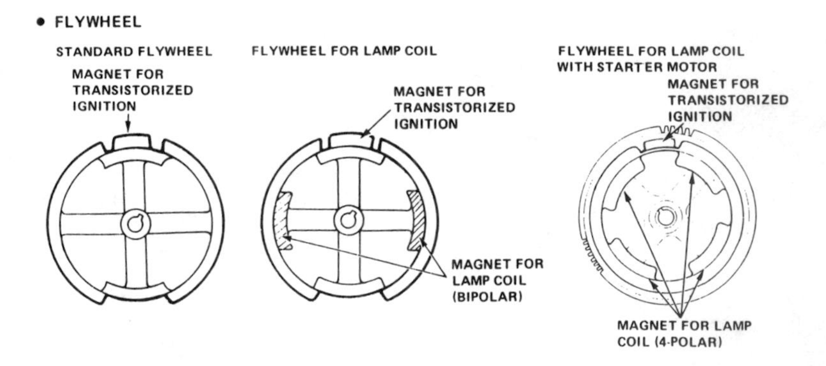

Fig. 5–60. Flywheel configurations for GX240 and GX340 engines. *(Courtesy Honda)*

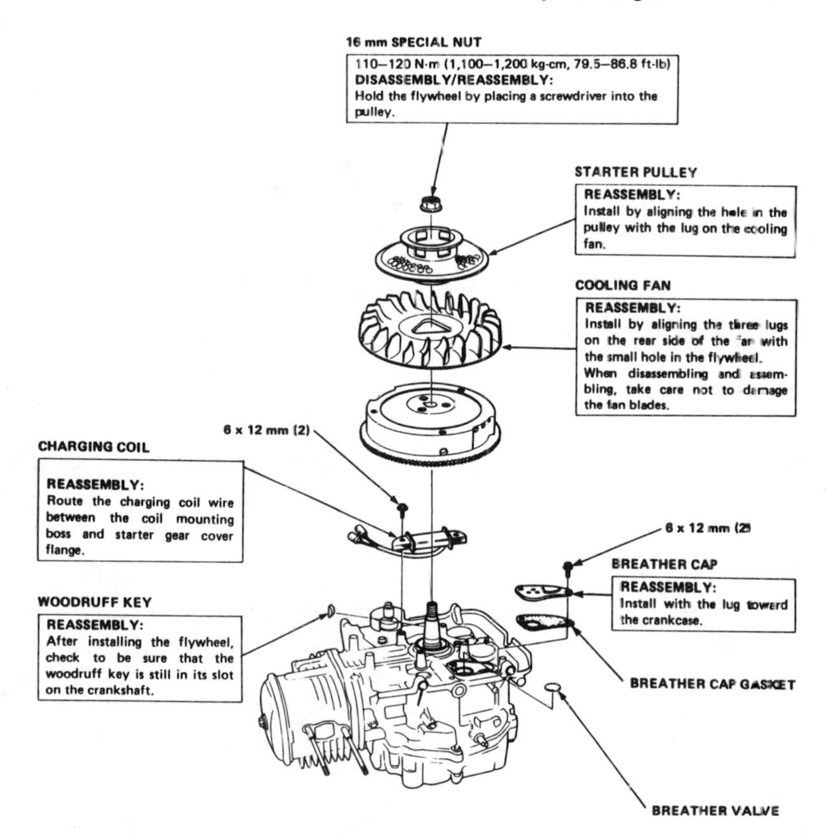

Fig. 5–61. Location of the charging coil on the GXV270 and GXV340 engines. *(Courtesy Honda)*

Chapter Highlights

In this chapter you should have learned:

1. That magnetism is an invisible force.
2. That natural and artificial magnets are used today.
3. That magnets have poles.
4. That artificial magnets can be produced by magnetic materials.
5. That permanent magnets are made from substances such as hardened steel and certain alloys.

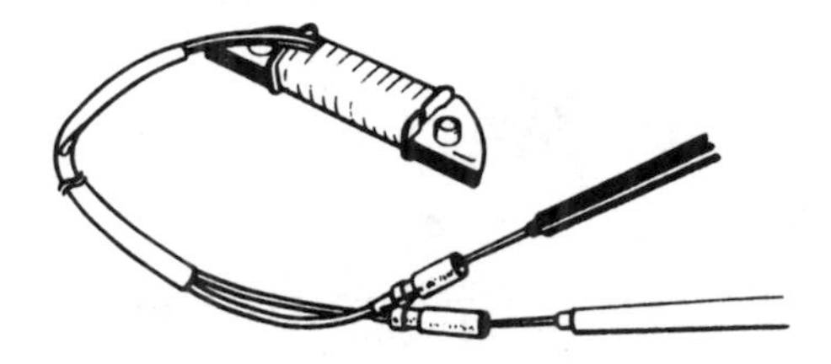

Fig. 5–62. Checking out the charging coil on the GXV270 and GXV340. *(Courtesy Honda)*

6. That a material with low reluctance is fairly easy to magnetize.
7. That the ability of a material to retain some residual magnetism is called retentivity.
8. That reluctance is magnetic resistance.
9. That the magnetic force surrounding a bar magnet is not uniform.
10. That like poles repel each other and unlike poles attract each other.
11. That motors and meters use magnetism.
12. That magnetic lines of force are imaginary.
13. That certain laws govern the behavior of magnets.
14. That magnetic flux is related to the total number of lines of force leaving or entering a pole.
15. The magnets cannot be treated roughly or heated.
16. That there are six ways to produce an emf.
17. That charged bodies react similarly to electrons.
18. That chemical reactions can be used to produce electricity.
19. That a primary cell cannot be recharged.
20. That a secondary cell can be recharged.
21. That thermocouples can produce electricity be being heated.
22. That silicon can be used to make solar cells.
23. That the piezoelectric effect can produce electricity by pressure.
24. That a horseshoe magnet also has two poles.

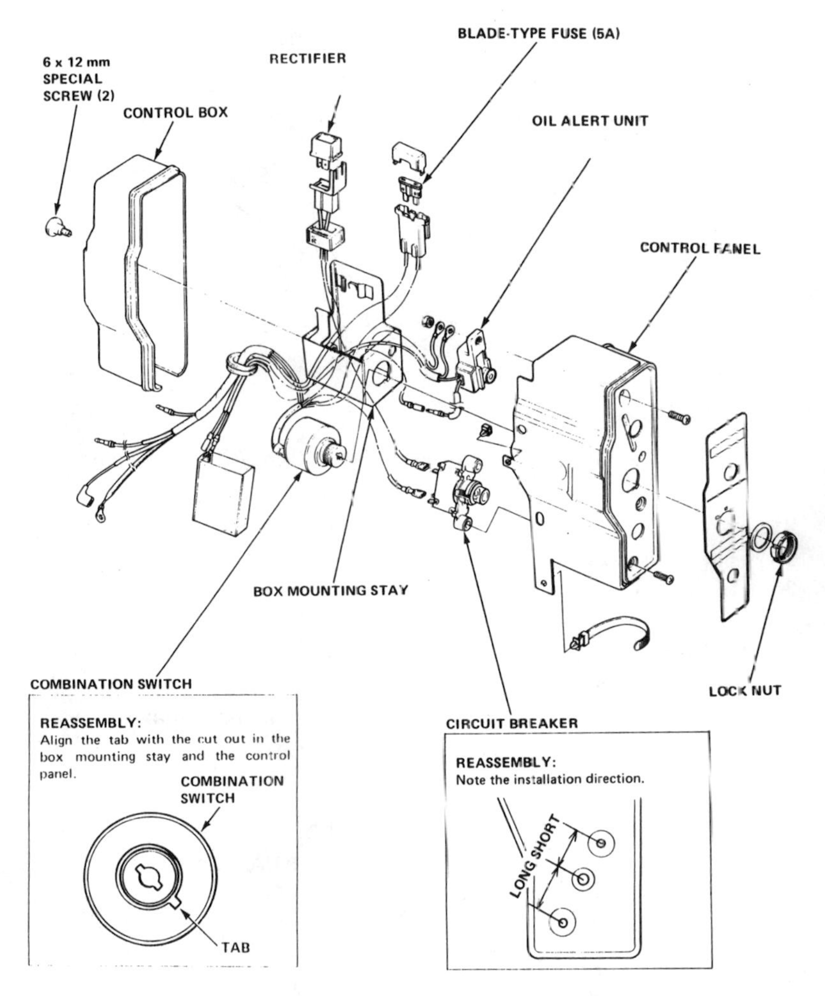

Fig. 5–63. Exploded view of control box for GX240 and GX340 engines. *(Courtesy Honda)*

25. That an electron generates a magnetic field when it moves.
26. That electricity produced by magnetism is used to provide a spark at the end of the gasoline engine's spark plug.
27. That 1 ampere is 1 coulomb in motion past a given point in 1 second.

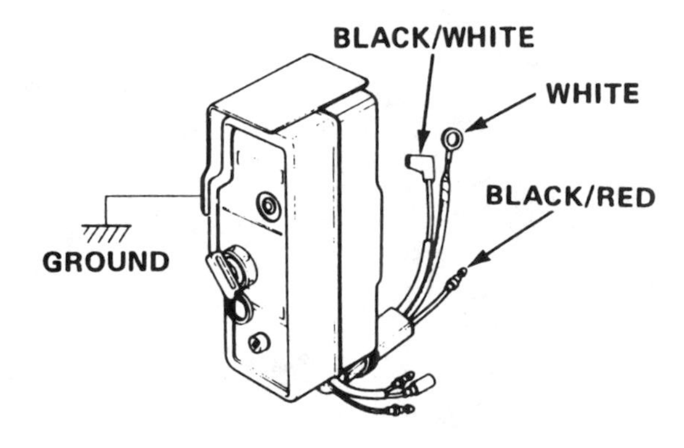

Fig. 5–64. Combination switch wires identified. *(Courtesy Honda)*

28. That a milliampere is one-thousandth of an ampere.
29. That a microampere is one-millionth of an ampere.
30. That the volt is the unit of measurement of electrical pressure, or voltage.
31. That the ohm is the unit of measurement for the electrical resistance of a circuit or device.
32. That the watt is the measurement unit for electrical power.

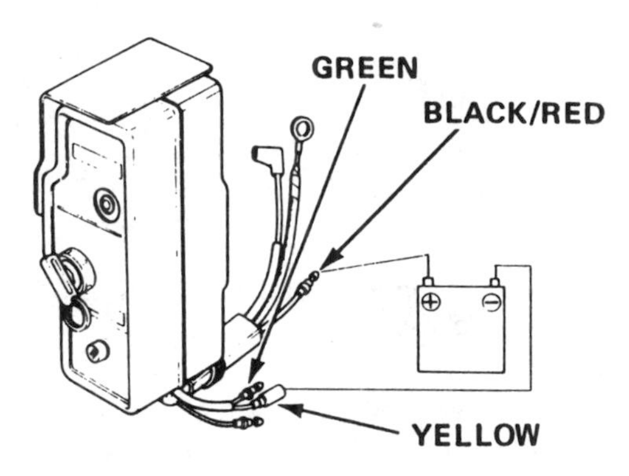

Fig. 5–65. Checking out the oil alert unit on the GX240. *(Courtesy Honda)*

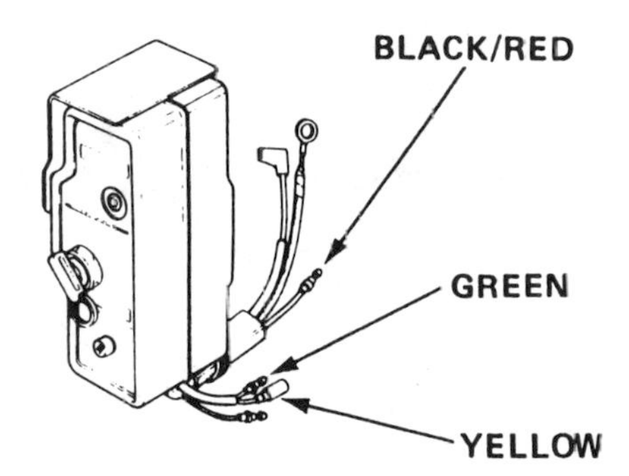

Fig. 5–66. Checking out the oil alert unit on the GX340. *(Courtesy Honda)*

33. That emf, potential difference, and voltage are the same.
34. That a kilohm is 1000 ohms.
35. That a megohm is 1,000,000 ohms.
36. That power is the rate at which work is done.
37. That work is done whenever a force causes motion.
38. That $P = E \times I$.
39. That $P = I^2R$.
40. That $P = E^2/R$.
41. That capacitor discharge ignition (CDI) systems are very efficient and reliable.
42. That the SCR is triggered to discharge the capacitor into the transformer, providing the high voltage for the spark.
43. That there are no components located under the flywheel in the CDI system.
44. That the CDI unit is enclosed in an epoxy to make a module.
45. That CDI units are not always interchangeable with other engines.
46. That the proper air gap setting between the magnets and the laminations is 0.0125″, the same for both CDI and breaker point ignitions systems.
47. That the oil alert system can save an engine by alerting the operator before the engine runs too low on oil.

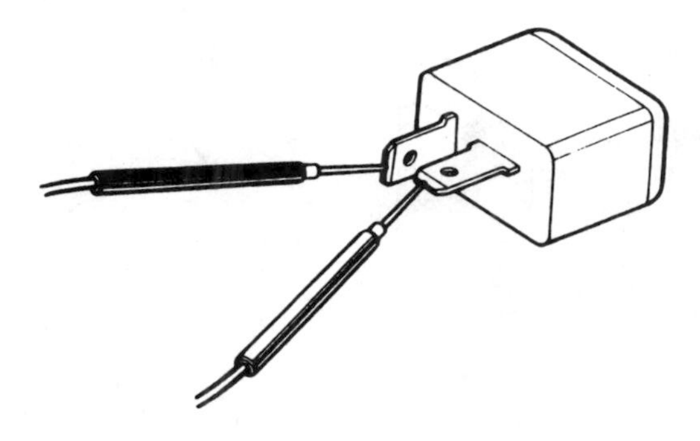

Fig. 5–67. **Checking out the rectifier.** *(Courtesy Honda)*

48. That an ohmmeter is used to check the oil alert system.

49. That the engine stop switch on the Honda engines can be checked with an ohmmeter.

50. That lamp coil kits are available for both 6-volt and 12-volt bulbs.

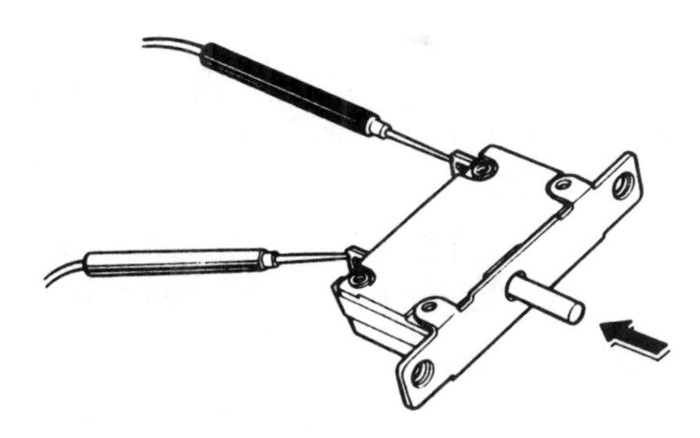

Fig. 5–68. **Checking out the circuit breaker.** *(Courtesy Honda)*

(With electric starter/without oil alert)

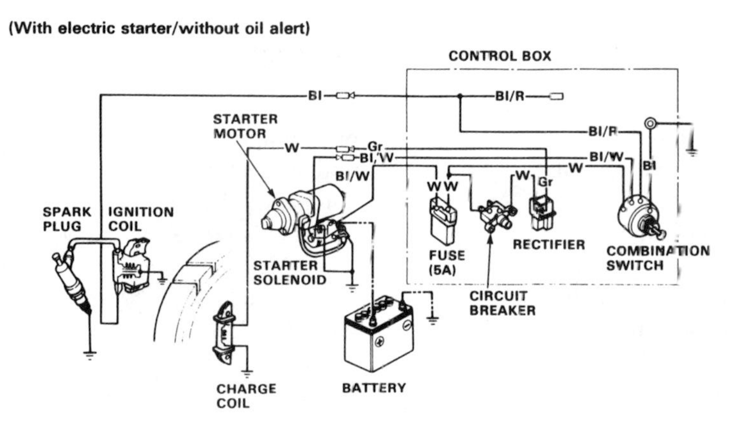

Fig. 5–69. Electrical diagram for Honda GX120K1 and GX160K1 with electric starter, but without oil alert. *(Courtesy Honda)*

(With electric starter and oil alert)

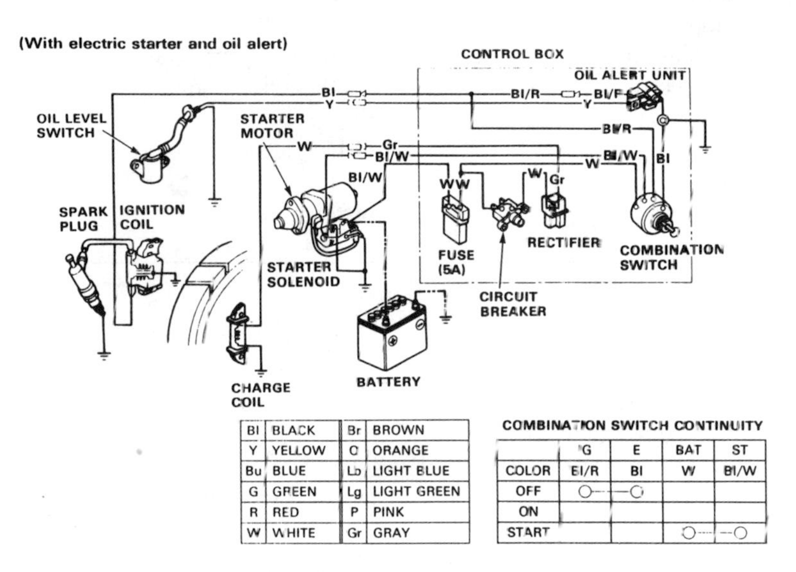

Bl	BLACK	Br	BROWN
Y	YELLOW	O	ORANGE
Bu	BLUE	Lb	LIGHT BLUE
G	GREEN	Lg	LIGHT GREEN
R	RED	P	PINK
W	WHITE	Gr	GRAY

COMBINATION SWITCH CONTINUITY

	IG	E	BAT	ST
COLOR	Bl/R	Bl	W	Bl/W
OFF	○—	—○		
ON				
START			○—	—○

Fig. 5–70. Electrical diagram for Honda GX120K1 and GX160K1 with electric starter and oil alert. *(Courtesy Honda)*

Vocabulary Checklist

- ammeter
- amperes
- capacitor discharge system
- CDI
- charged body
- circuit breaker
- combination switch
- control box
- coulomb
- diode
- emf
- engine stop switch
- filament
- flux density
- friction
- galvanic cell
- infinity
- kilohm
- laminations
- lamp coils
- lodestone
- magnetic flux
- magnetic materials
- magnetism
- megohm
- microampere
- milliampere
- nonmagnetic material
- north-seeking pole
- ohm
- oil alert buzzer
- permeability
- photoelectric cell
- piezoelectric effect
- polarity
- primary cell
- rectifier unit
- reluctance
- retentivity
- SCR
- secondary cell
- silicon cell
- solid state ignition system
- south-seeking pole
- thermocouple
- visual check
- watt
- wattage rating

Review Questions

1. What is magnetism?
2. How are artificial magnets produced?
3. What is another name for a natural magnet?
4. What are six characteristics of magnetic lines of force?
5. What is magnetic flux?
6. List six ways to produce electricity.
7. What is the difference between a primary and a secondary cell?
8. How many poles does a horseshoe magnet have?

9. The amount of voltage induced in a conductor depends on what three things?
10. What is the difference between a milliampere and a microampere?
11. What does the term *electromotive force* mean?
12. What is the difference between a kilohm and a megohm?
13. What is the unit of measurement for power?
14. What are the three formulas used for calculating power?
15. What is a magneto?
16. What is the primary voltage of the ignition coil of a small gasoline engine?
17. What is reluctance?
18. What is the purpose of breaker points in an engine's ignition system?
19. What is the usual size of the condenser used in small engines?
20. What does an oily spark plug indicate?
21. What does a normal plug look like?
22. If the plug is burned and/or pitted, what does this indicate?
23. What is a capacitor discharge system?
24. What is impulse coupling?
25. What role does the pawl play in impulse coupling?
26. What is meant by *solid state?*
27. What does *CDI* stand for?
28. How does the CDI system work?
29. Why is the solid state system more reliable?
30. What is a diode?
31. What is an SCR?
32. What is meant by the term *module?*
33. Where is the solid state electronic ignition system located on the engine?
34. How much of an air gap is specified between the laminations and the magnet?
35. What is the only on-engine check that can be performed to determine if the ignition system is working?
36. How does the oil-alert system operate?
37. Where is the oil level switch located?
38. Why is the engine stop switch necessary?
39. What is meant by the term *infinity* when used on an ohmmeter?
40. What instrument do you use to check the oil alert system?

CHAPTER 6

Starting and Stopping Systems

In order to use an engine it is necessary to start it. There are three methods used to start small gasoline engines. The manual start, the mechanical start, and the electrical start methods are used in various forms. It is the aim of this chapter to describe these three methods of starting small gasoline engines.

Manual Starting

There are two manual start systems used with small gasoline engines. They are the *rope* type and the *crank-and-kick* type. Each is distinct and different.

In order to start an engine from a no-run condition, it takes a lot of energy. There are springs to be compressed, oil to be moved—hard to do if it is cold and the oil is thick—and there is the up-and-down movement of the piston inside the cylinder. All these, plus the friction of the tight bearings and other friction within the engine, demand enough energy to cause sufficient momentum to get the engine to fire the first time to provide its own momentum.

In order to start an engine, the proper fuel-air ratio is needed. You usually assume that the carburetor is working properly before pulling the rope to start the engine. In some cases a primer is needed to make sure enough gasoline is present in the carburetor for the rich mixture needed for starting. The choke has to be set correctly and the ignition must be properly timed.

Starting the engine calls for the proper settings for the throttle and carburetor. Fig. 6–1 shows how the control should be set to the proper position. In other words, move the engine control to the *FAST* position. Next, grasp the starter handle as shown in Fig. 6–2. Pull rapidly outward on the cord. This causes the flywheel to spin fast enough to generate a voltage sufficient to jump the spark plug gap. Return the starter handle slowly to the engine. Repeat if necessary.

IMPORTANT: This engine may feature an automatic choke. In case of flooding, move the control to *STOP* and pull the starter rope six or seven times. Then move the control to *FAST* and start the engine. If the engine continues to flood, rotate the carburetor needle clockwise one-eighth of a turn to obtain a learner mixture (Fig. 6–3).

The carburetor needle valve has been adjusted so that the mixture is proper for average operating conditions. Normally, no changes are necessary. However, if the engine does not start easily, a minor adjustment will compensate for the differences in fuel, temperature, or altitude.

If the engine does not start easily when cold, rotate the needle valve counterclockwise one-eighth of a turn to produce a richer mixture.

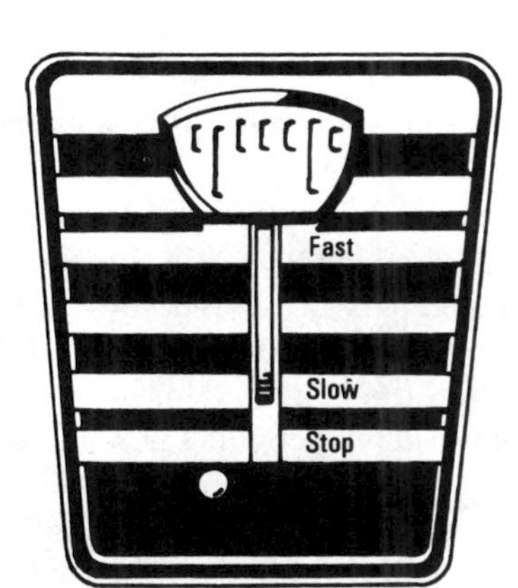

Fig. 6–1. Throttle control should be pushed forward to *FAST* position to start the engine.

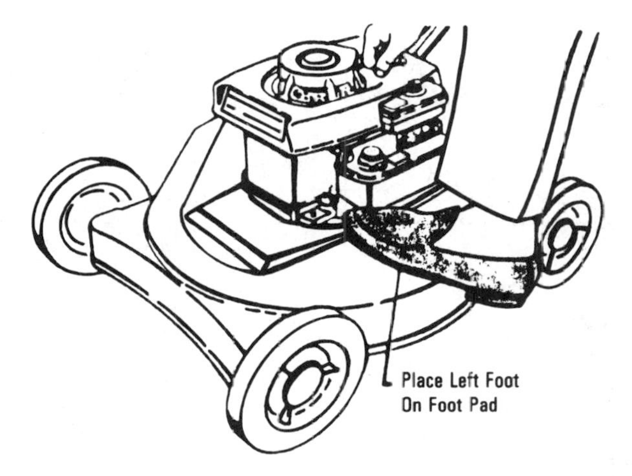

Fig. 6–2. If the engine is mounted on a lawnmower, put the left foot on the lawnmower and pull the crank rope with the right hand. Don't place your foot under the mower; injury may occur from the blade once the engine starts.

If the engine does not start easily when warm, rotate the needle valve clockwise one-eighth of a turn to produce a leaner mixture.

Various rewind-type manual starters are shown in Fig. 6–4 through 6–7. There are slight differences since they represent different years of production.

The starting operation should not be too long in duration. It takes about two times (normally, no more than twelve) to see if the engine will respond, unless it is flooded.

Once the engine starts, the starter should be disengaged. This prevents damage to the starter. The manually operated starter has to disengage to protect the engine operator. If it did not disengage, the high speed of the running engine would wear the starter excessively.

Rope starters are used on lawnmowers and outboard motors. Some of the simpler ones have a rope that is wound around the pulley attached to the flywheel. The pull on the rope causes the flywheel to spin. The pulley is notched so the end of the rope (with a knot in it) can be attached. Once it is pulled straight out, the end of the rope pulls free also.

The *direct rewind starter* is nothing more than a piece of rope

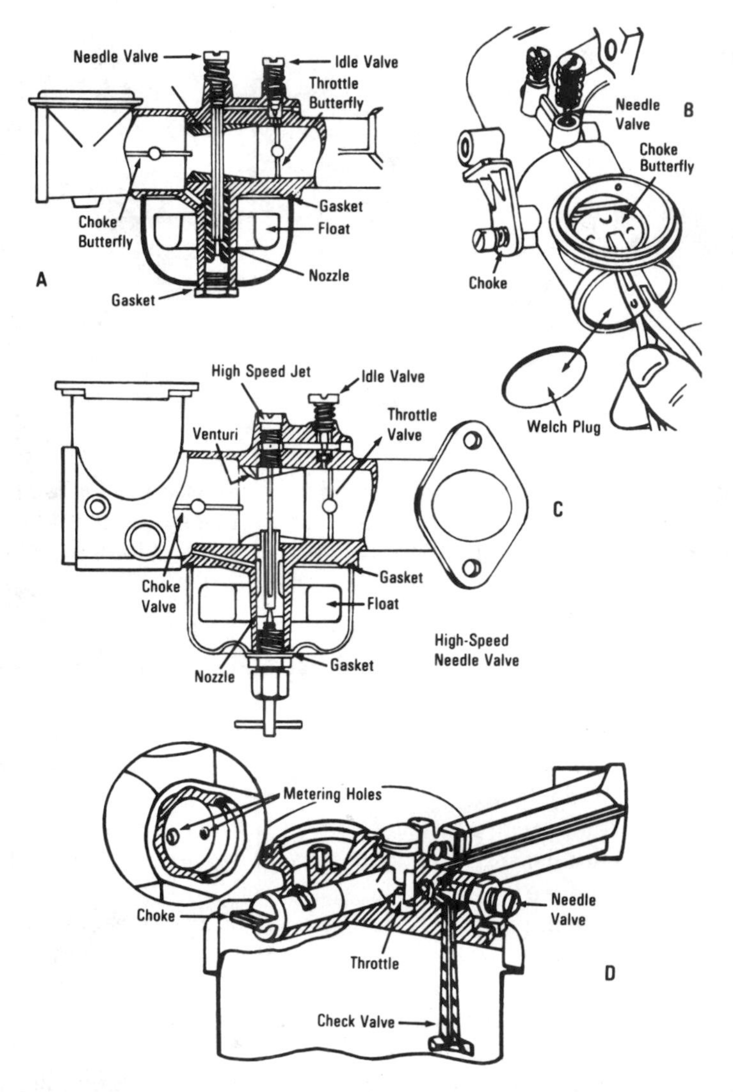

Fig. 6–3. The needle valve is located on the various carburetors at different points. Three of the Briggs & Stratton carburetors are shown here. (A) is the typical B & S small "one-piece" float type of carburetor. (B) shows the same carburetor as in (A), but gives the actual location of the needle valve adjustment. (C) is a cross-sectional view of the B & S larger one-piece float type of carburetor. (D) is the cutaway view of the typical suction (Vacu-jet) carburetor. Note the location of the needle valve.

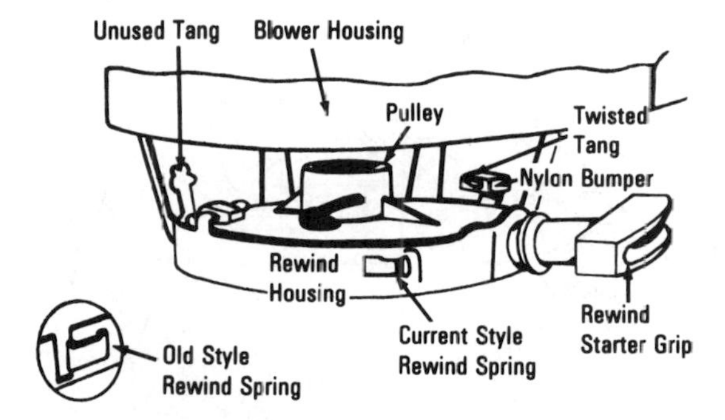

Fig. 6–4. Rewind starter assembly. This is the Briggs & Stratton old-style model series: 60000, 80000, 92000, 100000, and 110000.

mounted so that it retracts once it is pulled. This type has a clutch arrangement so the starter is disengaged once the engine starts (Fig. 6–8). If necessary, the sealed clutch can be disassembled by using a screwdriver or wedge to pry the retainer cover from the housing. Make sure the screen is screwed on before starting the engine.

The *geared rewind starter* has more leverage at start since the crankshaft is gear-driven. The rope is held by a threaded spool. A rewind spring is inside the spool.

Manual crank-type starters are used on multicylinder engines. It

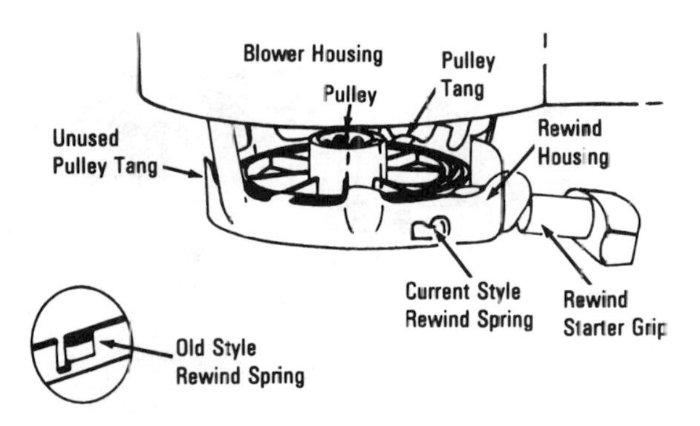

Fig. 6–5. Rewind starter assembly. This is the Briggs & Stratton model series 60000, 80000, 92000, 100000, and 110000. Slight changes have been made since those shown in Fig. 6–4.

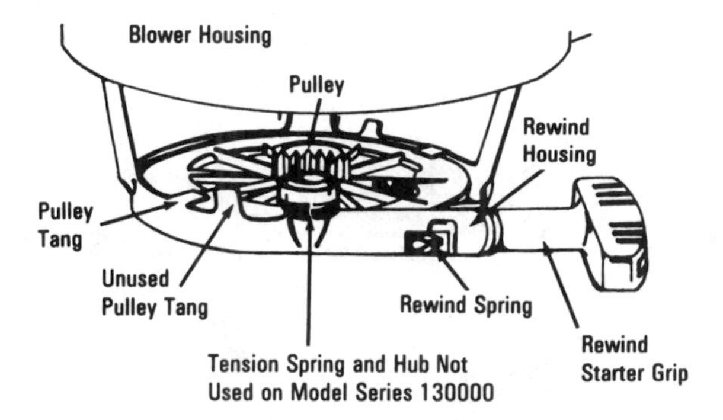

Fig. 6–6. Other modifications resulted in this rewind starter for Briggs & Stratton model series 130000, 140000, 170000, 190000, 220000, and 250000.

can be dangerous to use this type of cranking. The engine can backfire and cause enough torque on the hand crank to cause broken bones. This is usually an emergency type of cranking. The electric starter usually is engaged in the same crankshaft extension as the hand crank (Fig. 6–9).

Another type of manual starter is the *kick starter.* It is used on motorcycles or other small gasoline engines that are located so that the leg can be used as the power for cranking. This type of starter works

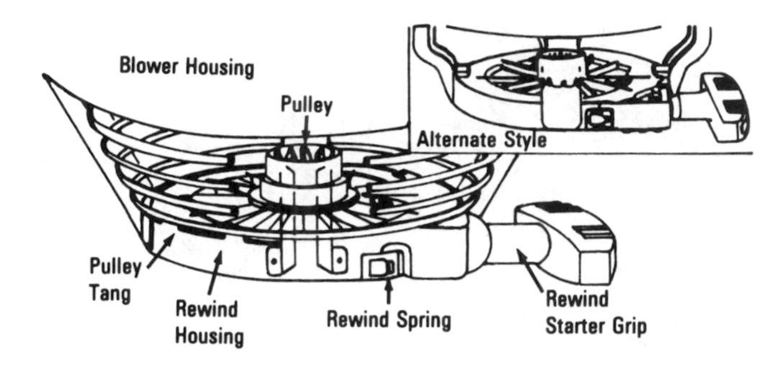

Fig. 6–7. Changes in the original present this model series with a newer type of starter. This one is used on Briggs & Stratton model series 140000, 170000, 190000, 250000, and 300000.

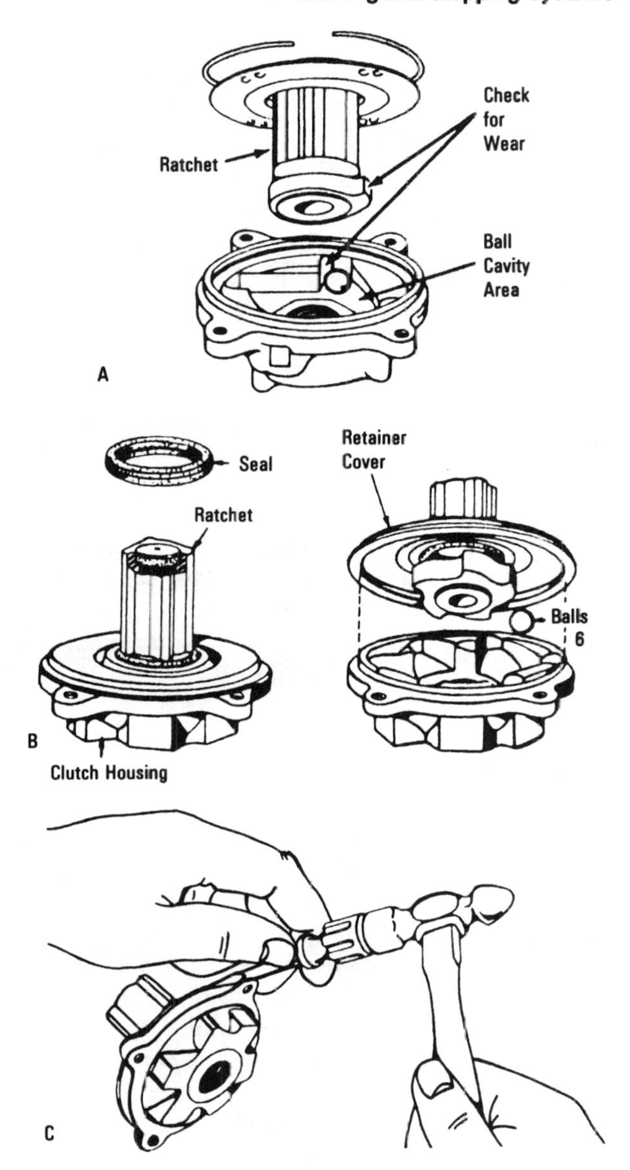

Fig. 6–8. (A) Old-style starter clutch used by Briggs & Stratton. (B) Sealed clutch assembly as now used on Briggs & Stratton models. (C) Disassembling the sealed clutch.

Fig. 6–9. A hand-crank starter as used on older-model automobiles and larger-size gasoline engines for light duty around the farm, home, and business.

through gears to drive the crankshaft and magneto. In some cases, the magneto isn't used since a battery furnishes the electrical power. The handle or foot rest of the kick starter usually folds back out of the way once the engine starts and the crank has been released. Note the arrangement of the springs and how the gear meshes with the crankshaft gears. The driver fits into the slot and causes the gear to turn. Once the engine starts, the driver is disengaged. A spring causes the crank to return if it does not start the first time or when the engine starts and the foot is removed from the crank (Fig. 6–10).

Mechanical Starting

There are a number of windup-type mechanical starters for small gasoline engines. The mechanical windup type is an example. A heavy spring is used to exert the necessary cranking pressure on the flywheel to spin it around. This is a replacement for the rewind type. However, the operation is basically the same. The spring is held in tension by a ratchet as the operator turns the handle to wind up the spring.

Once the spring is wound, a release-button or lever is provided to release the starting cycle. Fig. 6–11 shows the old and newer type of starter assemblies. These types of windup starters have either control knob or control lever releases. The control knob type is used with the unsealed four-ball clutch. The control lever is used with a sealed six-ball clutch. When working with this type of starter, caution is advised as the wound spring can cause bodily damage.

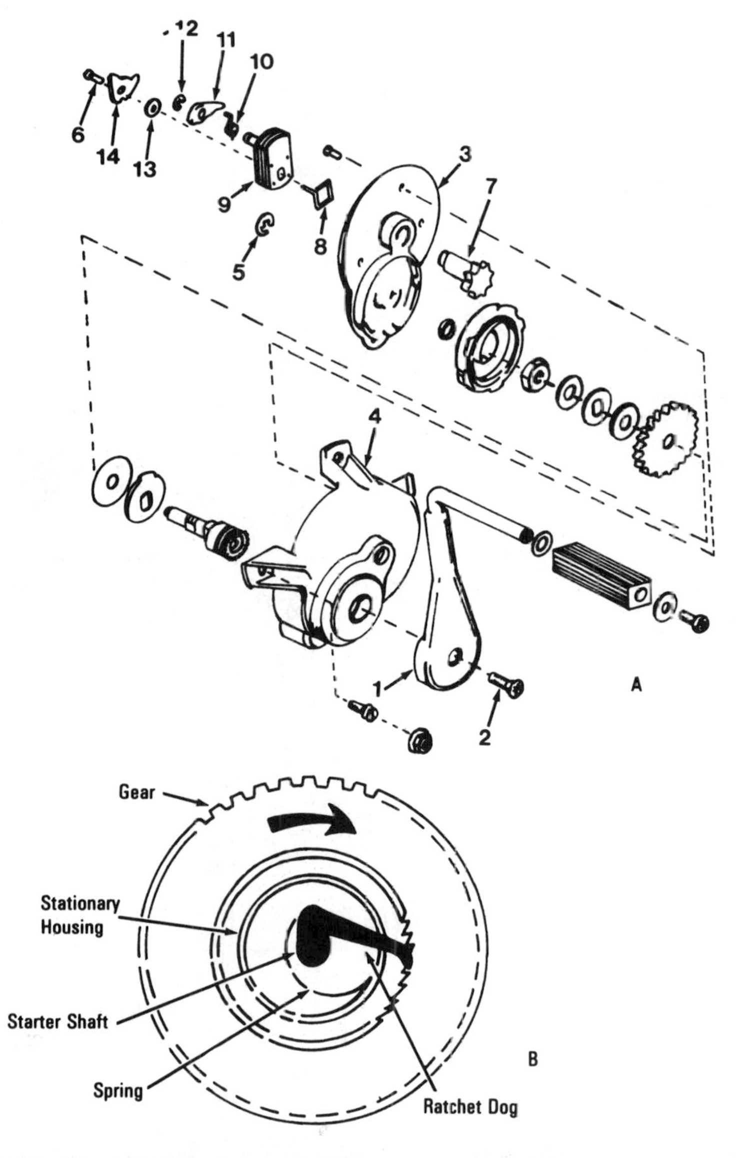

Fig. 6–10. (A) One type of kick starter as used on some small gasoline engines. (B) Another type of kick starter using a ratchet dog and a stationary housing. Direction of the gear movement is very important to the proper operation of the ratchet dog. *(Courtesy Tecumseh)*

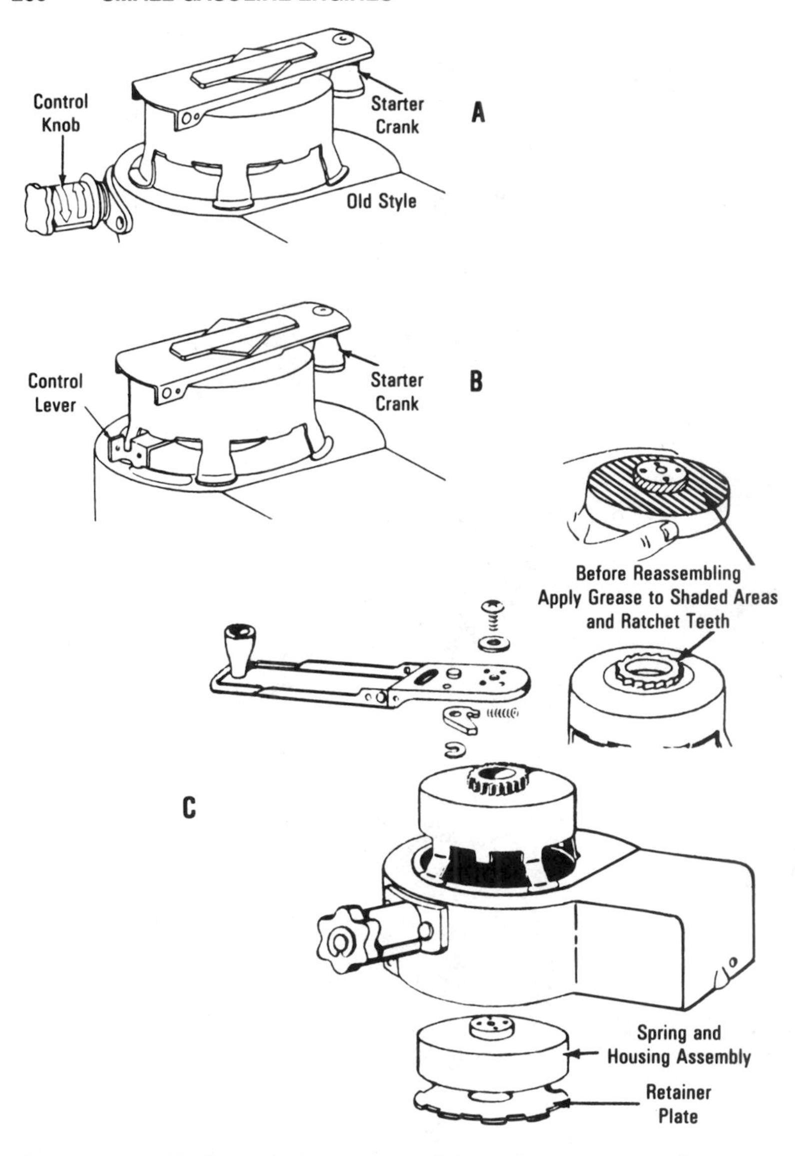

Fig. 6–11. (A) The windup starter is another type used to start the small gasoline engine. This is the old-style starter assembly. (B) The lever starter, control starter assembly. (C) Exploded view of the windup starter. Replacement of the starter spring is shown. *(Courtesy B & S)*

The Vertical Pull Starter

Some of the 92000 series of Briggs & Stratton engines use the vertical pull starter. The eight on the end of the model number 92008 will indicate that the starter is the vertical pull type (Fig. 6–12). This requires a different type of arrangement to cause the starter to work properly with a vertical pull and to transfer the results of the pull into horizontal rotary motion.

The starter consists of a plastic gear that meshes with the gear teeth on the flywheel. Once the rope is pulled upward, the plastic gear comes in contact with the bottom portion of the flywheel. There it engages the flywheel gear teeth to make the rest of the energy from the

Fig. 6–12. The 92000 series of Briggs & Stratton engines have a pull-up starter if the last number in the series is an 8 (92008). *(Courtesy B & S)*

upward pull transfer directly to rotate the flywheel. The flywheel is attached to the crankshaft, which in turn is attached to the connecting rod that is connected to the piston. The camshaft is driven by the crankshaft motion so it operates the valves. This movement of the flywheel also causes the magneto to operate since the magnets are embedded in the flywheel and come in close to the armature and coil of the magneto. Therefore, a pull upward causes the entire engine to swing into action.

Removing and Disassembling the Vertical Pull Starter—This starter is held onto the engine by only two screws. One of the screws also anchors the gas tank to the engine. It is longer than the other screw, which is used only to anchor the starter mechanism to the engine. Remove the two screws and hold the starter as shown in Fig. 6–13.

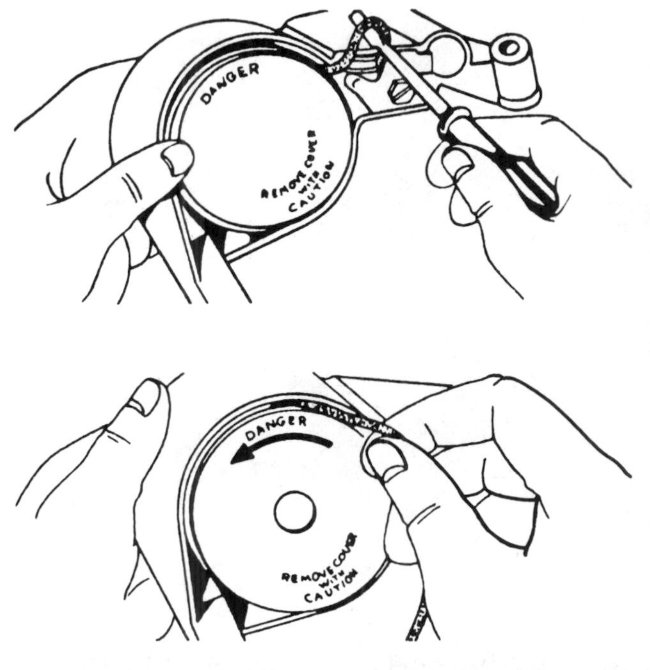

Fig. 6–13. Providing slack for the disassembly of the vertical pull starter before removing the cover. *(Courtesy B & S)*

Before you service the starter, relieve the tension from the rope. Use a screwdriver to lift the rope up approximately 12" (30.5 cm). Wind the rope and pulley counterclockwise two or three turns (Fig. 6–13). This releases the tension from the starter spring. The cover on the starter is clearly marked "Remove Cover with Caution." The spring can cause problems if released too suddenly, so follow the directions carefully. Wear a safety mask or goggles when working with this starter.

Once you have relieved the tension on the rope, take a screwdriver and remove the cover as shown in Fig. 6–14. After the cover is removed you can unscrew the anchor bolt and remove the anchor and bolt (Fig. 6–15). Inspect the starter spring for kinks or damaged ends. If the starter spring is to be replaced, *carefully* remove it from the housing at this time.

Remove the rope guide and make a note of the position of the link *before* removing the assembly from its housing (Fig. 6–16). The rope pulley and pin may be replaced if worn or damaged.

In order to handle the rope used to rewind starters more efficiently it is a good idea to make a rope-inserting tool as shown in Fig. 6–17. It can be used to remove the rope from the pulley (Figs. 6–18 and 6–19).

Next remove the rope from the grip (Fig. 6–20). If the pulley or gear is damaged, replace it with a new assembly. Clean all dirty or oily parts and check the link for proper friction. The link should move the

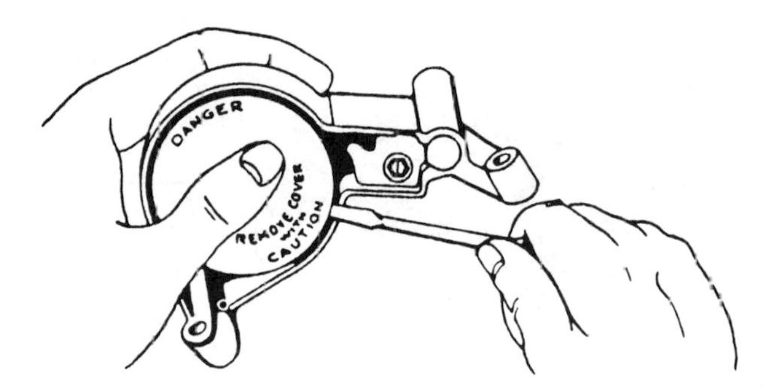

Fig. 6–14. Removing the cover with the prying action of a screwdriver. *(Courtesy B & S)*

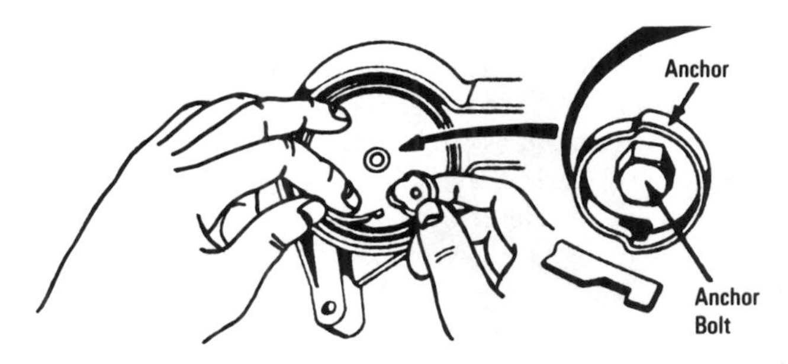

Fig. 6–15. Removing the anchor for the spring. Note the position of the anchor before removing it. *(Courtesy B & S)*

gear to both extremes of its travel. If not, replace the link assembly as shown in Fig. 6–21.

Reassembling the Vertical Pull Starter—In this reassembly we will assume that you are replacing the parts of the starter so that you will have a complete procedure just in case any part of the starter has to be replaced.

Install a new spring by hooking the end in the pulley retainer slot and winding the spring until it is coiled in the housing (Fig. 6–22). Keep in mind that when installing a new rope you should check the parts list to be sure the correct diameter and length rope is used. Oth-

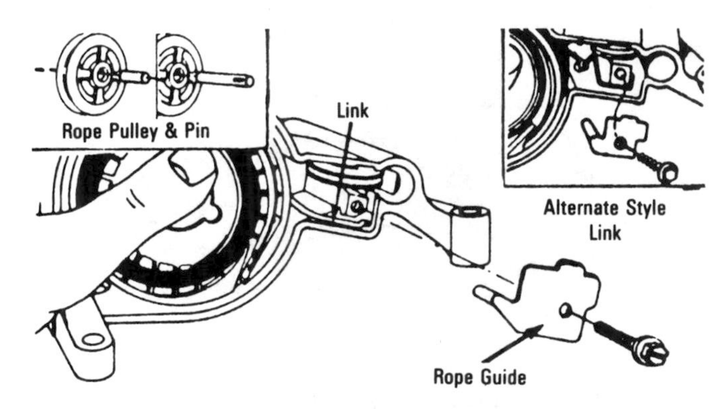

Fig. 6–16. Removing the rope guide. *(Courtesy B & S)*

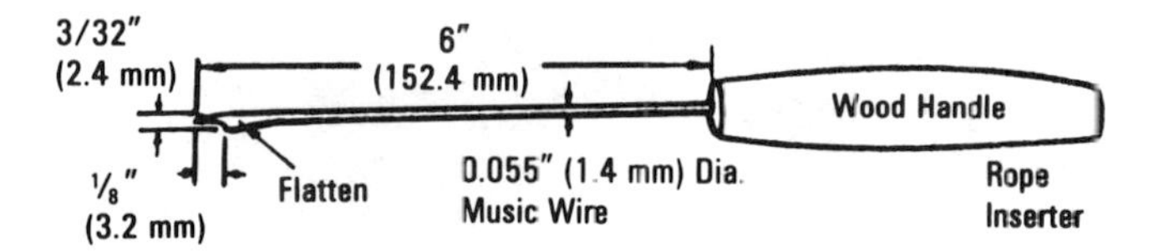

Fig. 6–17. Making a rope-inserting tool can be done with these plans. *(Courtesy B & S)*

Fig. 6–18. Using long-nose pliers to remove the rope from the starter pulley. *(Courtesy B & S)*

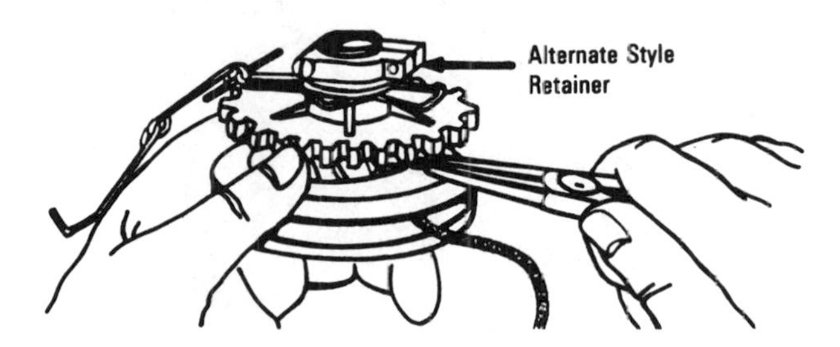

Fig. 6–19. Another way to remove the rope from the pulley is by using the alternate-style retainer. *(Courtesy B & S)*

Fig. 6–20. Removing the insert and then the rope from the grip. If the rope knot has been heat-sealed, you may have to cut off the knot. Don't forget to measure the length of the rope if you are installing a new one. *(Courtesy B & S)*

erwise you will wind up with a short rope with not enough room to wind it if the diameter is too large. Smaller-diameter ropes may not have the strength to take the abuse that a strong person exerts when there is an upward pull on the starter.

Thread the rope through the grip and into the insert (Fig. 6–23). Tie a small, tight knot. Test the knot to make sure it doesn't pull out when a force is applied to it. Heat-seal the knot to prevent it from loosening. Pull the knot into the pocket and snap the insert into the grip (Fig. 6–23). In some cases there may not be an insert, so make sure the

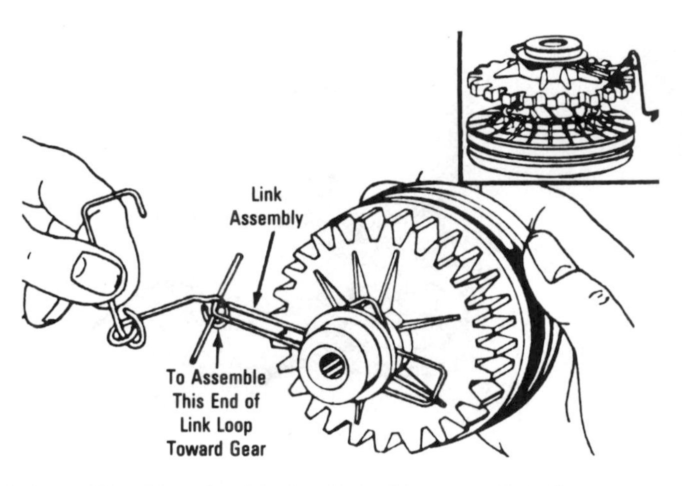

Fig. 6–21. Checking the friction link. *(Courtesy B & S)*

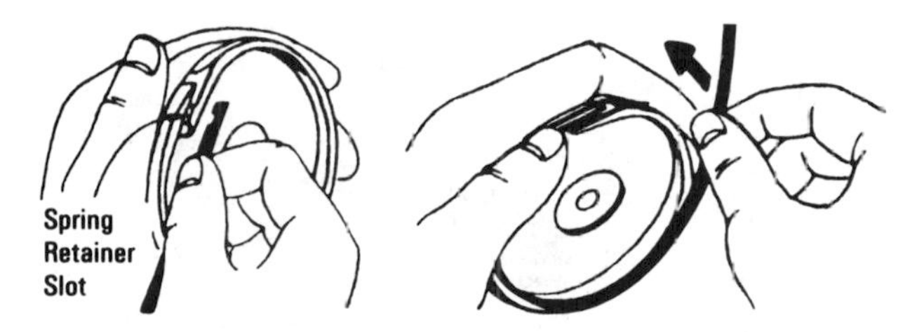

Fig. 6–22. Installing a new spring. Note the spring retainer slot. *(Courtesy B & S)*

knot on the rope end is large enough not to slide through the hole in the grip. The length of the rope should have been checked to make sure the replacement is the same length as the old rope.

Insert the rope through the housing and into the pulley. Now is the time to pick up your rope-inserter tool and use it. Tie a small knot, heat-seal, and pull it tight into the recess in the rope pulley. The rope must not interfere with the gear motion (Fig. 6–24). Note the $3/16''$ (4.8-mm) tail on the rope after the knot. Both the end that goes into the pulley and the end that went into the grip and insert earlier should have these tails.

Next, install the pulley assembly in the housing. The rope has not, at this time, been wound around the pulley. Make sure the link is in the pocket or hole of the casting as shown in Fig. 6–25. Install the rope guide and place the screw in tightly to hold the guide in properly.

To retract the rope see Fig. 6–26. Rotate the pulley in a counter-clockwise direction. Do this until the rope is fully retrieved or wound up to where the grip is touching the starter housing (Fig. 6–26). To tighten the screw, hook the free end of the spring to the spring anchor.

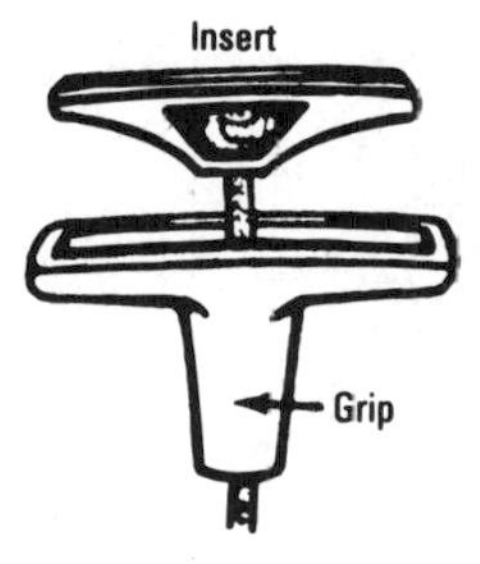

Fig. 6–23. Installing the rope in the insert and grip.

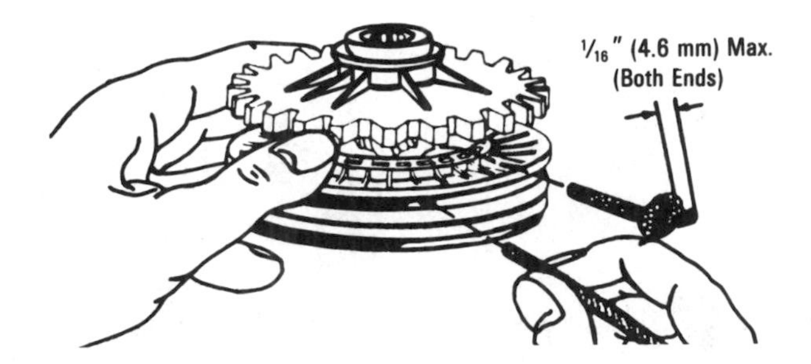

Fig. 6–24. Starting the rope in the pulley. Note that the end of the rope should allow for a small tail. *(Courtesy B & S)*

Install the screw. Torque the screw to 75 to 90 inch-pounds (8.5–10.2 Nm). Lubricate the spring with a small amount of engine oil or lubricant (Fig. 6–27).

You have almost finished the job of reassembly. Snap the cover in place. Wind the starter spring by pulling the rope out approximately 12" (30.5 cm). Wind the rope and pulley two or three turns in a clockwise direction. This should achieve the proper rope tension (Fig. 6–28).

Replace the starter on the engine by inserting the starter correctly and aligning the holes for the two screws. Tighten the screws and pull up on the starter to see if it engages the flywheel.

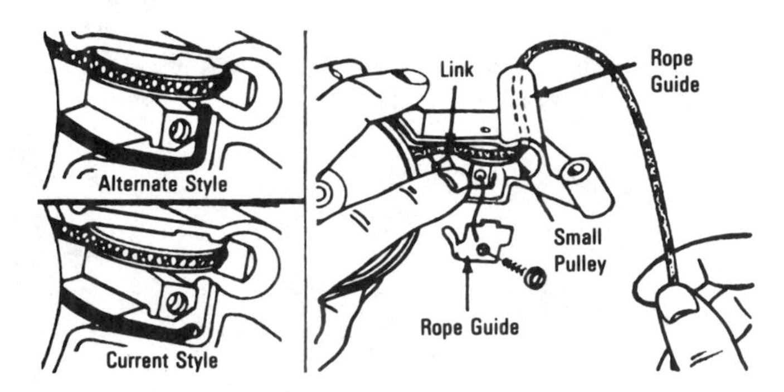

Fig. 6–25. Installing the pulley assembly. Two styles are shown. Note the rope passage over the small pulley. *(Courtesy B & S)*

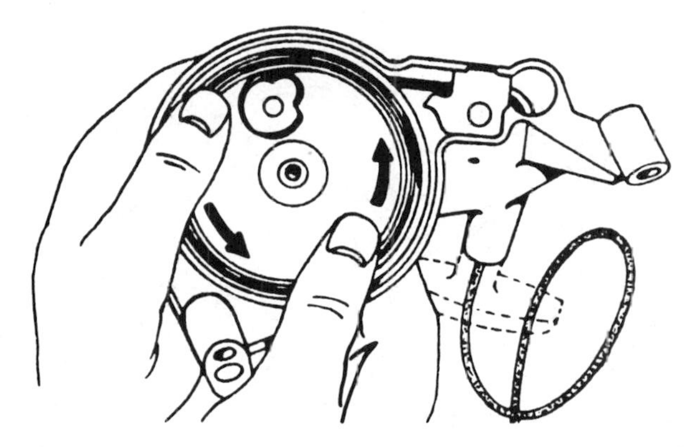

Fig. 6–26. Rotate the pulley to retract the rope up to the starter housing. *(Courtesy B & S)*

Electrical Starting

The electric starter is nothing more than an electric motor designed to turn the engine over and get it started. The starters used operate on either 12 volts DC or 120 volts AC (Fig. 6–29). The 12-volt starter is designed to work with small batteries mounted on the equip-

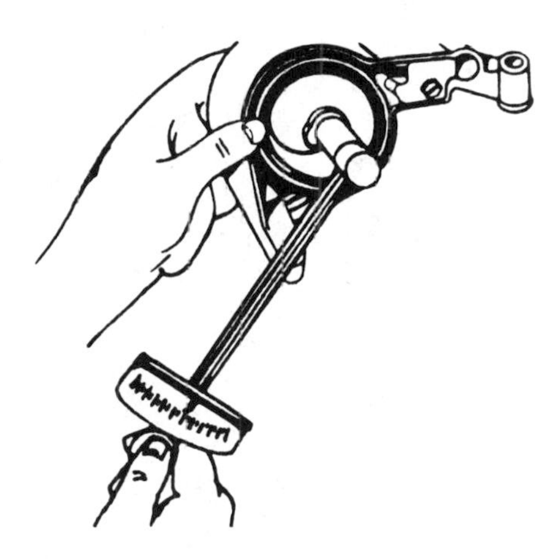

Fig. 6–27. Torque the screw to 75 to 90 inch-pounds to hold the pulley in place. *(Courtesy B & S)*

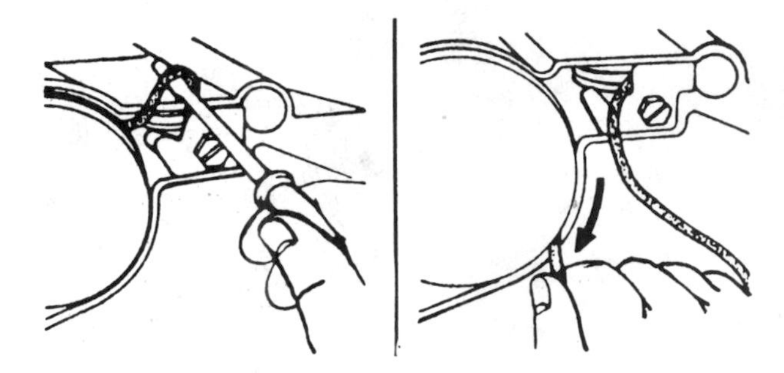

Fig. 6–28. Add tension to the rope so it will rest against the end of the starter housing when not in use. *(Courtesy B & S)*

ment that the engine drives. The 120-volt starter is mounted on equipment that will have access to house voltage. The snow blower and heavier-duty lawnmowers may use electric starters.

Gear-type engagement motors are similar to automobile starters. When the starter motor is activated, the helix on the back gear shaft drives a pinion gear into engagement with a ring gear attached to the engine flywheel. This causes the engine to turn over. Once the engine starts, the starter gear retracts and the starter can be removed from its circuit. Proper operation depends on the pinion gear moving freely on the helix (Fig. 6–29).

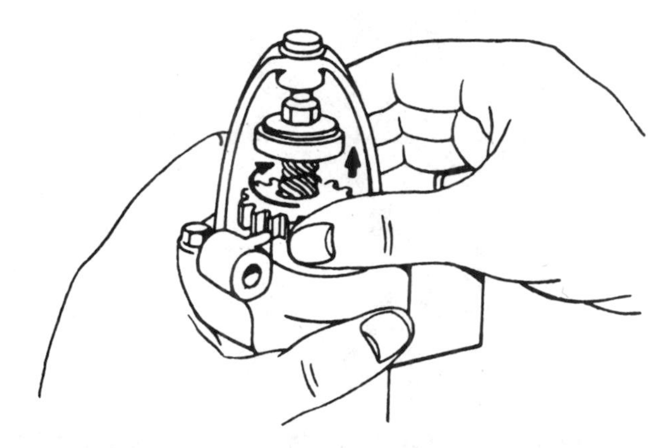

Fig. 6–29. The electric starter motor drive. *(Courtesy B & S)*

12-Volt Starter

The 12-volt starter is a DC electric motor. It is a series-wound type so it will have maximum torque for starting and for turning over the engine from a dead position.

Fig. 6–30 shows the 12-volt starter. The identification is located on the side. There are a number of manufacturers of the starter. Briggs & Stratton, American Bosch, Mitsubishi, and Motor Products produce starter motors for 12-volt operation. A battery is needed to start the motor. In some cases a ni-cad (nickel-cadmium) system is used to turn over the starter. Fig. 6–31 shows the ni-cad battery arrangement located on a lawnmower handle. This nickel-cadmium battery can be recharged with a typical 120-volt outlet providing power to a stepdown transformer and diode arrangement to produce DC for the battery charging.

Other engines may use a standard automobile-type lead acid cell battery. Some lawnmowers and small engine-equipped machines have their own generators for charging batteries while engines are running.

Battery condition has a lot to do with the problems associated with the 12-volt starter. If you have a fully-charged battery, then you may want to look to the starter itself for a source of trouble. Some of the possible troubles are:

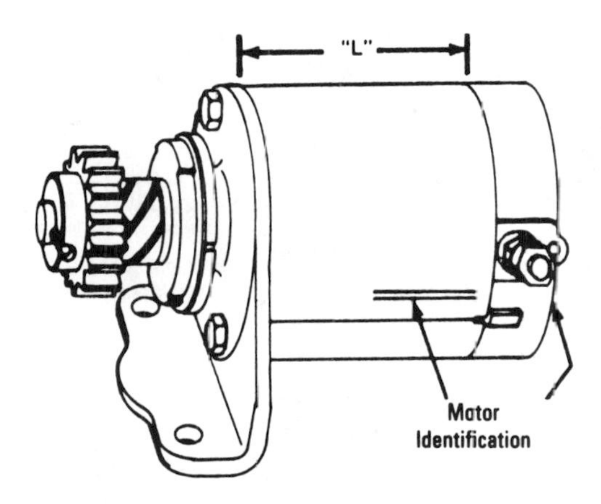

Fig. 6–30. A 12-volt DC starter motor used on Briggs & Stratton engines.

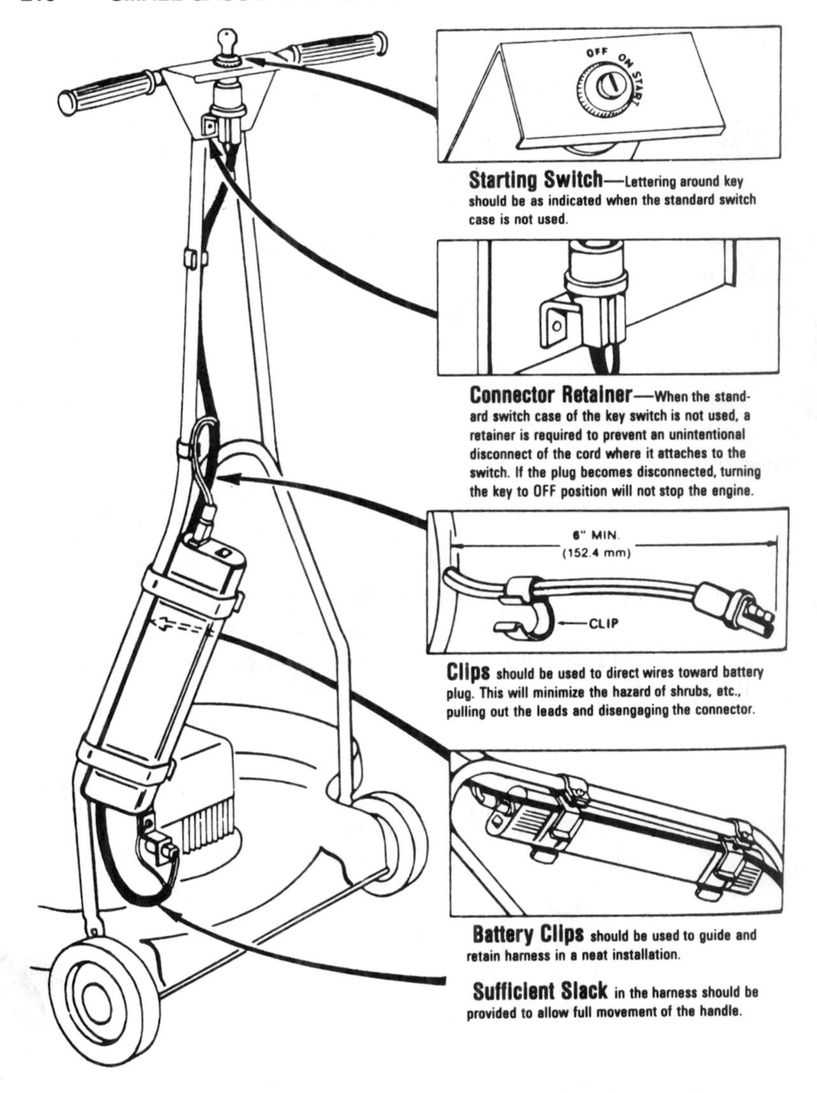

Fig. 6–31. The ni-cad electric starter with a key switch makes starting the small gasoline engine as simple as turning a key. The battery is nickel-cadmium and can be recharged. *(Courtesy B & S)*

- Dirty or worn starter motor commutator.
- Dirty or worn bearings.
- Weak magnet in the motor of a permanent-magnet-type motor.
- Worn brushes or weak brush springs.
- Wrong oil viscosity for the temperature expected.
- Bad or loose connection to the motor terminal.

The starter motor is located on the side of the engine so the pinion gear can engage the flywheel of the engine (Fig. 6–32).

120-Volt Starter

The 120-volt starter is similar to the 12-volt type, except it is made to operate on AC instead of DC. The armature is the same. The only difference is the end cap. The 120-volt motor has a rectifier assembly that rectifies or changes the 120 volts to DC and lowers it to the 12-volt motor voltage (Fig. 6–33). You can tell the difference in the motors by the 120-volt plug and wire coming from the starter (Fig. 6–34). The distance *L* shown in Fig. 6–34 indicates the 120- or 12-volt starter on the Briggs & Stratton models. If *L* is 3½″ (88.9 mm), then it is a 120-volt type. If *L* measures 3¹⁄₁₆″ (77.788 mm) or 3¾″ (95.25 mm), it is a 12-volt type. American Bosch usually incorporates the 110 or 12 in its motor identification label. (The 110 refers to 110 to 120 volts. Previously, 110 volts was standard; now 120 volts is the available voltage.) Mitsubishi has its own number system that does not easily identify the starter type. However, the Motor Products motor is made for 12-volt operation only. It is fairly obvious which motor is 12 volts and which is 120

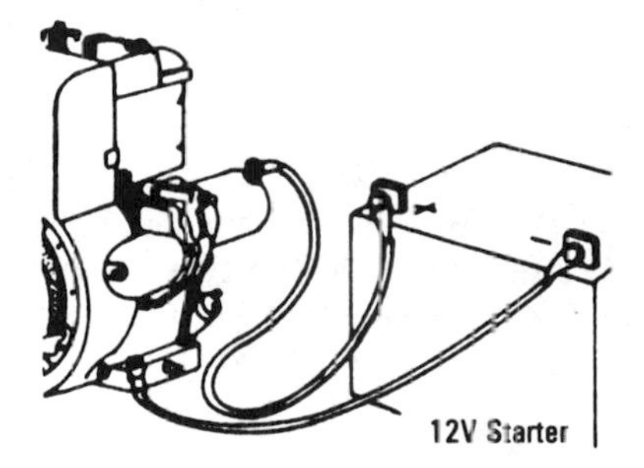

Fig. 6–32. Note how the starter is connected to the 12-volt battery. *(Courtesy B & S)*

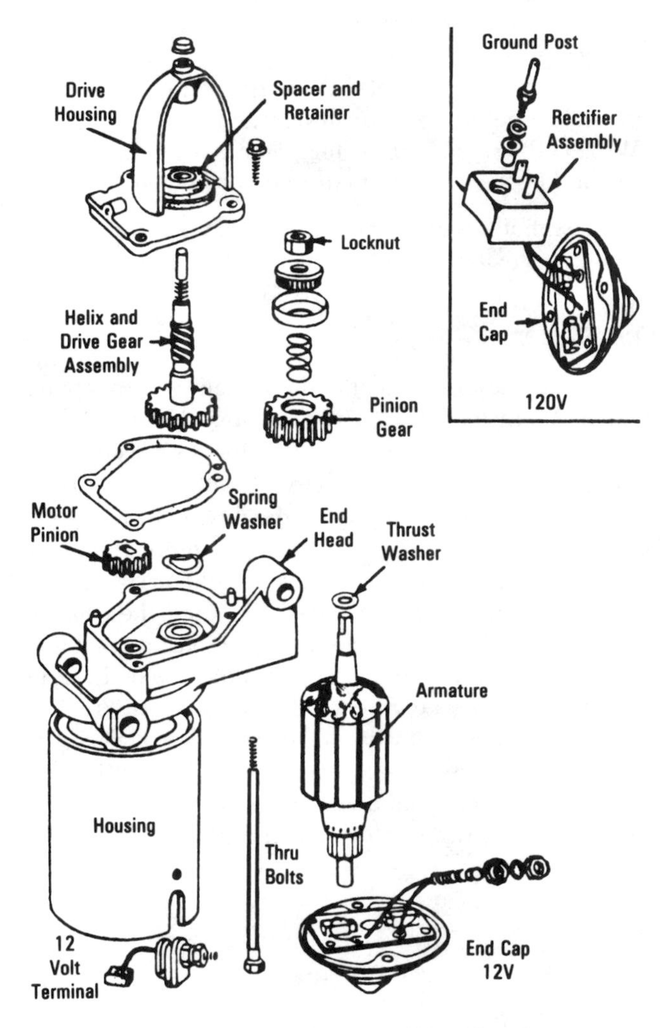

Fig. 6–33. Exploded view of the 12- and 120-volt starter motors. Note that the 120-volt end cap has a rectifier assembly that changes the 120 volts AC to 12 volts DC for use with the starter motor. The only difference in the two starters in the end cap. The rest of the motor is the same for both voltages. *(Courtesy B & S)*

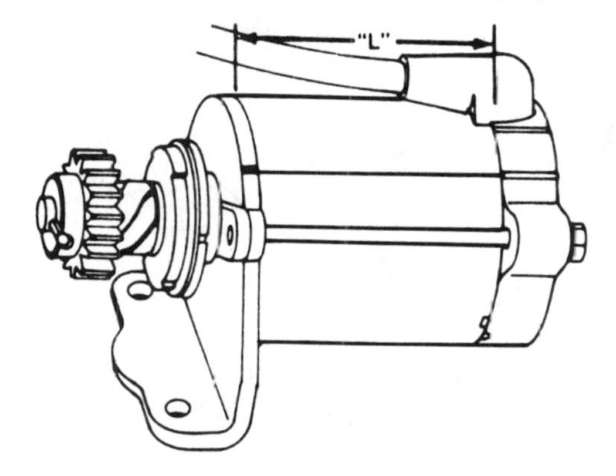

Fig. 6–34. The typical Briggs & Stratton 120-volt starter motor. Note that the difference is the wire coming from the motor.

volts because of the plugs associated with them. See Fig. 6–35 for further identification.

The 120-volt starter is equipped with a three-prong plug for safety. The longer prong in this plug is connected to the starter motor housing. When the starter motor is plugged into the three-wire cord

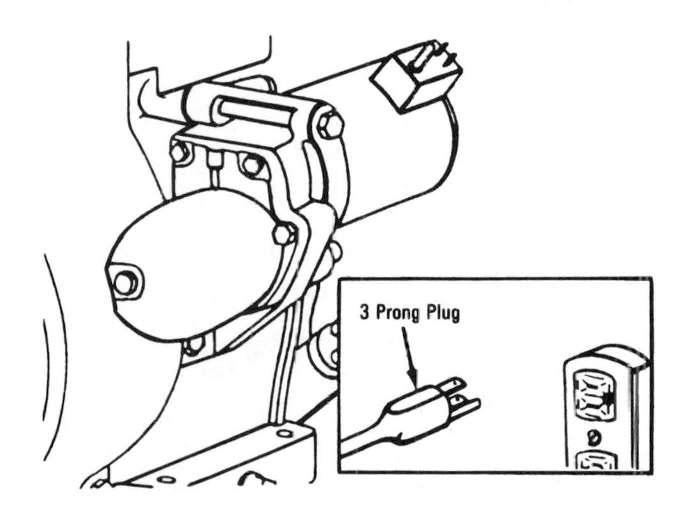

Fig. 6–35. The 120-volt starter motor is plugged into the wall socket. *(Courtesy B & S)*

supplied by the manufacturer, and the cord is plugged into a properly grounded receptacle, it will protect the user from shock should the starter motor insulation fail for any reason.

If an extension cord is used, it too should have a three-wire plug and receptacle. Do not use an extension cord over 25′ (7.62 m) long. Fig. 6–36 shows the 120-volt, gear-drive starter.

The starter engages the flywheel of the engine in order to crank it. Fig. 6–37 shows how the ring gear is attached to the flywheel for electric starter operation.

If the motor used as a starter has windings inside the housing, see Fig. 6–38. It is a series-wound motor and not a permanent-magnet type. The permanent-magnet type has a ceramic-type magnet attached to the housing and will have *no wires* attached.

Belt-Drive Starters, 120-Volt Type

Some starters are belt-drive arrangements. They use a belt to connect the starter motor to the flywheel of the engine to be cranked. There are two positions for the motor to be mounted (Figs. 6–39 and 6–40). They are adjusted until there is about ¼″ (6.35 mm) movement up and down with thumb pressure on the belt. Belts can be adjusted by

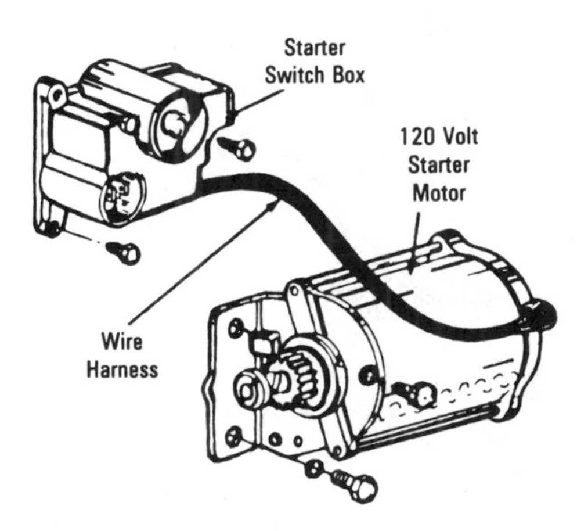

Fig. 6–36. The 120-volt, gear-drive starter motor. *(Courtesy B & S)*

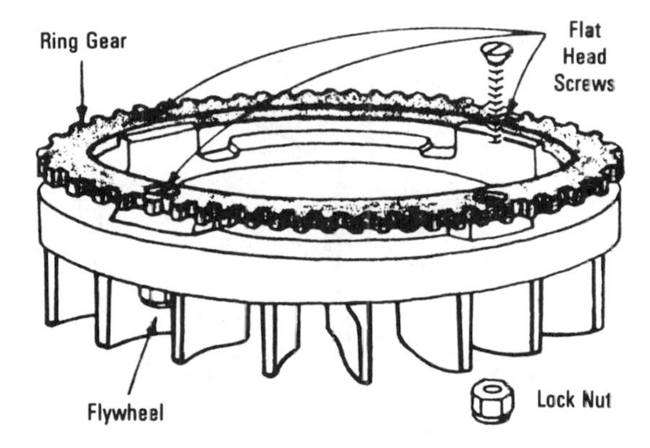

Fig. 6–37. Location of the ring gear on the flywheel for gear-drive starter. *(Courtesy B & S)*

loosening the two nuts shown in Fig. 6–41 and sliding the starter motor in the slots. Tighten the nuts to 15 to 20 inch-pounds (1.7 to 2.2 Nm).

In Fig. 6–41, illustrations *B, C,* and *D* have the distance *A* emphasized. This distance should be between 3⁄32″ and 1⁄8″ (2.38 to 3.175 mm) to be correct. If the dimension while cranking is less than 3⁄32″, the starter motor must be adjusted away from the engine. If more than 1⁄8″, the motor must be adjusted toward the engine.

Belt-Drive Starters, 12-Volt Type

This Briggs & Stratton electric starter automatically engages a belt clutch and cranks the engine when a 12-volt battery is connected between the terminal on the starter and the engine cylinder. When the engine starts, the belt clutch automatically disengages the starter motor from the engine.

Equipment driven by the engine should be disconnected before you try to start the engine. The starter is made large enough to turn over an unloaded engine.

To adjust the belt on this type of starter, take a look at Figs. 6–42 and 6–43. Loosen nut A and B slightly so the starter motor can just be moved by hand. Move the starter motor away from the engine as far as possible. Rock the engine pulley back and forth and at the same time slowly slide the starter motor toward the engine until the starter motor

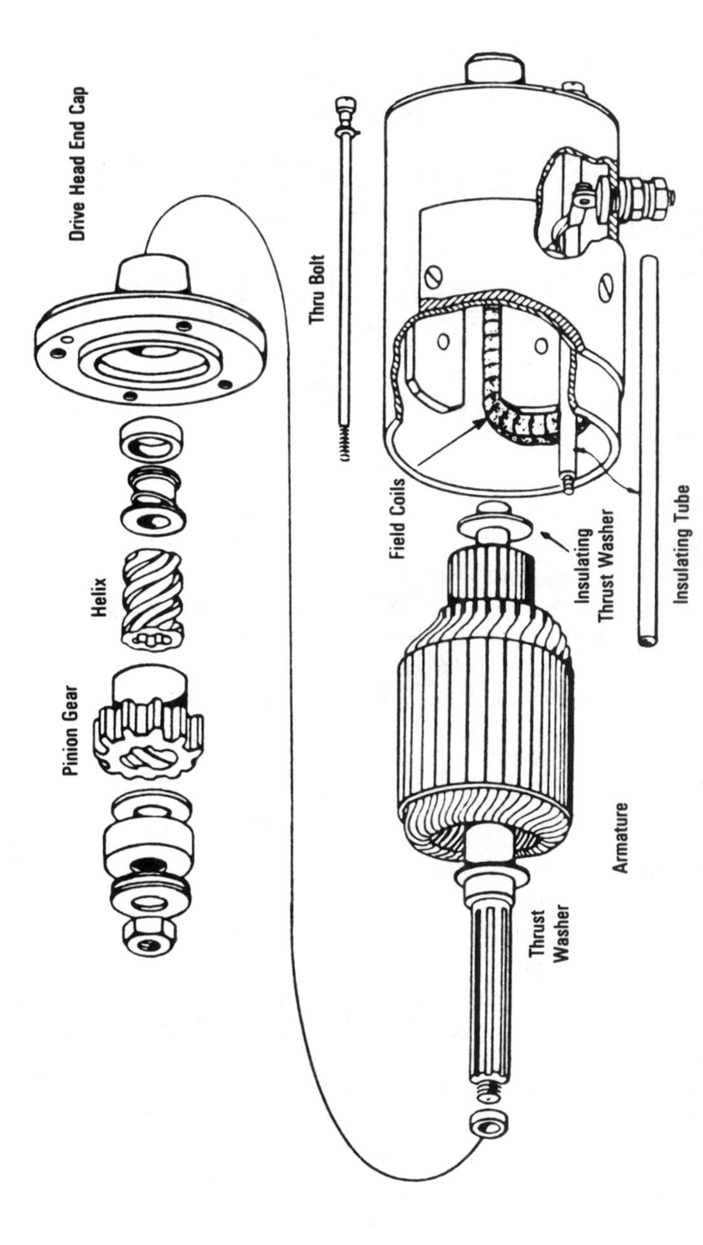

Fig. 6–38. Exploded view of the starter motor. Note that this one shows the field coils. In newer motors the field coils are replaced by permanent magnets. This is a series wound DC motor. *(Courtesy B & S)*

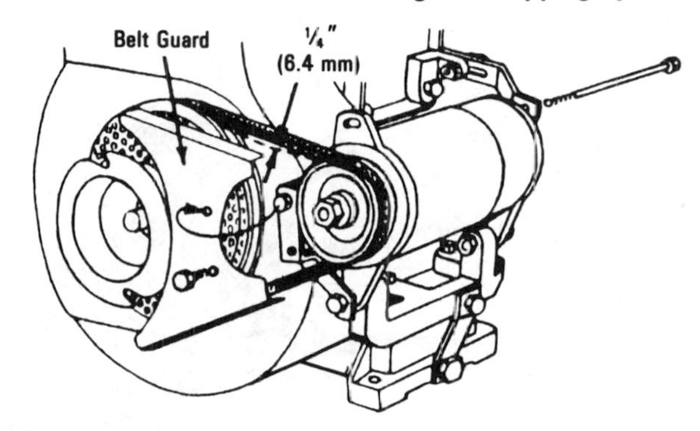

Fig. 6–39. The 12-volt starter generator (low position). *(Courtesy B & S)*

pulley stops being driven by the V belt. Move the starter motor another $\frac{1}{16}$" (1.6 mm) toward the engine. Tighten nuts A and B.

Electrical connections for the starter are shown in Fig. 6–44. A starter switch is wired in series with the + wire from the starter to the battery. This way the starter motor only turns over when the starter button or switch is pressed.

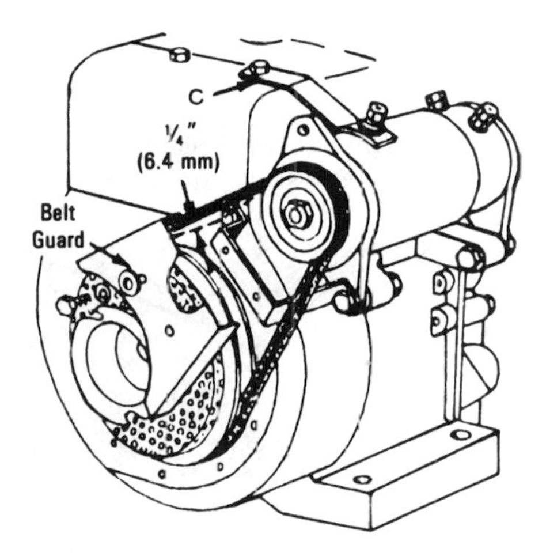

Fig. 6–40. The 12-volt starter generator (high position). *(Courtesy B & S)*

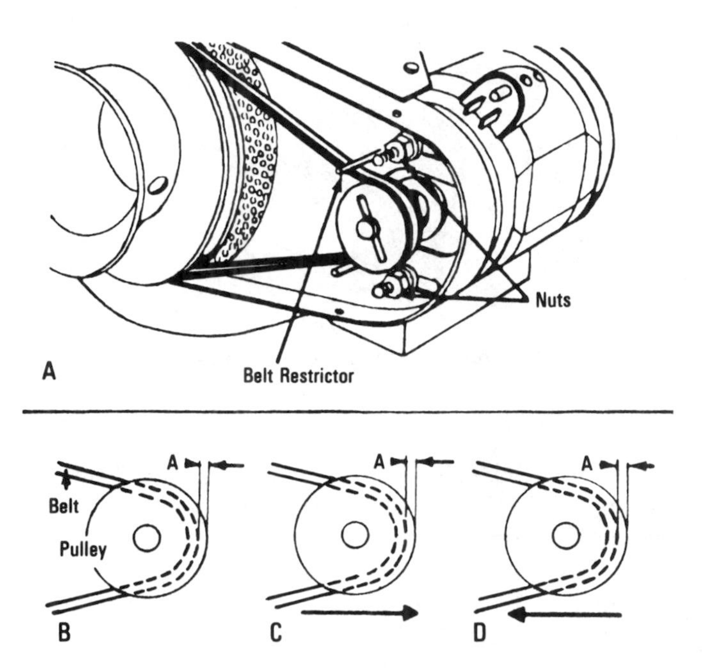

Fig. 6–41. Belt adjustment for the belt-driven starter generator. *(Courtesy B & S)*

Starter Generator

On some engines you will find a complete electrical system. A battery charger is included in the system. An electric motor can be con-

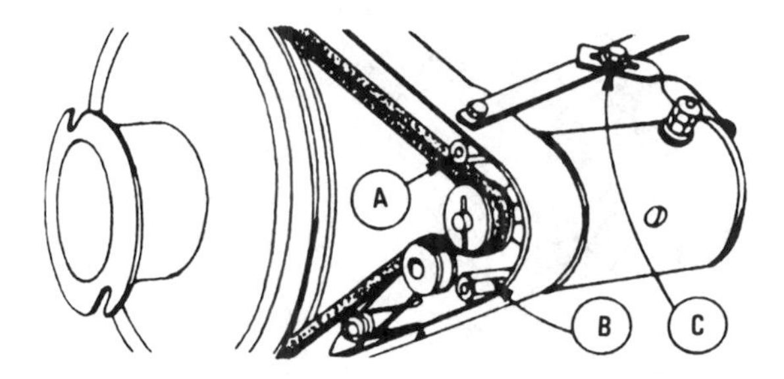

Fig. 6–42. Belt-adjustment locations. *(Courtesy B & S)*

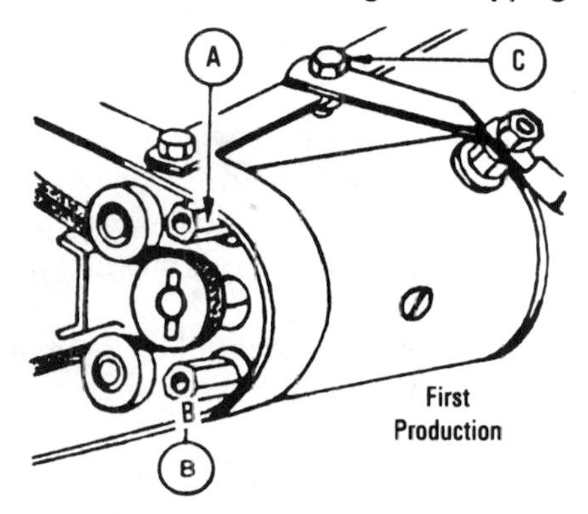

Fig. 6–43. Belt-adjustment locations. *(Courtesy B & S)*

verted to a generator by simply driving it instead of having electricity applied to it to make it run as a motor. Once the engine has started, the engine can be used as a source of power to drive the generator.

Stopping Systems

Stopping the Engine and Blade

Stopping the engine and the rotating lawnmower blade in an emergency is of great importance. In fact, since June 30, 1982 all lawn-

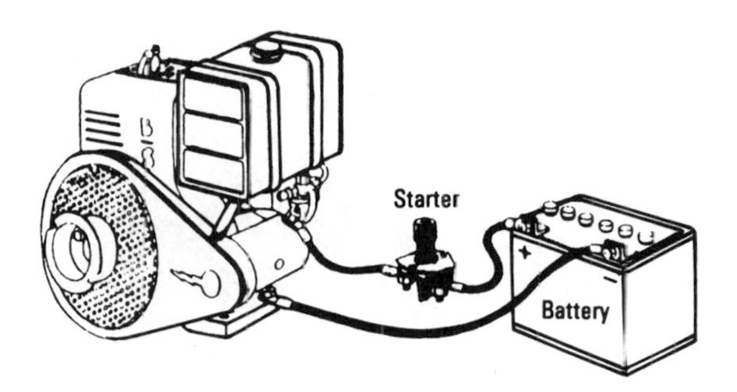

Fig. 6–44. Connecting a starter switch in the line from the battery to the starter motor on the engine. *(Courtesy B & S)*

mowers must have an approved method of stopping the rotating blade once the control has been released.

Tecumseh uses two methods of meeting compliance standards that have become law. Tecumseh's inside-edge flywheel brake system provides safety by shutting down the engine and lawnmower blade within seconds after the operator releases an engine/blade control at the handle of the lawnmower.

To stop the engine, a brake pad is applied to the inside of the flywheel. At the same time the ignition system is grounded out (Fig. 6–45). To restart the engine, the pad has to be pulled away from contact with the flywheel (Fig. 6–46). By pulling up on the brake control lever, a cable transmits power to the mechanism that supports the brake shoe. This action pulls the brake pad away from the inside edge of the flywheel and opens the ignition ground switch.

When removing the flywheel, make sure the brake pressure is removed from the flywheel. This makes for an easier removal of the flywheel.

Reassembling the Flywheel—The brake lever is compressed with an alignment pin (Fig. 6–47). Check the brake pad for contamination. It should be less than 0.060″ at the narrowest point and not contaminated. If it is not clean or has less than 0.060″ clearance, replace it.

Make sure the grounding pin is in its correct position. Install the

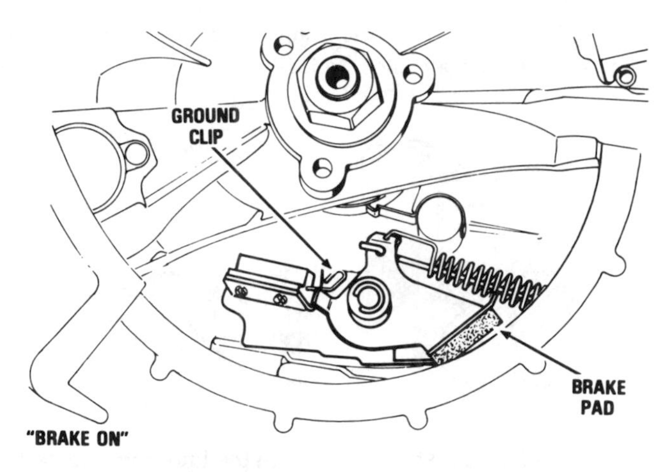

Fig. 6–45. Inside edge system—brake on. *(Courtesy Tecumseh)*

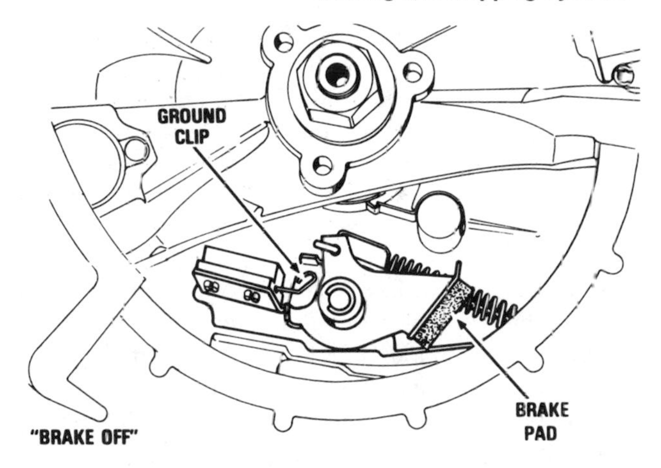

Fig. 6–46. Inside edge brake system—brake off. *(Courtesy Tecumseh)*

flywheel. Be sure that the ground wire to the grounding clip does not touch the flywheel.

Torque the flywheel nut to 22–27 foot-pounds.

Replacing the Brake Pad—To replace the brake pad, remove the flywheel. Remove the pin to aid in releasing the spring tension. Release

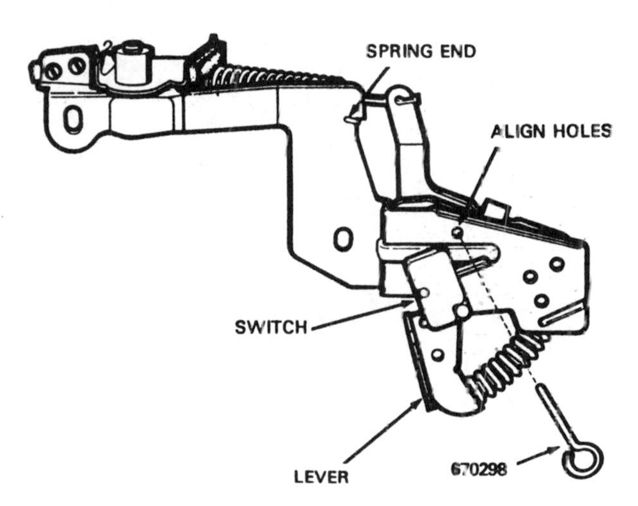

Fig. 6–47. Alignment hole and tool. *(Courtesy Tecumseh)*

the spring tension by pulling the short end of the spring with a needlenose pliers and unhook it from the bracket (Fig. 6–48). Pull the "E" clip from the brake pad shaft (Fig. 6–49). Pull the pad lever from the shaft and unhook the link. Clean the pivot post then apply a thin film of general purpose grease to the post. *Do not get grease on the brake pad!* If grease is on the pad, replace it and the lever with new ones. Reassemble the brake mechanism in the reverse order of the disassembly.

If the flywheel brake surface is damaged due to metal contact with the brake lever, the flywheel must be replaced.

Brake Mechanism Installation—If the brake assembly is removed during service to the engine, reassemble the brake mechanism in the full-down position on the mounting holes (Fig. 6–50). Re-torque the screws to 90 inch-pounds.

When attaching the cable make sure the screw end does not block the lever action of the mechanism (Fig. 6–51). Be very aware of this possible problem if you are not using a standard service part screw.

Electric Start Systems

The brake start mechanism may be used with either of two options for starting:

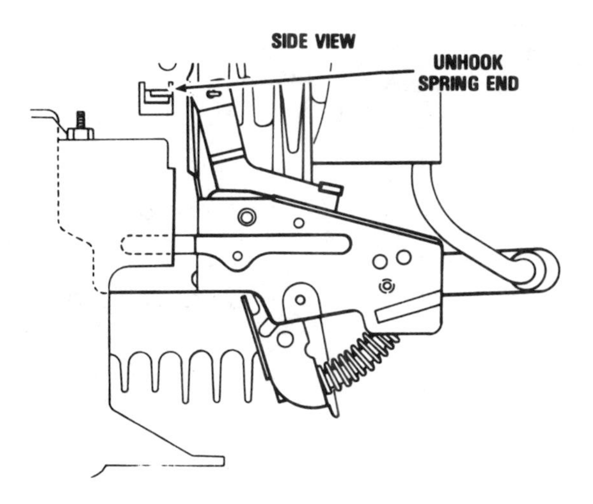

Fig. 6–48. Location of spring end. *(Courtesy Tecumseh)*

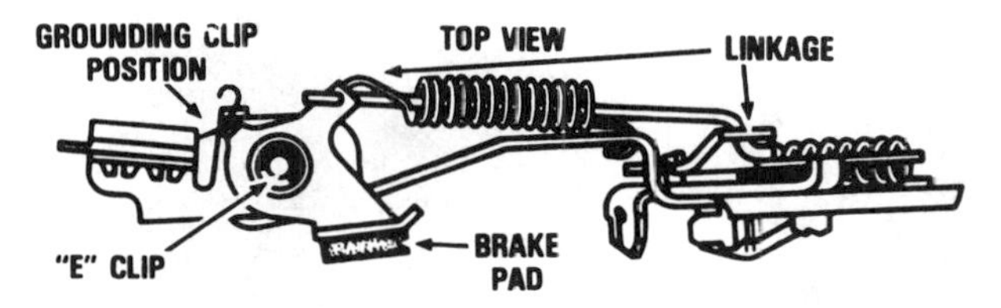

Fig. 6–49. Brake mechanism bracket. *(Courtesy Tecumseh)*

1. Manual rope start, and
2. 12-volt starter system.

Each system requires the operator to start the unit from behind the mower handle in the operator zone area. The electric starter system also provides a charging system for battery recharge when the engine is turning.

Figs. 6–45 and 6–46 show the *Brake On* and *Brake Off* positions of the brake shoe and the ignition ground wire that shorts out the ignition system.

To start the engine, the brake control must be applied so the brake pad is pulled back from the flywheel and the ignition shorting switch is opened. On the non-electric start systems, the recoil starter rope must be pulled to start the engine. On electric start systems, the starter is

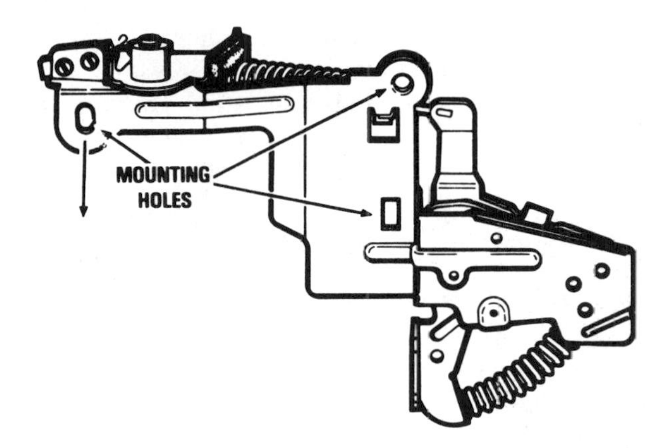

Fig. 6–50. Mounting holes for the brake mechanism. *(Courtesy Tecumseh)*

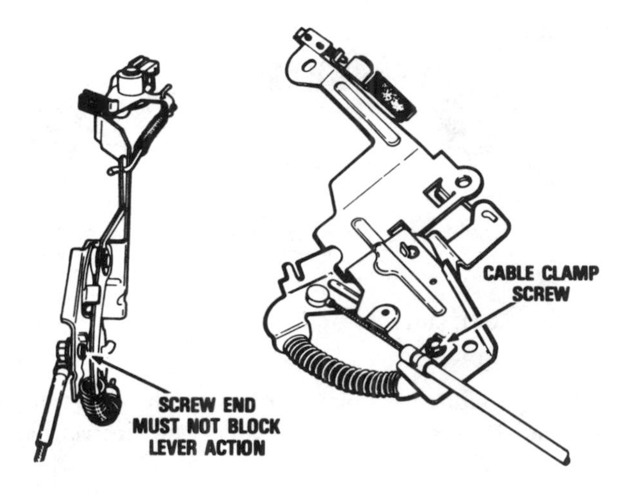

Fig. 6–51. Control cable conduit clamp screw location. *(Courtesy Tecumseh)*

energized to start the engine. This may be accomplished by a key switch or by two-motion control (Fig. 6–52).

All wiring beyond the connectors on the engine is supplied by the equipment manufacturer. Check all terminals and connections for corrosion and adequate contact, and all wiring should be checked for damage and the proper size. Figs. 6–52 and 6–53 show the two-motion control and the key switch control circuits.

Battery—Check the battery for proper operation. The charging system on the engine maintains the battery during normal use. When the battery is low, use the 120-volt auxiliary charger that is supplied by the equipment manufacturer.

Flywheel Removal—Follow the same procedure already described for the recoil starter engine.

Flywheel Installation—When you install the flywheel, be sure that the ground wire to the grounding clip does not touch the flywheel (Fig. 6–46). Torque the flywheel nut to 35 foot-pounds.

Replace the brake pad if it is contaminated or will not fit within 0.060″ (Fig. 6–54).

Brake Mechanism—If the brake assembly is removed during service to the engine, reassemble the brake mechanism in the full down position

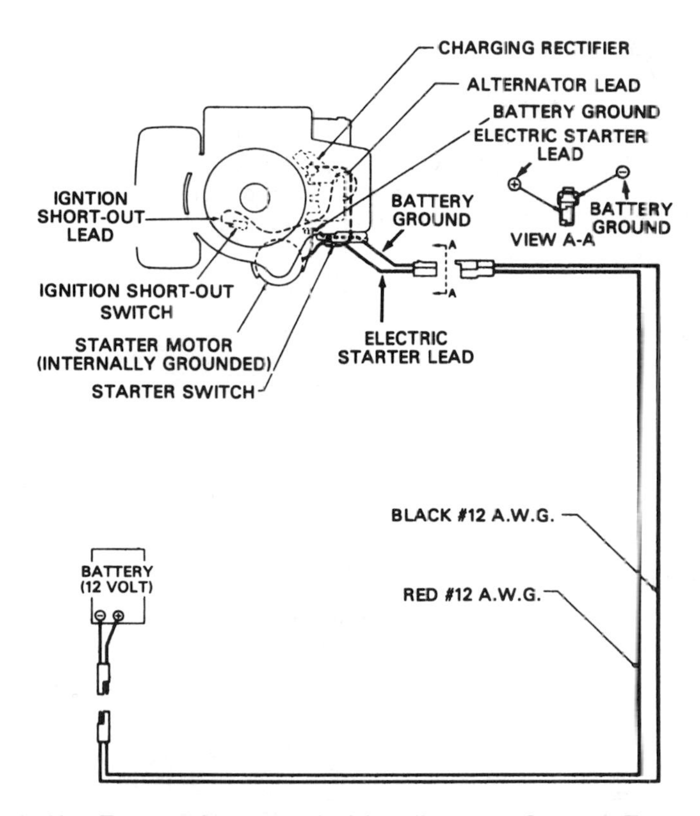

Fig. 6–52. Two-motion control wiring diagram. *(Courtesy Tecumseh)*

on the mounting holes shown in Fig. 6–55. Re-torque the screws to 90 inch-pounds.

Control Switch—Note how the control cable conduit clamp screw is located in Fig. 6–56.

The brake lever must close the switch before the starter can be engaged.

Disconnect the battery from the circuit before making checks!

To perform a continuity check on the switch, use a continuity light or meter. Remove the starter wire from the starter terminal of the switch. With one of the continuity unit's probes inserted in the brake

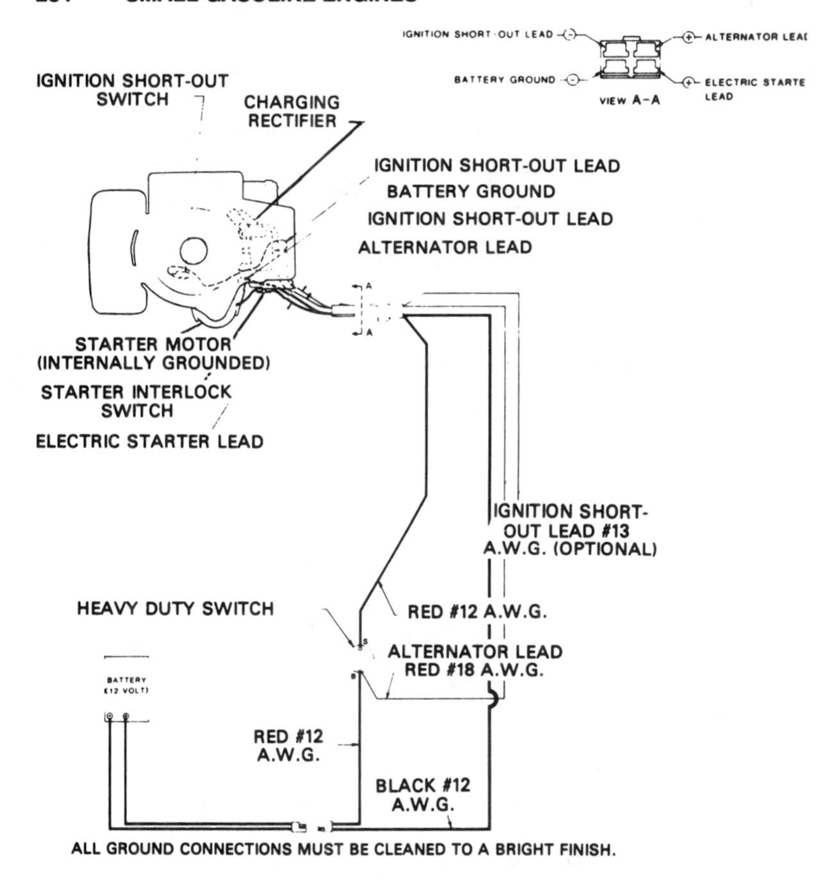

Fig. 6–53. Keyswitch control wiring system. *(Courtesy Tecumseh)*

start mechanism's terminal red wire connector and the other lead to the starter terminal (on switch), and press the switch button. The light on the meter should indicate continuity. If not, replace the switch. If continuity exists without pressing the switch button, replace the switch.

Replacing the Switch—Carefully grind off the heads of the rivets that hold the switch in place (Fig. 6–57). Remove the rivets from the back side of the brake bracket. Use the self-tapping screw to make threads in the bracket. Install the switch to the brake bracket in the proper position and secure the switch to the brake bracket with the machine screws. Be careful—overtightening the screws could break the switch.

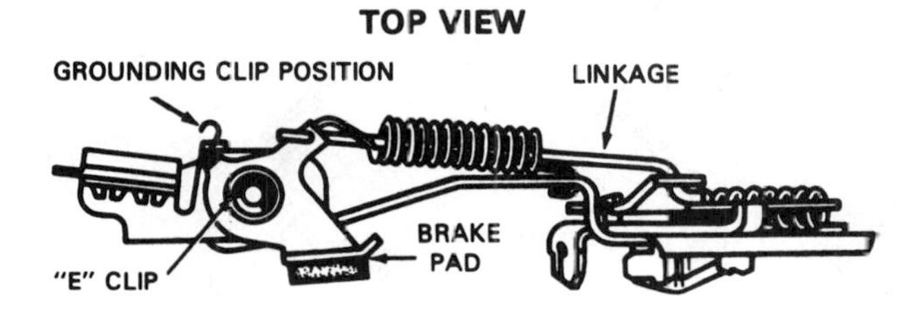

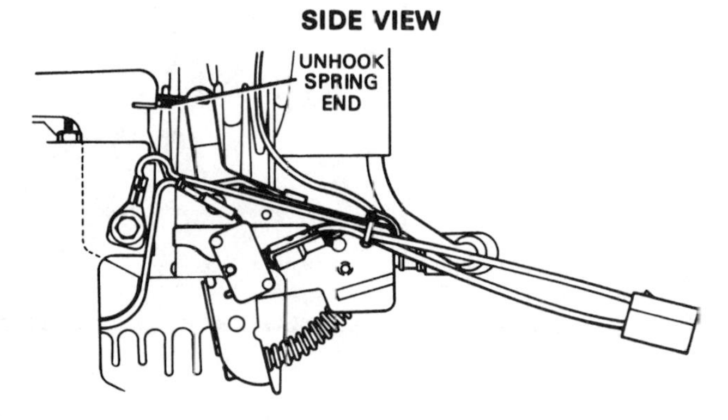

Fig. 6–54. **Top and side views of brake pad bracket.** *(Courtesy Tecumseh)*

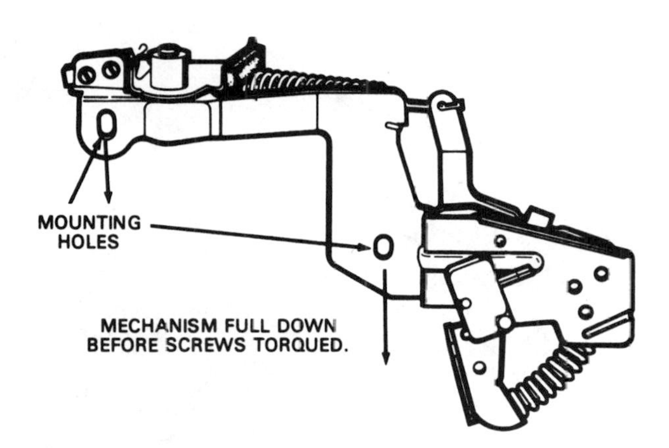

Fig. 6–55. **Brake mechanism bracket.** *(Courtesy Tecumseh)*

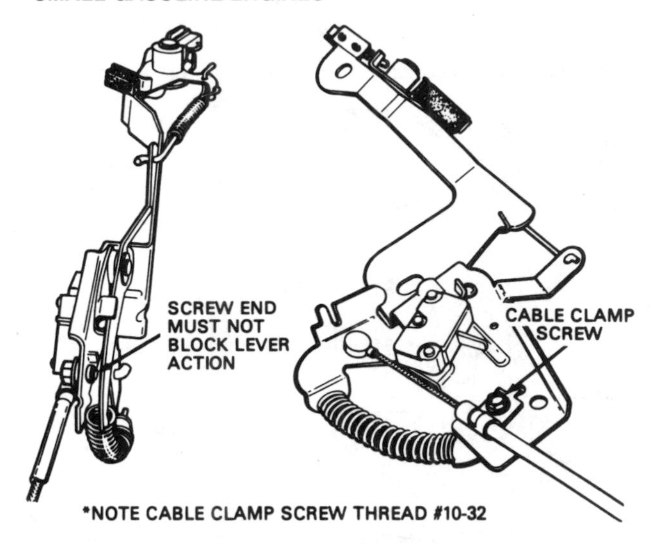

Fig. 6–56. Control cable conduit clamp screw location. *(Courtesy Tecumseh)*

Systems with Alternators—The charging system consists of a single alternator coil mounted to one side of the solid state module (Fig. 6–58). The output lead must not be allowed to dangle loose during engine operation since even intermittent arcing to ground will damage the diode.

To *set the air gap,* loosen all the lamination screws (Fig. 6–58) and

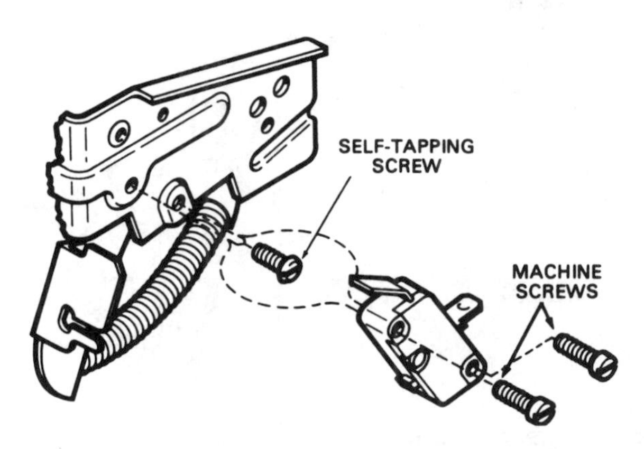

Fig. 6–57. Mounting of the interlock switch. *(Courtesy Tecumseh)*

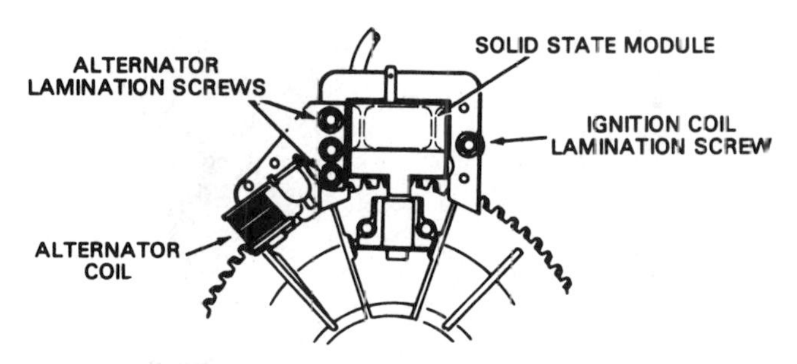

Fig. 6–58. One-coil alternator charging system. *(Courtesy Tecumseh)*

turn the flywheel so the magnet is centered across from the solid state module. Using the air gap gage, check the gap and adjust for 0.0125″. Tighten the ignition coil screw on the right side. Rotate the flywheel and perform the same procedure for a 0.0125″ air gap between the alternator coil, the laminations, and the magnet. Then tighten all alternator screws.

Alternator Check.—Disconnect the battery from the circuit before making an alternator check. Remove the starter wire from the starter terminal of the switch. Connect one probe of a continuity meter to the lead of the brake start mechanism's terminal red wire connector, the other probe to an unpainted surface on the engine for ground (Fig. 6–59). Now reverse the meter probes. If continuity exists in both probe positions, the alternator assembly must be replaced. It means the diode is shorted. Continuity should exist in *only one* of two probe positions. If there is no continuity in either position, the assembly must be replaced. It means there is an open diode. Keep in mind that diodes allow current to flow in one direction only while an ohmmeter has a battery with its polarity being present at the probes. The alternator's AC output has to be rectified and changed to DC before it can be used to charge the battery. The most efficient way to change AC to DC is to use semiconductor diodes. They are reliable, small, and inexpensive.

Bottom Surface Flywheel Brake System

The bottom surface flywheel braking system does the same thing the inside surface system did—it stops the lawnmower engine and blade in a matter of seconds.

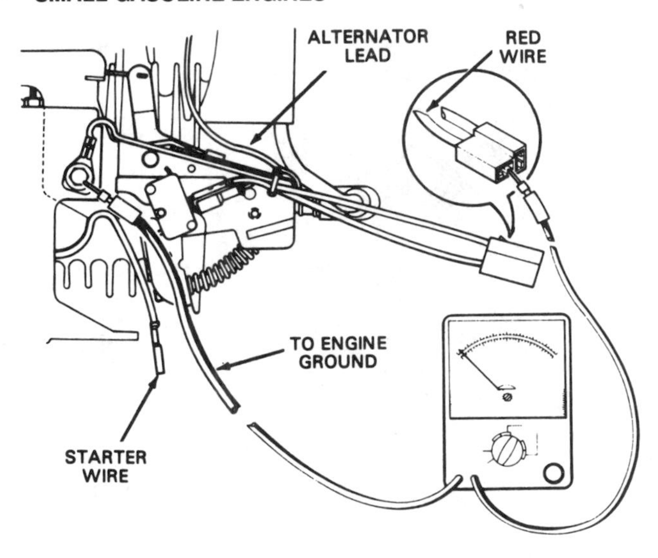

Fig. 6–59. Checking the electrical system. *(Courtesy Tecumseh)*

In the stop position, with the control released, the torsion spring rotates the brake lever as shown in Fig. 6–60, forcing the brake pad against the underside of the flywheel. It also shorts out the ignition, and opens the starter interlock switch.

In order to restart the engine, the control must be applied. This action pulls the brake pad away from the flywheel, opens the ignition short-out switch and closes the starter interlock switch (Fig. 6–61). This will allow the engine to be started by energizing the starter with a starter switch.

Fig. 6–62 shows the wiring for the electric start engine. Note the location of the parts and how the wires are color coded. The control cable conduit must be assembled against the stop in the bracket as shown in Fig. 6–63. If more adjustment is needed, it can be done on the handle. Make sure the bottom of the lever completely depresses the button on the starter interlock switch when the control is fully applied. The cable must provide enough travel so the brake will contact the flywheel. Some slack should exist in the cable adjustment to compensate for brake pad wear.

Flywheel Removal—Before the flywheel is removed or replaced, the

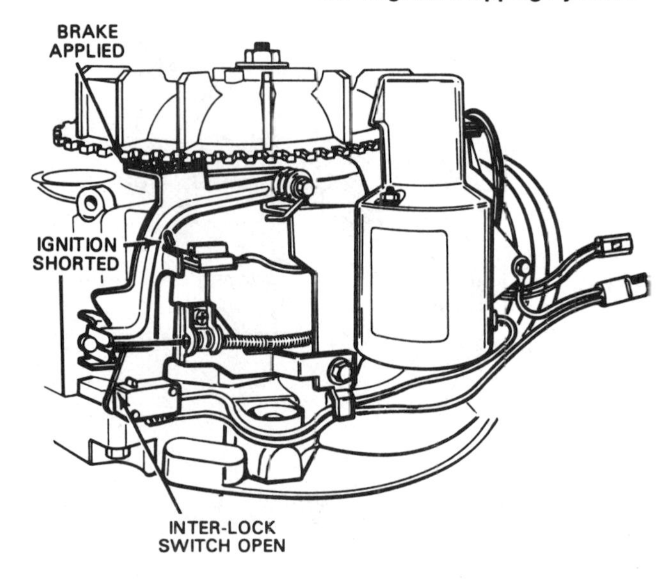

Fig. 6–60. Bottom surface brake system—brake on. *(Courtesy Tecumseh)*

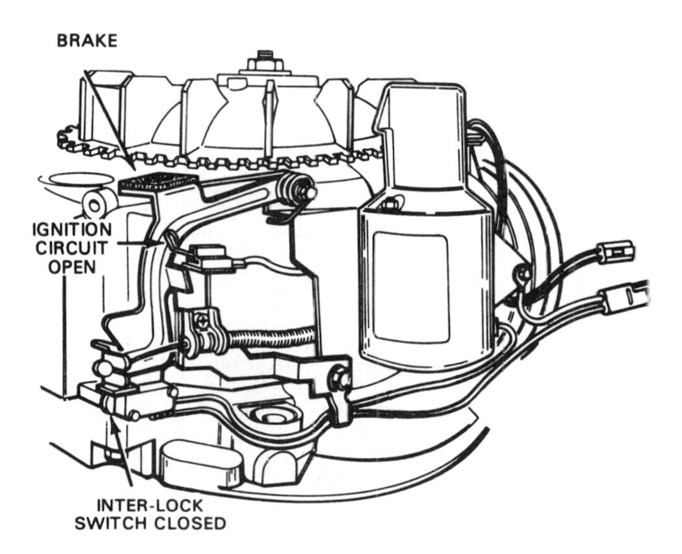

Fig. 6–61. Bottom surface brake system—brake off. *(Courtesy Tecumseh)*

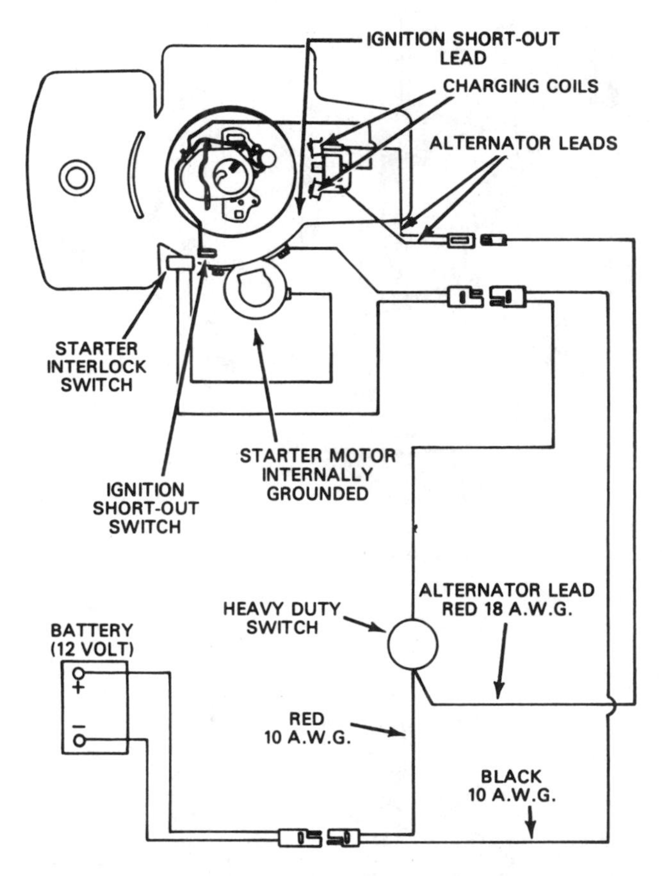

Fig. 6–62. Wiring diagram of a bottom surface brake system. *(Courtesy Tecumseh)*

tension of the torsion spring must be released. To remove the torsion spring, firmly grasp the short end of the spring with hose clamp pliers and unhook it from the bracket (Fig. 6–64). The spring must be released before the top starter bolt is removed or the threads in the cylinder block will be damaged.

Replacing the Brake Pad—After the tension has been released on the

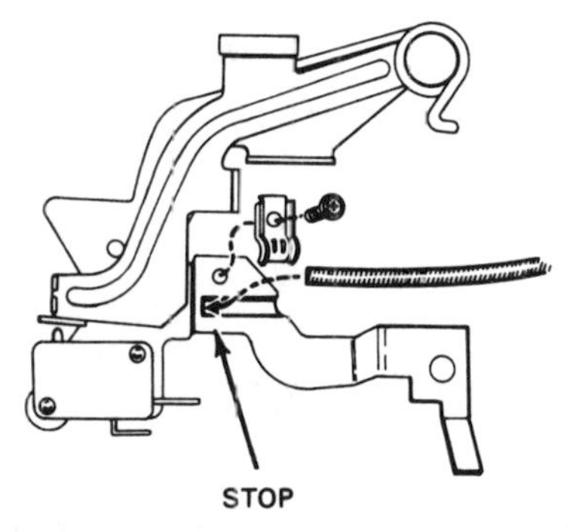

Fig. 6–63. Control cable conduit mounting. *(Courtesy Tecumseh)*

torsion spring, the bracket may be removed. The brake pad is bonded to the brake lever and must be replaced as an assembly when necessary.

Reassembling—When reassembling, be sure to torque the bracket screws to 60 to 70 inch-pounds. Install the ignition ground lead. Connect the wires to the starter interlock, reassemble the spring, make certain the long end is under the lever-pad assembly.

Electrical Circuit Checking—Disconnect the battery from the circuit before proceeding. The brake lever must close the interlock switch before the starter can be engaged. To check the interlock switch, use a simple continuity checker. Continuity should exist between the two terminals when the switch button is depressed and *no continuity* should be indicated when the switch is released.

Two-Coil Alternator—This charging system consists of two coils mounted on the outside legs of the coil laminations. The output lead must not be allowed to dangle loose during engine operation since even intermittent arcing to ground will damage the diodes (Fig. 6–65). To check the system, remove the plastic connector housing by inserting a small-bladed screwdriver or similar device to release the spade connector (Fig. 6–66). Clip off the connector and strip the wire ends. Make a simple continuity check as follows: connect one probe of a continuity meter to one lead of the alternator. The other probe goes to an unpainted surface on the engine for ground. Now, reverse the meter probes. If there is continuity in both probe positions, the alternator

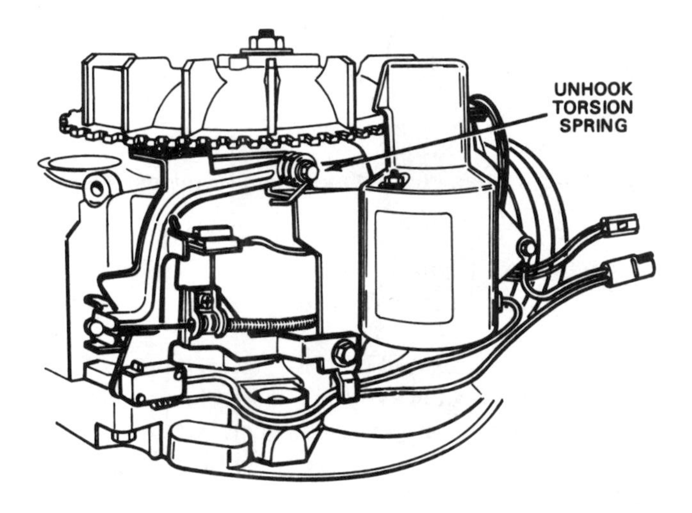

Fig. 6–64. Torsion spring location on a bottom edge brake system. *(Courtesy Tecumseh)*

assembly must be replaced. It means the diodes are shorted. Continuity should exist in *only one* position. If there is no continuity in both probe positions, the assembly must be replaced. It means the diodes are open. Perform the same tests on other alternator leads. If there is a defective reading, replace the unit.

Replace the spade connector (part no. 610885). Solder if necessary. Reinstall the plastic connector housing.

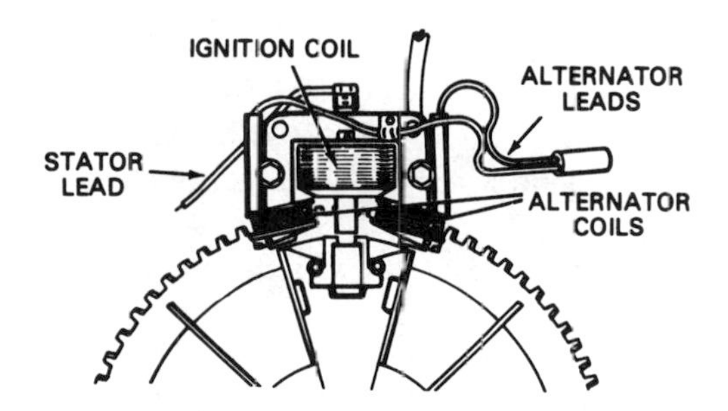

Fig. 6–65. Two-coil alternator charging system. *(Courtesy Tecumseh)*

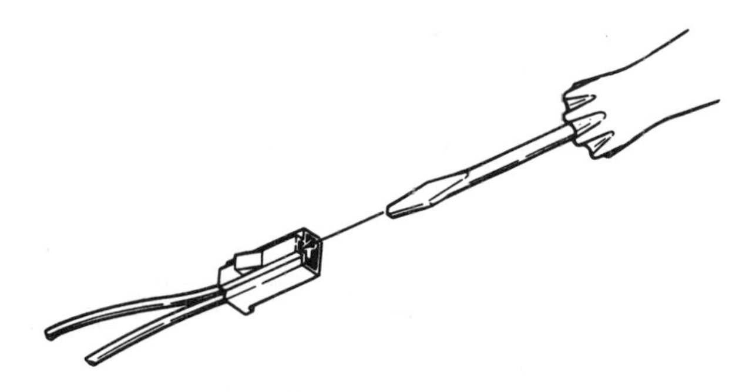

Fig. 6–66. Removing the plastic connector housing with a screwdriver. *(Courtesy Tecumseh)*

Chapter Highlights

1. There are two manual start systems for small engines.
2. Carburetor adjustments may be needed to get an engine started.
3. The direct rewind starter is nothing more than a piece of rope mounted so that it retracts once it is pulled.
4. More leverage at start is possible with a geared rewind starter since the crankshaft is gear-driven.
5. The kick starter is used where the foot can reach the starter.

6. There are a number of windup mechanical starters for small gasoline engines.
7. The vertical pull starter is gear-driven.
8. The electric starter requires a battery and an electric motor.
9. The only difference between the 120-volt AC starter and the 12-volt starter is the end where the diode is used to rectify the AC for the 120-volt starter.
10. Battery condition has a lot to do with the problems associated with the 12-volt starter.
11. Some starters are belt-drive arrangements.
12. Some electric start engines have a generator to charge the battery.
13. Alternators are used to generate power for charging the battery.
14. A rectifier unit is needed to change alternator output AC to DC for battery charging.
15. The brake control controls the brake pad and shorts out the engine ignition system.
16. Make sure you disconnect the battery from the circuit before making checks with an ohmmeter.
17. There are one-coil and two-coil alternators.

Vocabulary Checklist

alternator
auxiliary charger
belt adjustments
belt-drive starter
brake control
brake mechanism
brake pad
carburetor needle valve
circuit checking
control switch
direct rewind starter
electric start system
flywheel installation
flywheel removal
foot pounds
geared rewind starter
kick starter
manual starter
nickel-cadmium battery
reassembly
rich mixture
rope starter
spade connector
torque
two-coil alternator
vertical pull starter

Review Questions

1. What are the three methods used to start a small engine?
2. What is the procedure for starting an engine for the first time?
3. How does a rope starter differ from a direct rewind starter?
4. What is a geared rewind starter?
5. How does a kick starter work?
6. What's the difference between a manual starter and an electric starter?
7. How many types of electric starters are there?
8. List at least six troubles that can be found with electric starters.
9. Why should an extension cord used for an electric starter have a three-prong plug?
10. Explain the difference between belt-drive and gear-drive starters.
11. How is the engine stopped in a very short time?
12. What's the important caution to be aware of when replacing the brake pad?
13. What are the two options with which the brake start mechanism can be used?
14. What happens when the brake control is pulled back?
15. What is an important step to take when installing the flywheel?
16. Why does the output of the alternator have to be changed to charge the battery.
17. What does the rectifier unit do?
18. Other than stopping the blade, what does the bottom surface flywheel brake system do?
19. What is meant by the term *continuity*?
20. What does intermittent arcing do to the diodes in the two-coil alternator system?

CHAPTER 7

Fuel Systems and Carburetion

There are two types of combustion, external and internal. The internal combustion engine is the one studied here. That means the engine uses the fuel's energy to produce power. The steam engine is an example of an external combustion engine. The energy or fuel is used to turn water into steam, and then the steam is used to create mechanical energy or power. The power is generated by the engine, but the source of the energy to create the power is the heat from the fuel that was burned somewhere other than inside the steam engine (externally).

The internal combustion engine actually burns or uses the energy *inside* the engine itself. Both the two-cycle and the four-cycle engines are internal combustion types. The energy stored in the gasoline is released when burned inside the combustion chamber of the engine.

Heat generated inside an internal combustion engine may reach as high as 4000°F (2200°C). This heat is in most cases excess and not used to any advantage. One of the big problems of this type of engine is the dissipation of the heat. This is done either by water cooling or air cooling the engine combustion chamber.

A number of fuel systems are used in making the internal combustion engine work. The fuel systems make sure the fuel is at the right place at the right time. Carburetion is also necessary to make sure that the fuel is vaporized and combined with air in the correct proportions to create combustion when properly ignited.

Carburetion is the process of mixing fuel and air. However, the fuel has to be vaporized first. The carburetor is a device used to vaporize and meter the fuel-air mixture. Proper operation of an internal combustion engine depends on the carburetion process (Fig. 7–1).

This chapter will deal with the fuel systems needed for small gasoline engines and with the carburetion needed to keep the engines running properly.

Principles of Combustion

Before discussing combustion, let us take a closer look at the materials that will be combined to produce energy for making the small

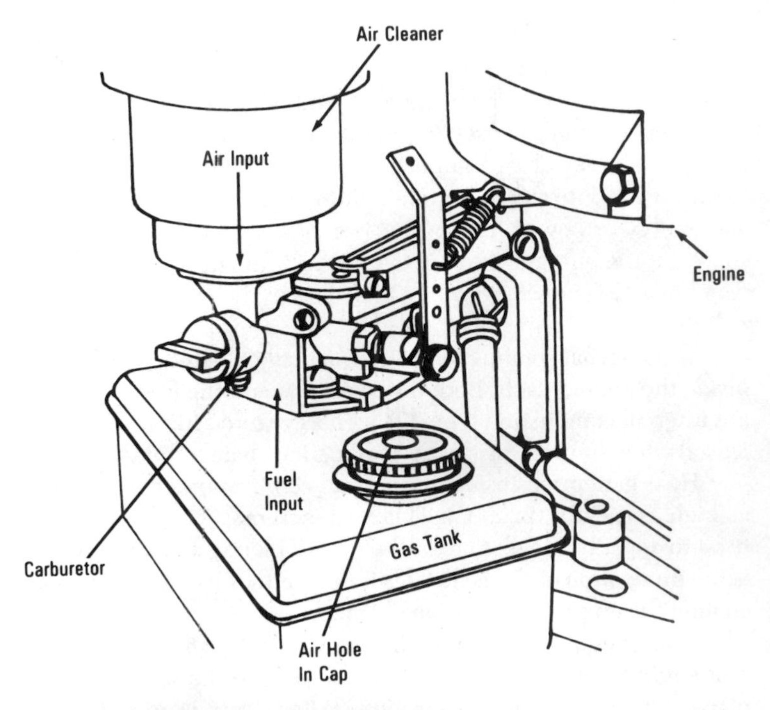

Fig. 7–1. Air and fuel are mixed in the carburetor. *(Courtesy B & S)*

gasoline engine operate and produce mechanical energy. We have to take a closer look at the fuel (gasoline) to understand the operation of the combustion process.

As you already know, gasoline evaporates easily. Fumes are given off when gasoline is exposed to the air (Fig. 7–2). If you strike a match in the presence of the fumes, you have rapid burning or an explosion. That means the gasoline, which is a liquid, has changed from a liquid to a gas in evaporating. Many substances change from one state to another. Water, for instance, changes from a liquid into a solid when it freezes. It also changes to a gas when it is boiled. That means water can take on three different states—liquid, gas, or solid—depending on the temperature. All three of these states of matter will be valuable to us in explaining how internal combustion engines operate (Fig. 7–3).

If the temperature is high enough, anything can be vaporized. Steel is vaporized by an atomic explosion. Any gas can be changed to a liquid or solid. It takes a lot of heat or cold to change substances from one state to the other. Most substances will remain in one state and have to be processed in most instances to be changed to another state. In this case we are interested in how gasoline changes from a liquid to a vapor.

Gasoline Vaporization

Gasoline is a petroleum product. Petroleum is made of hydrocarbons. Hydrocarbons are made of different groupings of hydrogen and

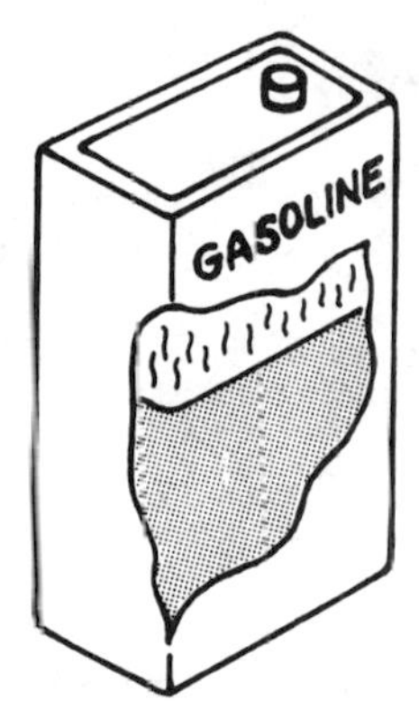

Fig. 7–2. The fumes above the gasoline in the half-full container represent a mixture of gasoline and air that is easily ignited.

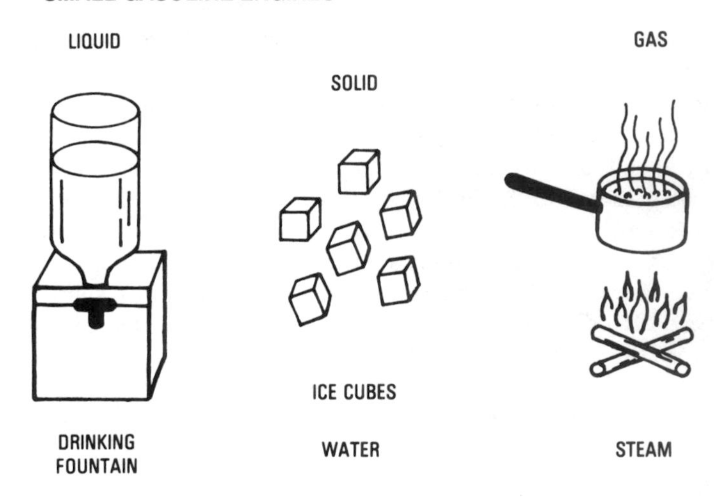

Fig. 7–3. Water can be found in three states—liquid, solid, or gas.

carbon atoms. Petroleum molecules will oxidize (combine with oxygen) easily when heated. That makes the gasoline mixture's explosive qualities greater than those of dynamite (Fig. 7–4).

In order to make gasoline, petroleum is boiled or heated until

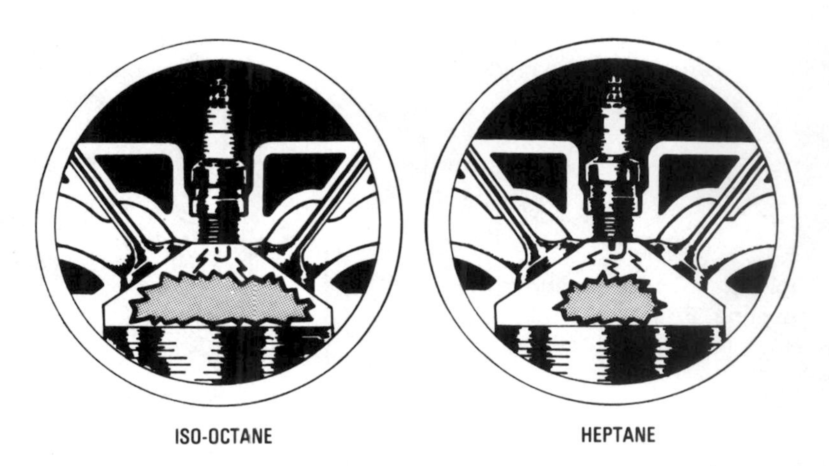

Fig. 7–4. Hydrocarbons such as heptane and iso-octane are plentiful in gasoline, but are mixed to improve ignition and power.

some of the products are vaporized or boiled off. These vapors are collected and cooled back into a liquid form. Gasoline is one of the products produced by heating petroleum until it boils off some of its most volatile components. When petroleum is heated, it separates into gasoline, naphtha, benzine, kerosene, gas oils, lubricating oils, fuel oils, petroleum jelly, paraffin, and petroleum coke. These are but a few of the products of petroleum. Many other products are made from petroleum, such as plastics, medicines, and dyes.

Gasoline will evaporate if left in an open container. Once it completely evaporates, it leaves a residue. This residue is a gummy substance that can clog small holes and cause problems in any carburetor. Fresh gasoline, or gasoline that has been kept in a sealed container, is best for use in any engine.

Gasoline can be manufactured from coal, natural gas, and vegetation. However, certain standards are set nationally concerning where the gasoline will vaporize under pressure and temperature conditions. Manufacturers design their gasolines to be used in certain areas where they know the climatic conditions.

Commercial gasolines have two liquid hydrocarbons in them. These hydrocarbons are *heptane* and *iso-octane*. If too much heptane is included in a gasoline relative to the iso-octane, the fuel will not burn evenly. This can cause problems with ping or knock in an engine (Fig. 7–5). The greater the compression ratio, the more iso-octane is needed to make the engine operate properly. Iso-octane aids the fuel-burning process. It evenly distributes burning in the combustion chamber. Gasolines are rated according to their octane content. Most *regular* gasoline has an octane rating of 89, and *unleaded* is rated at 87. Most small gasoline engines have a compression ratio of 6 to 1 (6:1) and can easily use the regular or even lower-octane gasolines.

High-octane gasolines are sold as *premium*. They cost more because of the content of iso-octane in relation to the less expensive heptane. Hydrocarbons will leave carbon deposits inside the engine when burned. They also give off carbon particles in the exhaust gases.

Some gasolines have *tetraethyl lead* and *ethylene bromide* added to improve the burning of heptane. This aids in the elimination of the *ping* or *knock* caused by low-octane gasolines. Unleaded gasolines do not have the tetraethyl lead or ethylene bromide. Unleaded gasoline can cause exhaust valves to heat more in a small gasoline engine if it is operated for any period of time.

Fig. 7–5. A fuel charge has to burn evenly to eliminate engine ping or knock that can damage the engine parts.

Gasoline Properties

The gasoline used by internal combustion engines needs to be able to vaporize at the correct time. Complete combustion takes place when there is enough oxygen to combine with the carbon during the burning or oxidation process. If the correct amounts of oxygen combine with the carbon, the by-product of the combustion process will be carbon dioxide (CO_2). If enough air is not available to allow oxygen to combine with the burning hydrocarbons, then carbon monoxide (CO) will be produced. Carbon monoxide is very dangerous to humans (Fig. 7–6). It is odorless and colorless and can cause death if enough accumulates within a confined area. Most gasoline engines produce large amounts of carbon monoxide when running. Be sure the internal combustion engine is operated only in a well-ventilated space. At least make sure the exhaust gases are vented outside the inhabited space.

Perfect combustion can occur only when there is enough oxygen to combine with the carbon. This means that the gasoline has to be vaporized and mixed with air before it is fed into the combustion chamber and ignited. That is the job of the carburetor (Fig. 7–7).

Vapor Lock—Gasoline can vaporize before it reaches the carburetor. If this happens in the fuel lines, the vapor will block the flow of gaso-

One Carbone Atom +
Two Oxygen Atoms = CO_2

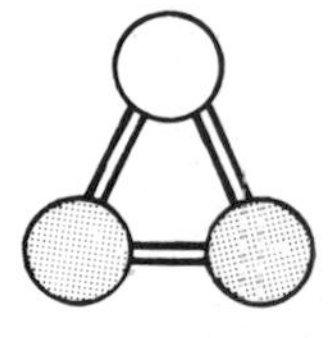

CO_2
CARBON DIOXIDE

One Carbon Atom +
One Oxygen Atom = CO

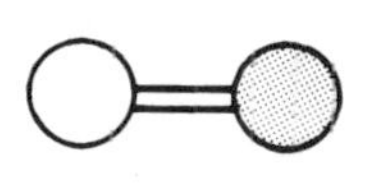

CO
CARBON MONOXIDE

Fig. 7–6. Carbon dioxide is harmless, but carbon monoxide is deadly when breathed.

line and a condition known as *vapor lock* is created. This happens sometimes in hot weather (Fig. 7–8).

In your handling of gasoline, make sure you keep it clean and free of dirt, grass clippings, and other contaminants. Contaminants such as kerosene, lubricating oil, and other solubles can create problems in starting and operation of the engine.

Don't allow gasoline to stand too long in a container. Remember especially to drain fuel tanks for winter storage. Drain snowblower engine tanks for summer storage. Gummy deposits may clog the carburetor of an engine. It will have to be cleaned and adjusted when the engine is brought out of storage.

A good grade of *regular* unleaded gasoline is usually recommended by the manufacturer. This is also true of the two-cycle engines where oil is mixed with the gasoline. However, check the manufacturer's recommendations to make sure.

Fuel—Tecumseh Products Company strongly recommends the use of fresh, clean, *unleaded* regular gasoline in all Tecumseh engines. Unleaded gasoline burns cleaner, extends engine life and promotes good starting by reducing the buildup of combustion chamber deposits. Leaded gasoline or gasohol containing no more than 10% ethanol can be used if unleaded is not available. Newer engines have been built with the ability to withstand the use of unleaded gasoline.

Never use gasoline containing *methanol*, gasohol containing more

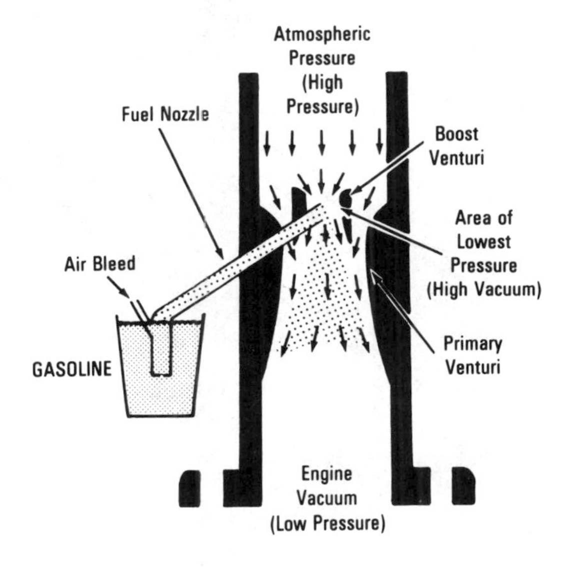

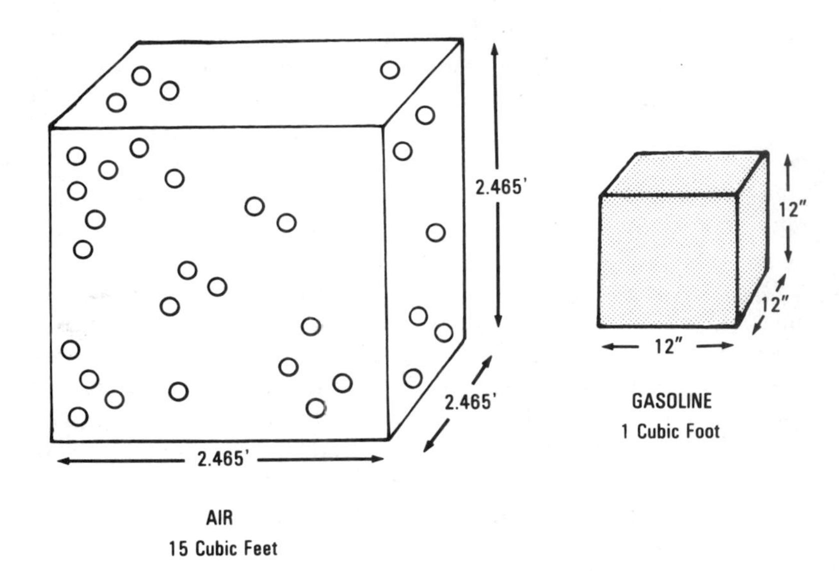

Fig. 7–7. A carburetor mixes fifteen parts of air to every part of gasoline. Any number of devices are used to do the mixing, but the carburetor still is the most popular and inexpensive.

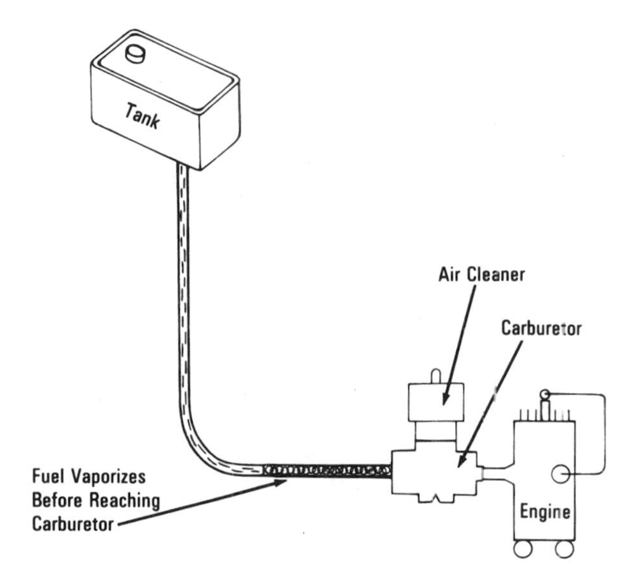

Fig. 7–8. Gas bubbles forming in the gasoline line before it reaches the carburetor can block the flow of gasoline to the carburetor. This is called vapor lock and happens mainly in the summer or at higher temperatures. Some refineries improve their product for the summer months to eliminate vapor lock.

than 10% ethanol, gasoline additives, premium gasoline, or white gas because the engine/fuel system can be damaged.

Storage—Never store an engine for 30 days with fuel in the tank indoors or in enclosed, poorly-ventilated enclosures. This is especially important where fuel fumes may reach an open flame, spark or pilot light as on a furnace, water heater, clothes dryer, or similar device.

If the engine is unused for 30 days or more, prepare it properly:

1. Remove all gasoline from the *carburetor* and *fuel tank* to prevent gum deposits from forming on these parts and causing possible malfunction of the engine.

You can run the engine until the fuel tank is empty and the engine stops due to lack of fuel. If you decide to drain the carburetor or tank use an approved container and do it outdoors away from any open flame.

Disconnect the fuel line at the carburetor or fuel tank. Be very careful not to damage the fuel line, fittings or fuel tank. Drain any remaining fuel from the system.

NOTE: If *gasohol* has been used, complete the draining of the system and then put a small amount of unleaded (or regular leaded) gasoline into the fuel tank and repeat the draining instructions.

If the oil has not been changed recently, this may be a good time to do it.

Remove the spark plug and pour one (1) ounce (0.029 liter) of engine oil into the spark plug hole. Crank the engine over slowly several times.

Avoid the spray from the spark plug hole when cranking the engine over slowly.

Replace the spark plug.

Clean the engine by removing any clippings, dirt, or chaff from the exterior of the engine.

Fuel Systems

The basic parts of a fuel system are a tank, a carburetor with air cleaner, and an intake manifold (Fig. 7–9). However, there are other parts that make for better operation of specialized engines.

Fuel Tank—The fuel tank is nothing more than a small container for holding the gasoline or fuel in storage until used. A number of sizes and shapes are available for small gasoline engines (Fig. 7–10).

The feed system for getting the fuel from the tank to the carburetor is also important. There are three ways of doing this. One involves gravity. In this case the tank is mounted above the carburetor and gravity will cause the gasoline to flow from the tank down to the carburetor. The system may use a vacuum to cause the fuel to flow into the carburetor. This vacuum may be obtained by a number of methods. See the operation of carburetors later in this chapter for a more detailed explanation. There is also the forced-feed type of system to get the fuel from the tank to the carburetor or wherever the gasoline will be mixed with the air. This usually involves a fuel pump. Small pumps are used for small gasoline engines and will be discussed later.

Carburetor—Another essential component of the fuel system for a gasoline engine is the carburetor. This device causes the fuel to be

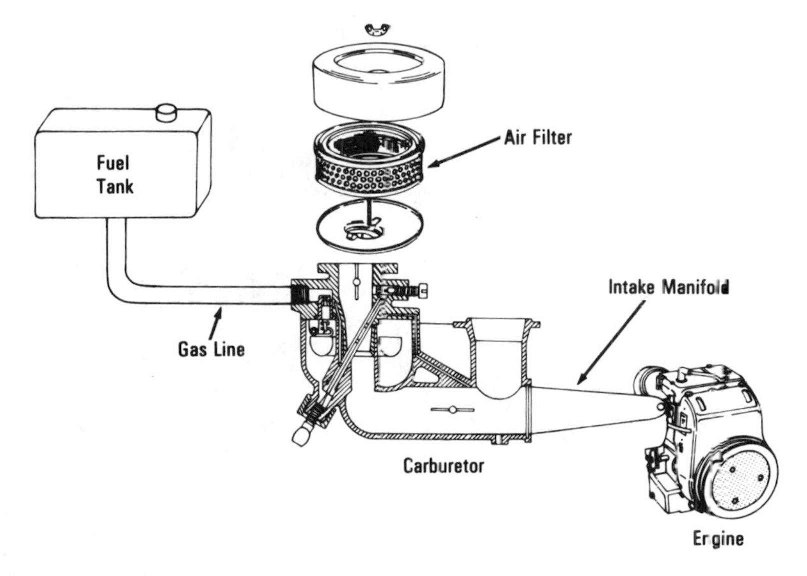

Fig. 7–9. The fuel system for a small gasoline engine. Note the air filter and fuel tank locations.

vaporized and mixed with air. The proper ratio of air to fuel is important. This means that the carburetor will have to be able to maintain this ratio during engine operation. Therefore, this is a very critical component in the operation of an internal combustion engine.

Air Filter—Since large volumes of air are needed to mix with the gasoline vapor, an air filter is very essential. This device cleans the air so that it will be free of objects or contaminants that may clog the small holes in the carburetor.

Intake Manifold—The manifold takes the gasoline and air mixture from the carburetor to the combustion chamber. This should be capable of preventing the vaporized gasoline from condensing back to a liquid state.

Fuel Tanks

Fuel tanks serve one purpose: to store the gasoline or other fuel until it is needed by the engine. There are a number of types of gaso-

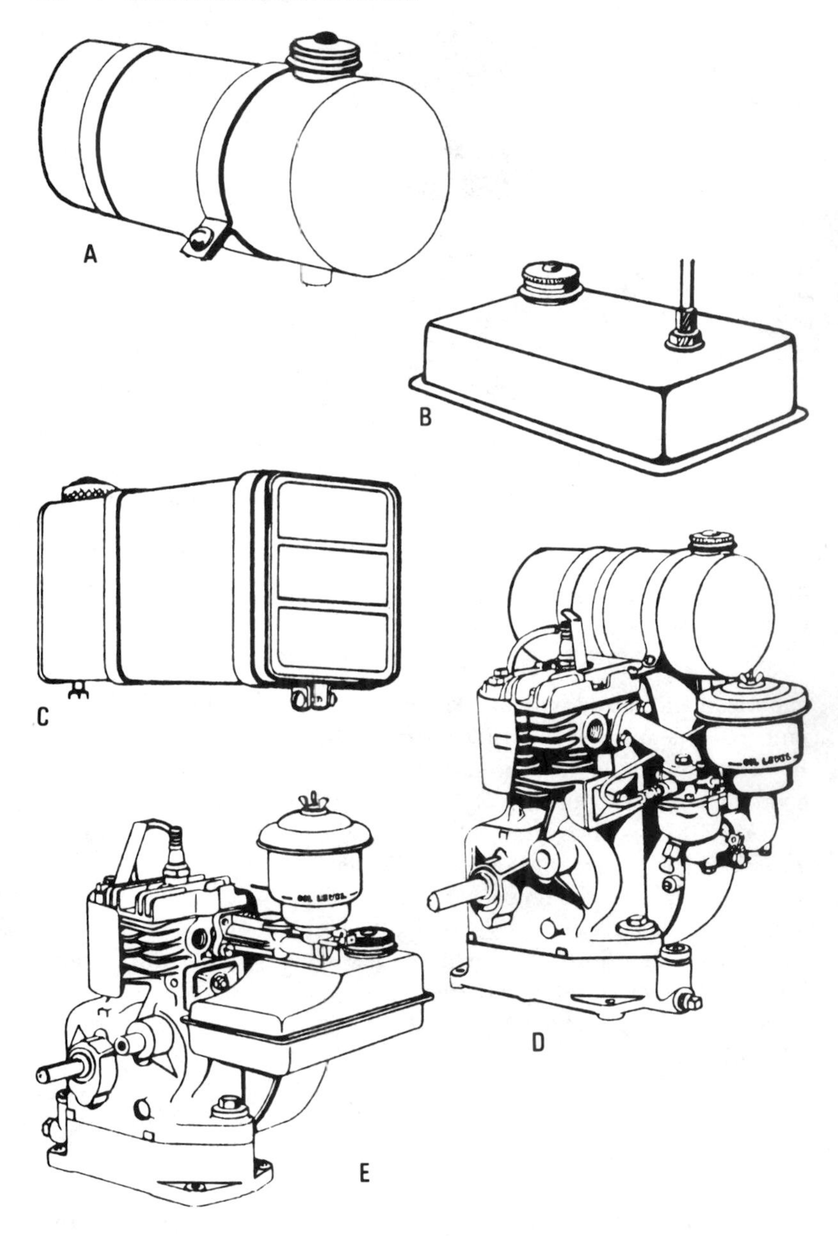

Fig. 7–10. Fuel tanks come in a variety of shapes and sizes.

line tanks used for small gasoline engines (Fig. 7–10). Some tanks have a screen fitted into the tank inlet to strain the gasoline as it is put into the tank. The tank cap is vented. The outlet for the fuel is usually screened to make sure that grass clippings and other foreign matter that may have gotten into the tank do not reach the carburetor.

Some larger tanks will have a draincock for cleaning the fuel system. A fuel shutoff is also available on some of the larger tanks (Fig. 7–11). Some gas tanks will have a gage that measures the amount of gasoline left in the tank. This is usually a float arrangement with an indicator on the outside of the tank (Fig. 7–12).

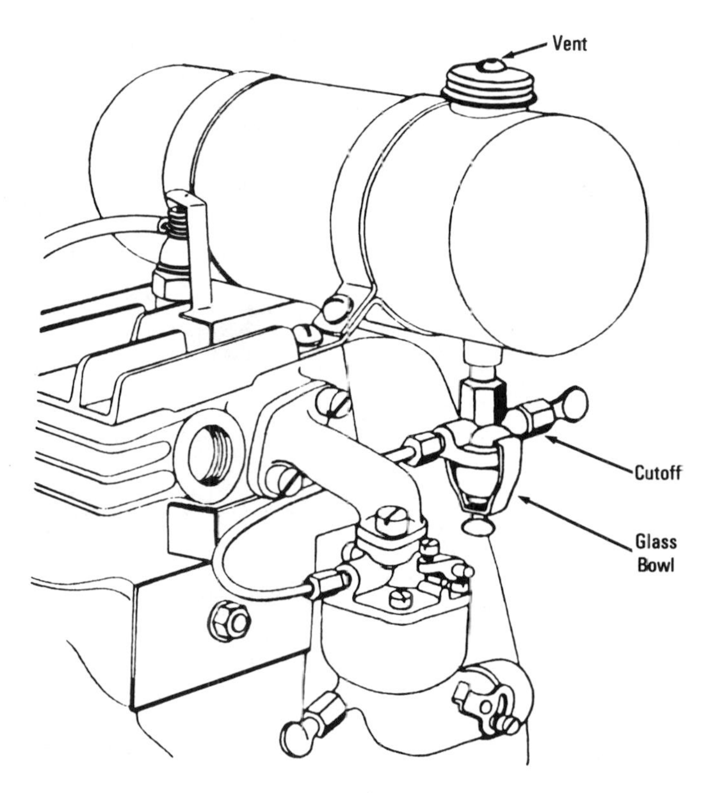

Fig. 7–11. Some tanks have a cutoff valve and a glass bowl that can be removed for cleaning. *(Courtesy B & S)*

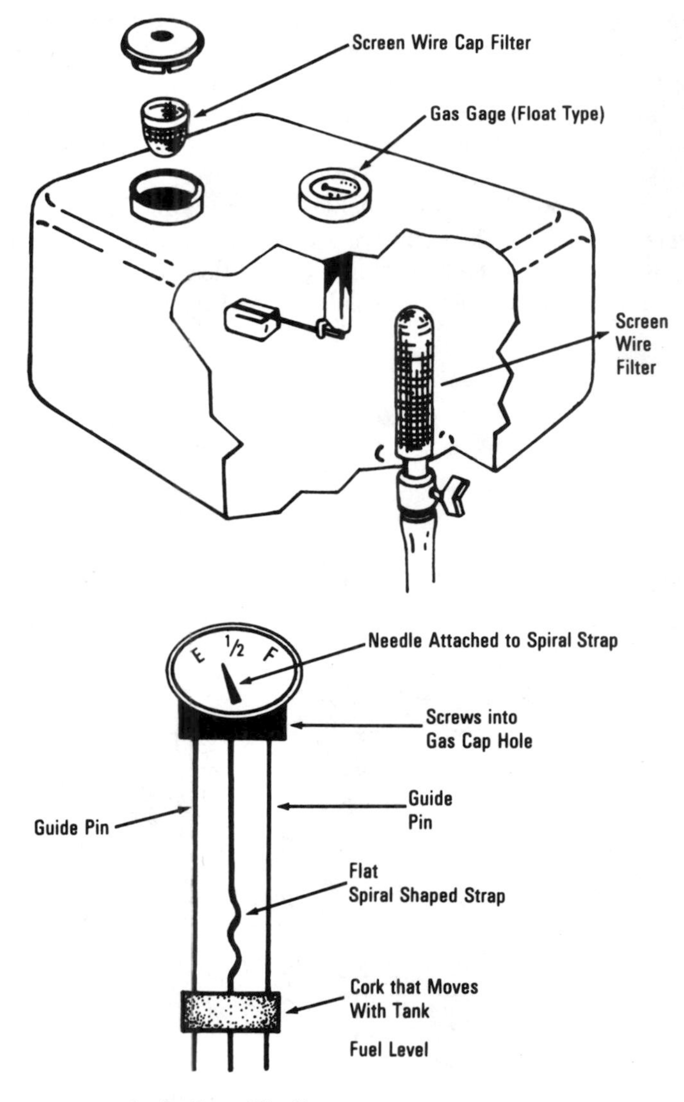

Fig. 7–12. A float-actuated, fuel-level indicator is available for those who need to know the fuel level.

Fuel Feed Systems

There are four types of systems used to move the gasoline from the tank to the carburetor. They are the gravity type, suction-feed system, forced-feed system, and standard vacuum system.

Gravity System—In the gravity-feed system, the tank is above the carburetor. Fuel flows down by gravity. Notice an air vent hole in the carburetor bowl so that air can flow out as fuel flows in (Fig. 7–13). If one or both of these holes becomes clogged or plugged, the flow of fuel stops and so does the engine.

As the fuel enters the bowl, it raises the carburetor float (Fig. 7–14). The float, in turn, raises the needle in the float valve. When the needle touches the seat, it shuts off the fuel flow. The position of the float at this time is called the float level. The float level in general should be high enough to afford an ample supply of fuel at full throttle and low enough to prevent flooding or leaking.

To set the level on the carburetor, invert the upper body. The float and the body cover should be parallel (Fig. 7–15). If not, bend the tang on the float to obtain this position. The actual distance on the small

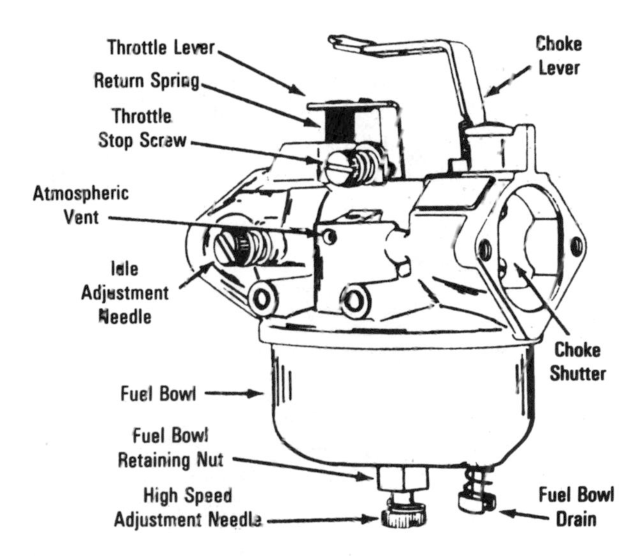

Fig. 7–13. Atmospheric vent location on a gravity-feed system's carburetor. *(Courtesy Tecumseh)*

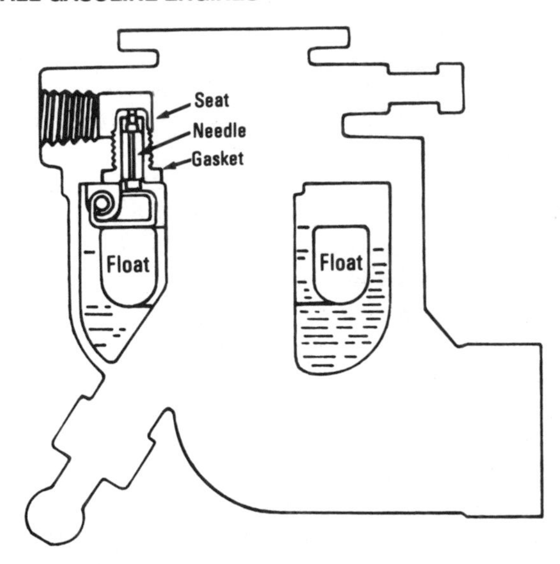

Fig. 7–14. The float level controls the action of the needle against the seat. *(Courtesy B & S)*

carburetors is 5⁄16″ (7.94 mm) between the float and the gasket. On larger models it is 3⁄16″ (4.76 mm). It is seldom necessary to measure this distance. The float level is not as critical as on some carburetors. Remember, however, that there should be one gasket between the float valve seat and the carburetor. No gasket or two gaskets will

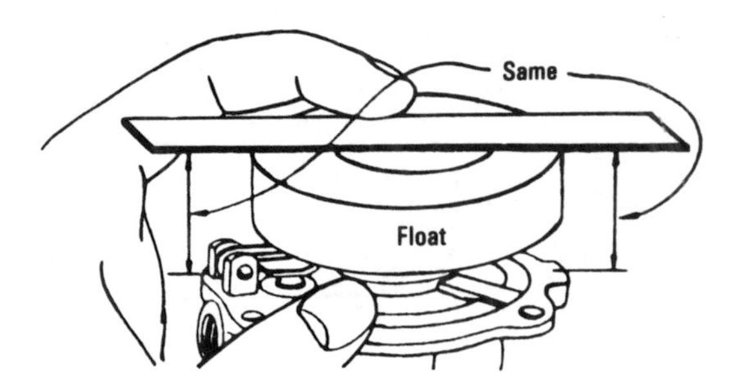

Fig. 7–15. The float adjustment means the two distances shown should be the same. *(Courtesy B & S)*

change the float level. Once the fuel has reached the carburetor, it seeks its own level in the bowl. This level is well below the discharge holes (Fig. 7–16). Note the discharge holes in the venturi. This is the location of the greatest air velocity. As the piston in the cylinder moves down with the intake valve open, it creates a low-pressure area that extends down into the carburetor throat and venturi. Two things then start to take place: The air pressure above the fuel in the bowl pushes the fuel down in the bowl and up in the nozzle to the discharge holes. At the same time the air rushes into the carburetor air horn and through the venturi, where its velocity is greatly increased (Fig. 7–16).

The nozzle extending through this air stream acts as an airfoil. This creates an area of still lower pressure on the upper side. This allows the fuel to stream out of the nozzle through the discharge holes into the venturi, where it mixes with the air and becomes a combustible mixture ready for firing in the cylinder.

A small amount of air is allowed to enter the nozzle through the

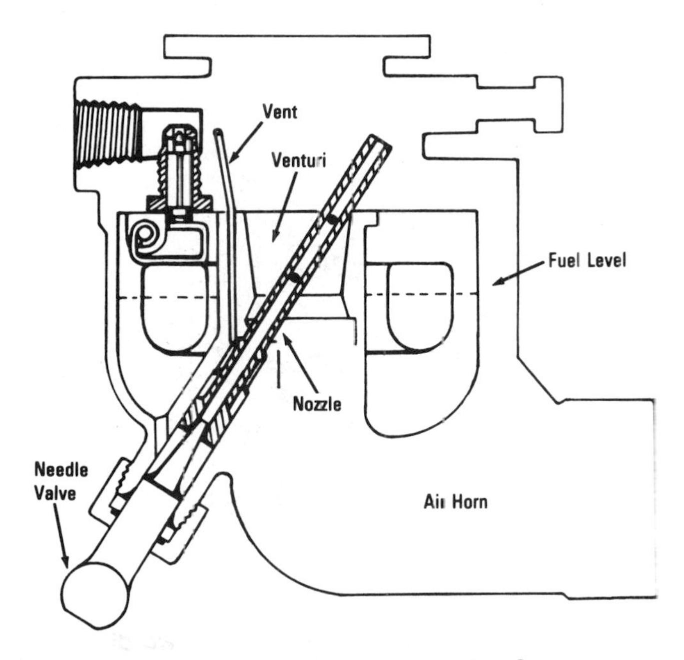

Fig. 7–16. Operation of carburetor. *(Courtesy B & S)*

vent. This air compensates for the difference in engine speed and prevents too rich a mixture at high speed.

If an engine ran at only one speed, it would be easy to design a carburetor for it. However, since a range of speeds is called for in most engines, the carburetor must be designed to take into account the highest and lowest speeds as well as those in between. Since smooth economical operation is desired at varying speeds along with starting at all times, some additions must be made to the carburetor. These will be discussed later in the chapter under carburetion principles.

Suction-Feed System—In the suction-feed system, the gas tank is *below* the carburetor. This means the fuel most flow upward, against gravity. Therefore, the force of atmospheric pressure must be used to get the gas from the tank to the carburetor above it (Fig. 7–17).

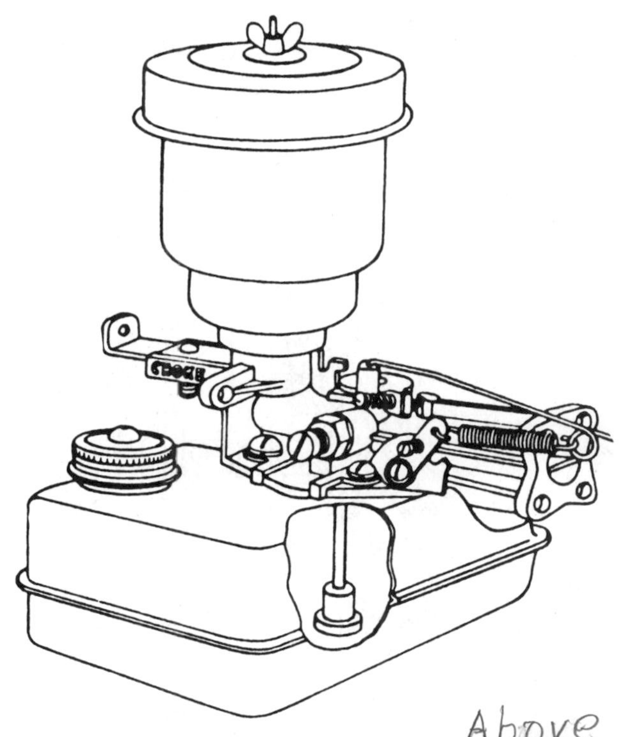

Fig. 7–17. Suction-feed carburetor is located below the fuel tank. *(Courtesy B & S)*

The fuel tank cap has a vent hole. This is to allow the pressure in the tank to remain constant. Before adjusting the carburetor, pour in enough fuel to *half* fill the tank. The distance the fuel has to be lifted will affect the adjustment. At half full we have an average operating condition and the adjustment will be satisfactory if the engine is run with the tank full or nearly empty.

As the piston goes down in the cylinder with both the intake valve and the throttle open, a low-pressure area is created in the carburetor throat. A slight restriction is placed between the air horn and the carburetor throat at the choke. This helps to maintain the low pressure.

The difference in pressure between the tank and the carburetor throat forces the fuel up the fuel pipe, past the needle valve, through the two discharge holes. The throttle is relatively thick, so we have in effect a venturi at this point, thus aiding vaporization. A spiral is placed in the throat to aid acceleration and also to help keep the engine from dying when the throttle is opened suddenly.

The amount of fuel at operating speed is metered by the needle valve and seat. Turning the needle valve in or out changes the setting until the proper mixture is obtained. This adjustment must always be made while the engine is running at operating speed, not at idle speed. While the needle valve may look like an idle valve due to its position, it is a true high-speed, mixture-adjusting valve (Fig. 7–18).

Since no accelerator pump is used on this carburetor and since these engines are used on lawnmowers where rapid acceleration is needed, the mixture should be rich. Turn the needle valve inward until the engine begins to lose speed, indicating a lean mixture. Then open the needle valve (turn outward) past the point of smooth operation until the engine just begins to run unevenly. Since this setting is made without load, the mixture should operate the engine satisfactorily under load.

These carburetors do not have an idle valve, but the mixture at idle speed is controlled in a different way. As the throttle closes to idle, the leading edge takes a position between the two discharge holes. The larger of the discharge holes is now in the high-pressure area, and the flow of fuel through it will cease. The small hole will continue to discharge fuel, but the amount will be metered by the hole size and will be in proportion to the reduced air flow. For this reason it is important that the small discharge hole be the proper size. The needle valve will allow much more fuel to pass than should go through the small dis-

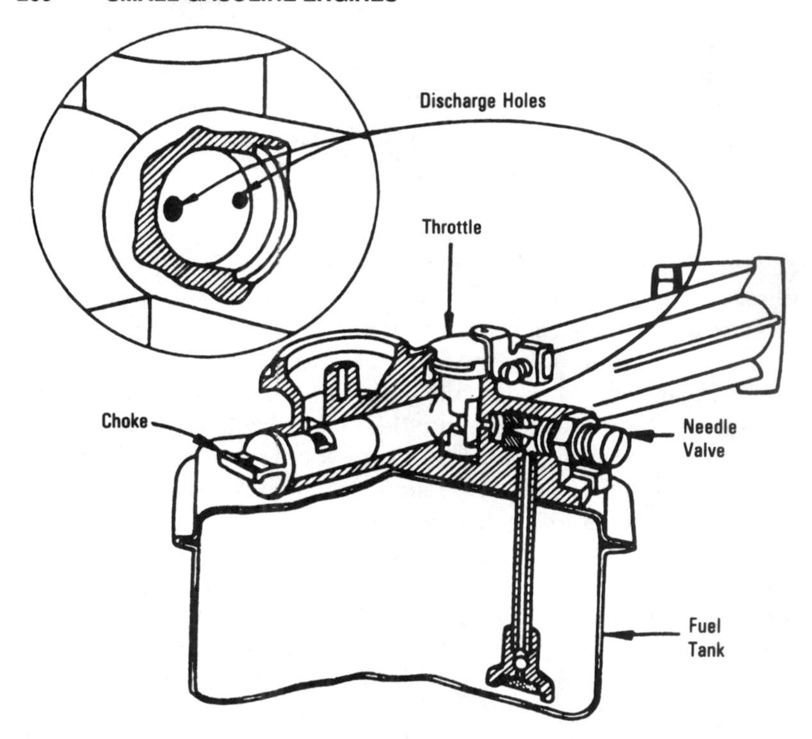

Fig. 7–18. Needle valve location on the carburetor for a suction-type fuel system. *(Courtesy B & S)*

charge hole. A #68 drill bit can be used as a plug gage to check the small hole. A #56 drill bit can be used to check the larger hole. This can be done with the needle valve and seat removed.

You will notice that a small section is milled out of the throttle where it meets the discharge hole. This concentrates the flow of air past the hole and ensures good vaporization.

The idle-speed adjusting screw should be set to obtain an idle speed of 1750 rpm. This may seem fast to people accustomed to auto engines, but it is necessary in order to have fast acceleration. It also helps cooling and lubrication. A slight unevenness may be noticed at idle speed, but this is normal and no readjustments of the needle valve should be made.

The choke is the sliding plate mounted at the small space between the air horn and the carburetor throat. The choke is pushed in to close the air intake for starting, but should be pulled out as soon as the engine starts. Due to the rich mixture, very little choking is necessary. The use of this choke should be understood fully. Many complaints of engine trouble prove to be nothing more than failure to use the choke properly.

The latest engines with suction-feed carburetors incorporate a ball check in the fuel pipe, which ensures a steady flow of fuel to the needle valve and discharge holes (Fig. 7–19). The ball drops each time the vacuum or suction ends. This prevents the gasoline in the carburetor from draining back into the tank.

Standard-Type Vacuum System—In this type of system, an auxiliary tank is located above the carburetor. The manifold vacuum sucks the air out of the auxiliary tank and creates a vacuum. The vacuum tank then sucks fuel out of the regular fuel tank and into the auxiliary tank. This means the fuel actually flows uphill to the vacuum tank in an attempt to fill the vacuum created by the manifold vacuum, which was, in turn, created by the piston moving downward inside the cylinder. A difference in pressure from 3 to 12 pounds per square inch (psi) is created. That means the difference between atmospheric pressure and the pressure created by the piston moving downward is about 3 psi when the throttle is wide open and about 12 psi when the throttle is closed. Enough pressure differential is created to keep the vacuum tank filled. A vent hole in the tank top allows air to enter and keeps the tank from overfilling.

The standard-type vacuum system is seldom used on small gasoline engines. It is mentioned here to illustrate some of the possibilities in the fuel systems for a gasoline engine.

Forced-Feed System—The forced-feed fuel system uses a fuel pump. This pump moves the gasoline from the storage tank to the carburetor. Some two-cycle engines have the fuel pump made into the carburetor body.

Fuel Pumps

Carburetor built-in fuel-pump systems are used on the two-cycle engines (Fig. 7–20). Note the location of the pump element in the carburetor.

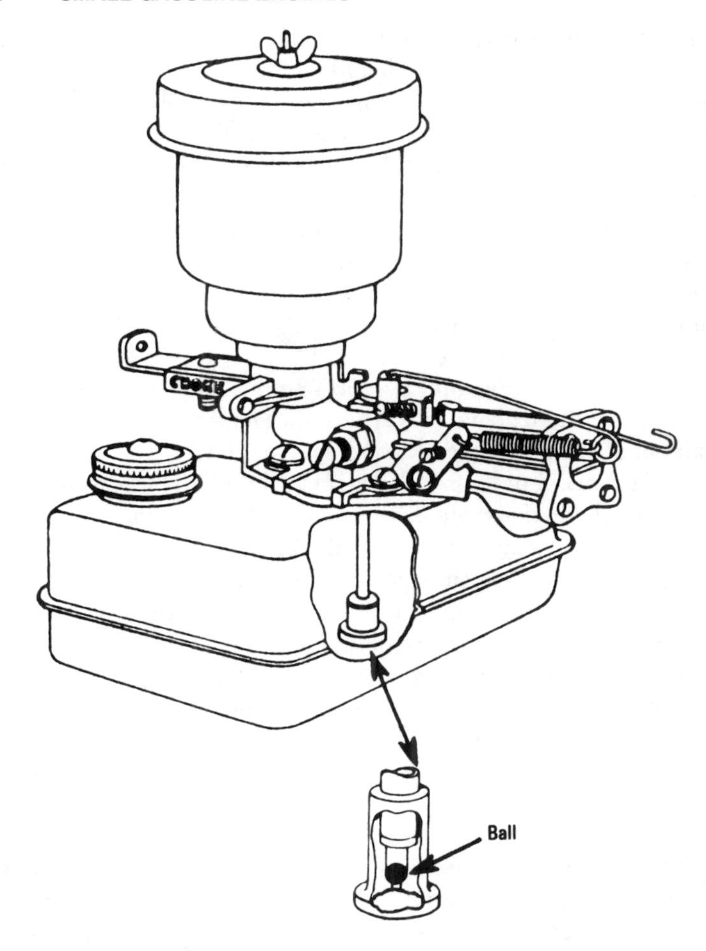

Fig. 7–19. The ball drops when suction ceases. This prevents the gasoline from dropping back into the tank. *(Courtesy B & S)*

Operation of the fuel pump in the carburetor can be explained using Fig. 7–21 through 7–25. In Fig. 7–21 rapid deflation (*1*) draws fuel through the passage (3) and the inlet fitting check valve (*4*) and into the pump element cavity (2). The pump is operated by crankcase pulsations.

In Fig. 7–22, rapid inflation (*1*) forces fuel out of the element cav-

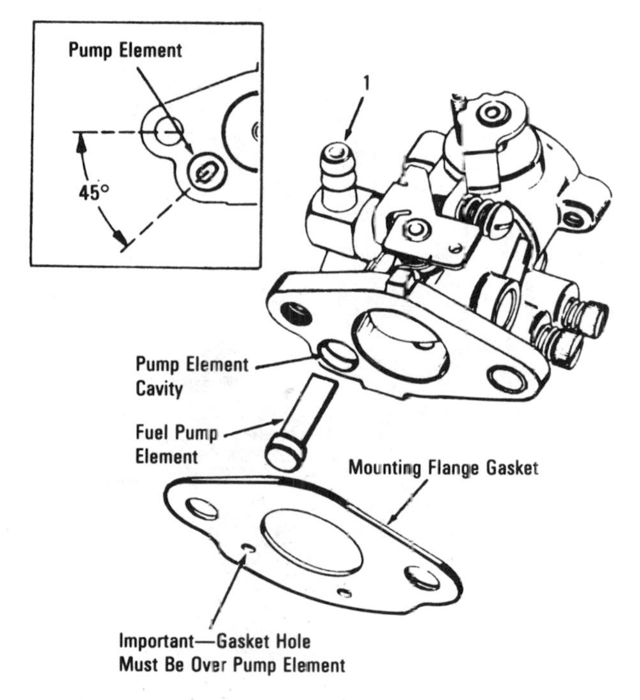

Fig. 7–20. Location of the pump element in the fuel-pump carburetor. *(Courtesy Tecumseh)*

ity (2), through the passage (3), and blocks it at the inlet fitting check valve (4). The body check valve (5) opens to allow fuel passage into the carburetor.

The built-in fuel-pump element (Fig. 7–23) is inserted into an opening in the mounting flange end of the carburetor body, with the slot opening at a 45° angle. Keep in mind that the hole in the gasket must be directly over the pump element in order for the crankcase pulsations to operate the pump element.

The engine piston (Fig. 7–24), moving out of the crankcase area, creates a partial vacuum that collapses the fuel-pump element in the carburetor. On the outside of the element, suction opens the inlet flap, allowing a supply of fuel to flow from the tank and lines into the cavity created by the deflating pump element. Suction pulls the outlet flap

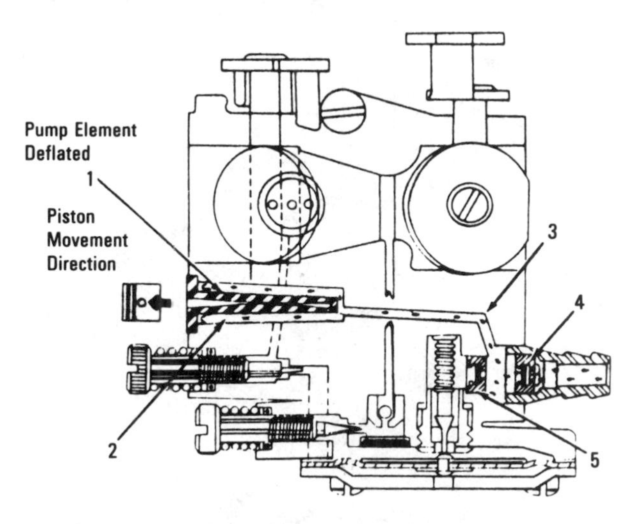

Fig. 7–21. Operation of the fuel pump in the carburetor—rapid deflation of pump element. *(Courtesy Tecumseh)*

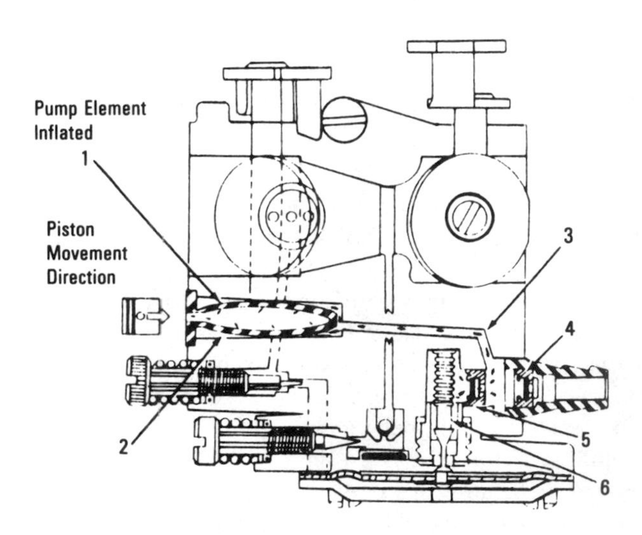

Fig. 7–22. Operation of the fuel pump in the carburetor—rapid inflation of pump element. *(Courtesy Tecumseh)*

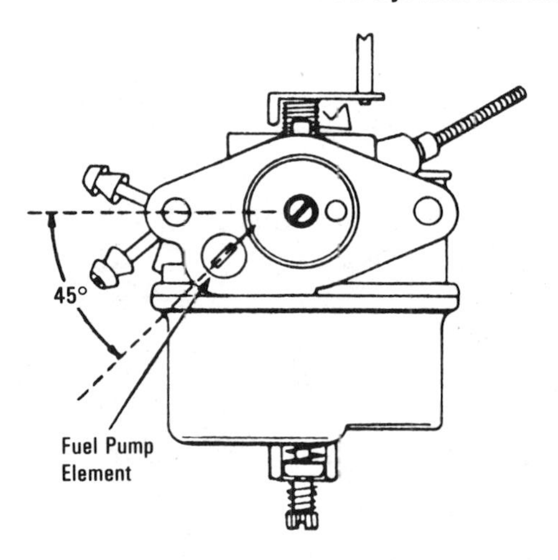

Fig. 7–23. Proper mounting position of the fuel-pump element. *(Courtesy Tecumseh)*

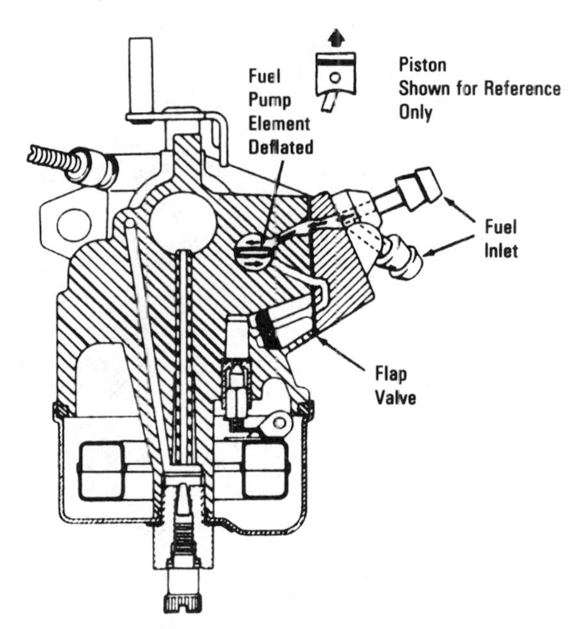

Fig. 7–24. Operation of the fuel pump in the carburetor—fuel pump element deflated. *(Courtesy Tecumseh)*

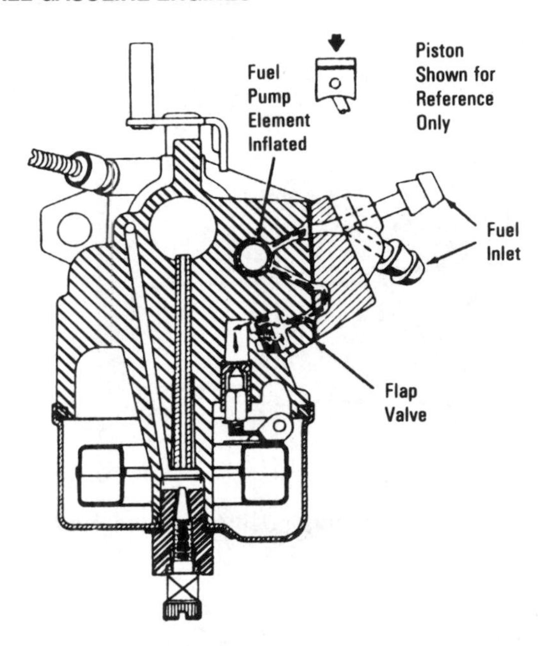

Fig. 7–25. Operation of the fuel pump in the carburetor—fuel pump element inflated. *(Courtesy Tecumseh)*

closed. This seals the outlet port so that fuel isn't pulled from the area of the inlet needle seat.

With the piston stroke downward (Fig. 7–25), the crankcase pressure enlarges the pump element, forcing fuel out of its cavity. This pressurized fuel acts against the outlet flap valve, opening it. This allows a head of pressurized fuel to be transmitted to the inlet needle and seat port. The inlet valve is pressed against the inlet port. This seals it so that pressurized fuel does not escape back into the fuel tank and lines.

The valve cover assembly contains the inlet tube (there are two inlet tubes in an outboard engine) from the fuel supply. Fig. 7–26 shows the location of the inlet fitting, the fuel-pump flap valve cover, and the throttle post and lever.

Impulse Fuel Pump

Another type of fuel pump used on two-cycle engines is the impulse type. The pump is operated by changing pressures created by

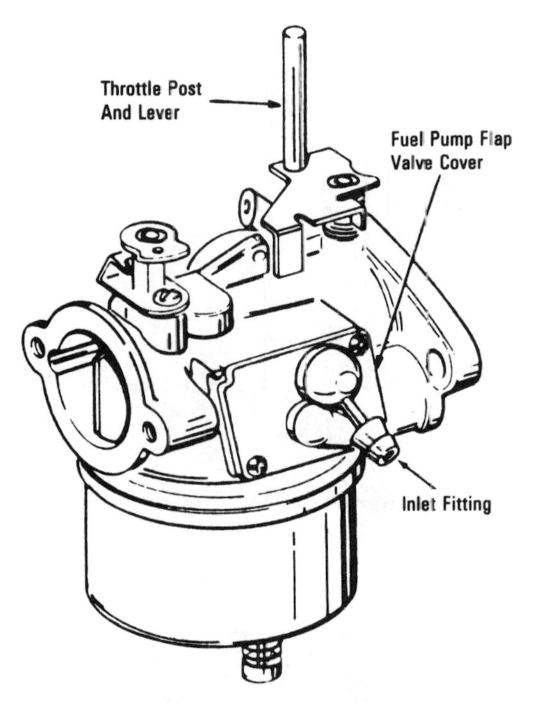

Fig. 7–26. Location of the fuel-pump flap valve cover. *(Courtesy Tecumseh)*

the engine piston movement and transmitted through the pulse line (Fig. 7–27). This pump is mounted at the carburetor inlet port. It has connections from the fuel supply and a pulse line to the engine crankcase.

The pump is operated by changing pressures in the crankcase. A decrease in pressure (suction) in the pulse chamber allows atmospheric pressure to push a fuel charge into the pump chamber from the fuel supply. The direction of fuel flow is controlled by flap valves. A suction action causes the fuel outlet flap to seal and the fuel inlet flap to open. Both flap valves have a coil spring to overcome before they can open to allow fuel to flow (Fig. 7–28).

An increase in pulse chamber pressure (above atmospheric) acting on the diaphragm forces fuel out of the fuel-pump chamber. Initially, the fuel fills a booster chamber, then fills the passageway to the inlet

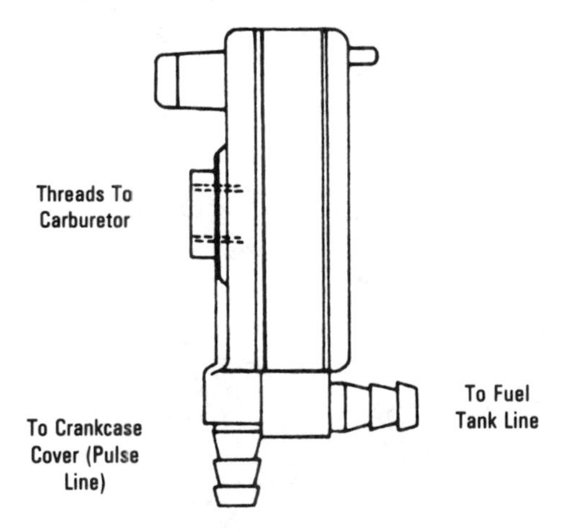

Fig. 7–27. Impulse fuel pump for two-cycle engines. *(Courtesy Tecumseh)*

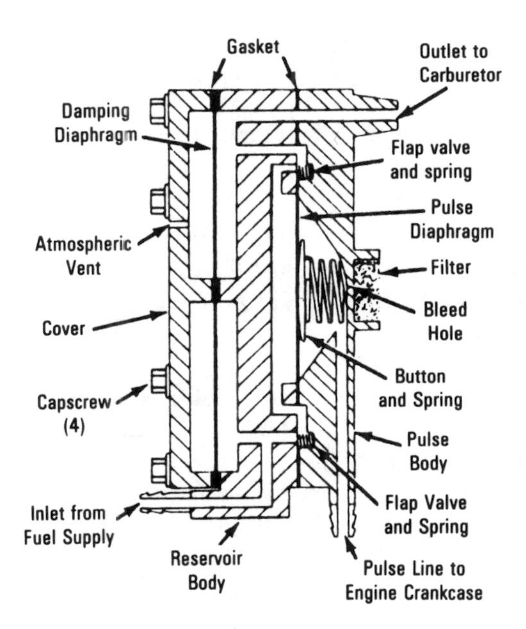

Fig. 7–28. Parts of the impulse fuel pump. *(Courtesy Tecumseh)*

needle and seat area of the carburetor. Pressurized fuel can't escape back to the fuel supply because the flap valve at that port seals. The fuel outlet flap valve is forced open and the fuel flows in that direction.

Once fuel pressure is established from the pump to the carburetor, each pulse of fuel pressure from the pump chamber expands a diaphragm in the booster chamber. On the suction stroke, the expanded booster diaphragm continues to exert pressure against the fuel to the carburetor, keeping this pressure constant. A small bleed hole in the pulse chamber acts to dampen extremes in pressure to that area (Figs. 7–29 and 7–30).

Fig. 7–31 shows an exploded view of the fuel pump. One of the advantages of the pulse- and vacuum-operated fuel-pump system is that the engine can be operated in any position. If the float-type carburetor is used, the engine must be kept horizontal or have no more than a 45° tilt.

Eccentric-Operated Fuel Pump

This type of fuel pump is operated by an arm that rides on an eccentric, or cam, lobe (Fig. 7–32). Reciprocal movement of the lever pulls the diaphragm down against the spring, then allows the spring to push the diaphragm back up. Downward movement of the diaphragm

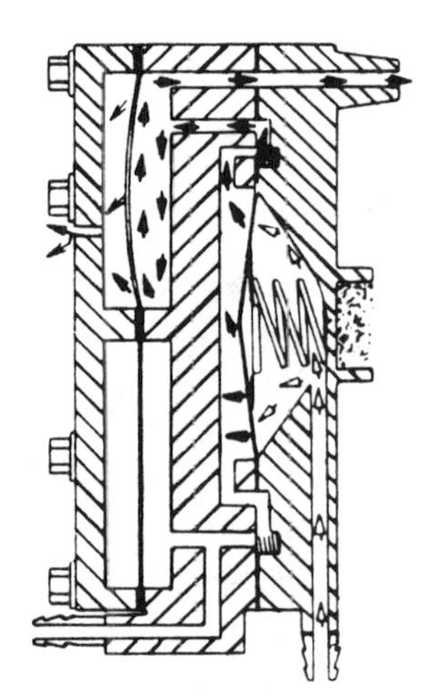

Fig. 7–29. Operation of the impulse fuel pump. *(Courtesy Tecumseh)*

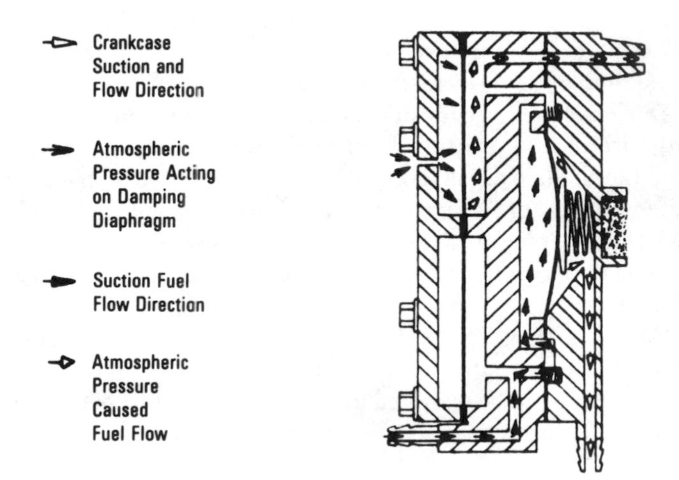

Fig. 7–30. Operation of the impulse fuel pump. *(Courtesy Tecumseh)*

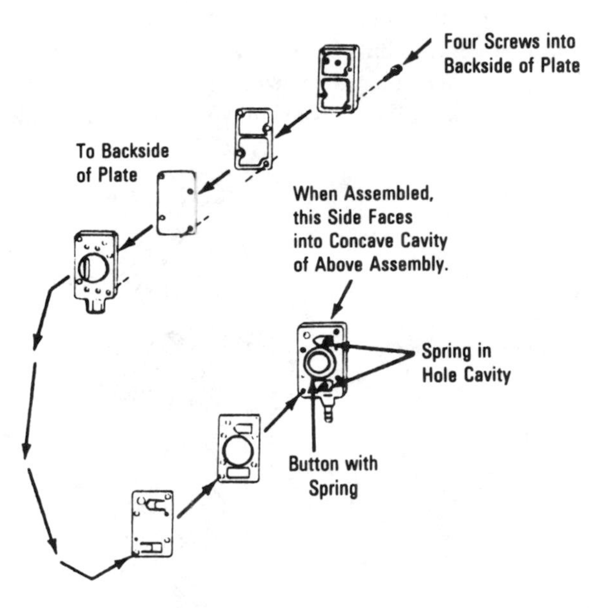

Fig. 7–31. Exploded view and assembly pattern for the impulse fuel pump. *(Courtesy Tecumseh)*

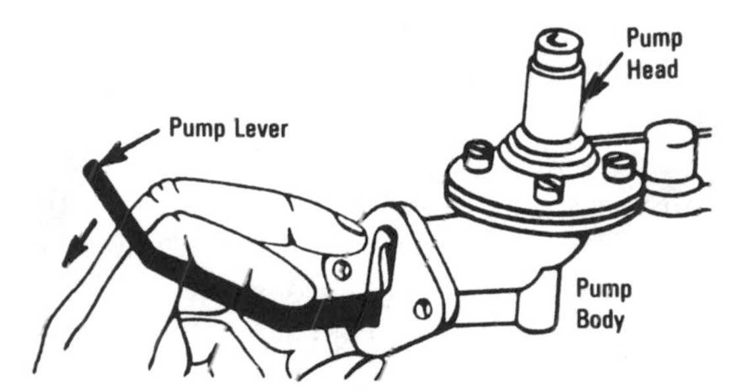

Fig. 7–32. Push-rod arm or lever for operation of the fuel pump. *(Courtesy Tecumseh)*

creates low pressure in the pump chamber (Fig. 7–33). A spring-loaded valve allows fuel to flow into the pump chamber. When the diaphragm moves upward, it pushes the gasoline out the other spring-loaded valve to the carburetor. The valve on the left is closed so that the gasoline has

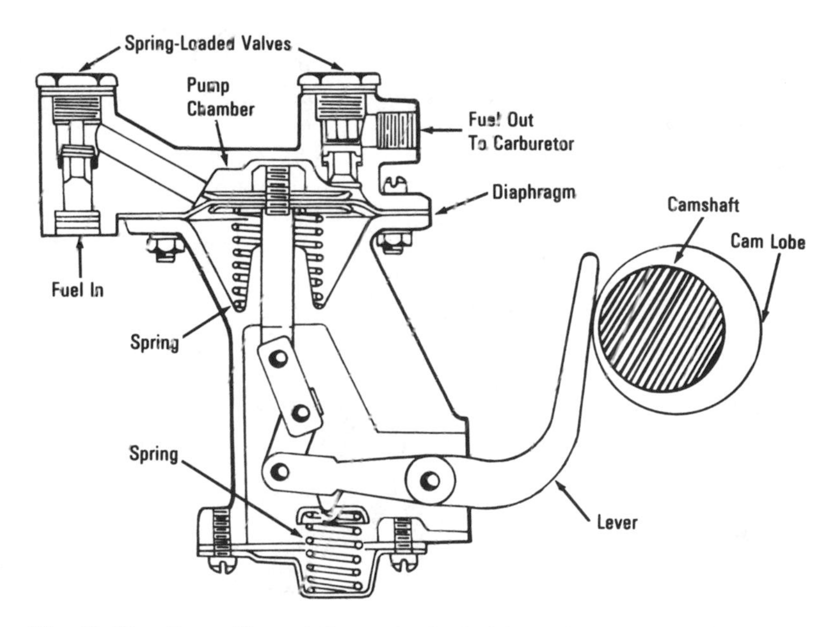

Fig. 7–33. Operation of the camshaft-driven fuel pump. *(Courtesy Tecumseh)*

to move out through the valve at the right of the carburetor. It also has been designed to allow gasoline to flow in only one direction. Once the carburetor becomes filled with gasoline, and is not in need of any more, the chamber of the pump is filled. The back pressure created by the valve in the carburetor closes off the fuel line from the carburetor to the fuel pump. This means the spring mounted under the diaphragm is unable to push the diaphragm upward. However, the lever arm and the rest of the pump continues to operate without restriction as the cam lobe continues to rotate and operate the arm (Figs. 7–33 and 7–34).

The diaphragm is made of a flexible material that will not permit gasoline to flow through it. The diaphragm is held in place by a cover plate exerting pressure around its entire circumference. The pump in Fig. 7–34 has four screws to hold the pump head in place over the diaphragm.

The Fuel Pump Add-On

The fuel pump may also be installed on four-cycle engines that normally use gravity feed. Figs. 7–35, 7–36, and 7–37 show how a fuel pump may be added to take advantage of the crankcase vacuum. In some cases these pumps are factory installed when specially ordered. In most instances, however, the pumps are added in the field.

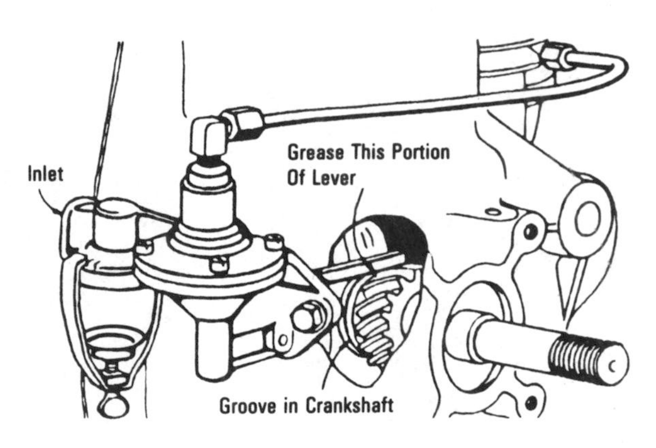

Fig. 7–34. Location of the pump lever on the crankshaft. *(Courtesy B & S)*

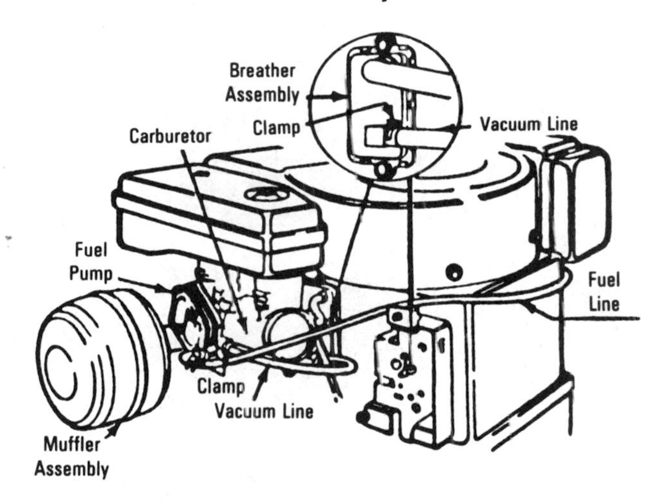

Fig. 7–35. Pump on the carburetor, breather valve location. *(Courtesy B & S)*

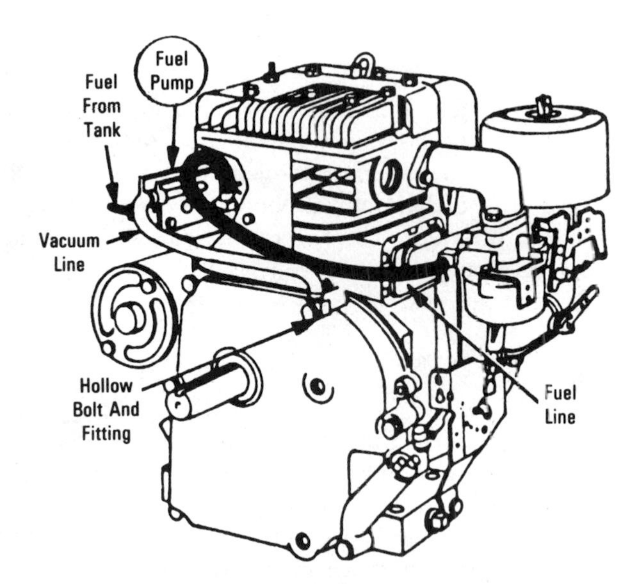

Fig. 7–36. Pump on bracket, vacuum through hollow bolt. *(Courtesy B & S)*

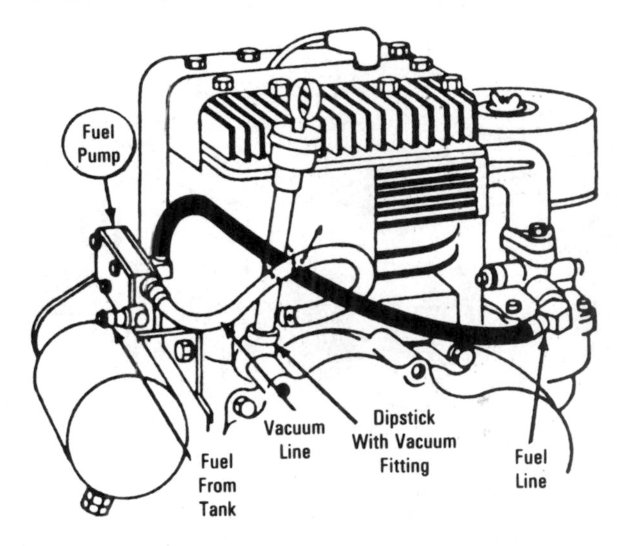

Fig. 7–37. Vacuum from dipstick. *(Courtesy B & S)*

The Priming Pump

Fig. 7–38 shows a primer bulb instead of a choke valve being used to get the engine started. The bulb is compressed and causes a rich charging mixture to flow to the carburetor venturi and intake manifold for easier starting. The carburetor float bowl is vented through the primer tube to the flexible bulb. The vent is closed when the operator's finger depresses the primer bulb.

Air Cleaners

Since air is fifteen times as great as fuel in terms of volume, it is very important that the air be clean. The carburetor holes are very small and the fuel is filtered by screens in the line, in the tank, and sometimes in the tank filler cap. This ensures clean fuel, but the mixing of the air without clogging particles is needed. The air cleaner usually consists of some type of filter. The filter may be either dry or wet.

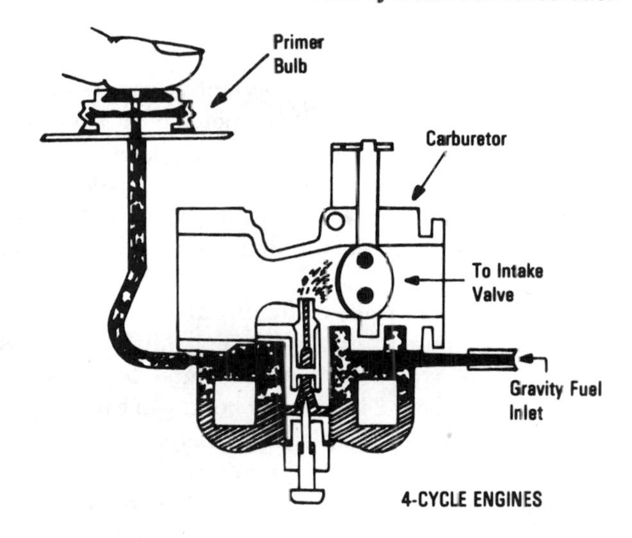

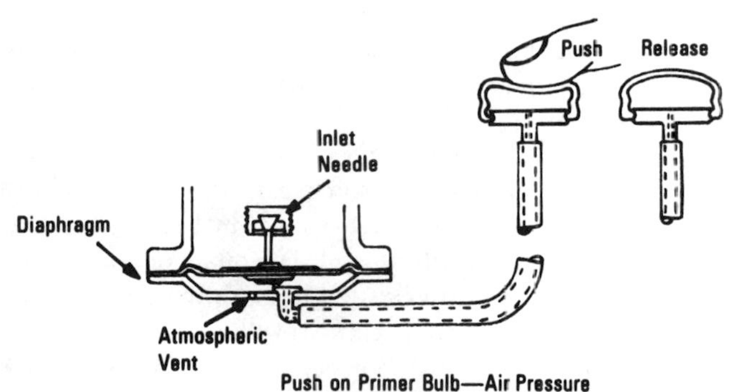

Push on Primer Bulb—Air Pressure
Lifts Diaphragm, then Leaks off
Through Atmospheric Vent Hole.

2-CYCLE ENGINES

Pressing the bulb forces air pressure against the diaphragm. As the diaphragm lifts, the inlet needle is lifted off its seat and the fuel within the reservoir is forced up through the passages into the air horn. A one-way valve in the body prevents the fuel from being forced back up into the fuel tank.

Fig. 7–38. Priming pump.

Wet Filters

The air entering the engine is important in engine performance and engine life. Power will decrease 3½ percent for every thousand feet above sea level. Power will also decrease 1 percent for every 10°F above the standard temperature of 60°F (15.55°C). In addition, the ambient temperature is important in the cooling of the engine. Ambient temperature is the temperature of the air immediately surrounding the engine.

The great trouble, however, is the dirt and grit carried into the engine through the carburetor. When you remember that a typical engine operating at 3600 rpm uses about 390 cubic feet of air an hour entering at the rate of about 24 miles per hour, you can imagine what happens if the air cleaner does not work properly. A great percentage of engines used on lawnmowers and agricultural implements are operated where conditions are extremely dusty. If the air cleaner does not have enough oil, or if it is so dirty that it can absorb no more, the grit in the air enters the engine.

Dirt can work down the cylinder walls, past the rings, and mix with the oil. This produces an abrasive that causes rapid wear of rings, cylinders, and bearings.

While the filter may be dry or wet, most filters are made of oil-saturated foam. The older models or those used in extremely dusty atmospheres will use the wet filter such as that shown in Fig. 7–39. Note how the air is forced into contact with the oil. The oil level in the filter is very important. Too much oil can cause it to be drawn into the engine and a smoky, hard-starting engine is the result.

Fig. 7–40 shows how the oil-bath air cleaner is constructed; the oil-level line is clearly marked. For many years the oil-bath air cleaner (Fig. 7–39) was considered the best, but recently Briggs & Stratton developed the oil foam "no spill" air cleaner (Fig. 7–41). This cleaner uses a polyurethane element. The important patented feature is that it is sealed. Other cleaners are made with a polyurethane element, but some are merely blocks of material with no seals of any kind, thus allowing the air and dirt to bypass the element. The cleaner in Fig. 7–41 uses the edges of the element as gaskets so that all the air must pass through the element (Fig. 7–41B).

Another advantage of the "no spill" feature of the air cleaner is that oil will not spill if the engine is tilted. If the element becomes loaded

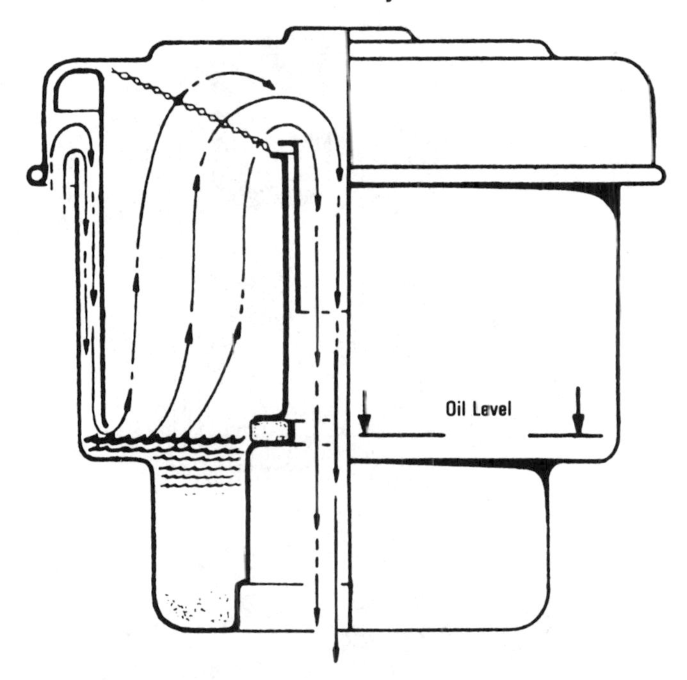

Fig. 7–39. Older type of oil-bath filter. It tilts and spills oil when the engine is slanted. *(Courtesy B & S)*

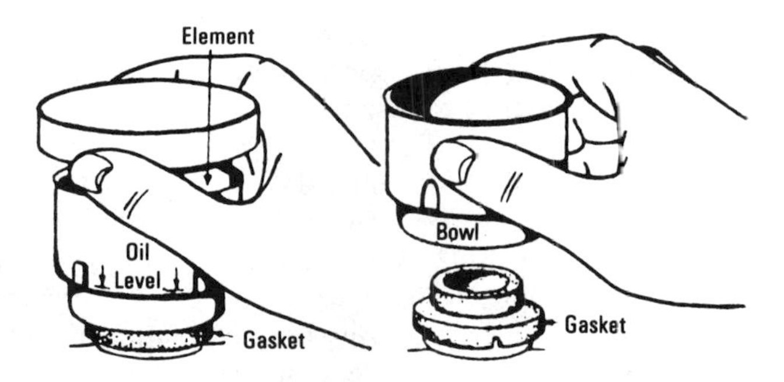

Fig. 7–40. Oil-bath air cleaner. The gasket was made of material to ensure a tight seal. *(Courtesy B & S)*

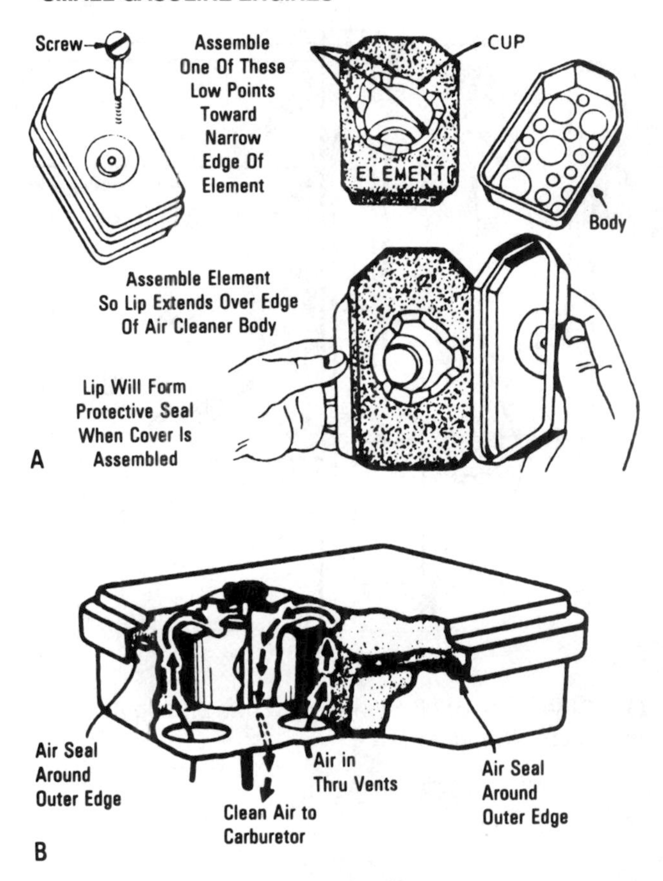

Fig. 7–41. (A) Oil-foam air cleaner. Note how the element has a lip extended over the edge of the air-cleaner body. (B) Cutaway view of a working foam air cleaner. *(Courtesy B & S)*

with dirt, the air supply will be shut off and the engine will lose power or stop entirely. Then the element can be cleaned, reoiled, and reinstalled as good as new (Fig. 7–42). Slight variations of this type of filter are shown in Fig. 7–43. Fig. 7–43C shows a foam precleaner element which can be washed in detergent, rinsed, dried, saturated with oil, and replaced. The only difference here is the shape of the foam material. Other manufacturers also use the foam with oil saturation as a filter element. Fig. 7–44 shows how various shapes of the polyurethane element are used to fill air cleaners.

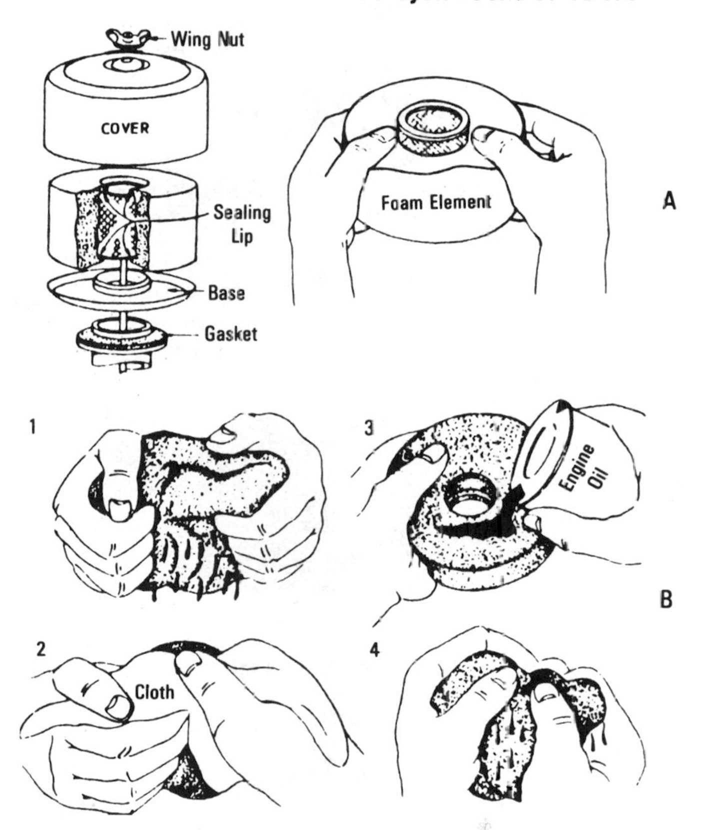

Fig. 7–42. (A) Oil-foam air cleaner. (B) Remove air cleaner carefully to prevent dirt from entering the carburetor. Take the air cleaner apart and clean by washing the foam element in kerosene or liquid detergent and water to remove the dirt. Wrap the foam in cloth and squeeze dry. Saturate the foam with engine oil. Squeeze to remove excess oil, then reassemble parts and fasten to the carburetor securely with a screw or wing nut. *(Courtesy B & S)*

Dry Filters

Dry filter elements are not saturated with oil. They are usually made with a paper element that has microscopic holes to trap dust and dirt. These have to be taken out and occasionally tapped lightly (every 25 to 100 hours depending on the atmospheric conditions) to remove

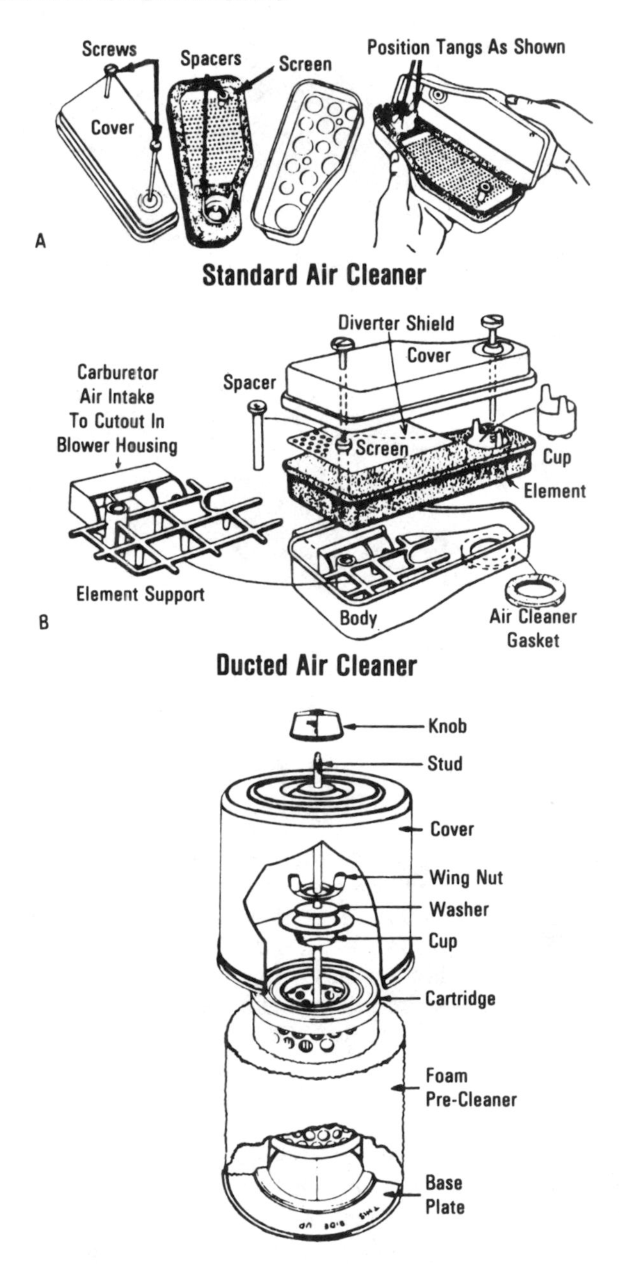

Fig. 7–43. Oil-foam, air-cleaner variations. *(Courtesy B & S)*

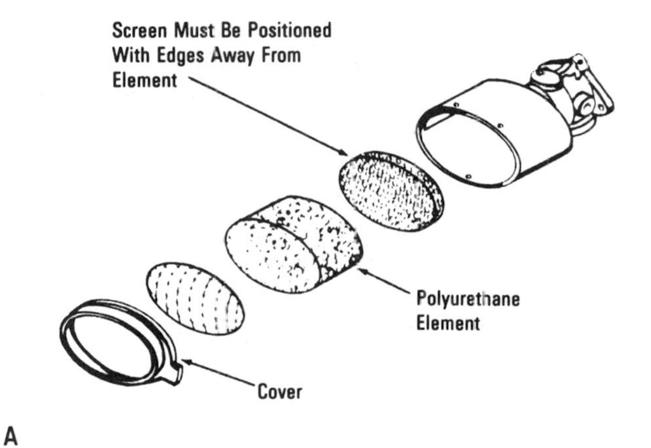

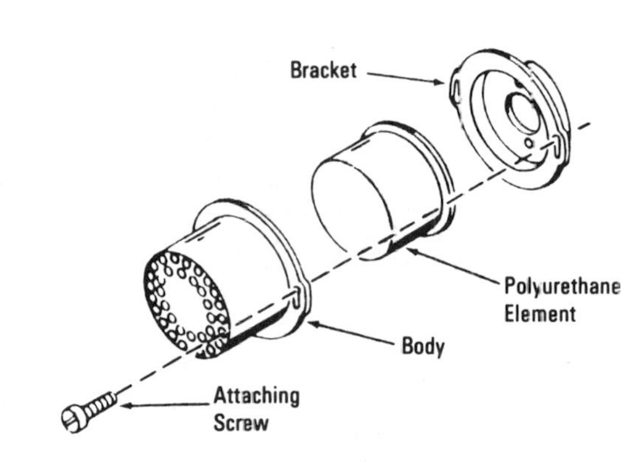

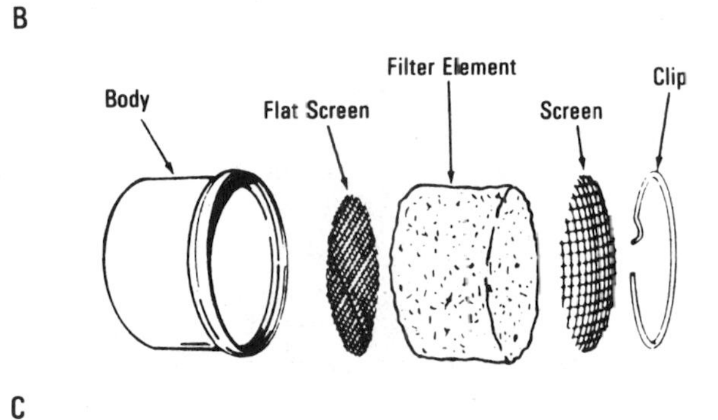

Fig. 7–44. How various polyurethane element shapes fill air cleaners. *(Courtesy Tecumseh)*

the larger particles. When the element becomes very clogged, it may be necessary to wash the element and dry it. In some cases the air hose has to be turned on it from the inside out to remove the particles. Generally speaking, a light tapping will remove most particles. However, follow the manufacturer's recommendations for a thorough cleaning every 100 hours or yearly. Some paper elements cannot tolerate water, so check before scrubbing with soap and water.

Fig. 7–45 shows some of the dry elements used to clean the air for small and larger-size engines. Replacement elements are available from local suppliers of engine parts.

Principles of Carburetion

The basic purpose of a carburetor is to produce the correct ratio of air and fuel for the ignition system to ignite and explode. The explosion produces power when it causes a piston to be moved from its topmost location to its lowest location. The power is harnessed through the piston, connecting rod, crankshaft and power take-off. So, as you can see, the proper ratio of air and fuel is important to the functioning of a gasoline engine.

If the carburetor is to make for a smooth-running engine at various speeds and under varying load conditions, it needs to be designed with these variables in mind. The price or cost of the carburetor in relation to the rest of the engine is also important.

Principles of the Venturi and the Airfoil

In order for a carburetor to operate properly, it must take into consideration the principles of the venturi and the airfoil. Atmospheric pressure must be taken into account in designing a carburetor. Atmospheric pressure, which may vary slightly due to altitude or temperature, is a constant force that tends to equalize itself in any given area (Fig. 7–46).

The carburetor uses the atmospheric pressure to advantage. The carburetor creates its own low-pressure areas and thus obtains movement of either air or intervening fuel. The greater the difference in pressure between the two areas, the greater the distance the fuel can be raised.

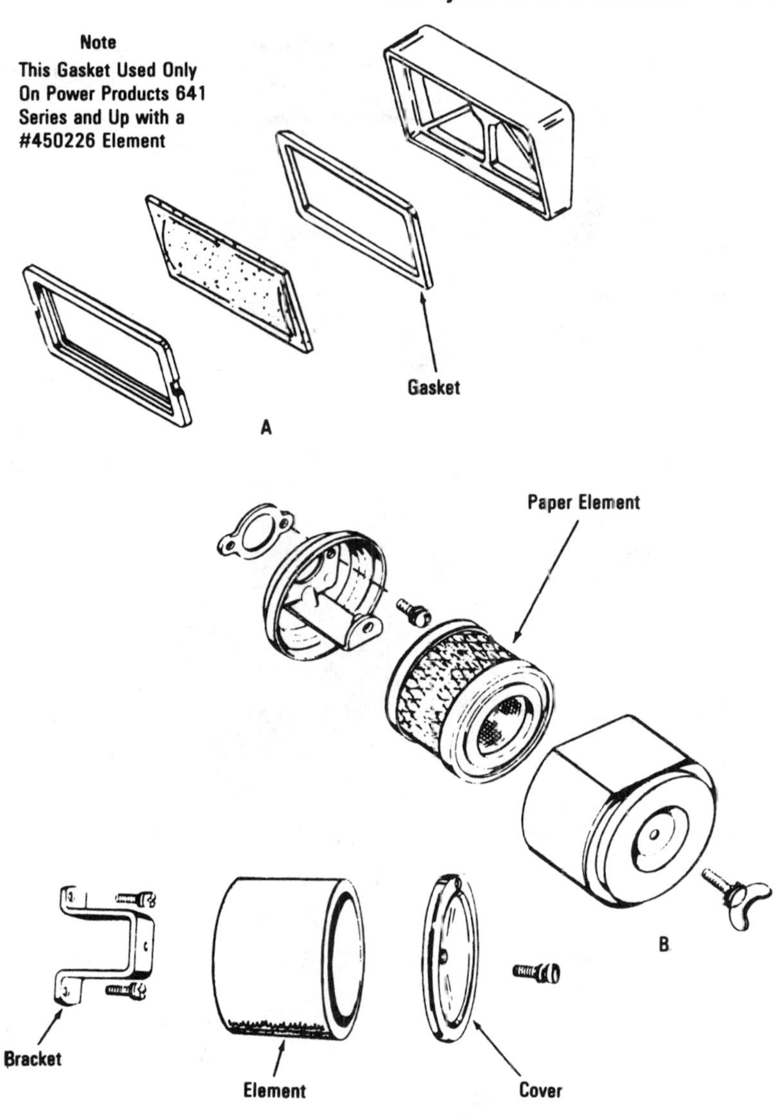

Fig. 7–45. Types of dry air cleaners. (A) Felt-type air cleaner (dry type). (B) Fiber-element air cleaner (dry type). (C) Paper-element air cleaner (dry type). Can be washed, dried, and reused. *(Courtesy Tecumseh)*

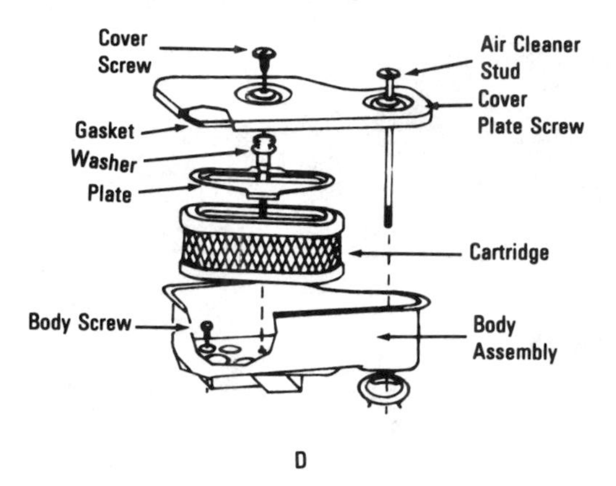

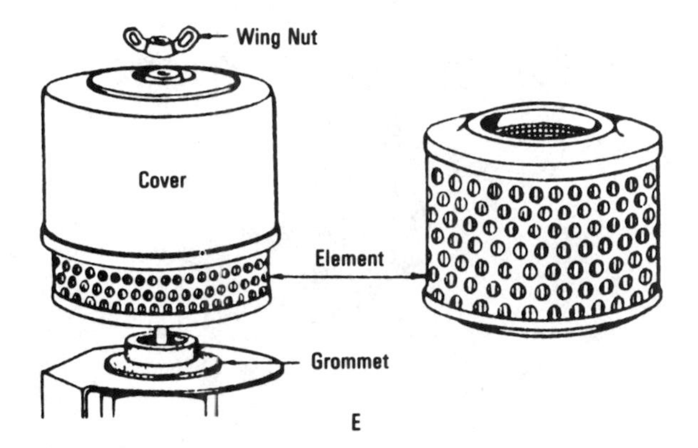

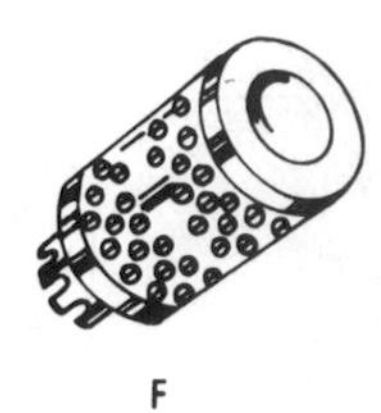

Fig. 7–45. (Cont'd). (D) Cartridge air cleaner (dry type). *(Courtesy B & S)* **(E) Cartridge air cleaner (dry type).** *(Courtesy B & S)* **(F) Cartridge air cleaner (dry type).** *(Courtesy Sears)*

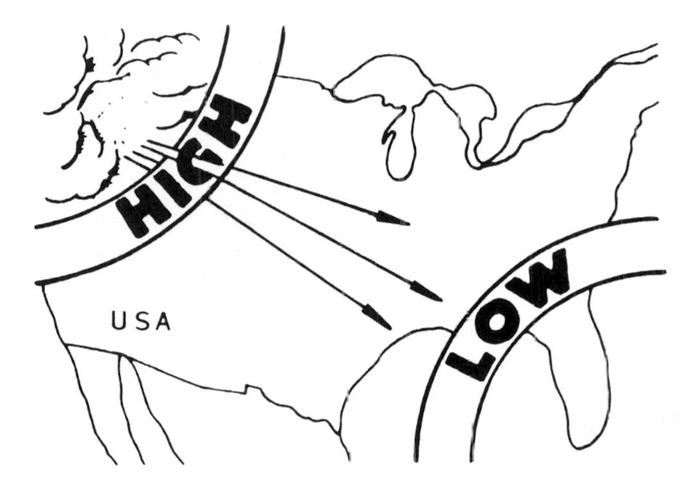

Fig. 7–46. Air moves from high- to low-pressure areas.

In the following discussion the term *vacuum* or *suction* will be used, for convenience, when *difference in pressures* would be more technically accurate.

The Venturi

When air or water is suddenly forced through a constricted area, it has to accelerate to maintain the volume of flow. This can be seen in Fig. 7–47. That means the air is moving faster through a small hole than through a larger hole or pipe, if the same amount of air is to move through the smaller space. This faster movement of air through a smaller space creates what is called the *venturi effect.*

What does all this have to do with carburetion? It is the principle on which carburetion is based. Controlling the flow of air makes a difference when the air and fuel are to be mixed in the proper proportions.

The Airfoil

The airfoil also has to do with the control of air flow. When air is at rest or still, it has the same pressure on all sides of an object—such as

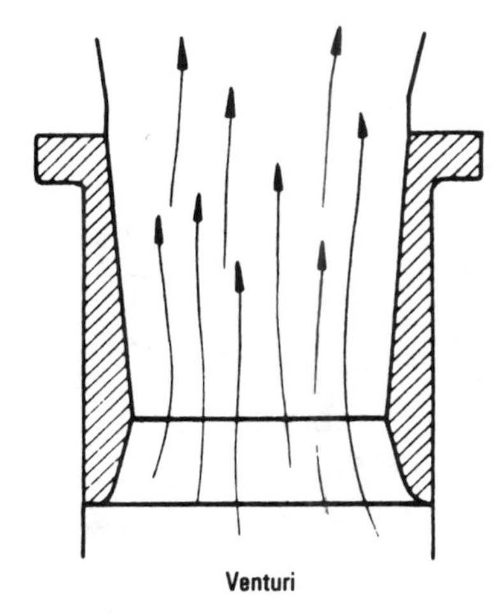

Fig. 7–47. Venturi effect caused by air speeding up to pass through a narrow opening.

a tube within a tube (Fig. 7–48). However, once the air starts to move past the stationary tube, the pressure on the downward side will be lower than that on the upward side (high side).

This phenomenon creates a vacuum at one side of the stationary object, or a low-pressure side. This can be used in the design of a car-

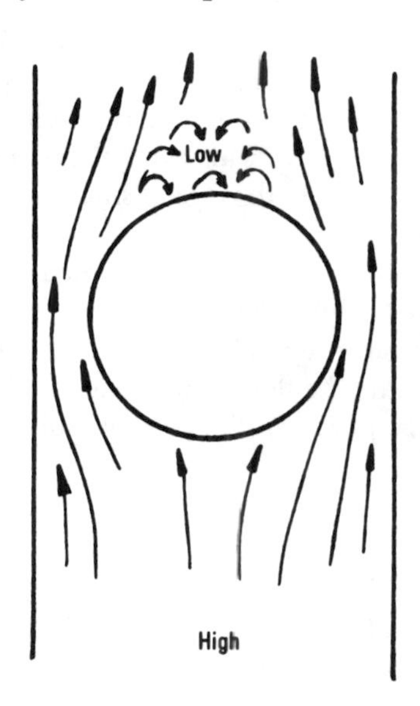

Fig. 7–48. Low-pressure area created by an obstruction or the creation of airfoil *(Courtesy B & S)*

buretor where fuel is sucked into an air stream. Take a look at the different types of carburetors that make use of this principle.

Gravity-Feed Carburetors

In the gravity-feed carburetor, the tank is mounted above the carburetor. The fuel will flow by the force of gravity. An air vent hole is placed in the tank cap so that air can be drawn in as fuel flows out. A vent hole in the carburetor bowl allows air to flow out as fuel flows in (Fig. 7–49).

If either of the holes is plugged, the fuel will cease to flow and the engine will stop running.

As the fuel drops down into the carburetor bowl, it raises the float (Fig. 7–50). The float in turn raises the needle in the float valve. When the needle touches the seat, it shuts off the fuel flow. This is referred to as the level for the float, or float level.

The float level should be high enough to supply enough fuel at full throttle and low enough to prevent flooding or leaking. Flooding also causes the hard starting of an engine.

The level of the gasoline in the carburetor bowl, or the float level, has to be adjusted. This can be done by making sure the float has the same measurements on both sides as shown in Fig. 7–50B. The opera-

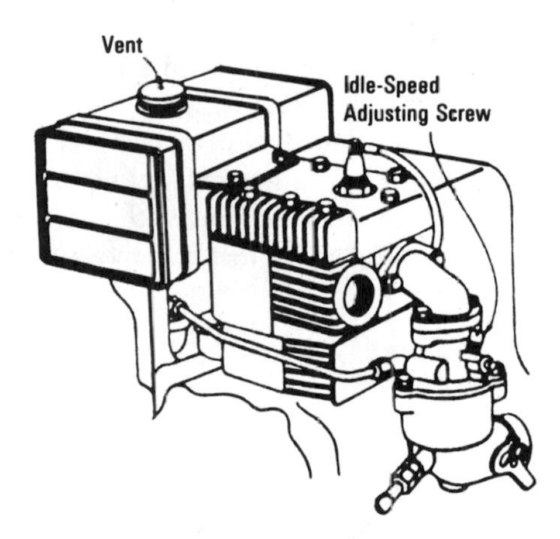

Fig. 7–49. Note the vent in the gasoline tank cap. *(Courtesy B & S)*

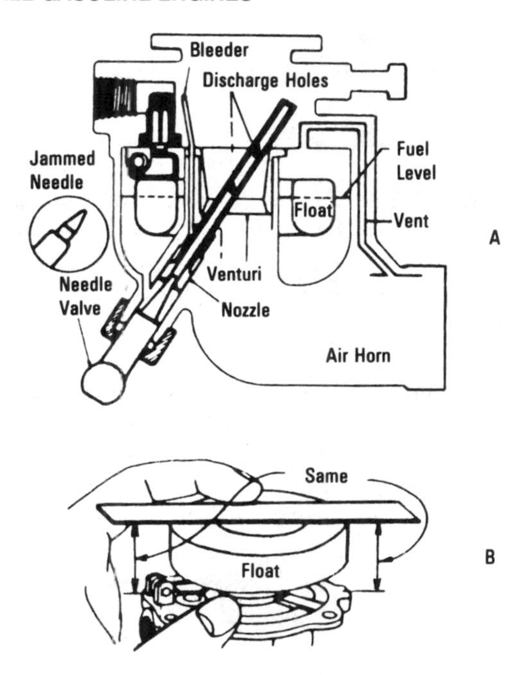

Fig. 7–50. (A) Jammed needle can cause more gasoline to flow than normal, making adjustment very difficult. (B) Adjustment of float level. *(Courtesy B & S)*

tions of the venturi, air horn, and the nozzle in Fig. 7–50A have already been discussed earlier in the chapter.

Keep in mind that the ideal mixture of air and fuel is 15:1. Remember that an engine operating under heavy load requires a richer mixture than under light load. In order to regulate the mixture, a threaded needle is used. It has a tapered point that projects into the end of the nozzle (Fig. 7–50A).

Carburetor Adjustment

To adjust the carburetor for maximum power, run the engine at the desired operating speed. Then turn in the needle valve until the engine slows down. This indicates a lean mixture. As the needle is moved inward it restricts the amount of gasoline allowed to flow past its needle nose. Less fuel means a leaner mixture. Note the position of the needle valve adjustment end. Now adjust the needle valve in the opposite direction (outward). This creates more space between the needle valve point and the seat and allows more fuel to flow past the needle nose. The engine speeds up. If you go too far, it will flood. Then you will notice the engine slow down. Back off on the adjustment until the engine appears to run at its fastest speed. Adjust the engine so that it is now halfway between the fastest speed (with the rich mixture) and the slow speed (with too lean a mixture). This should be about the correct running speed. If the engine is adjusted to the most economical point—lean mixture—it will not necessarily be correct for the engine. Too lean a mixture causes overheating, detonation, and short valve life. In this type of engine there is no accelerator pump. The mixture must therefore be richer so that the engine will not stop when the throttle is suddenly opened. Engines that run at constant speeds (such as generator power sources) can be slightly leaner than those whose use requires changes in speed.

The Throttle

The throttle is a flat disc, called a butterfly. It is mounted on a shaft and placed in the carburetor throat above the venturi (Fig. 7–51).

The throttle does not affect the carburetor air flow when it is wide open. As the throttle is closed, it reduces the flow of air through the carburetor. This, in turn, causes the engine to lose power and speed. At the same time it allows the pressure in the area below the butterfly to increase. This means that the difference between the air pressure in the carburetor bowl and the air pressure in the venturi is decreased.

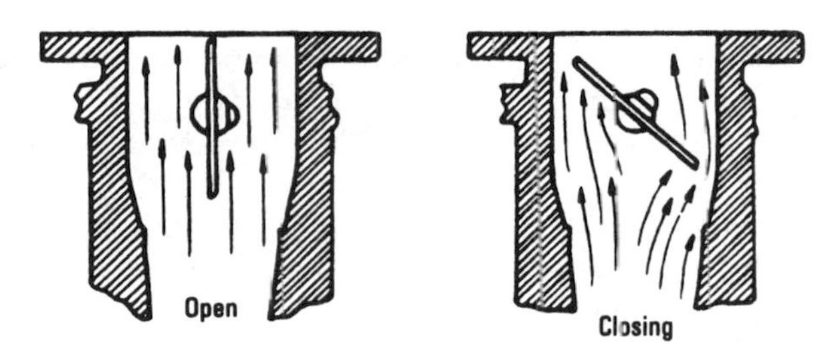

Fig. 7–51. Throttle effect on air flow in a carburetor. *(Courtesy B & S)*

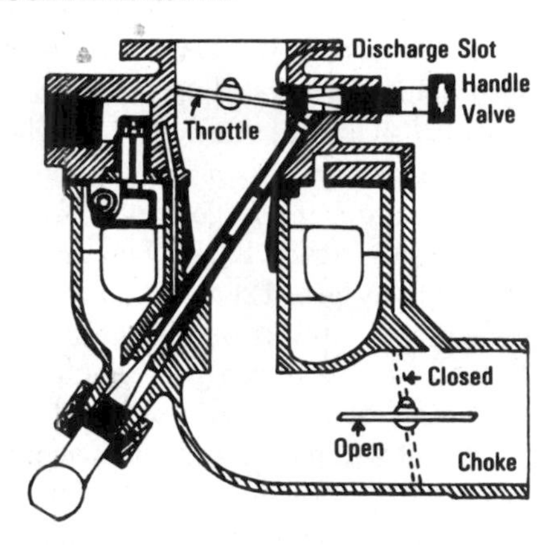

Fig. 7–52. Throttle and choke locations in a carburetor. *(Courtesy B & S)*

Movement of the fuel through the nozzle is slowed down. Thus, the proportion of fuel and air remains approximately the same. As the engine speed slows down to idle, this situation changes (Fig. 7–52).

At idle speed the throttle is practically closed. This means very little air is passing through the venturi. That causes the pressure in the venturi and in the float bowl to be about the same. The fuel is not forced through the discharge holes. That means the mixture tends to become too lean.

To supply fuel for the idle, the nozzle is extended up into the idle valve chamber. It fits snugly in the upper body to prevent leaks. Because of this tight fit, the nozzle must be removed before upper and lower bodies are separated during disassembly or the nozzle will be bent.

The idle valve chamber leads into the carburetor throat above the throttle. Here the pressure is low. This causes the fuel to rise in the nozzle past the idle valve and flow into the carburetor through the discharge slot.

The amount of fuel is metered by turning the idle valve in or out until the proper mixture is obtained. Here again, if the needle is screwed in too far, damage to the valve can result.

Adjustment of the idle valve is similar to that of the needle valve, but it should be made *after* the needle valve has been adjusted. The idle speed is not the slowest speed at which the engine will run. On small engines it is 1750 rpm. On larger engines the idle speed may be as low as 1200 rpm. Use a tachometer to adjust the speed precisely.

The idle-speed adjusting screw is located on the throttle shaft. Turn it until the desired idle speed is obtained. Hold the throttle closed. Turn the idle valve inward until the speed decreases. Then turn it outward until the speed increases and again decreases. Then turn the idle valve to a point midway between the two settings. Usually the idle-speed adjusting screw will have to be reset to the desired idle speed.

Starting the Engine in Different Temperatures

The next problem is starting the engine in different temperatures. Different fuels are also a variable to be taken into consideration in making adjustments. A butterfly is mounted on a shaft and is placed in the air horn. With this *choke* we can close, or almost close, the air horn and get a low-pressure area in the venturi and throat (Fig. 7–52).

This means a rush of fuel is obtained from the nozzle with a relatively small amount of air. Even with low vaporization, this extra-rich mixture will give easy starting. Only a portion of the fuel will be consumed while choking, and a large portion will remain in the cylinder. This raw gasoline will dilute the crankcase oil and may even cause scuffing due to washing away of the oil film from between the engine rings and the cylinder wall. For this reason, prolonged choking should be avoided.

Suction-Feed Carburetors

Suction-feed carburetors have the tank mounted *below* the carburetor. In fact, in most instances the carburetor is mounted on top of the tank. This is one of the more popular types of carburetors for lawnmower engines. Fuel will not flow upward due to gravity. That means some other method must be used to suck or draw up the fuel to the carburetor.

Fig. 7–53 shows the cutaway view of the tank and the location of the fuel pipe. A vent hole is found in the fuel tank cap. This allows the pressure in the tank to remain constant. Before adjusting the carburetor, pour in enough fuel to *half fill* the tank. The distance the fuel has to be lifted affects the adjustment. At half full we have an average op-

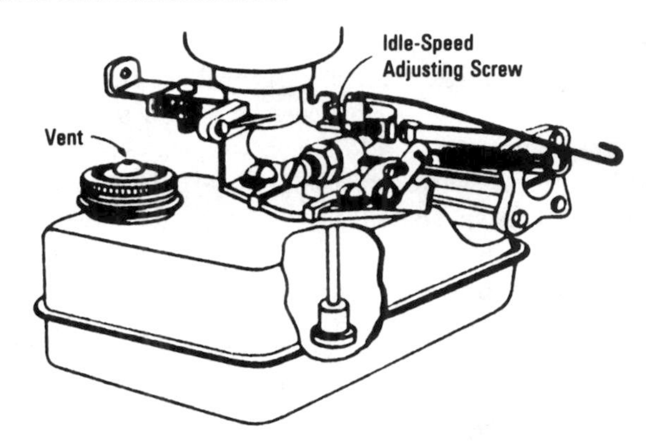

Fig. 7–53. Suction-feed carburetor. *(Courtesy B & S)*

erating condition. The adjustment is then satisfactory if the engine has to run with the tank full or nearly empty.

As the piston goes down in the cylinder (with the intake valve and the throttle open) a low-pressure area is created in the carburetor throat. A slight restriction is placed between the air horn and the carburetor throat at the choke. This helps maintain the low pressure.

Fuel is forced up the fuel pipe by a difference of pressure between the tank and the carburetor throat. The fuel passes through the fuel pipe, past the needle valve, and through two discharge holes.

The throttle is relatively thick. This thickness produces a venturi effect that aids in the vaporization of the fuel. A spiral is placed in the throat to help accelerate the engine. It also prevents its dying when the throttle is opened suddenly.

The amount of fuel at operating speed is metered by the needle valve and seat. Turn the needle valve in or out to change the setting until the proper mixture is obtained. Make the adjustment at operating speed, not at idle.

Operation of the choke and the idle adjustments have already been discussed previously in this chapter.

Pulsa-Jet Carburetors

This type of carburetor has a diaphragm-type fuel pump and a constant-level fuel chamber (Fig. 7–54).

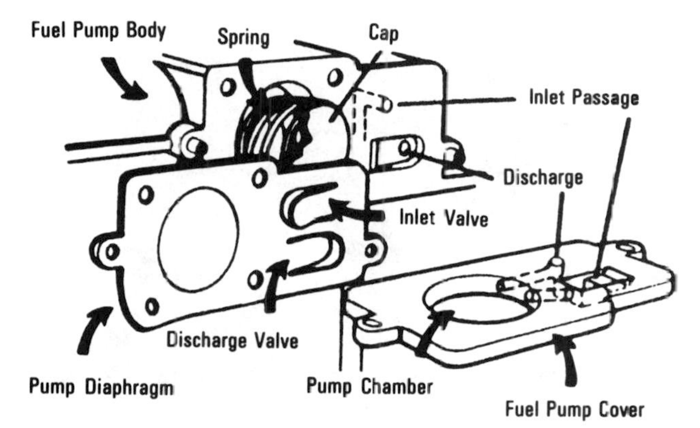

Fig. 7–54. Inlet valve and discharge valve location on a pump diaphragm for a Pulsa-jet carburetor. *(Courtesy B & S)*

The fuel tank, the fuel pump, and the constant-level fuel chamber serve the same functions as the gravity-feed tank, the float, and the float chamber of the conventional float-type carburetor.

The Pulsa-jet type of carburetor allows the proper amount of fuel to be delivered when needed. It can produce the same horsepower as the float-type carburetor. The Pulsa-jet provides a constant fuel level directly below the venturi as shown in Fig. 7–55. With this design very little *lift* is required to draw gasoline into the venturi. The venturi can be made larger. This permits greater volume of fuel-air mixture to flow into the engine. An increase in horsepower is the result.

A vacuum is created in the carburetor elbow by the intake stroke of the piston. Fig. 7–56 shows how this pulls cap *A* and pump diaphragm *B* inward and compresses spring *C*.

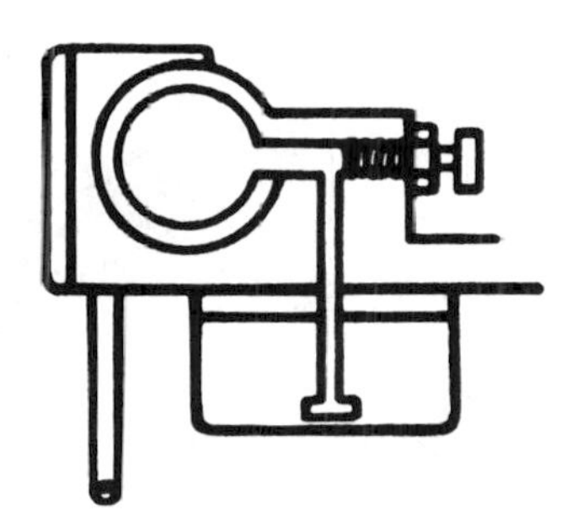

Fig. 7–55. Constant-level fuel chamber on the Pulsa-jet carburetor. *(Courtesy B & S)*

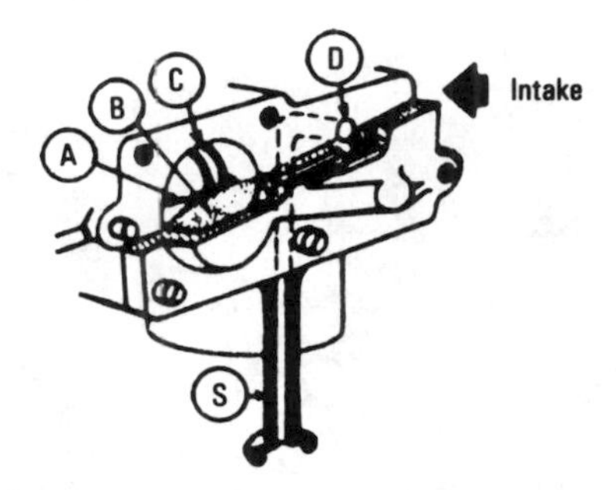

Fig. 7–56. Operation of the diaphragm that causes the pump to lift gasoline to the carburetor. *(Courtesy B & S)*

The vacuum (Fig. 7–56) created on the cover side of the diaphragm pulls gasoline up the suction pipe (*S*) and under intake valve *D*. The gasoline goes into the pocket created when the diaphragm moved upward.

When the engine intake stroke is completed, spring *C* pushes plunger *A* outward. This causes gasoline in the pocket above the diaphragm to move into valve *D*. This opens discharge valve *E*. The fuel is then pumped into fuel cup *F* (Fig. 7–57).

During the engine's next intake stroke the cycle is repeated. This pulsation of the diaphragm keeps the fuel cup full. Excess fuel flows back into the tank.

The venturi of the carburetor (Fig. 7–58) is connected to the intake

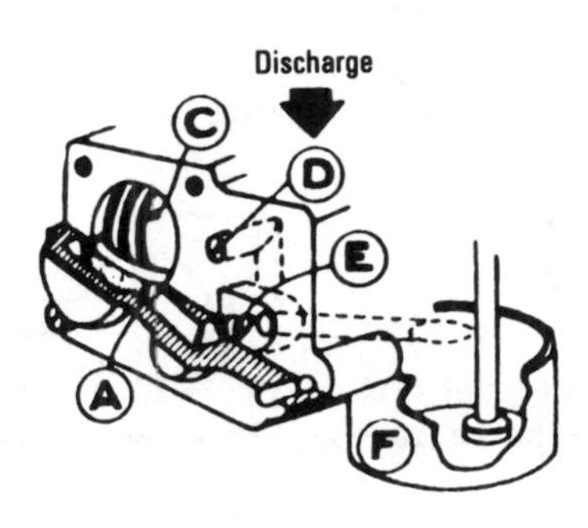

Fig. 7–57. Operation of the diaphragm to cause the pump to discharge fuel during the operation of the fuel cycle. *(Courtesy B & S)*

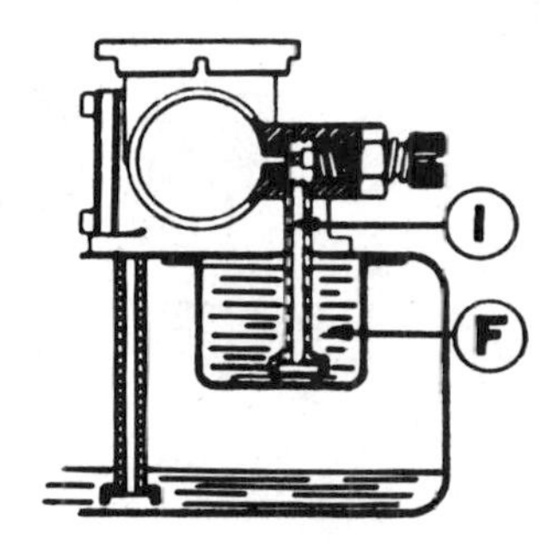

Fig. 7–58. Venturi created by the large pipe opening draws gasoline up through the fuel pipe and into the carburetor. *(Courtesy B & S)*

pipe (*I*), which draws gasoline from the fuel cup (*F*). A constant level is maintained in the fuel cup. The engine gets a constant air-fuel ratio no matter what fuel level exists in the main tank.

From this point, the carburetor operates and is adjusted in the same manner as is the suction-type carburetor. There is an exception, however. The fuel tank does not have to be half full as with the suction-type carburetor. The small cup is always full, so the tank can have any level of fuel during adjustments. There are no valve checks in the fuel lines. The flaps on the diaphragm serve as valves.

Governors

The purpose of a governor on a small gasoline engine is to regulate the speed under varying conditions. With a fixed throttle position, the engine can speed up if the load is lightened. If the load is increased, the engine slows down or even stops.

A governor, on the other hand, closes the throttle if the load is lightened. Or it opens the throttle to obtain more power if the load is increased.

Basically there are two types of governors—the *pneumatic,* or *air vane* (Fig. 7–59), and the *mechanical,* or *flyball weight* (Fig. 7–60).

Pneumatic Governors

The pneumatic governor is operated by the force of the air from the flywheel fins. When the engine is running, air from the flywheel

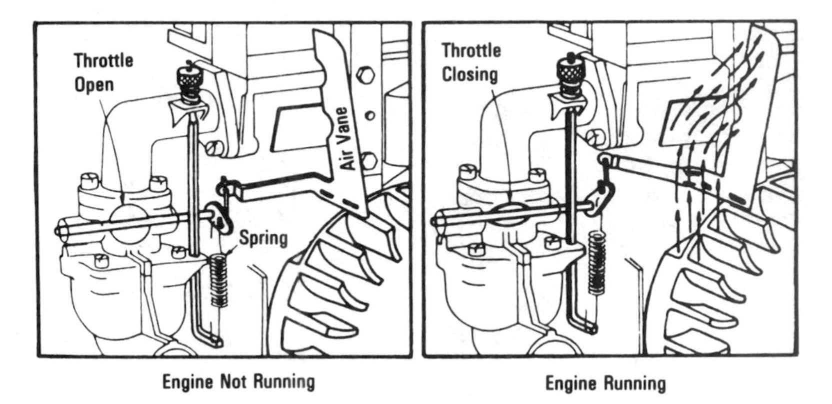

Fig. 7–59. Air vane type of governor for small gasoline engines. *(Courtesy B & S)*

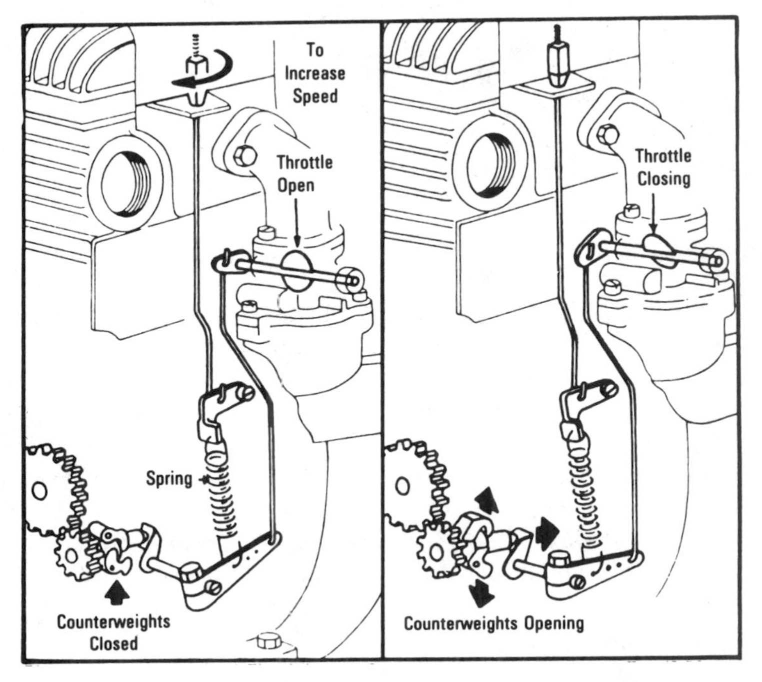

Fig. 7–60. Mechanical governor for small gasoline engines. *(Courtesy B & S)*

fins pushes against the air vane. The air vane is connected to the carburetor throttle by means of a link. The force and movement of these parts tends to close the carburetor and thus slow down the engine speed.

Opposed to this is the governor spring, which tends to pull the opposite way. This opens the throttle. This spring is usually connected to an adjustable control of some kind. This is so that the tension on the spring can be changed at the will of the operator. The engine speed is increased by increasing the tension of the spring. Decreasing the tension lowers the engine speed. The point at which the pull of the spring equals the force of the air vane is called the *governed speed.*

Two-Cycle Pneumatic Governors—The air vane on the two-cycle engine operates as follows. The engine speed increases, the air-flow rate increases. This tends to pivot the air vane against the tension of the governor spring (Fig. 7–61). This movement is transferred to the throttle lever and shaft to partially close the throttle. Two-cycle engines usually have a direct connection between the vane and the lever. As the throttle closes, the fuel flow to the engine is decreased. This means the engine speed decreases. This, in turn, decreases the amount of air that is directed against the air vane. The governor spring becomes controlled and pivots the air vane toward the fan. This movement is transferred to the throttle lever to open the throttle, increasing the fuel flow to the engine. As fuel flow increases, engine speed increases. In this manner engine speed is stabilized. The engine maintains a constant speed regardless of the engine load.

Mechanical Governor

The mechanical governor works in a manner that is similar to the pneumatic governor. However, instead of air blowing against the vane, the centrifugal force of the flyball weights opposes the governing spring (Fig. 7–60).

The operation is the same in both the mechanical and the pneumatic governor. As the load on the engine increases, the engine starts to slow. As soon as this happens, the centrifugal force of the flyball weights lessens. This allows the governor spring to pull the throttle

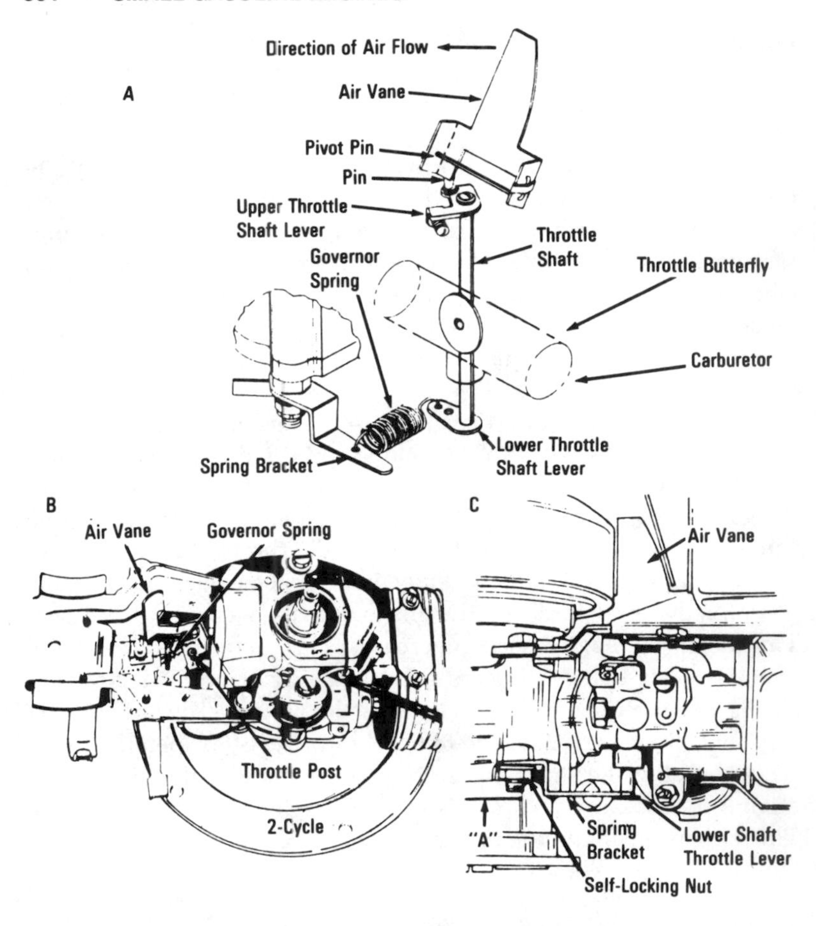

Fig. 7–61. (A) Two-cycle-engine, air-vane governor. (B) Location of the governor on the two-cycle engine. (C) Parts of the air-vane governor that can be adjusted. *(Courtesy Tecumseh)*

open wider, increasing the horsepower to compensate for the increased load. Thus, it maintains the desired governed speed.

If the load on the engine lessens, the engine starts to speed up. This increases the pressure of the centrifugal force and the spring will be stretched a little farther. This closes the throttle and reduces the

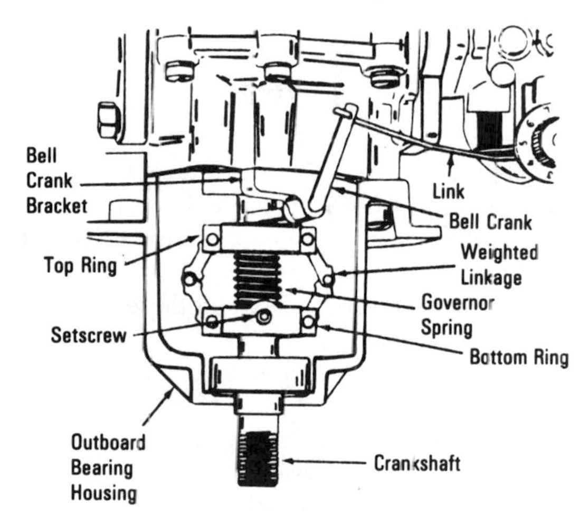

Fig. 7–62. Mechanical governor for a two-cycle engine mounted at the power take-off end of the engine. *(Courtesy Tecumseh)*

engine power. A properly functioning governor maintains this desired governed speed within fairly close limits.

In general, an engine that has good compression, carburetion, and ignition will operate efficiently. However, dirt or neglect can quickly ruin an engine. It should be the duty, therefore, of everyone who operates an engine to become familiar with all aspects of good maintenance procedures.

Two-Cycle Mechanical Governor—Fig. 7–62 shows the arrangement of the two-cycle engine's governor mounted on the power take-off end. The governor assembly is made up of two rings, a weighted linkage, and the governor spring. The lower ring is secured to the crankshaft, so that the governor assembly rotates with the crankshaft. As rotation speed increases, weighted linkages are thrown outward by centrifugal force. The top ring moves toward the bottom ring The governor bell crank contacts the top ring and follows it. This permits the throttle to close, slowing the speed of the engine. The centrifugal force decreases as a result of the slowed engine. This causes the governor spring to move the top ring upward. That, in turn, overcomes the throttle spring to open the throttle. This maintains a constant operating speed.

Chapter Highlights

1. Heat generated inside an internal combustion engine can reach 4000°F (2200°C).
2. Carburetion is the process of mixing fuel and air.
3. Gasoline has hydrocarbons of heptane and iso-octane.
4. Hydrocarbons will leave carbon deposits inside the engine when burned.
5. Perfect combustion can occur only when there is enough oxygen to combine with the carbon.
6. Gasoline can vaporize before it reaches the carburetor. This causes vapor lock.
7. Gummy deposits from fuel can clog the carburetor.
8. The intake manifold takes the gasoline from the carburetor, where it is mixed with air, and moves it to the combustion chamber.
9. There are four types of fuel-feed systems used to move gasoline from the tank to the carburetor.
10. The suction-feed system has the gas tank mounted below the carburetor.
11. In the standard-type vacuum fuel system, the auxiliary tank is located above the carburetor.
12. The forced-feed fuel system uses a fuel pump.
13. Some two-cycle engines have the fuel pump built into the carburetor body.
14. Another type of fuel pump used on two-cycle engines is the impulse type.
15. The eccentric-operated fuel pump is operated by an arm that rides on an eccentric, or cam, lobe.
16. A fuel pump may also be installed on four-cycle engines that normally use gravity feed.
17. The primer pump is used instead of a choke valve to get the engine started.
18. Air cleaners may be either the wet or dry type.
19. The carburetor uses atmospheric pressure to advantage.
20. The venturi effect means the air moves faster through a smaller space than a larger one.
21. The airfoil has to do with air at rest and the air in motion and the resulting pressure differences.

22. The float is used in gravity-feed carburetors to make sure the engine doesn't flood.
23. The ideal mixture of air and fuel is 15:1.
24. The butterfly is a flat disc located on a shaft in the carburetor throat above the venturi.
25. The choke can close, or almost close, the air horn in the carburetor to get a low-pressure area in the venturi and throat.
26. The gas tank cap has a hole to allow for atmospheric pressure to operate the fuel system.
27. The Pulsa-jet carburetor has a diaphragm-type fuel pump and a constant-level fuel chamber.
28. The purpose of a governor on a small gasoline engine is to regulate the speed under varying load conditions.
29. The pneumatic governor is operated by the force of the air from the flywheel fins.
30. Leaded gas or gasohol with no more than 10% alcohol can be used in most engines. Later model engines call for clean unleaded regular gasoline.
31. Drain the gas tank if the engine is stored for more than 30 days.

Vocabulary Checklist

air filter
benzine
butterfly
choke
combustion
ethanol
evaporating
forced-feed system
gasohol
gravity system
heptane
hydrocarbon
idle valve
impulse fuel pump
intake manifold
iso-octane
kerosene
methanol
naphtha
nozzle
paraffin
petroleum
pneumatic governor
priming pump
suction-feed system
tachometer
vacuum system
vapor lock
venturi effect
volatile

32. All gasoline from the carburetor and gas tank should be removed for storage.

Review Questions

1. What are the two types of combustion?
2. How hot does the inside of an internal combustion engine get?
3. What is a hydrocarbon?
4. Why should gasoline be vaporized to make it work?
5. What are heptane and iso-octane?
6. Why are tetraethyl lead and ethylene bromide added to gasolines?
7. What is the difference between regular and unleaded gasoline?
8. What is vapor lock?
9. When does vapor lock occur?
10. Why don't you want gasoline to stand over a month in a container?
11. What are the basic parts of a fuel system?
12. What does a carburetor do?
13. What does an air filter do?
14. What is an intake manifold?
15. What are the four types of systems used to move gasoline from the tank to the carburetor?
16. How do you set the carburetor level?
17. Where is the gas tank located in reference to the carburetor in a suction-feed system?
18. What is an auxiliary tank? Where is it used?
19. What is an impulse fuel pump?
20. How does an eccentric-operated fuel pump work?
21. What is the purpose of a primer pump?
22. Where are wet air filters used?
23. Where are dry air filters used?
24. What is the venturi effect?
25. What is a butterfly? Where is it located?
26. What is the purpose of a choke?
27. What is a needle valve?
28. Explain how the Pulsa-jet carburetor works.
29. What is a pneumatic governor?
30. How does the mechanical governor work?
31. What is an air vane? How does it operate?

32. Why should you drain the gasoline tank if the mower is stored for more than 30 days?
33. What is the procedure used to store an engine?
34. Why isn't methanol recommended for use as a fuel?
35. What is gasohol?

CHAPTER 8

Exhaust and Cooling Systems

In most cases, the small gasoline engine uses an air-cooled system for dissipating heat generated by combustion. In order to get rid of this excess heat, fins are necessary to provide greater surface area for contact with air. Air is used to move the heat from the surface of the engine to a location away from the heat generator.

In some engines a water-cooled system is used, for which a radiator, a fan, and, in most instances, antifreeze are necessary. This cooling system is completely different from that of the air-cooled engine. Both types will be discussed here in relation to the small gasoline engine.

The exhaust system is rather simple and effective in a small gasoline engine. It consists of a pipe nipple and a muffler in the majority of designs. In some cases there is an exhaust manifold. The basic operational parts of the exhaust systems used in small engines as well as the parts of the cooling system will be discussed in this chapter.

Principles of Operation of the Water-Cooled System

Most automobile engines use a water cooling system (Fig. 8–1). The water-cooled system is also used on small gasoline engines used as outboard motors. This is primarily because using water is very convenient (Fig. 8–2).

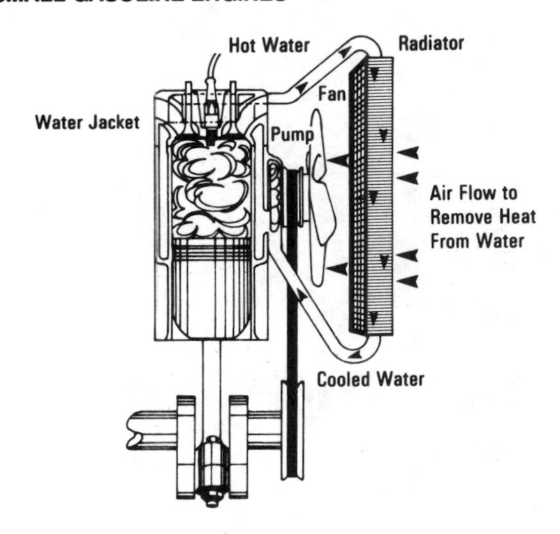

Fig. 8–1. Closed water-cooled system.

Fig. 8–2. Two-cylinder, two-cycle outboard motors. *(Courtesy Chrysler)*

Water cooling is more efficient than air cooling primarily because water is a better conductor of heat than is air. Water cooling uses both convection and conduction to accomplish the job of keeping an engine running at the proper temperature. However, there are some disadvantages to this type of system. There is the necessity for a radiator, a water pump, hoses, and antifreeze. The possibility of problems with rust, freezing, and clogging of the radiator cores is always present. Maintenance takes more time than with an air-cooled system.

One of the advantages of an outboard motor is the lack of a radiator. The water is taken in and exhausted as it is used (Fig. 8–3). In this type of engine the water pump becomes the main concern of the cooling system.

Thermostatically controlled cooling systems increase engine efficiency. Engines operate best at design temperatures. These have been selected by the manufacturer and are best known by the temperature of the thermostat (140°F to 145°F) installed in the engine. This means

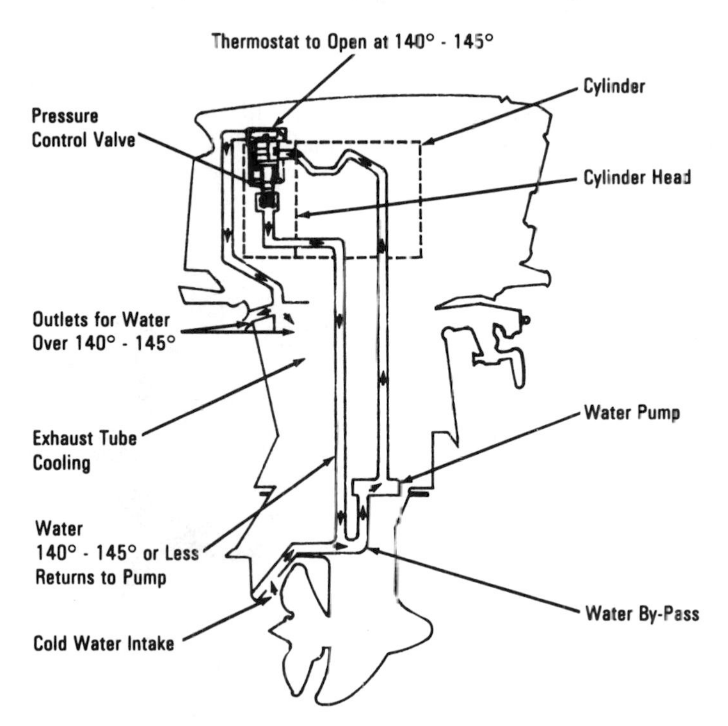

Fig. 8–3. Water-cooling route for an outboard motor.

that the proper temperature is maintained no matter what the speed of the engine.

The thermostat is closed when the engine is started. The pressure-control valve bypasses the water back to the water pump to overcome the pressure-control-valve setting. This continues until the engine becomes hot enough to bring the temperature of the water up to the opening point of the thermostat. Fig. 8–3 shows the opening point to be 140°F to 145°F (60°C to 63°C).

Once the thermostat opens, the hot water is passed to the atmosphere, where it cools slightly before falling back to the river, lake, or source.

Fig. 8–4 shows the water-pump impeller made of synthetic rubber, which acts as a water circulator at high speed and as a displacement pump at low speed.

Principles of Operation of the Air-Cooled System

Most small gasoline engines rely on air for cooling. They have been mounted on large pieces of metal that will conduct the heat to the surrounding air, or they may have fins that expose more surface area to

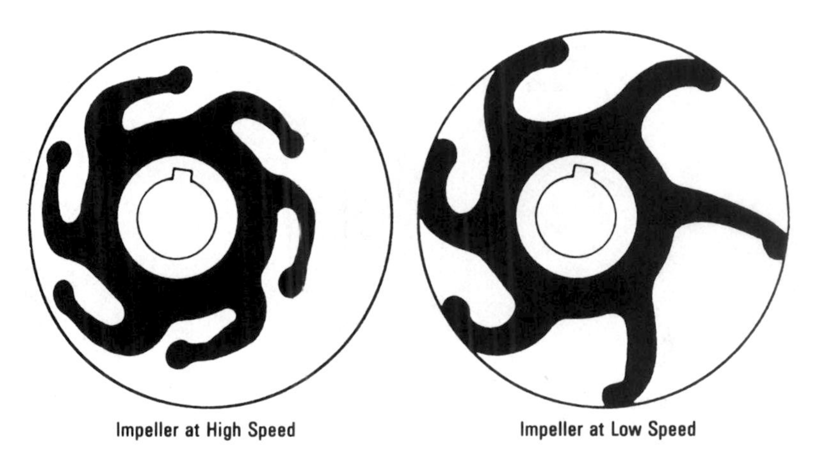

Fig. 8–4. Water-pump impellers at low and high speed.

the air. Air is not a very good heat conductor. However, designers of the small engines have provided for the elimination of most of the heat generated by combustion. There are two types of air-cooled engines. One is the *open-draft type,* which has a fan mounted on the end of the crankshaft. The fan pulls air over the exposed engine parts. This provides cooling. Most engines rely on a slight breeze to move the heat away from the engine. This breeze may be created by the movement of the engine as on a lawnmower or by a fan built into an engine.

In this exposed, or open-draft, type of engine, the fins are very effective in removing heat from the combustion chamber to the air (Fig. 8–5). Aluminum is used in most small engines as it is very effective in dissipating heat.

The second type of air-cooled engine has a shroud and flywheel fan and is the *forced-draft type* (Fig. 8–6). The only difference between this type of cooling and the exposed type is the use of the shroud and fan. The shroud channels the air to where it is needed for cooling purposes. The fan generates the breeze or air movement needed for cooling.

Fig. 8–7 shows how the shroud is in place and the air is drawn in

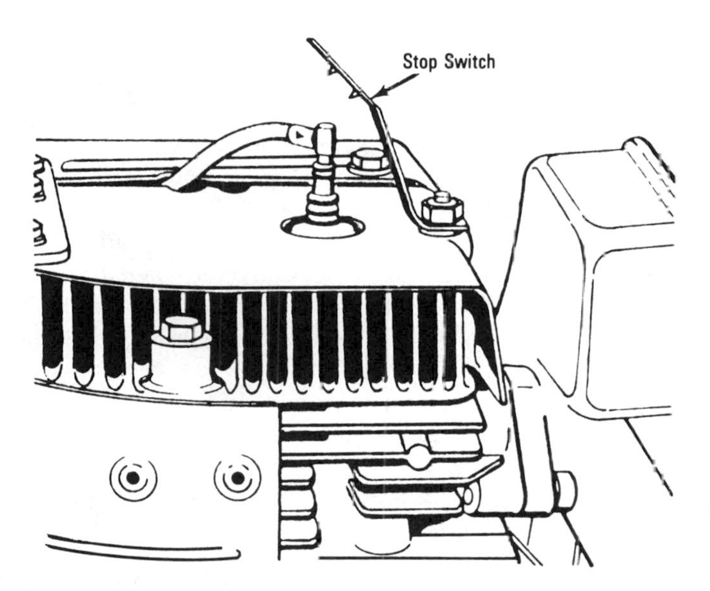

Fig. 8–5. Cooling fins for an air-cooled engine. *(Courtesy B & S)*

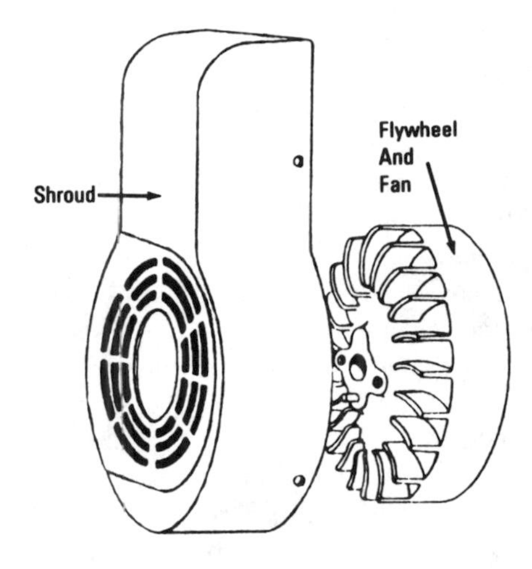

Fig. 8–6. Flywheel fan and shroud for a small gas engine.

at the top of this vertical crankshaft engine and exhausted downward through the fins.

Cooling System Maintenance

In order to keep the engine from overheating, it is necessary to see to it that the fins are not blocked. This happens occasionally when the small engine is used as a lawnmower. Grass and clippings get jammed in the fins located under the shroud.

In Fig. 8–8 you can see where grass particles, chaff, or dirt may clog the air-cooling system. This is especially true when the engine has been in prolonged service cutting dry grasses. Regular maintenance of the engine includes removing the shroud and cleaning the grass and dirt particles from the fins.

The restriction of the air flow over the fins can result in an overheated engine with some severe consequences.

The Exhaust System

There are two parts to an exhaust system: the exhaust manifold, which collects the exhaust and channels it to the muffler, and the muffler itself.

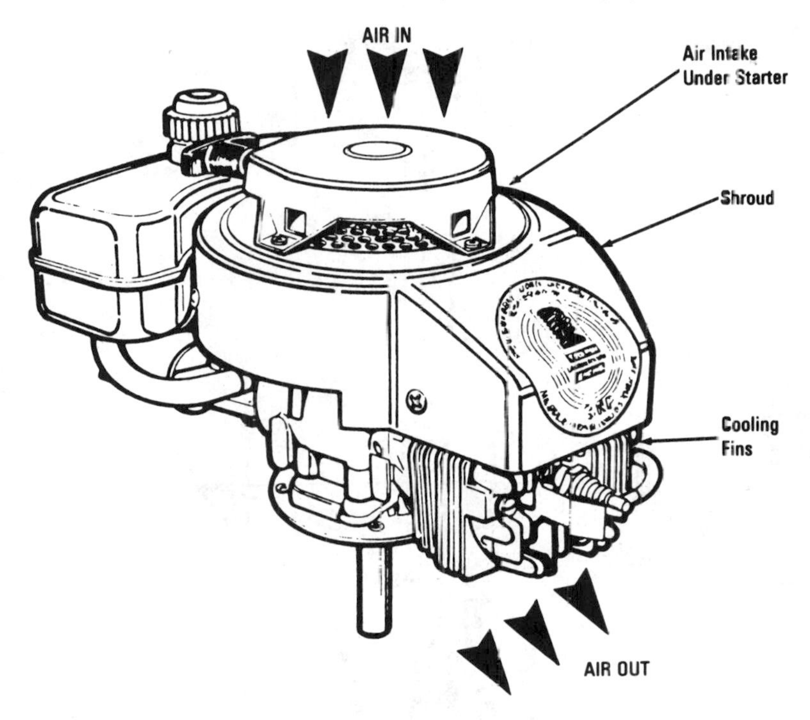

Fig. 8–7. Air intake and exhaust for an air-cooled engine. *(Courtesy Tecumseh)*

Exhaust Manifolds

The basic purpose of the exhaust manifold is to collect the exhaust gases. This is not necessary on most small engines since the heat from combustion is not used for any other purpose and it is exhausted to the surrounding air. As can be seen from Fig. 8–9, the exhaust manifold does not exist in the small gasoline engine. There is a threaded hole in the engine casting for a pipe nipple to be screwed on. A muffler fits onto the other end of the pipe nipple for a complete installation.

Multicylinder engines usually have an exhaust manifold. In automobile engines, the manifold heat is used to increase the temperature of the intake mixture so that the gasoline will be more vaporized for more efficient combustion.

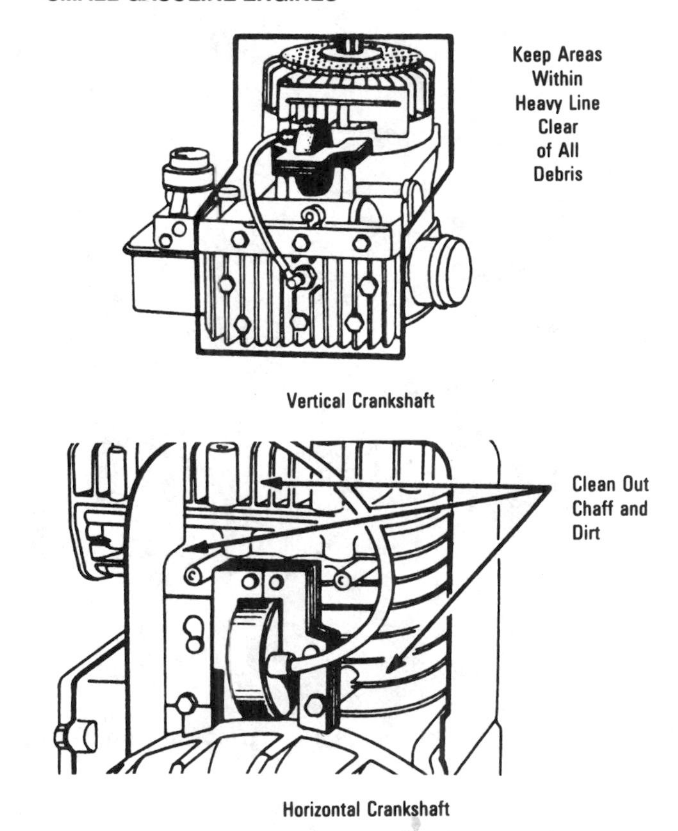

Fig. 8–8. Maintenance of the cooling fins is important in keeping air flow to its greatest advantage. *(Courtesy B & S)*

Mufflers

Gasoline engines make a loud noise when operating. They are so noisy that some type of muffling of the sound is necessary. The noise is created by the expansion of the hot gases, by the noise of the engine transmitted through the exhaust, and by the gases striking cooler air on leaving the engine. This is the thunderclap phenomenon.

Mufflers are devices used to muffle sound. This can be accomplished in a number of ways. The muffler is designed to operate with a specific type of engine. It absorbs the noise generated by the engine

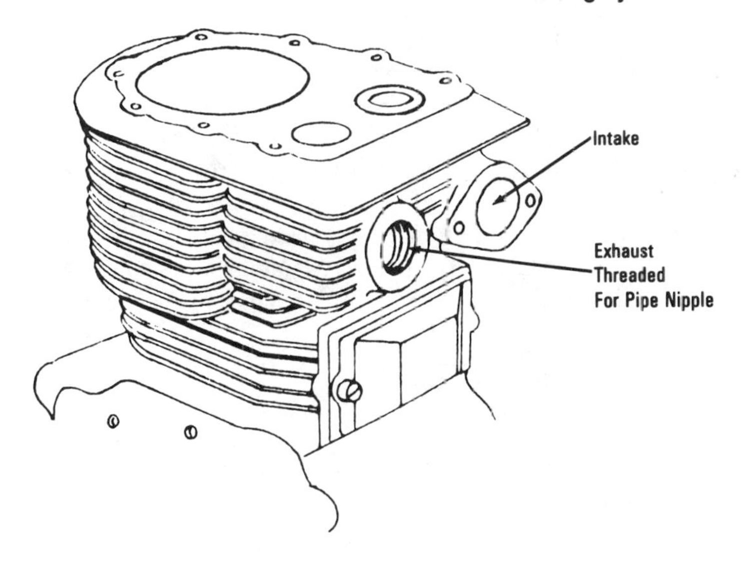

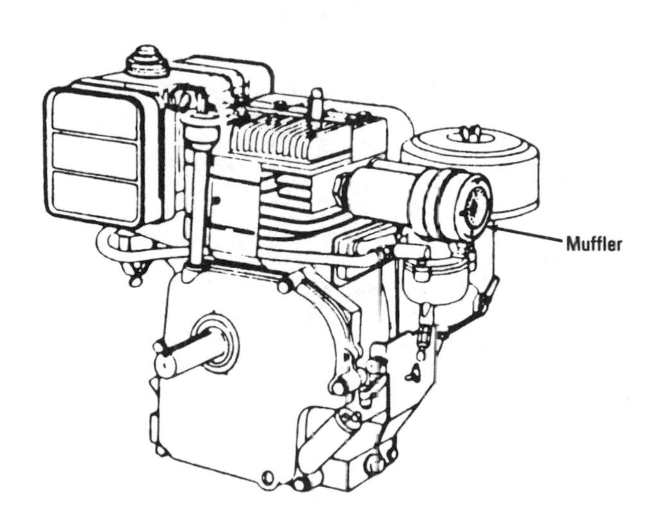

Fig. 8–9. Location of the exhaust system on a small gas engine. *(Courtesy B & S)*

and the expanding gases and delays the hot gases slightly so they hit the cooler outside air in a steady stream instead of immediately. This has a tendency to lessen the thunderclap effect.

Mufflers have to be designed so that they present a minimum of back pressure on the engine. Excess back pressure can affect efficiency.

Mufflers are designed for engines according to where the engine is to be used. For instance, if you use an engine for a lawnmower, you want it to be quieter than if you used it for a pump placed near a well

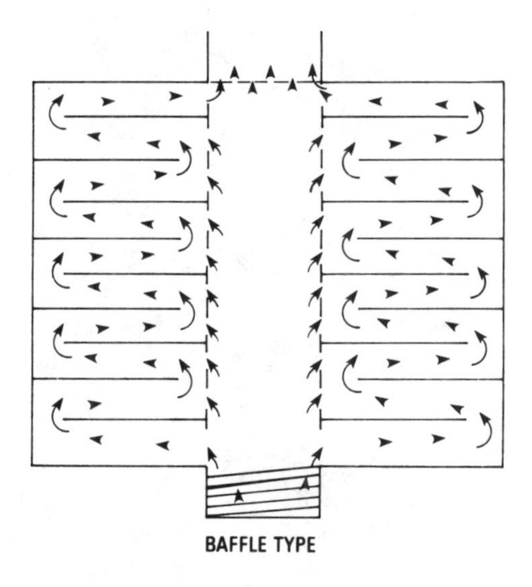

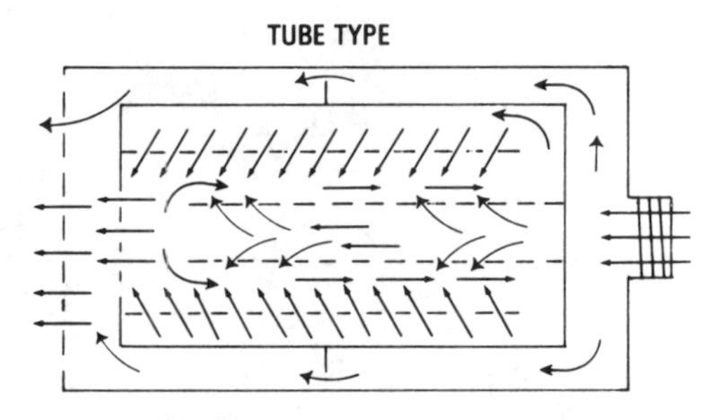

Fig. 8–10. Two types of small-engine mufflers.

and away from the house. In outboard motors, the exhaust is usually routed underwater. The water has its own muffling effect.

Fig. 8–10 shows two types of mufflers. The *baffle type* allows the gases to expand and cool before being ejected into the atmosphere. The *tube type* is the more modern type, though both types are still used. The tubes inside the tube type have holes to allow for the expansion of the gases. The delayed gases can escape when the exhaust stroke of the engine is over. That means the pressure decreases and the straight-through path is not utilized. Baffling the noise can be very important in the operation of a small engine since as much as 20 percent of the engine's power can be consumed by the muffler. In some cases where the wrong muffler is used or the muffler becomes clogged, it can consume as much as 50 percent of the engine's power.

A two-stroke engine generates more noise than the four-stroke-cycle engine. That means there has to be a different muffler for each type of engine. Fig. 8–11 shows some of the mufflers available for the small gasoline engine. Those in the top row are for the bolt-on types. Two bolts hold the muffler to the engine. Those in the bottom row of Fig. 8–11 show the mufflers that screw into an engine and the pipe nipple. These mufflers are available for replacement use.

Exhaust deflectors are important on some engines. If equipment will be affected by the exhaust gases, it is best to deflect exhaust. Fig.

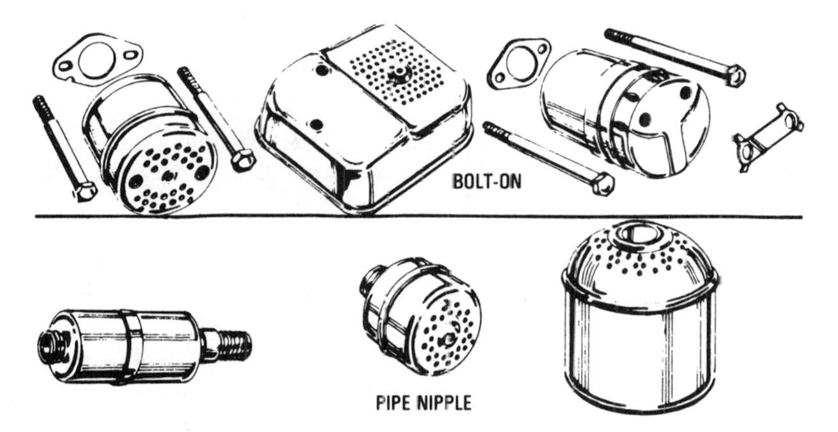

Fig. 8–11. Bolt-on and screw-on mufflers for small gas engines. *(Courtesy Frederick Manufacturing)*

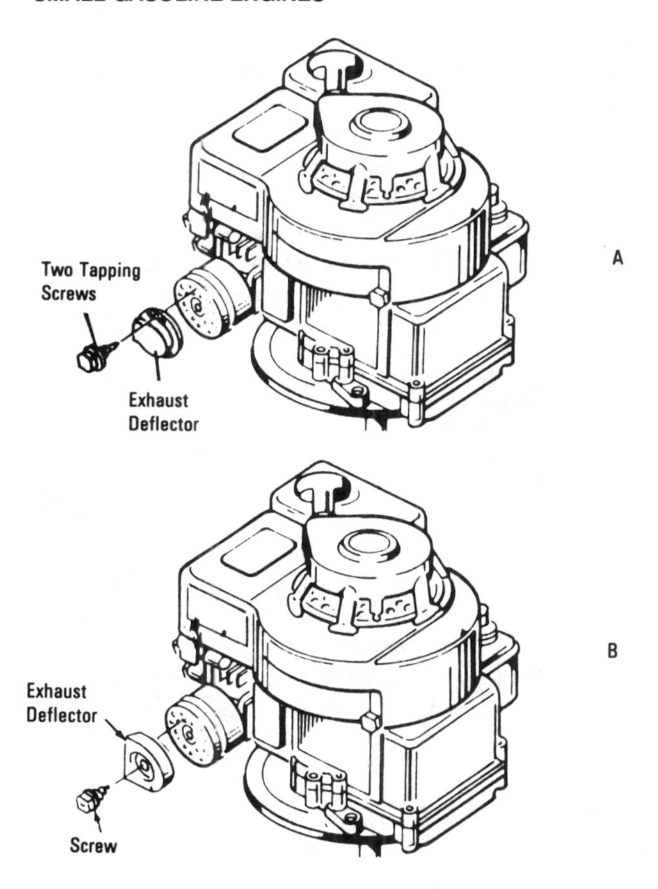

Fig. 8–12. (A) Two tapping screws are used to hold an exhaust deflector onto a muffler. (B) A single screw is used to attach an exhaust deflector onto a small-engine muffler.

8–12 shows several ways a small piece of metal is used to serve as a deflector. It may use a single screw for attachment, or it may be held in place with several sheet-metal screws.

Various shapes of mufflers are available for small gasoline engines. Honda makes small engines for use on lawnmowers. They have a different-shaped muffler, but they do the same job as the basic type shown here. They allow for the expansion of the hot gases before they reach the atmosphere to create a thunderclap.

Spark arrestors are placed on some small gasoline engines so that they will not start fires if they are used in the woods or near dry grass or leaves. The aim is to prevent the emission of small red-hot pieces of carbon that can become a source of fire.

Chapter Highlights

1. Air is used to move heat from the surface of the engine to a location away from the heat source.
2. In the water-cooled engine, the cooling system is much more complicated than in the air-cooled system.
3. The exhaust system in a small gasoline engine is fairly simple.
4. The water-cooled system is used to cool outboard motors.
5. A water-cooled system requires a water pump.
6. A thermostat is needed for a water-cooled system.
7. A radiator cools the water in a closed system.
8. The open-draft system uses a fan to cool the engine.
9. The forced-draft system uses a shroud and flywheel fan to cause forced draft around the engine fins.
10. Fins must not be blocked if air cooling is to work properly.
11. The exhaust manifold and the muffler make up the exhaust system.
12. Mufflers eliminate the thunderclap effect created by an internal combustion engine.
13. Mufflers are designed for engines according to where the engine will be used.
14. Spark suppressors are used to make sure carbon sparks are not emitted from a running engine when it is used near a dry area susceptible to fire.
15. The two-stroke engine makes more noise than a four-stroke engine.

Review Questions

1. Water-cooling systems on small or large engines use two methods of dissipating heat. What are they?

Vocabulary Checklist

antifreeze
closed system
conduction
convection
displacement pump
exhaust deflector
exhaust manifold
forced-draft air-cooled engine
maintenance
open-draft air-cooled engine
pipe nipple
radiator
spark arrestor
thermostatically controlled
thunderclap phenomenon

2. Which type of system, air-cooled or water-cooled, takes the greatest amount of maintenance?
3. What is a thermostat? Is one used on air-cooled engines?
4. What is the closed system of cooling?
5. List the two types of air-cooled engines.
6. What type of engine cooling uses a shroud and flywheel fan?
7. How do air-cooled systems become inoperative?
8. What is an exhaust manifold?
9. Why is a muffler needed for small gasoline engines?
10. What are the two types of mufflers?
11. What is an exhaust deflector?
12. What are spark arrestors? Why are they needed?

CHAPTER 9

Lubrication Systems

In order to reduce friction and improve engine efficiency, some type of lubrication has to be used. This lubrication may have a number of designations. It has a viscosity rating and a rating for the cleaning factor.

In this chapter we will take a look at how a lubricant is selected, classified, and used. There are a number of ways to lubricate an internal combustion engine. These methods will be discussed and analyzed.

Classification of Lubricants

Lubricants are needed to reduce friction. Two surfaces moving over one another create friction that can decrease the efficiency of a machine and generate heat. A lubricant must be able to reduce the friction and conduct heat to another location to be dissipated.

Anything that reduces friction can be called a lubricant. The tallow of a candle has been used. The grease from any number of animals was used to lubricate the wheels of covered wagons. Even ice, which is slippery, can be used as a lubricant in some cases. However, in today's high-technology world the job of reducing friction has been closely studied, and some definite rules and standards have been established. Standards make it easier to order to obtain the proper lubrication for mechanical devices.

One of the important factors to be considered in an engine lubricant is the temperature. A lubricant for use in an engine must be able to take the temperatures without breaking down or failing in its job.

Some materials have good lubricating qualities and others are very poor in reducing friction. For a comparison of the various materials that can be used as lubricants, take a look at Table 9–1. Here you will find some of the common materials that can be used for various lubricating jobs. You may not be able to use some of them, however, because of their very poor lubricating qualities.

Oils are measured in *viscosity,* which is the internal friction between molecules of a liquid or gas. The friction developed in a bearing in which the rubbing surfaces are separated by a fluid film (oil) depends directly on the viscosity of the fluid. Viscosity also impedes the flow of oil through the bearing. You may be able to obtain #30 or #40 oil. The #40 oil is thicker than the #30. Some multiple-viscosity oils are also available for engines. They are labeled 10W-30 or 10W-40. That means they have a viscosity of 10 for low temperatures and a viscosity of 40 for high temperatures. With these oils you don't have to change the oil in an engine for the winter and summer months, for the oil will adjust to the temperature.

The viscosity rating indicates the ability of the oil to flow at various temperatures. Notice in Table 9–2 that the Society of Automotive Engineers (SAE) has labeled the viscosity of their oils with numbers running from SAE 10 to 60. Some special oils are now available with a viscosity rating of SAE 5. All oils flow at higher temperatures, but the ability to flow at low temperatures is extremely important in some climates or locations. The SAE number also indicates how easily the oil will pour from a can. The lower the number, the thinner the oil and the more easily it will pour. There are other factors used to classify or rate lubricants. Engineers rate the lubricants as to their viscosity index, flash and fire points, carbon residue, sulfur content, neutralization number, color, and saponification number. The only ones you need to worry about with a small gasoline engine are the SAE numbers ranging from 5 through 60. The manufacturer will indicate which is acceptable for use in its engine.

Saponification indicates the foaming action of the oil. The less foaming the better, since the foam can cause lack of flow and reduce the level of the lubricant below required limits. Saponification is defined as the act or process of conversion into soap, or the hydrolysis of

Table 9–1. Lubricating Qualities and Viscosities

Material	Viscosity at 70°F in Centipoise
Honey	1500
SAE 50 oil	800
Glycerin	500
SAE 30 oil	300
SAE 10 oil	70
Ethylene glycol	20
Kerosene	2
Water	1
Gasoline	0.4
Air	0.018
Hydrogen	0.009

any ester into the corresponding alcohol and acid. Lubricants usually have waxy substances added to resist saponification.

If a manufacturer specifies a specific grease or oil, you should make every effort to obtain that product. The manufacturers have worked with the engine under various conditions of operation to determine the exact qualities of lubricant needed for the long life of the engine. In most cases they will specify a particular grade of oil or grease. You are then left to choose the brand you wish to use.

Table 9–2. Recommended SAE Viscosity Grades

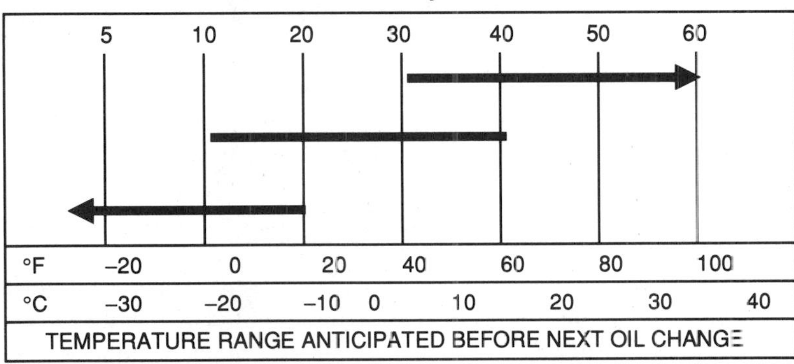

(Courtesy B & S)

Selection of Lubricants

Since the manufacturers' recommendations are to be followed, we will start with those of Briggs & Stratton engines, since they represent the largest number in operation. Any high-quality detergent oil having the American Petroleum Institute classification "For Service SC, SD, SE or MS" can be used in Briggs & Stratton engines. Detergent oils keep the engine cleaner and retard the formation of gum and varnish deposits (Table 9–2).

Summer	*Winter*
Above 40°F (4.4°C), use SAE 30. If not available, use SAE 10W-30 or SAE 10W-40	Under 40°F (4.4°), use SAE 5W-20 or SAE 5W-30. If not available, use 10W or SAE 10W-30. Below 0°F (−17.8°C), dilute 10 percent with kerosene.

The oil recommendations are the result of extensive testing. No special additives should be used.

Oil changes should occur after each twenty-five hours of engine operation. Changes should occur more often under dirty operating conditions. In the normal running of any engine, small particles of metal from the cylinder walls, pistons, and bearings will gradually work into the oil. Dust particles from the air also get into the oil. If the oil is not changed regularly, these foreign particles cause increased friction and a grinding action, which shorten the life of the engine. Fresh oil also assists in cooling, for old oil gradually becomes thick and loses its effectiveness as well as its lubricating qualities.

Table 9–3 shows the capacities for the various sizes of Briggs & Stratton engines.

Two-Cycle Engines

Two-stroke-cycle engines require a mixture of oil and gasoline. The ratio of oil to gasoline is extremely important. The manufacturer specifies this ratio in writing somewhere on the engine in plain view. A typical two-cycle gas/oil mixing chart is shown in Fig. 9–1. The manu-

Table 9–3. Capacity Chart

Basic Model Series	Capacity	
Aluminum	**Pints**	**Liters**
6, 8, 9, 11 cu. in. Vert. Crankshaft	1¼	0.6
6, 8, 9 cu. in. Horiz. Crankshaft	1¼	0.6
10, 13 cu. in. Vert. Crankshaft	1¼	0.8
10, 13 cu. in. Horiz. Crankshaft	1¼	0.6
14, 17, 19 cu. in. Vert. Crankshaft	2¼	1.1
14, 17, 19 cu. in. Horiz. Crankshaft	2¾	1.3
22, 25 cu. in. Vert. Crankshaft	3	1.4
22, 25 cu. in. Horiz. Crankshaft	3	1.4
Cast Iron		
9, 14, 19, 20 cu. in. Horiz. Crankshaft	3	1.4
23, 24, 30, 32 cu. in. Horiz. Crankshaft	3	1.9

(Courtesy B & S)

facturers recommend that you follow the fuel and engine oil requirements listed in the owner's manual and disregard conflicting instructions on oil containers.

For the best performance, use regular-grade leaded gasoline. Unleaded automotive gasoline is an acceptable substitute, with two-cycle or outboard oil rated SAE 30 or 40. The terms *two-cycle* and *outboard* are used by various companies in various ways to designate oil specifications they have designated for use in two-cycle engines.

Caution: Multiple-weight oils (10W-30, 5W-30, etc.) are *not* recommended for mixing with gasoline for use in a two-cycle gasoline engine. Use only the oil marked for two-cycle or outboard engines.

Fuel Mixture—Mix the oil and gasoline in a clean container. Cleanliness of the fuel and oil is essential for proper engine performance. Make sure that gasoline and oil are stored in clean, covered, rust-free containers. Dirt in fuel can clog small ports and passages of carburetors, causing engine failure. Use fresh gasoline only. Gasoline that stands for a long period of time can develop a gum that results in fouled spark plugs and clogged fuel lines, carburetors and fuel screens. Dirty oil causes engine wear. When servicing engines showing indica-

Gas/Oil Mixing Chart for Two-Cycle Engines

See Owner's Manual and Operating Instructions for Individual Engine		
	Gasoline	**Amount of SAE 30 oil to be added**
Fuel	1 Gallon or 4 liters	⅜ pint or 6.4 oz. or 200 cm^3
Mixing	2 Gallons or 8 liters	¾ pint or 12.8 oz. or 400 cm^3
Table	5 Gallons or 20 liters	1 quart or 32 oz.* or 1 liter (1000 cm^3)

*These ratios of oil and gasoline produce a 20:1 mixture.

Ratios will have to be changed for 25:1, 50:1 or 16:1. See owner's manual for proper oil measurements for these ratios. cm^3 is also written as cc.

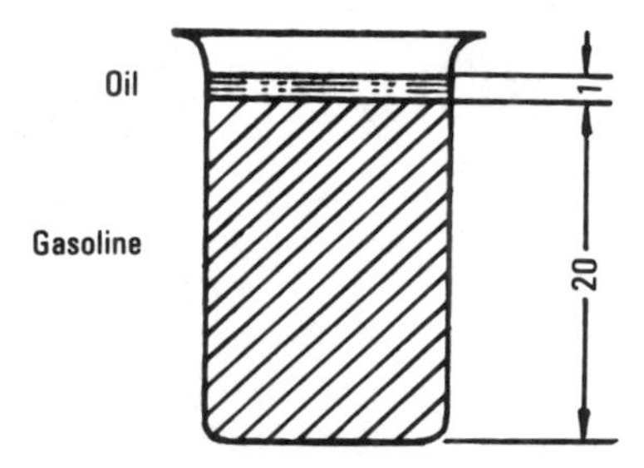

Fig. 9–1. Mixing chart for 20:1 for a two-cycle engine.

tions of dirty gasoline or oil, report the condition to the engine owner, cautioning him or her against continued use of contaminated fuels or lubricants.

Filling the Specifications—To make sure the mixing of oil and gasoline is properly accomplished, the container should be filled about one-quarter full with gasoline. Add oil as indicated from the chart or by the manufacturer, shake the container vigorously, and then fill the remainder of the container with gasoline. Once the oil and gasoline are mixed, they do not separate.

Various ratios of two-cycle engines are: 20:1, 25:1, 50:1, and 16:1. Since there are many different ratios, make sure you are using the right one for your engine.

Components of Lubrication Systems

The two-cycle engine and the four-cycle engine differ greatly in their lubrication systems. The two-cycle engine has little or no lubrication system since the gasoline is mixed with the oil. However, in some cases there is oil injected into the engine and mixed inside the crankcase. The four-cycle engine has a complete system for lubricating all moving parts inside the engine.

Two-Cycle Engine

Fig. 9–2 shows the typical two-stroke-cycle engine with the fuel mixture containing oil coming in contact with the moving parts of the piston, crankshaft, and connecting rod. The oil is in the form of a mist and will settle on the metal to form a film for lubrication purposes. Some of the oil is also drawn up into the combustion chamber, where it is burned along with the gasoline. That is one reason for the puffs of blue smoke coming from the two-cycle exhaust when it changes speeds rapidly. The spark plug has to be cleaned more often in a two-cycle engine since carbon deposits collect rapidly from the burning of oil.

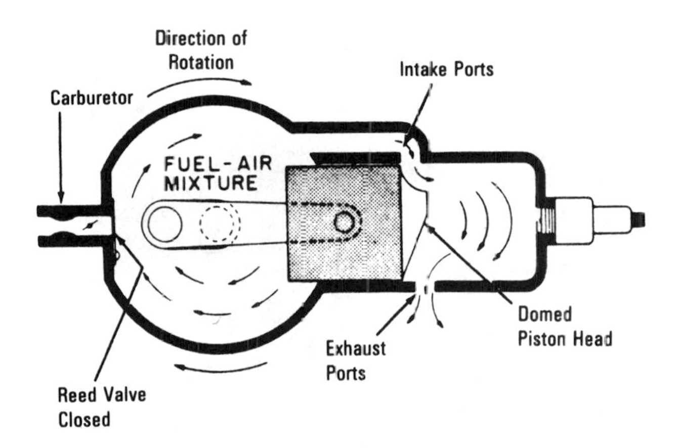

Fig. 9–2. Typical two-stroke-cycle engine with the fuel mixture containing oil coming in contact with the moving parts of the engine. *(Courtesy Tecumseh)*

Although the gasoline may represent fifty parts to every one part of oil, there is still enough oil to lubricate the engine's moving parts properly.

Some larger two-cycle engines have a separate oil and gasoline tank (Fig. 9–3). The oil is injected into the crankcase at the proper moment to mix with the air-gasoline mixture. The oil tank is separate and the oil pump is synchronized with the fuel injection from the gasoline tank.

Four-Cycle Engine

Oil has four purposes. It cools, cleans, seals, and lubricates. There are two methods the Briggs & Stratton engines use for lubrication. They employ the *driven splash-oil slinger* or the *connecting-rod dipper* method.

Since clean oil is essential for good lubrication, it is recommended that the oil be changed after the first five hours of operation. After this change the oil should be changed after every twenty-five hours. It

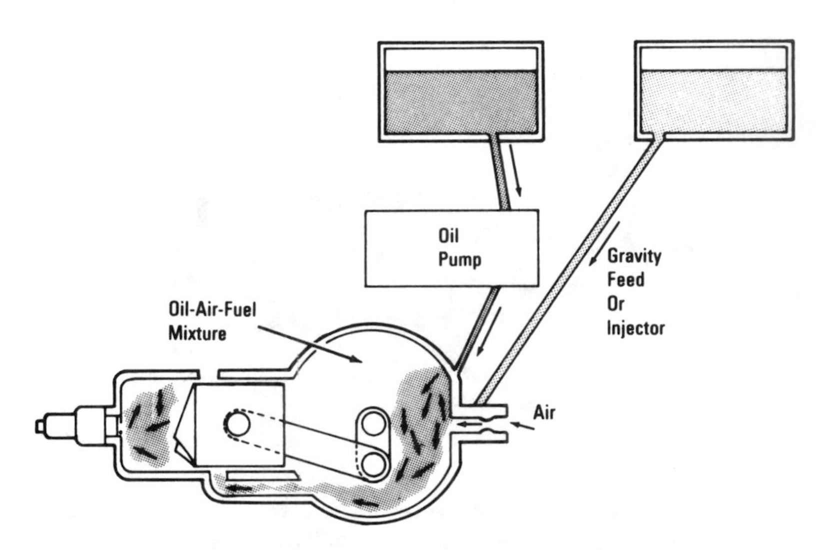

Fig. 9–3. Fuel injection and oil injection methods for a two-cycle engine.

should be changed more often if the engine operates in a dirty atmosphere.

With proper care, oil will last indefinitely. However, the oil may be contaminated with acids, water, dust, metal particles, carbon, and blow-by from the combustion chamber. These can cause damage to the engine if left suspended in the oil. In time, a sludge will form in the crankcase. This sludge should be removed by changing the oil.

How long the engine operates without internal damage depends a great deal on the condition of the lubricating oil. Engine life can be prolonged, all other factors equally considered, if the engine oil is kept clean and uncontaminated.

Oil Change—In order to change the oil, it is necessary to find the drain plug. Fig. 9–4 shows the location of the drain plug for the horizontal- and vertical-crankshaft engines. After draining the oil while the engine is still warm, replace the drain plug. Remove the oil-fill plug or cap and refill with new oil of the proper grade. Replace the oil-fill plug or cap. Check the oil level regularly or at least every five hours of engine operation.

In the 6:1 gear-reduction models you can check by referring to Fig. 9–5. Remove the oil plug in the lower half of the gear cover every hundred hours of operation to check the oil level. Add SAE 10W-30 oil at the upper oil plug until oil runs out of the lower hole. Replace both plugs. Keep in mind that the filler plug has a vent hole and must be replaced in the *top* opening.

Fig. 9–6 shows how to locate the drain plug on other gear-reduction models. The drain plug at the bottom of the gear case should be removed and the oil drained every hundred hours of operation. To refill, remove the oil-check plug and oil-fill plug and pour oil (same

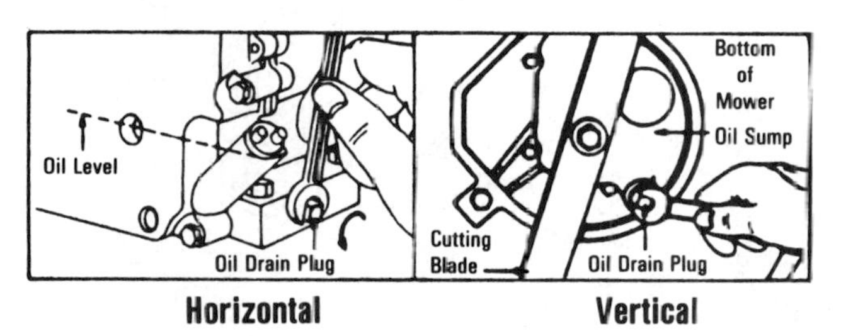

Fig. 9–4. Drain plug locations for an oil change on a four-cycle engine *(Courtesy B & S)*

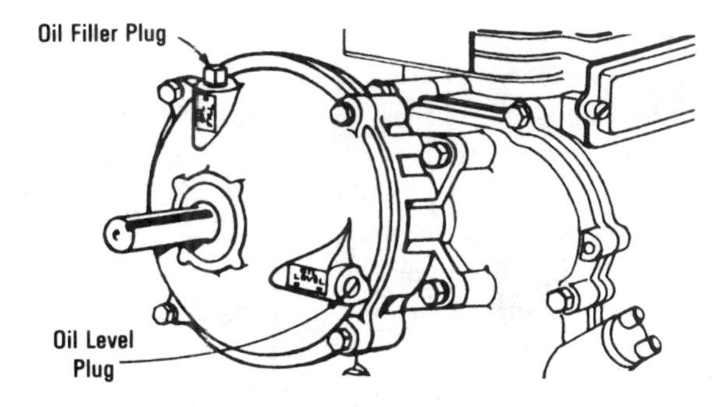

Fig. 9–5. Filler plug and oil check level for a gear-reduction model engine. *(Courtesy B & S)*

grade as used in the crankcase) into the filler hole until it runs out the level check hole. Replace both plugs. The oil-fill plug has a vent hole and must be installed on top of the gear case cover.

Cast-iron engines with a gear-reduction box must be checked also for oil level in the gearbox. The reduction gears are lubricated by the engine crankcase oil. Remove the drain plug from the gear case cover to drain the oil remaining in the gear case (Fig. 9–7).

Some engines have an extended oil fill and a dipstick (Figs. 9–8 through 9–12). When installing the extended oil-fill and dipstick as-

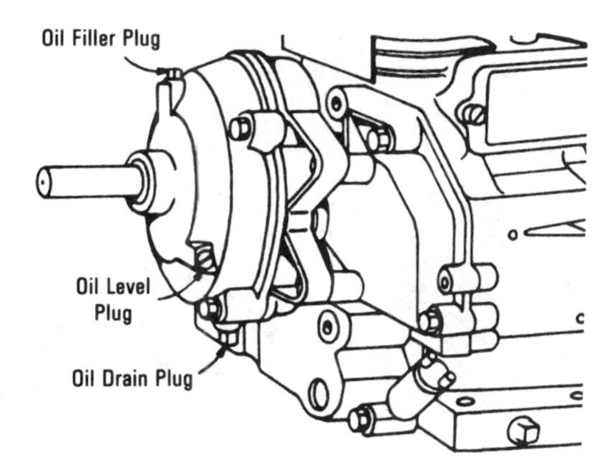

Fig. 9–6. Gear reduction on an aluminum engine. The oil-filler plug and oil level as well as oil drains are shown. *(Courtesy B & S)*

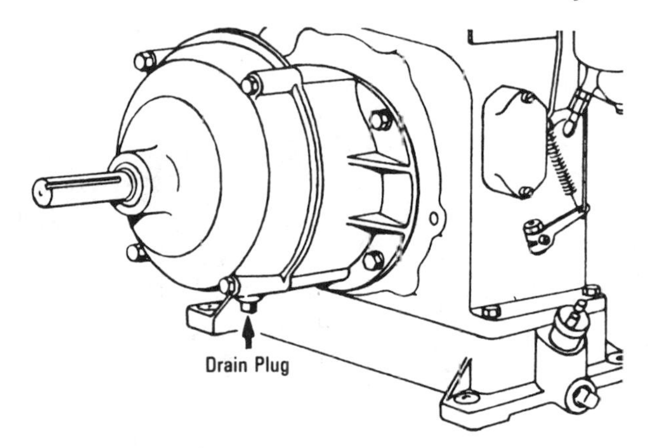

Fig. 9–7. Drain plug for the cast-iron engine. *(Courtesy B & S)*

sembly, the tube must be installed so the O-ring seal is firmly compressed. This type of dipstick and extended oil-filler tube can be added to any model. Fig. 9–8 shows how the installation is accomplished by pushing the tube downward toward the sump, then tightening the

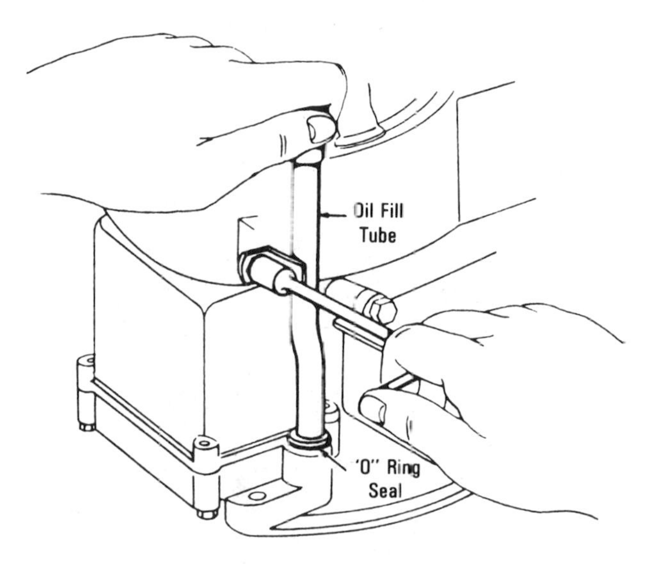

Fig. 9–8. Extended oil-fill and dipstick attachment. *(Courtesy B & S)*

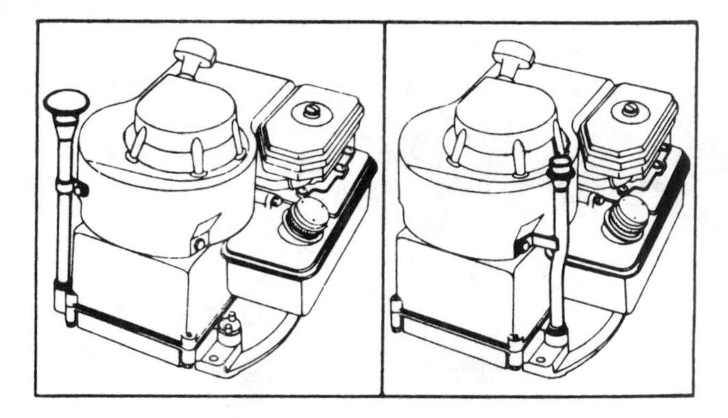

Fig. 9–9. Other models with dipstick and extended oil-filler tube. *(Courtesy B & S)*

blower housing screw, which is used to secure the tube and bracket. When the cap and dipstick assembly is fully depressed or screwed down, it seals the upper end of the tube.

A leak at the seal between the tube and the sump or at the seal at the upper end of the dipstick can result in a loss of crankcase vacuum. This can discharge smoke through the muffler.

Do not overfill the sump or crankcase with oil when the extended

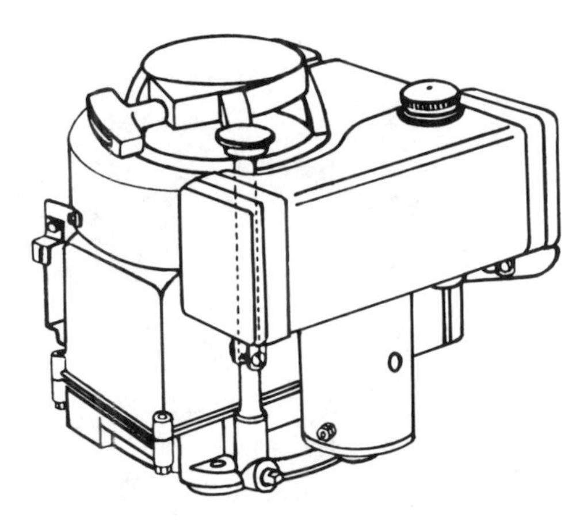

Fig. 9–10. Location of the fill tube and dipstick. *(Courtesy B & S)*

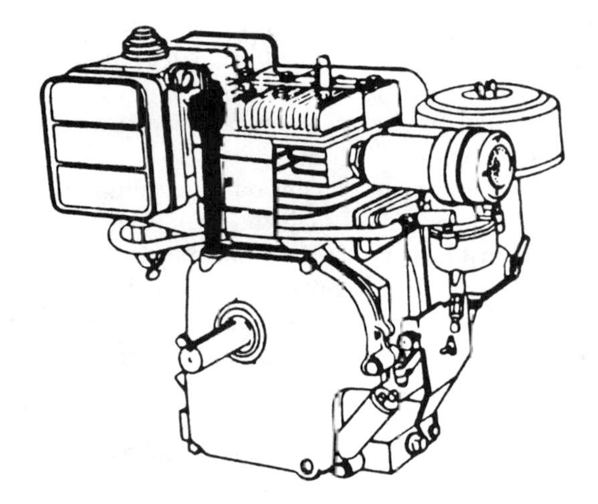

Fig. 9–11. Location of the fill tube and dipstick. *(Courtesy B & S)*

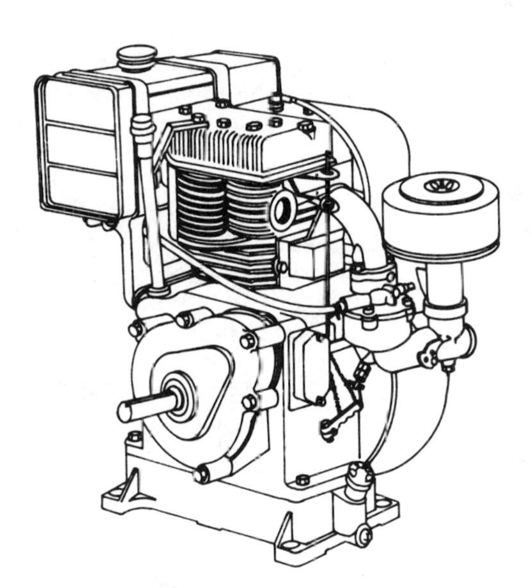

Fig. 9–12. Location of the fill tube and dipstick. *(Courtesy B & S)*

filler and dipstick is used. The dipstick is marked "DO NOT OVERFILL." Excessive oil will cause a smoking condition as the engine attempts to discharge the surplus oil.

Fig. 9–9 through 9–12 show the various locations of dipstick filler tubes on different engines.

Breathers—In order to prevent oil from being forced out of the engine at the piston rings, oil seals, breaker plunger, and gaskets, a partial vacuum must be maintained in the crankcase.

The breather is used to maintain a vacuum in the crankcase (Fig. 9–13). The breather has a fiber disc valve which limits the direction of air flow caused by the piston moving back and forth. Air can flow out of the crankcase, but the one-way valve blocks the return flow. That means a vacuum is created in the crankcase.

In the single-cylinder engine, the breather is usually located at the valve chamber. This is the area where the valve are accessible for repair. There are two types of breathers. One is the open type and the other is the closed type. The open-type breather just ventilates the crankcase to the atmosphere. This was the case with automobile engines before the PCV valve was installed in 1967. A breather cap was placed on some of the oil-filler tubes in earlier models of small engines.

The closed-type breather allows air to pass out of the crankcase, but not back in. Breathers can be seen in Figs. 9–14 and 9–15. The fumes from the crankcase when the engine is operating are drawn out of the crankcase and recirculated into the combustion chamber by way

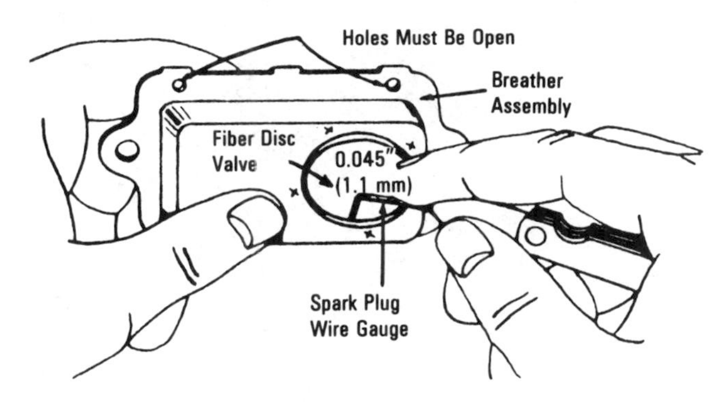

Fig. 9–13. Checking the breather for proper clearance. *(Courtesy B & S)*

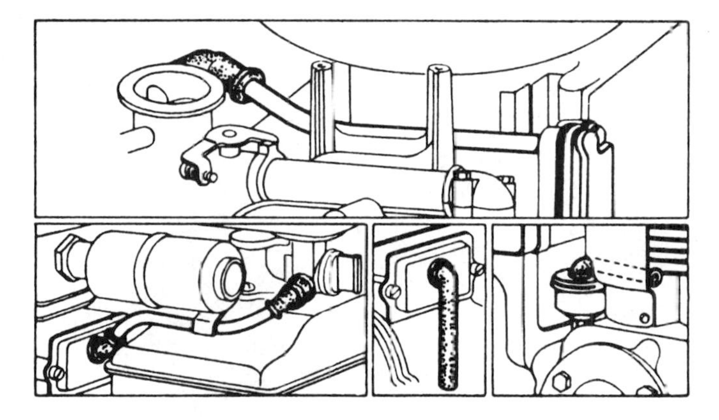

Fig. 9–14. Location of the breather assembly on B & S models.

of the carburetor. This gives complete combustion of the fumes and keeps down the pollutants created by these fumes.

Breather Checks—If the fiber disc valve is stuck or binding, the breather cannot function properly. It should be replaced. A 0.045″ (1.1-mm) wire gage should not enter the space between the fiber disc and the body (Fig. 9–13). You may use a spark plug gage wire to do the checking. The fiber disc valve is held in place by an internal bracket which will be distorted if pressure is applied to the fiber disc valve. Therefore, do not apply force when checking with a wire gage.

If the breather is removed for inspection or valve repair, a new gasket should be used when replacing the breather. Tighten the screws securely to prevent oil leakage.

Most breathers are now vented through the air cleaner to prevent dirt from entering the crankcase and contaminating the oil. Check to be sure venting elbows or the tube are not damaged and are sealed properly.

The breather filters are usually steel wool or plastic foam impregnated with oil. They should be cleaned or replaced if clogged.

Breathers are connected to the carburetor or vented in a number of ways.

Figs. 9–14 and 9–15 show how some of the arrangements are made. The flexible rubber tubing is connected directly to the atmo-

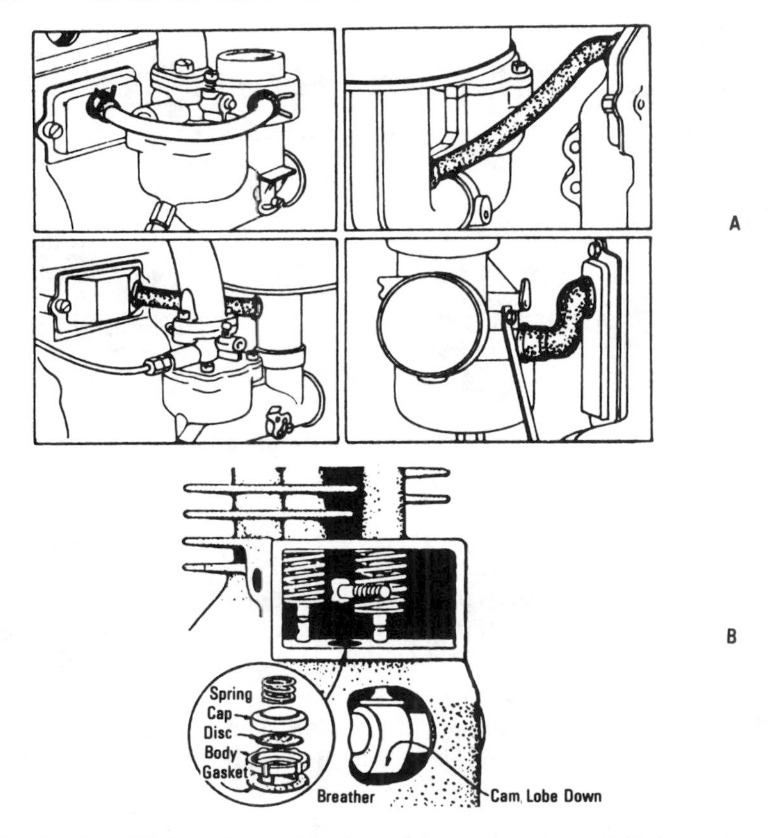

Fig. 9–15. (A) Location of the breather assembly on B & S models. (B) Some B & S engines have internal breathers. Insert shows the assembly sequence. *(Courtesy B & S)*

sphere in one of the assemblies in Fig. 9–14. This was the practice in some of the older engines.

Just keep in mind that one of the first things to check if the engine is burning oil is the breather. It may be clogged and the pressure building up in the crankcase is forcing oil past the rings into the combustion chamber. Once this has been eliminated as a cause of oil burning, check the other standard trouble areas.

Internal Lubrication

The cylinder walls and internal moving parts of the engine are lubricated by several methods. The *oil-pump method* squirts oil on the moving parts. The *splasher method* picks up oil with a dipper on the end of the connecting rod and splashes it all over the inside of the engine.

Fig. 9–16 shows one type of oil pump used on some horizontal-crankshaft models. The pump plunger is actuated by the camshaft and the oil is sprayed from the outlet in the plunger to lubricate the engine. Another type of plunger pump is similar, but oil is delivered by way of a tube connected to the pump (Fig. 9–17).

In Fig. 9–18, notice the different type of pump used for horizontal-crankcase engines.

There are a number of types of oil pumps: the piston type, which has a piston moving up and down in a cylinder with ball check valve plus a pressure relief valve, and the double-acting piston-type pump. Linkages that drive these pumps may be direct extensions of the piston shaft to ride on one of the cams that operate the valves or cam-operated rocker arms.

There is an ejector pump system that sprays oil onto the moving

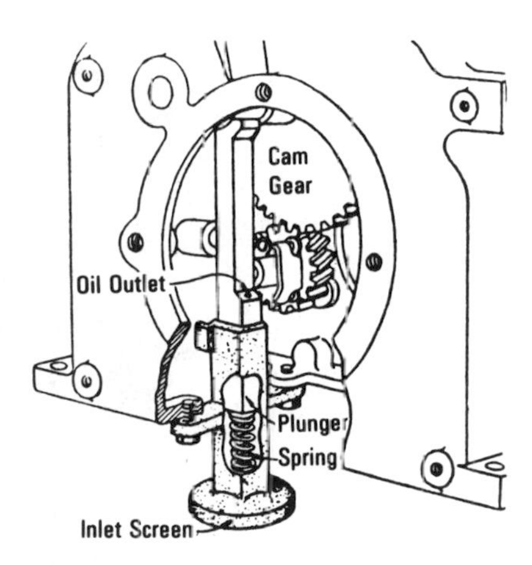

Fig. 9–16. Oil pump used on some horizontal-crankshaft models. *(Courtesy B & S)*

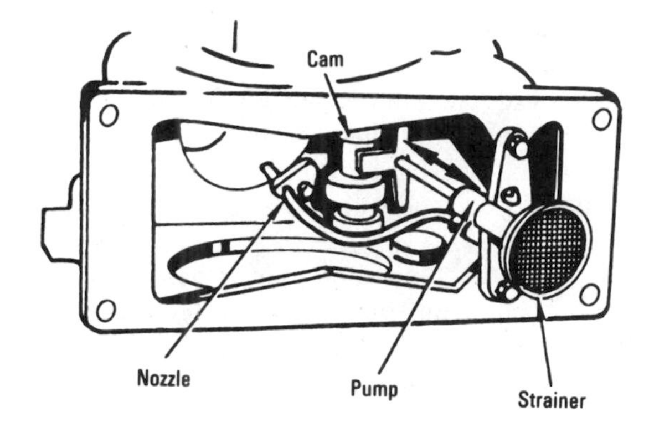

Fig. 9–17. Piston-drive oil pump with a tube to direct the oil spray.

parts of the engine. In most cases the spray is directed to hit the connecting rod where it connects to the crankshaft. This type of pump presents a spray mist of oil to lubricate all parts of the engine. This type may also use the oil dipper to splash oil around. Both methods ensure good lubrication. The dry-sump pump system is used on foreign engines and should be mentioned in passing so that you are aware of its

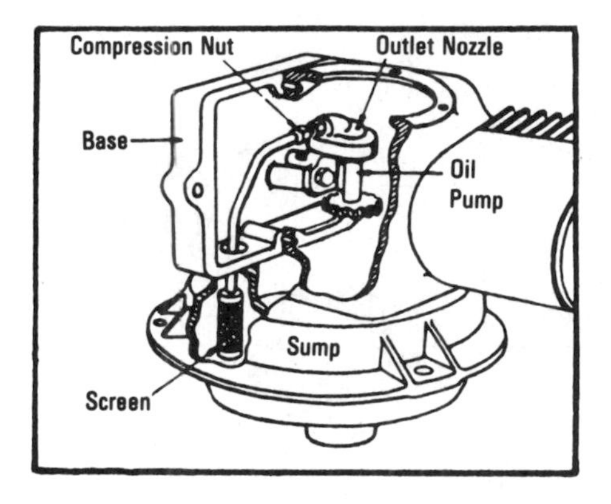

Fig. 9–18. Oil pump installed on some B & S horizontal-cylinder engines. *(Courtesy B & S)*

existence. The oil is not placed directly into the crankcase but is stored in the separate oil tank. It is usually placed outside the engine crankcase area. Oil is distributed by an oil pump. Once the oil drops back to the bottom of the crankcase, it is lifted from this sump area to the reservoir or tank by the pump. The pump then uses the reservoir area to obtain oil to spray on the moving parts of the engine.

Other types of oil pumps that may be encountered are the impeller, or vane, type and the gear type. These are not in widespread use today in small gasoline engines, but are mentioned here just in case you see one of these types in an old engine.

Oil Slingers

In the splash system of lubrication, the dipper dips into the oil reservoir of the engine. The oil reservoir is usually located in the crankcase. Fig. 9–18 shows the location of the slinger or dipper on the end of the connecting rod.

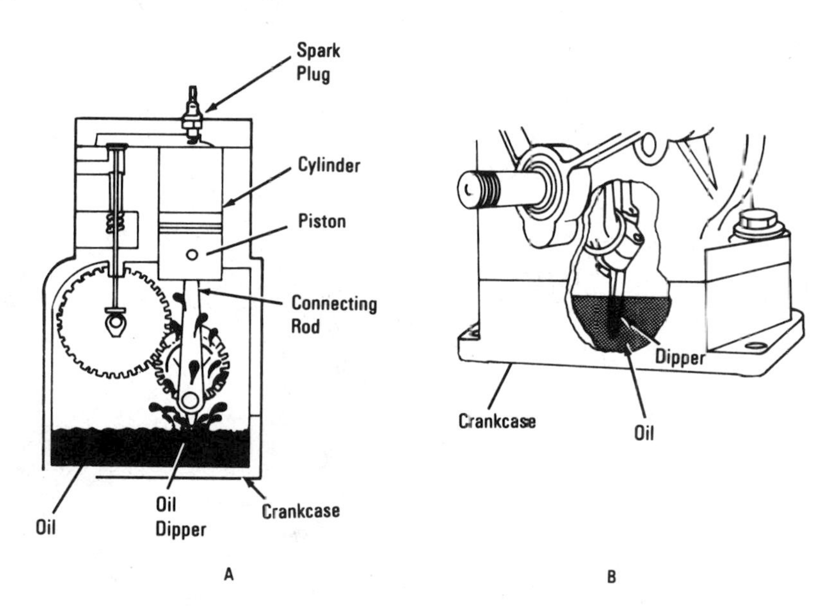

Fig. 9–19. (A) Dipper slings oil in the crankcase. (B) Location of the dipper during its operation.

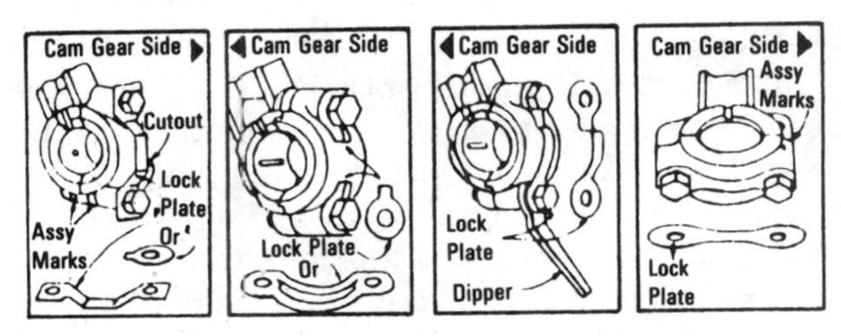

Fig. 9–20. Connecting-rod installation on horizontal-crankshaft engines. *(Courtesy B & S)*

Briggs & Stratton engines that are made of aluminum alloy and cast iron do not have oil pumps. The splasher or dipper operates as shown in Fig. 9–19. As the crankshaft rotates, it causes oil to be splashed on all areas of the moving engine components. Fig. 9–20 shows how the connecting rod is installed on horizontal-crankshaft engines. Note how the dipper is mounted on the connecting rod.

The oil slinger is driven by the cam gear. Old-type slingers using a die-cast bracket assembly have a steel bushing between the slinger

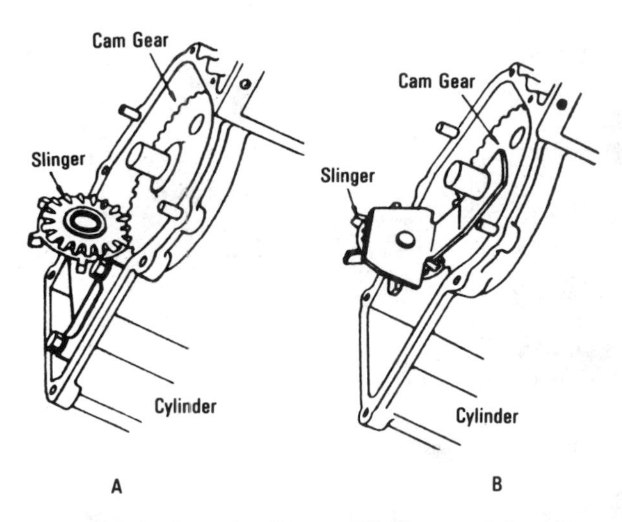

Fig. 9–21. (A) Old-style oil slinger. (B) Newer-style, stamped-steel slinger and bracket. *(Courtesy B & S)*

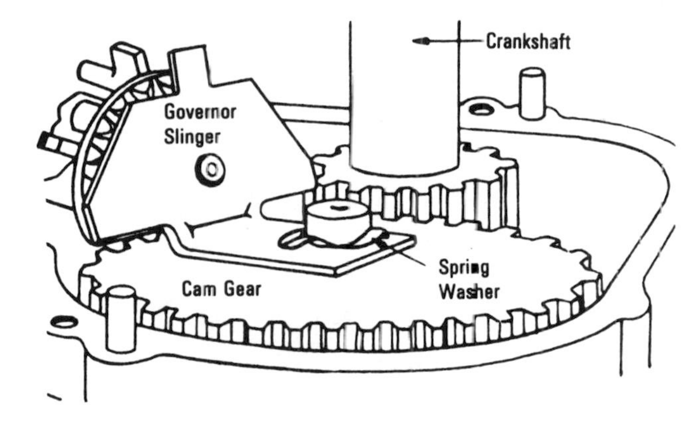

Fig. 9–22. Oil slinger and bracket with a vertical-crankshaft engine. The spring washer is not used on some models. *(Courtesy B & S)*

and the bracket. The bracket should be replaced if it is worn to a diameter of 0.490″ (12.4 mm) or less. Oil slingers and brackets are shown in Fig. 9–21. Replace the steel bushing if it is worn. A newer-style oil slinger is shown in Fig. 9–21B. Fig. 9–22 shows the oil slinger and bracket on a vertical-crankshaft engine. Replace the gears and slinger if worn. The spring washer is used only on some models. The newer-style stamped-steel slinger appears to be more efficient than the older type.

Oil holes are placed in proper locations in engine parts for lubrication purposes. As the oil that the slinger splashed onto the cylinder walls and on the connecting rod runs back toward the crankcase, it also flows into the oil holes (Fig. 9–23). This is true of the connecting rod bearings and the main bearings for the crankshaft. From this you can see the importance of having clean oil in the crankcase. Frequent oil changes, according to the manufacturer's recommendations, are essential for long, trouble-free engine operation.

Another reason for oil changes is the *blow-by* of combustion products. The components of the combustion chamber can easily get by the rings and down into the crankcase if the rings are worn or the cylinder sides are scored. This material can generate acids in the oil and produce particles that can clog the holes that provide lubrication for the crankshaft and the connecting rod bearings.

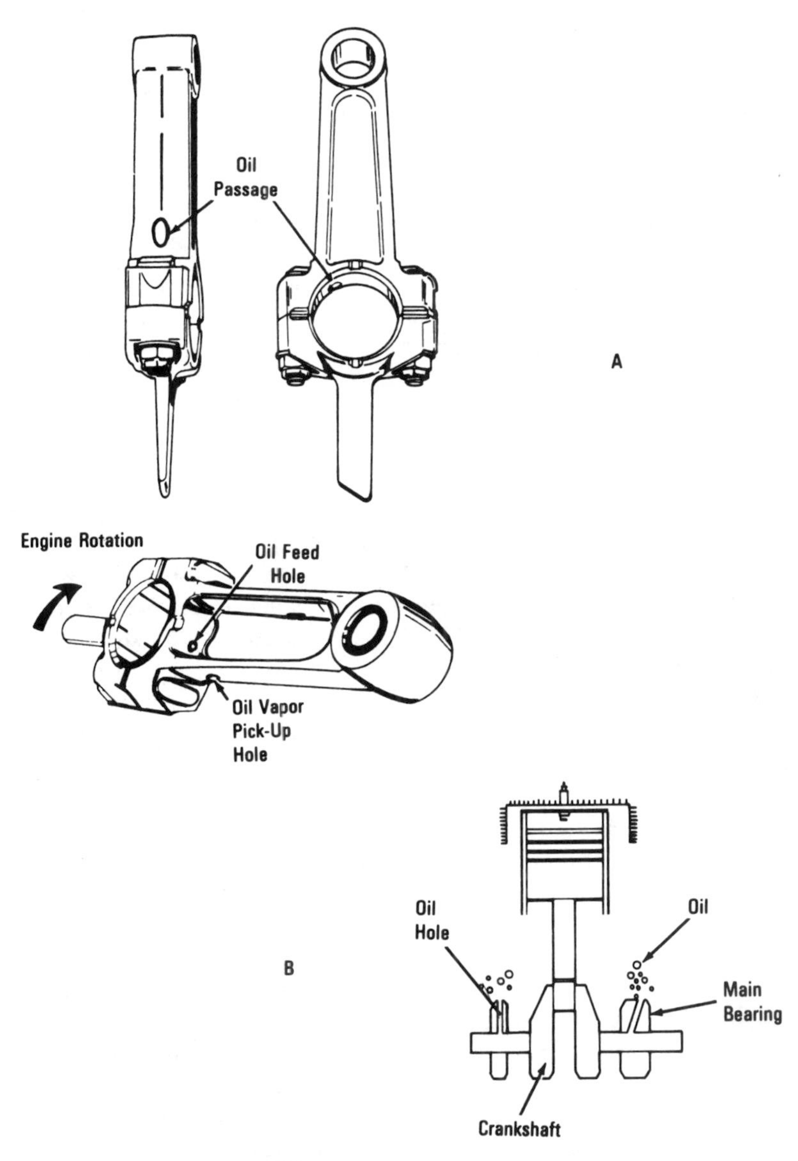

Fig. 9–23. (A) Oil holes in the connecting-rod assembly. (B) Oil hole locations in the main bearing assembly. *(Courtesy Tecumseh)*

Chapter Highlights

1. Lubricants are needed to reduce friction.
2. Anything that will reduce friction can be called a lubricant.
3. Viscosity is the internal friction between molecules of a liquid or gas.
4. Viscosity is used to indicate the ability of an oil to lubricate and eliminate friction by producing a film of oil.
5. Some materials have good lubricating qualities and others are very poor in reducing friction.
6. Any high-quality detergent oil having the American Petroleum Institute classification for SC, SD, SE, or MS can be used in small gasoline engines.
7. Oil changes should occur after each twenty-five hours of operation, or more often under dirty operating conditions.
8. Two-stroke-cycle engines require a mixture of oil and gasoline.
9. Gasoline that stands for some time can develop a gum.
10. Various ratios of two-cycle engines for gas to oil are 20:1, 25:1, 50:1, and 16:1, depending on the manufacturer.
11. Some larger two-cycle engines have separate oil and gasoline tanks. The oil is injected into the crankcase at the proper time to mix with the air-gasoline mixture.
12. Oil has four purposes: it cools, clean, seals, and lubricates.
13. The splash and the dipper methods of lubrication are used in small gasoline engines.
14. Oil will last indefinitely if properly cared for.
15. Gear-reduction-box engines use the same oil as the engine. The gearbox must be checked also when the engine oil is checked.
16. Some engines have a dipstick that extends outside of the engine.
17. Do not overfill the sump or crankcase when changing oil.
18. A partial vacuum must be obtained in the crankcase when the engine is operating. The breather valve does this.
19. Breathers are connected to the carburetor in a number of arrangements.
20. There is an ejector pump system for spraying oil onto the moving parts of the engine.
21. Impeller-, or vane-, type and gear-type oil pumps are used in small engines.

22. In the splash system of lubrication, the dipper dips into the oil reservoir of the engine.
23. The oil slinger is driven by the cam gear.
24. Oil holes are placed in proper locations in engine parts for lubrication purposes.
25. Another reason to change oil is the blow-by of combustion products.

Vocabulary Checklist

blow-by
bracket
breather
clog
component
connecting-rod dipper
dipstick
dry-sump pump system
fiber disc valve
foul
friction
gas-oil mixture ratio
horizontal-crankshaft engine
impedes
impeller
injected
lubricants
O-ring
plunger
rocker arm
saponification
splash-oil slinger
uncontaminated
vertical-crankshaft engine
viscosity

Review Questions

1. What is the purpose of a lubricant?
2. What is one of the most important factors to be considered in the selection of an engine lubricant?
3. What is viscosity? Where is the term used?
4. What is saponification?
5. Which would you use in the winter, SAE 10W-40 or SAE 20?
6. Why are gasoline and oil mixed for the operation of a two-cycle engine?
7. How do you mix the oil and gasoline for a two-cycle engine?

8. What are four different ratios used for fuel mix in a two-cycle engine?
9. How do four-cycle engines lubricate their moving parts?
10. Why is an oil change important?
11. How often should you change the oil in a four-cycle engine?
12. What is the purpose of a breather?
13. How do you check the breather for proper operation?
14. What type of engines have oil pumps?
15. What is an oil slinger? Where is it used?
16. What is blow-by?

CHAPTER 10

Power Drives

Small gasoline engines are used to drive many types of machines. Some of these machines have transmissions and power take-offs. There are a number of different types of power drives, transmissions, and mechanisms that drive various machines with the aid of a small engine. In this chapter we take a look at some of the basic principles that aid in the multiplication of the power of a small gasoline engine.

Simple devices can be used to vary speed and increase the ability of a small engine to do work that otherwise it would be unable to do. Transmissions not only transmit power from one place to another, but in most instances they also have mechanical advantage (MA). Power drives may be direct or indirect. The direct type can be illustrated by the lawnmower blade attached directly to the crankshaft of the engine. Other types are not so direct. They may have the load attached through gears, pulleys and belts, or chains and sprockets. All these devices will be discussed and examined in this chapter.

Physical Factors That Affect Machines

There are a number of physical factors that affect the operation of simple machines. A good grasp of these factors allows for a better understanding of the way transmissions, clutches, and compound machines work.

Friction, for instance, is a very important factor in the engine and in the application made of it in a clutch. Friction is necessary for a friction clutch to operate. But what is friction? *Friction is a force that opposes motion.* Motion is what we must have in the operation of small gasoline engines. The small gasoline engine is the source of motion for the machines we are working with in this unit.

Nature of Friction

When two surfaces slide over one another there is a force that opposes the motion. This is called friction. Some people believe that friction is caused by irregular surfaces that can drag. This is usually eliminated by using a film of oil on both surfaces to allow for smoother passage. If surfaces are polished, they have a tendency to move more freely over one another, or with less friction. Very smooth surfaces are called for when two metals are to be in contact with one another. However, if two metal surfaces are too smooth, they have an increase in friction. There is a point at which smoothness is not enough.

Friction may also be an aid to you. It helps when you need traction between wheels and the road surface. Ice has a tendency to lessen friction, and automobiles lose control on these slick surfaces. Friction clutches are also useful in the manufacture of small machines. You even use friction between the soles of your shoes and the floor you walk on. Applying the brakes on an automobile calls for friction to stop the car. The friction between the brake lining and brake drum as well as between the road surface and wheel surface make it possible to control the speed of a moving automobile. Friction holds a screw in wood and metal, and helps in many other ways.

Friction can also be a hindrance. We have to polish and lubricate bearings to eliminate friction. We have to reduce the friction in engines by various lubricants. We have to do any number of things to eliminate or reduce friction. Therefore, we should take a closer look at what types of friction there are and just what can be done about this force.

Sliding Friction—Some general statements can be made about sliding friction. The friction that causes difficulty in moving a book over a tabletop is sliding friction. Some apparent contradictions can be found in any examples used to illustrate sliding friction.

1. Friction acts parallel to the surfaces sliding over one another, and in the direction opposite to that of the motion.
   Example: Slide a book over a tabletop. Friction acts parallel to the surface of the tabletop. It also acts in the direction opposite to that in which the book is moved.
2. Friction depends on the materials and their surfaces.
   Example: Friction may act differently among different materials and also among different surfaces composed of the same material.
3. Sliding friction is less than starting friction.
   Example: Pull on any object to get it started. Note that it takes more force to get it to start sliding than it does to keep it sliding once it has moved.
4. Sliding friction is nearly independent of the speed of sliding.
   Example: Braking a car means that the harder you apply the brakes, the greater the stopping friction. However, there is a point at which friction decreases with an increase in velocity.
5. Friction is practically independent of the area of the contact between the surfaces.
   Example: Sliding a brick along a surface takes about the same amount of force whether the brick is flat, on its end, or on an edge.
6. Friction is directly proportional to the force pressing the two surfaces together.
   Example: It is easier to pull a 1-pound load in a box over a surface than to pull the same size box making the same surface contact with 100 pounds in it.

Coefficient of Friction—The coefficient of friction is the ratio of the force needed to overcome friction to the normal force pressing the surfaces together. The coefficient of friction varies with the nature of the surfaces in contact and with the degree of polish of the surfaces (Table 10–1).

Reducing Friction—The reduction of friction means an increase in the efficiency of the machine. Therefore, it is possible to reduce machine friction at least four ways:

1. *Polished bearings.* When a wheel is turned on an axle, it is easy to reduce the friction by polishing the axle and the bearing surface. The materials should be hard enough to resist scratching and wear.

Table 10–1. Selected Coefficients of Friction

Materials	Static	Sliding
Hard steel on hard steel	0.78 (dry)	0.42 (dry)
Steel on graphite	0.21 (dry) 0.09 (lubricated)	
Aluminum on mild steel	0.61 (dry)	0.47 (dry)
Brass on cast iron		0.30 (dry)
Glass on glass	0.94 (dry) 0.35 (lubricated)	0.40 (dry) 0.09 (lubricated)
Cast iron on cast iron	1.1 (dry) 0.2 (lubricated)	0.15 (dry) 0.07 (lubricated)
Bronze on cast iron		0.22 (dry) 0.077 (lubricated)
Steel on babbitt	0.42 (dry) 0.17 (lubricated)	0.35 (dry) 0.14 (lubricated)
Teflon on teflon or teflon on steel	0.04	0.04

2. *Antifriction metals.* Steel sliding on steel has a high coefficient on friction. The coefficient of friction is less when steel slides over bearings of lead and antimony called babbit. Some bearings are lined with the alloy to reduce friction. Other combinations of metals are now available for bearings.
3. *Ball bearings.* Steel sliding on steel has a very high coefficient of friction. When roller bearings are added, the coefficient is reduced to about one-hundredth of the steel-on-steel arrangement. Roller bearings are also used to reduce friction. Ball bearings and roller bearings are very good devices for reducing friction between two pieces of steel making direct contact. Of course, the proper lubrication is necessary to further reduce the friction (Fig. 10–1).
4. *Lubricants.* Fluid friction is substituted for solid friction when lubricants are used between sliding surfaces. Fluid friction is generally less than solid friction.

Metal-to-metal contact is everywhere when you look at a small gasoline engine. The piston makes contact with the cylinder walls; the

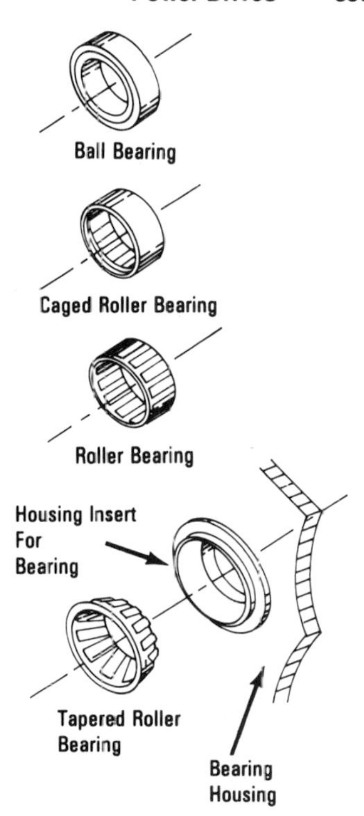

Fig. 10–1. Bearings are available in ball, caged roller, roller, and tapered roller configurations.

connecting rod makes contact with the crankshaft. There is need to reduce friction in these areas and others. Even the cam that touches the push rod for the valves makes metal-to-metal contact (Fig. 10–2). Proper lubrication and proper lubricants are needed. Proper bearing surfaces are also called for in the engine. All these things are necessary to reduce friction and improve the efficiency of the engine.

Basic Machines and Mechanical Advantage

There are six simple basic machines. We use machines to transform energy, to multiply force, and to transfer energy from one place to another. Connecting rods, crankshafts, drive shafts, and axles all are used to transfer energy from the combustion in the cylinder of an engine to the drive attached to the crankshaft.

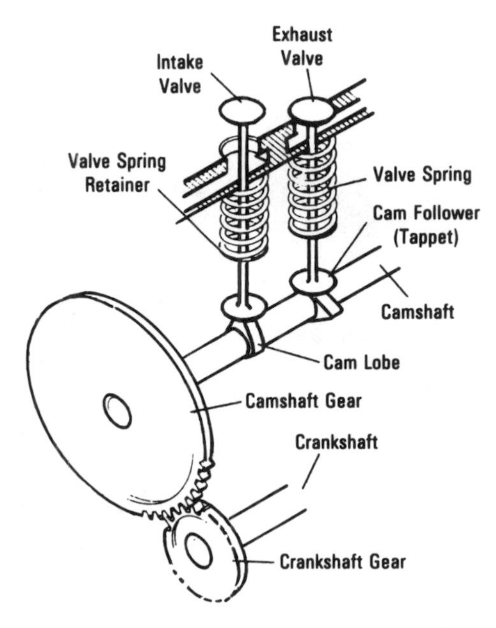

Fig. 10–2. Lubricants make it possible to have metal-to-metal contact without wearing out the part too rapidly. A thin film of oil protects the parts from actually coming in contact with one another.

We can modify machines to cause them to change velocity or speed and to change direction of a force. The best way to get at the workings of basic machines, and eventually the small gasoline engine, is to look at the six simple machines with which everyone is familiar from previous experiences.

Simple Machines

There are six basic machines. These are the *lever,* the *pulley,* the *wheel and axle,* the *inclined plane,* the *screw,* and the *wedge.* Other machines are either modifications of these simple machines or combinations of two or more of them.

Keep in mind that the wheel and axle are fundamentally levers. The wedge and screw are modified inclined planes.

Mechanical Advantage—There are two ways to define the mechanical advantage of machines: the ideal mechanical advantage and the actual mechanical advantage.

The *ideal mechanical advantage* (*IMA*) is the ratio of the distance the effort force moves to the distance the resistance force moves.

The *actual mechanical advantage* (*AMA*) is the ratio of the resistance force to the effort force.

Machine Efficiency—The efficiency of any machine is the ratio of its actual mechanical advantage to its ideal mechanical advantage, converted to a percentage.

$$\text{Efficiency} = \frac{AMA}{IMA} \times 100\%$$

The *AMA* is never greater than the *IMA* since all machines have to overcome friction within the machine itself. That means the efficiency of a machine is always less than 100 percent.

In order to obtain a better understanding of mechanical advantage and how machines make use of it, take a closer look at the six types of basic machines we mentioned earlier. Each has a mechanical advantage.

The Lever

A *lever* is a rigid bar that is free to turn about a fixed point called the fulcrum (Fig. 10–3). The fulcrum is a pivot point. Effort or force is exerted on one lever arm, and tends to rotate the lever in one direction. Resistance or some other force is exerted on the other lever arm, and tends to rotate the lever in the opposite direction. There are three classes of levers.

Class 1 Lever. A heavier weight can be lifted by a lighter weight. Note the fulcrum point and how the distance times the force applied is important (Fig. 10–3A).

Class 2 Lever. Note the location of the fulcrum and the amount of force or effort needed to lift the weight located in the middle of the lever (Fig. 10–3B).

Class 3 Lever. In this case the weight to be lifted takes five times the force of its weight to be lifted (Fig. 10–3C).

In each case the location of the fulcrum and the point at which the effort was exerted makes a difference in the amount of effort needed to

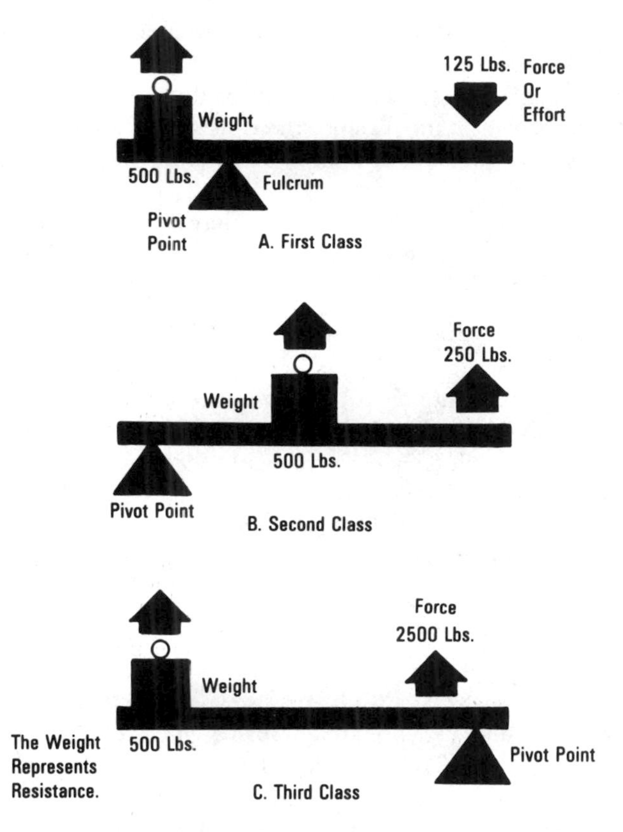

Fig. 10–3. The three classes of levers.

move the weight or opposition. This can be summarized as: *The product of the acting force (input) and its distance from the fulcrum is equal to the resistance force (load) times its distance from the fulcrum.*

Notice that in the first two classes of levers the mechanical advantage is greater than one, but in the third-class lever it is less than one.

The Pulley

A *pulley* is a wheel that turns readily on an axle, which in turn is mounted in a frame (Fig. 10–4). When one or more pulleys are enclosed in a frame, the machine is called a *block.* If a series of pulleys with rope or chain attached is combined, then the arrangement is

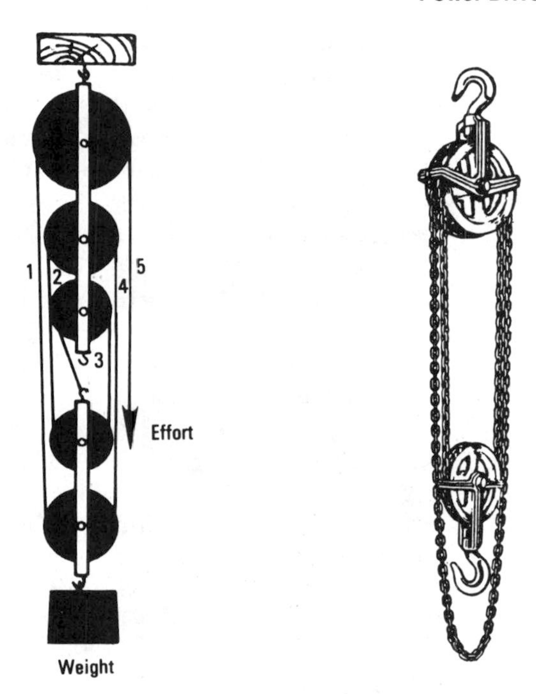

Fig. 10–4. Block and tackle or commercial chain hoist.

called the *block and tackle*. Note how the pulleys are arranged in Fig. 10–4 and how the force exerted is multiplied or the direction of the pulleys changes with each arrangement. The ideal mechanical advantage of a single movable pulley is two. Many different combinations of fixed and movable pulleys are possible. The mechanical advantage of a single fixed pulley is one. As you increase the number of pulleys, you increase the mechanical advantage. There is no mechanical advantage in a fixed pulley that does not move. Mechanical advantage is equal to the number of rope strands used to support the movable pulleys and the load. Count the number of movable pulleys and find the mechanical advantage.

The Wheel and Axle

The wheel and axle is really a Class 1 lever. *The wheel and axle is a wheel or crank rigidly attached to an axle.* This means that both the wheel and axle have the same angular velocity. The mechanical advantage comes from the two different circumferences—that is, the circumference of the wheel is larger than that of the axle. If you pull down on the rope attached to the outside or wheel portion, it will take less effort to lift the weight attached to the rope connected to the smaller-diameter axle (Fig. 10–5). In the wheel and axle, the axle may be the drive part while the wheel is the driven part. In that case the axle is the input for the force and the wheel is the output. That means the mechanical advantage can be figured by dividing the input by the output. For every revolution of the axle there is a revolution of the wheel; they are tied together. However, if the axle is small in diameter and the wheel has a larger diameter, then the wheel has a larger circumference or outside curve to make contact with the earth or roadbed. This means a small axle can turn a wheel of large circumference and cause it to cover much more ground than the smaller-circumference axle would if placed in contact with the ground. One application of the wheel and axle is the automobile, where the axle is driving the large-circumference wheel for a great mechanical advantage. It allows the engine to turn over fairly slowly while the wheel covers a much greater distance.

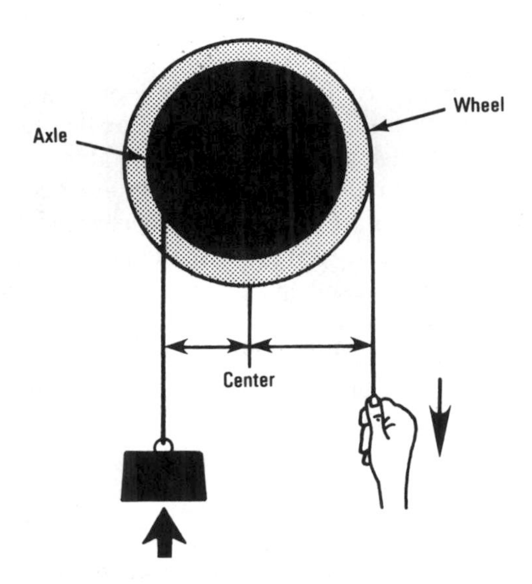

Fig. 10–5. Wheel and axle.

The Inclined Plane

The *inclined plane* is just what the title suggests: A plane or surface is inclined or raised. When you use an inclined plane (Fig. 10–6), it is possible to exert less force than normally thought necessary to lift an object. The object or weight is moved up a slanted surface rather than lifted straight up, or vertically. The inclined plane resembles a right triangle. The mechanical advantage is found by taking the length of the incline and dividing it by the vertical distance, or height.

The Wedge

The *wedge* is a double-inclined plane (Fig. 10–7). There is so much friction in using a wedge that an ideal mechanical advantage is of little significance. A wedge that is long in proportion to its thickness is easier to drive. The mechanical advantage depends on the ratio of its length to its thickness.

The Screw

The *screw* is derived from the inclined plane (Fig. 10–8). The screw is an inclined plane wound around a cylinder. The distance be-

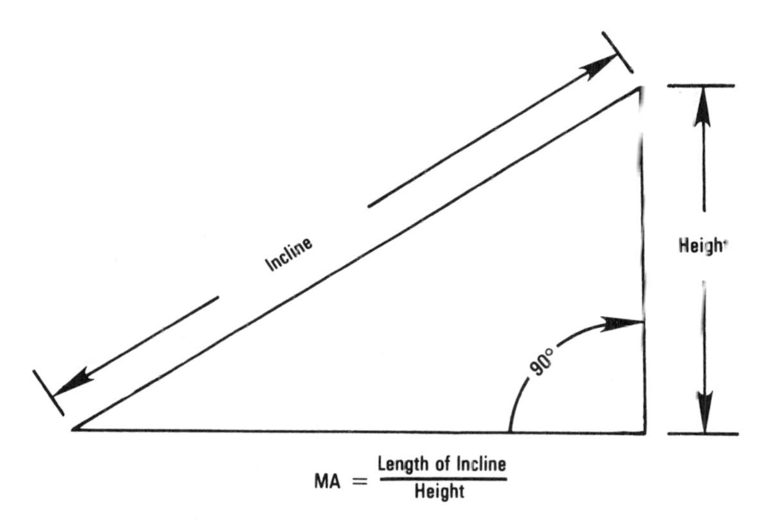

Fig. 10–6. Inclined plane.

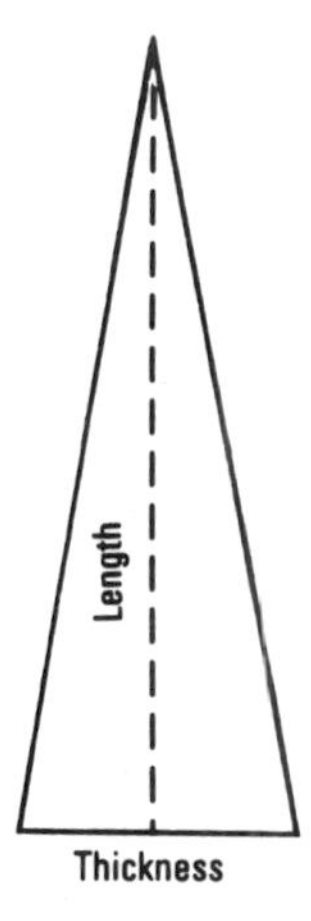

Fig. 10–7. A wedge is nothing more than two inclined planes.

tween threads is called the *pitch* of the screw. The effort force is often applied at one end of a lever that is set in the head of the screw or to a wrench attached to the head of the screw. That means the screw is turned one revolution with the effort force. It makes a complete circle. This means the circumference is equal to 2 pi (π) times the radius of the moving arm or screwdriver handle. The mechanical advantage is

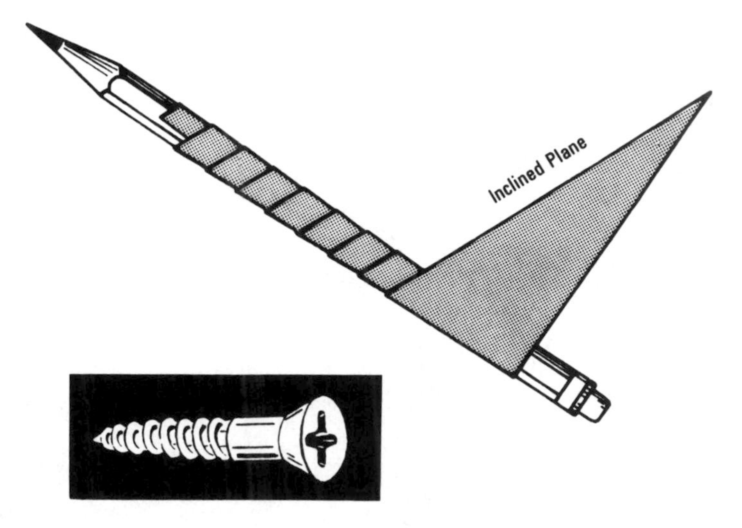

Fig. 10–8. The screw is a spiral that is similar to the inclined plane being wrapped around the pencil.

found by dividing the circumference ($2\pi r$) by the pitch of the screw thread. Bolts, nuts, and screws of all kinds are examples of this type of simple machine.

It is impossible to make any machine 100 percent efficient. That is because it is impossible to eliminate friction and because of the weight of the parts of the machine. For instance, there may be almost 100 percent efficiency when the lever is used. However, the weight of the arm still makes it less than 100 percent. The block and tackle is about 60 percent efficient. The inclined plane may reach 80 percent efficiency when the surfaces are very smooth. The wedge has friction such that it is almost impossible to ascertain its efficiency. The jackscrew has a high mechanical advantage, but friction reduces its efficiency to about 25 percent.

Examples of the Uses of Basic Machines

So far you have looked at the six simple machines. Each has an application in the small gasoline engine. Take, for instance, the crank. This is an example of the wheel and axle. The cam and follower is an example of the inclined plane being put to use. While there are many applications of screws everywhere you look in the manufacture or assembly of the engine and its component parts, some of the best examples are the screws used to hold the head to the cylinder block. The spring is an application of the lever machine. Many springs are used in the construction of a small gasoline engine. The pulley is evident in the many power take-offs from small engines. The small gasoline engine is all-inclusive, ranging from simple machines to compound machines. It contains parts of all types of machines and in turn produces power by the use of some combinations of simple machines. The next part of this chapter will deal with some of the compound machines. Combinations of gears are used to produce transmissions for small and large engines. Gears are made in different shapes and with different types of threads or teeth. Some of the gears used in small gasoline engines will be shown and discussed in the next portion of this chapter.

Compound Machines and Types of Transmissions

A compound machine is the combination of two or more simple machines. To find the mechanical advantage of a compound machine, you will to use the individual mechanical advantages of the various simple machines.

Gears

Gear wheels are used to vary the speed or direction of a twisting force (Fig. 10–9). Note that the direction of the large gear wheel is the opposite of that of the smaller gear wheel. That means direction can be changed by using a gear arrangement. Note that the gear ratio between the teeth in the two gear wheels is 2:1. That means the larger wheel has thirty-two teeth to the smaller one's sixteen teeth. Every time the large wheel is turned one complete revolution, the small gear wheel will rotate through *two* revolutions. That means that with this arrangement there is a *velocity* advantage. In order to obtain a greater speed, it is possible to change the gear ratio and produce the needed or required speed. If the small wheel is the driver and the larger gear wheel is driven, then the smaller wheel turns twice to move the larger one through one revolution. That produces a mechanical advantage and a gain in force, or twisting action.

Notice what can happens with a bicycle sprocket arrangement. The bicycle wheel has a gear with only seven teeth. The sprocket has twenty-eight teeth. This is a 28:7, or 4:1, ratio. That means the smaller

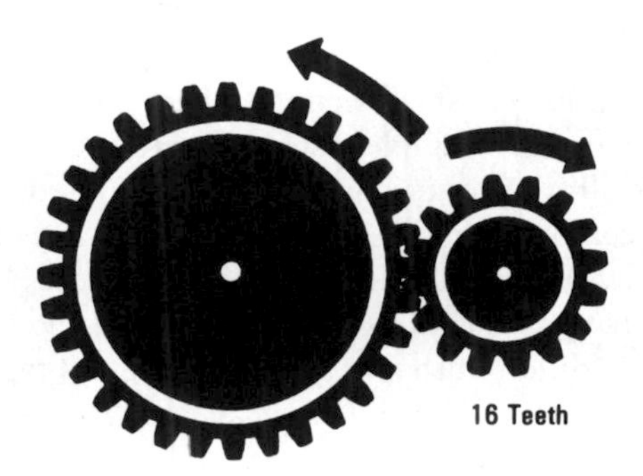

Fig. 10–9. Gears with a 2:1 ratio. There are thirty-two teeth on one wheel and sixteen on the other. The small wheel rotates twice for every rotation of the larger wheel.

gear rotates four times for every rotation of the sprocket gear that is driving it, creating a velocity advantage here. The mechanical advantage is in speed (Fig. 10–10). This same type of arrangement can be used in pulleys where a belt is used to connect the two pulleys. The smaller pulley can drive a slower device by putting a larger pulley on the device to be driven; or a large pulley can drive the small pulley at a very high speed (Fig. 10–11). One advantage of the sprocket, the pulley and belt drives, and the V-belt sheaves is that the drive and driven pulleys are both moving in the same direction. By twisting the belts or drive mediums the direction of rotation can be reversed (Fig. 10–12).

Worm Gear—A worm gear is nothing more than a screw that is meshed with a gear wheel. The effort is applied to the worm wheel. If the gear wheel has fifty teeth, then the worm will have to turn fifty times to make one revolution of the gear wheel (Fig. 10–13).

The main reason for using the worm gear is to turn the direction of rotation of the power source 90°. It allows some rather interesting machine applications and makes power transmission compact.

Clutches

Clutches are used in an apparatus where you want the drive power source to be separated from the driven element at various times. A

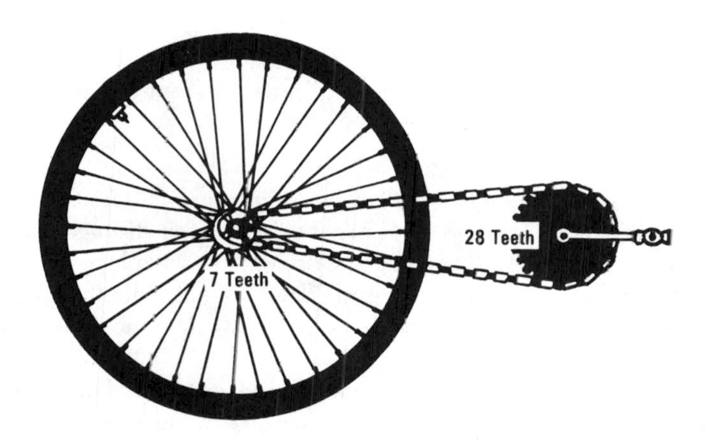

Fig. 10–10. The sprocket has twenty-eight teeth and the rear wheel has seven teeth. What is the mechanical advantage?

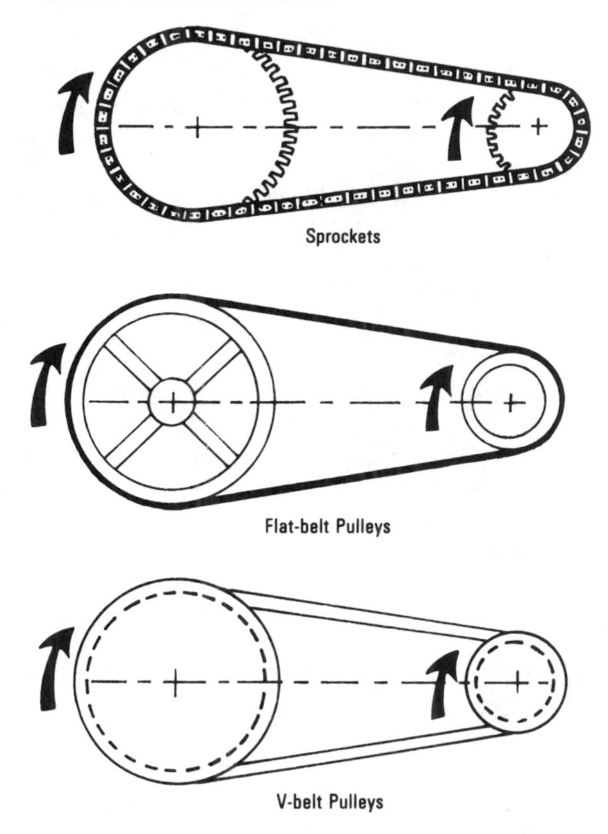

Fig. 10–11. Pulley arrangements. Three different types of connection devices—V belt, flat belt, and chain.

clutch is set between two parts of a power transmission machine. It is arranged so that the two parts may be engaged or disengaged whenever desired (Fig. 10–14A). Friction drives with a stepless ratio change allow output speeds to be controlled within small limits. A stepless ratio change means a smoother change without obvious steps or change points.

A narrow wheel on a friction drive can be adjusted on a threaded shaft to make contact with any radius of the drive disc. A typical clutch of this type can change the speed of the driven machine by changing the position of the smaller wheel on the drive plate (Fig. 10–14C). Au-

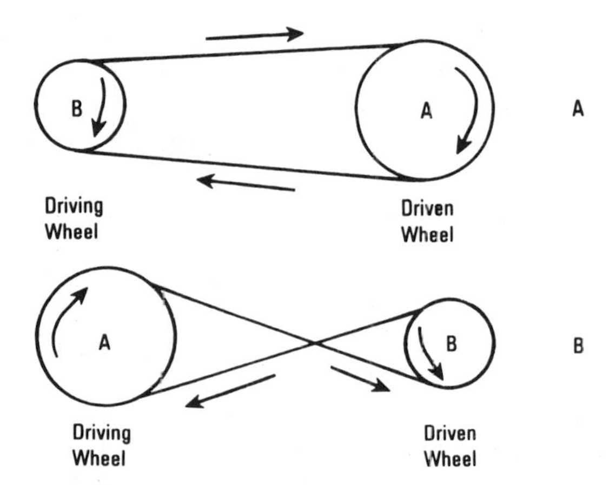

Fig. 10–12. By twisting the belt it is possible to reverse the direction of rotation.

tomatic friction clutches use centrifugal force to engage the mechanism (Fig. 10–14B). This type is used in weed cutters where a two-cycle engine at high speed turns a plastic cord to cut weeds. The engine must be revved up to about 3600 rpm before the centrifugal clutch engages the cutter head.

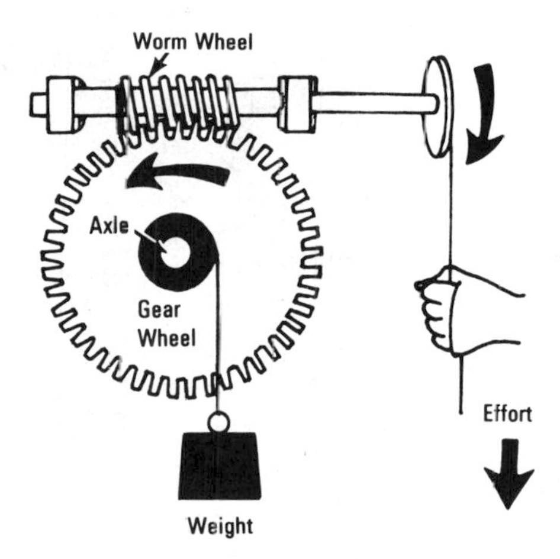

Fig. 10–13. The worm gear is a modification of the screw applied to the gear wheel.

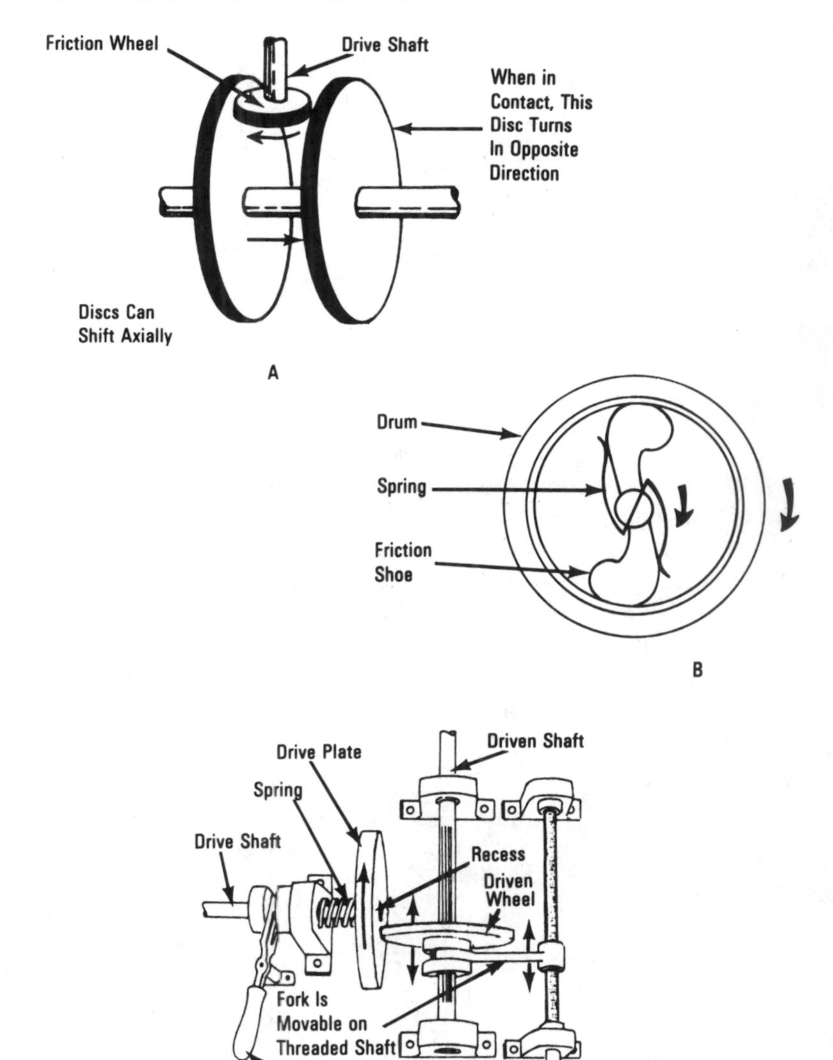

Fig. 10–14. (A) A friction wheel is attached to the drive wheel. The drive wheel contacts the discs. The discs rotate in opposite directions. (B) An example of a centrifugal friction clutch. The shoe is forced outward by the centrifugal force to make contact with the drum. (C) A friction wheel makes contact with the driving disc. (D) A direct-contact clutch catches the grooves in the opposite member and becomes attached directly. (E) An idler pulley is used to engage the belt and make it tight enough to drive the shaft.

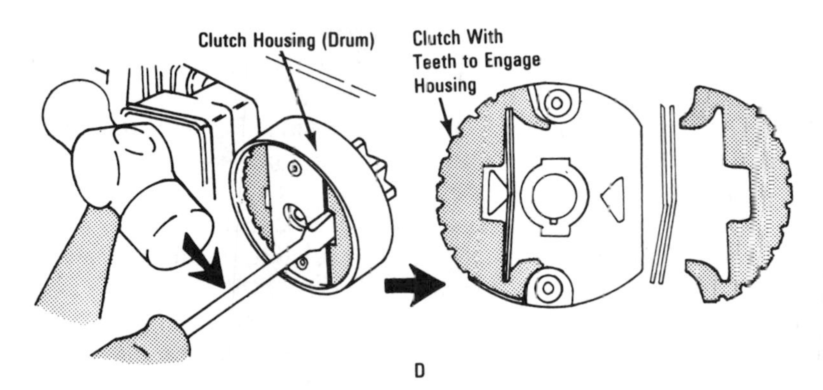
Clutch Housing (Drum)
Clutch With Teeth to Engage Housing

D

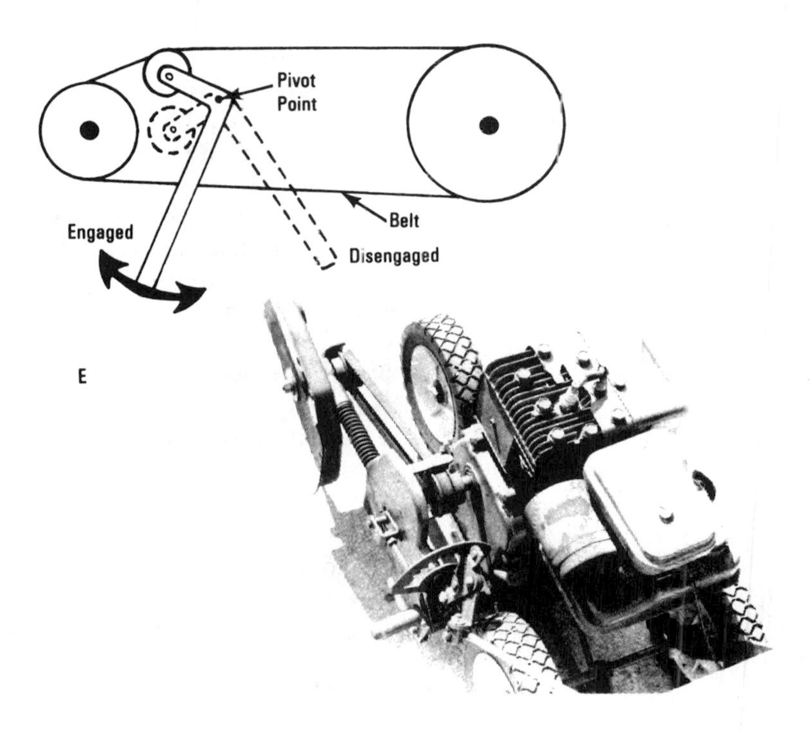
Pivot Point
Belt
Engaged
Disengaged

E

The variable-speed drive clutch arrangement can be used by lifting up the clutch wheel from the plate. If the reverse direction is needed, the small wheel can be moved over to the other side of the drive plate. This reverses the direction of rotation for the output shaft. The clutch wheel and drive plate arrangement is used in Ariens snowblowers.

Another type of clutch is used by Cooper in their lawn edger. The drive belt is tightened enough to cause the cutter blade to rotate by using the clutching method shown in Fig. 10–14E. The typical idler pulley declutching arrangement is used.

As you can see, friction plays a great role in any transmission of power and the control of that power to the point of use. The clutch is used to remove the power without stopping the engine each time. This will allow for the changing of gears or for idling the engine. There are a number of variations of each type shown here. Some use a rubber-tire wheel on a drive plate. Others may use centrifugal weights as wedges that engage the internal spline teeth of a drum. Or a simple cone clutch may slide into engagement when expanded. In most instances the clutches for small gasoline engines are very simple affairs. They are easily understood once you examine them and know that their function is to apply and remove power when needed. In some instances the adjustment of the moving arms or levers is critical. Check the manufacturer's recommendations for adjustments of the clutch wheel. For instance, the distance between the clutch wheel rubber tire and the clutch plate on an Ariens snowblower is only 0.02″ (0.5 mm) when in neutral.

Friction has been used to cause the clutch to operate. It is used in the power transmission of many machines. Any belt is an example of the most obvious use of friction. Friction keeps the belt gripping the pulleys without slipping.

Brakes use friction to stop a car or to slow it down. The amount of friction determines the action. Brakes may be operated by air, as on large trucks, or by hydraulic pressure, as in the automobile. In some cases the brake is actuated by mechanical means—a cable pulls a brake shoe in contact with a rotating surface. This is the case in the parking brake of an automobile and in some lever-operated brakes for lawnmowers. The brake operates by causing enough friction to overcome the momentum of the moving object.

Types of Clutches

Clutches are devices that allow for the removal of power from a machine while the engine is still running. This comes in handy when you are not ready to turn off the engine but need to remove something in the path of the lawnmower or cart. Any number of demands for clutching can be brought to mind when the applications of the small gasoline engine are enumerated.

Cone-Type Clutches—Cones are used for clutching purposes. One cone, either the inner or outer, is lined with a friction material to provide contact and nonslippage once engaged. One cone slips into the other to make contact between the drive and driven shafts (Fig. 10–15).

Positive-Drive Clutches—These are nonslip clutches that engage the power source directly with the driven shaft. This type is used in Ariens snowblowers to disconnect the engine from the auger or

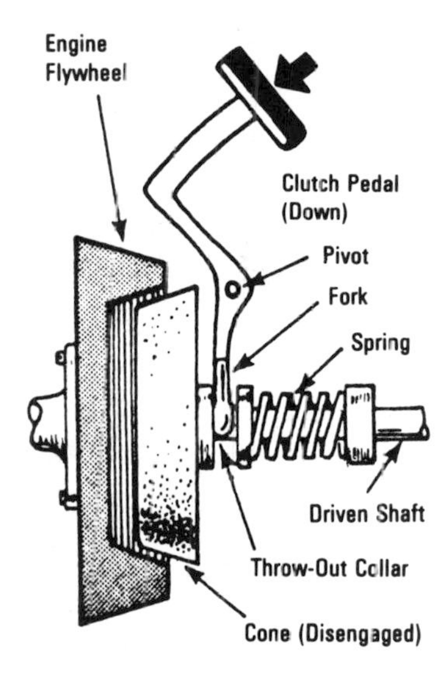

Fig. 10–15. A cone-type clutch. Old automobiles used it when the first models were designed. It still has applications with small engines.

wheels. The operation of this type of clutch depends on the shafts being at rest or turning at a very low rpm (Fig. 10–16).

Overrunning Clutches—The automobile uses the overrunning clutch. Some industrial applications are also evident in larger machines where the power has to be disconnected and the driven part stopped for maintenance or change. The ratchet-and-pawl overrunning clutch is used in industry and in some hand tools such as the wrench or the hand brace.

The roller overrunning clutch uses a spring-loaded roller and ramp. The roller turns freely when the clutch is operated in one direction. When reversed, the roller moves up the ramp to bind with the outer member of the clutch and to engage the load.

There are many variations of the overrunning clutch. The main feature is that it can be used only where there is a minimum of control needed.

Disc Clutches—The automobile uses a friction clutch that has a driven plate (the friction disc) splined to the clutch or transmission input shaft. The spline permits the disc to move back and forth on the

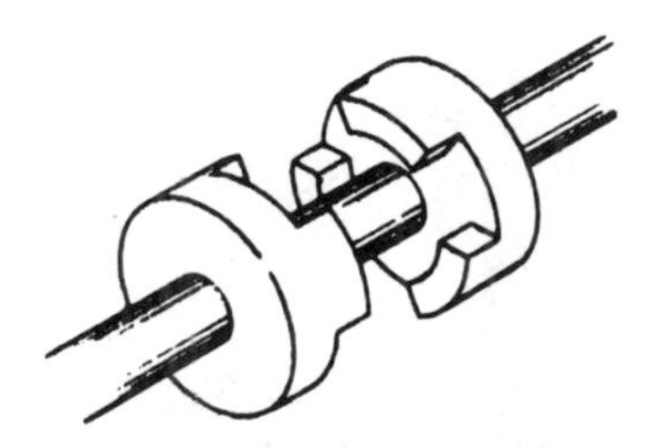

Power transmitted in either direction.

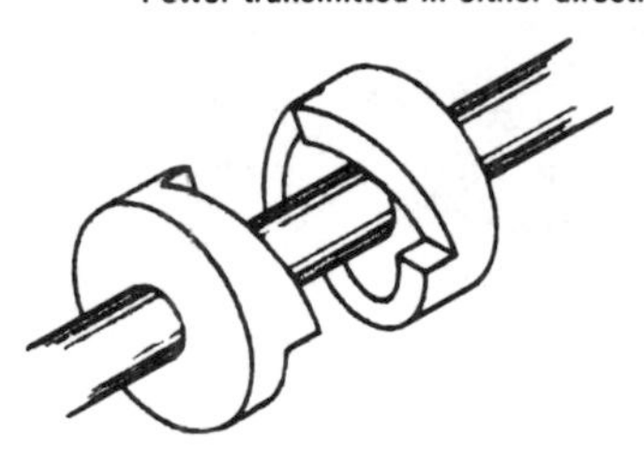

Power transmitted clockwise only.

Fig. 10–16. Directly coupled clutch. Note how the power can be coupled and transmitted.

shaft while causing the shaft to rotate with it. The flywheel of the engine is a polished surface that contacts the friction plate of the clutch when it is released. The friction plate is lined with a high-friction material, which can take the heat generated by contacting the rotating flywheel and coming up to speed.

Take a look at Fig. 10–17 for an example of how the clutch works.

Magnetic Clutches—Automobile air-conditioning compressors use this type of clutch—electromagnetism causes one plate to attract to another. This attraction creates a nonslip bond at normal speed. The clutches operate on direct current.

Centrifugal Clutches—Centrifugal clutches use centrifugal force and friction to operate. Centrifugal force tends to pull a rotating body away from its center of rotation. Friction exists between any two bodies in contact where one body is trying to move relative to the other body. There are two basic parts to this type of clutch. Take a look at Fig. 10–18 for a look at how the basic idea works.

As already mentioned, the centrifugal clutch comes in handy with the weed cutters that utilize a plastic cord. The speed of the engine must be such that the centrifugal force causes the clutch to engage and the string to be rotated at high speed for weed cutting.

Other Types—There are many other types of clutches: the overload release clutch, the overload clutch, the dry fluid clutch, the wet fluid clutch, and the electric clutch. The electric clutch may be the electromagnet disc type, the metal particle type, or the eddy current type. Each has a particular application and each has its ability to do a specific job.

Clutches are necessary to make use of any machine that is not a constant-speed type. A machine that needs the engine to run while various adjustments are made on the driven machine is an ideal candidate for the clutch. The type of clutch must be selected according to the job to be performed.

The clutch allows the engine to be uncoupled from the load while gears are shifted to change the speed of a machine. This changing of speeds is possible by the use of various gears in a transmission

Asbestos

At one time asbestos was used for clutch plates and brake pads for automobiles and motorcycles. Asbestos is one of a group of natural

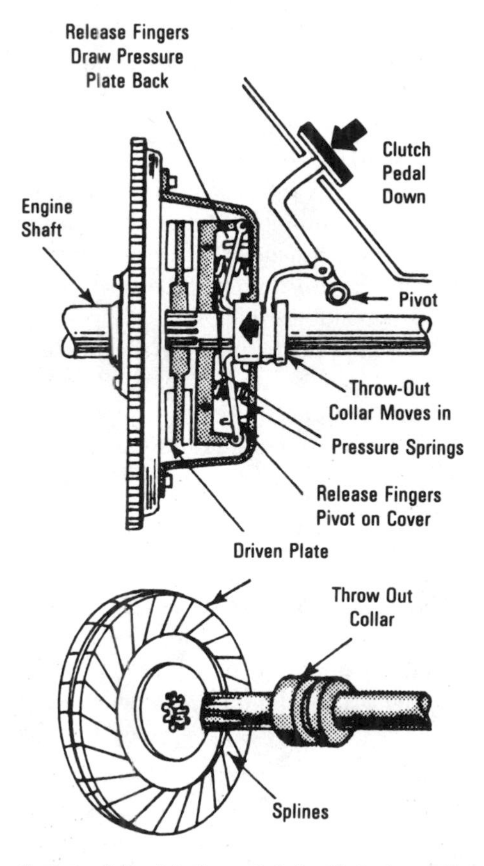

Fig. 10–17. Automobile friction clutch. Note how the driven plate is constructed with a high-friction material.

minerals made up of tiny fibers. Unless completely sealed in a product, asbestos can easily crumble into a dust of tiny fibers small enough to become suspended in the air and inhaled into the lungs, where they remain for many years.

Exposure to asbestos fibers over a period of time can cause lung cancer, mesothelioma (a cancer of the lung and abdominal lining), or asbestosis (a chronic lung condition that eventually makes breathing nearly impossible) fifteen to forty years later. Between 1900 and 1986

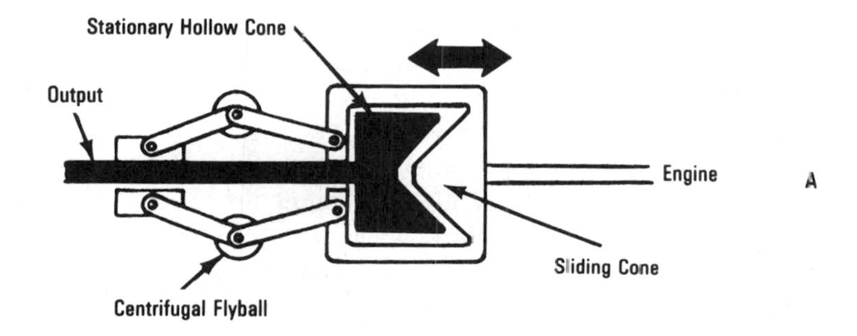

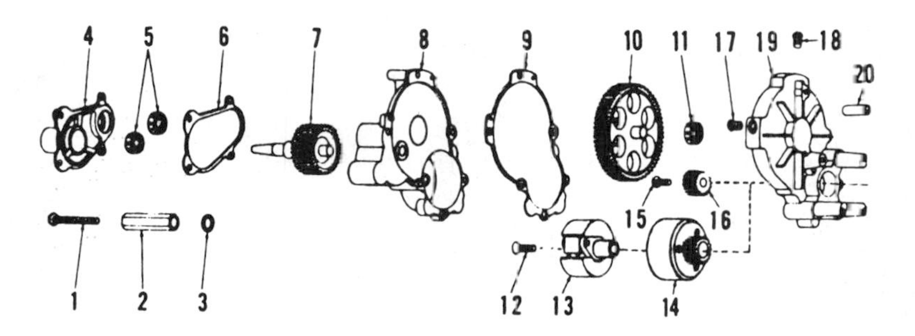

B

Exploded view of double reduction gear unit. On models without clutch (13 & 14), drive gear (16) is used.

1. Extension shaft screw
2. Extension shaft
3. Sealing ring
4. Reduction gear cover
5. Roller bearings
6. Gasket
7. Output gear
8. Gear housing cover
9. Gasket
10. Intermediate gear
11. Roller bearing
12. Phillips head screw
13. Clutch hub and shoe assy.
14. Clutch drum & drive gear
15. Phillips head screw
16. Drive gear
17. Plug
18. Plug
19. Gear housing
20. Spacer (reverse position gear case models only)

Fig. 10–18. (A) One method of using centrifugal force to cause a clutching action. As the speed of the engine increases, it turns the balls and they fly outward. As the balls move outward they cause the sliding cone to come in contact with the stationary cone. The stationary cone is attached to the output shaft. (B) One type of centrifugal clutch used in an O & R small gasoline engine.

over 31 million tons of asbestos were used in the United States for hundreds of purposes. In 1989, the Environmental Protection Agency ordered a ban on almost all uses of asbestos, such as brake linings, roofing shingles, and water pipes, by 1997.

Manufacturers of brake linings and clutch plates have looked elsewhere to obtain material of a high-friction nature. The Vesrah Company, Inc. of Chicago and Tokyo, Japan uses a synthetic fiber, *Kevlar*, by DuPont to make clutch and brake linings to replace asbestos. Formula One car racing now uses Carbon/Carbon brakes. Today, S45 friction material in motorcycle brake pads uses ceramics, developed in the United States for the space program as a heat shield. Motorcycles, some automobiles and ATVs produced in the world today have brake systems designed for sintered metal pads.

Mechanisms of Drives and Transmissions

A number of different types of drives are available to the person who needs to change the direction of rotation or the speed of rotation of a power source.

Drives

There are right-angle drives, gearbox drives, gear-shifting transmissions, and free-wheeling drives that can be used to transfer power from the source to the user.

Right-Angle Drives—This type of drive consists of two gears mounted so that they mesh at 90° (Fig. 10–19). The input shaft may have a gear with the same number of teeth and the same diameter as the output shaft. In this case all that is accomplished is the changing of direction of the input and output. The mechanical advantage of the arrangement is still one. A mechanical advantage can be obtained by having the two gears of different diameters (Fig. 10–19B). In most cases the gear drives are housed in appropriate containers with the proper seals, bearings, and lubrication fittings.

Gearbox Drives—The gearbox contains a number of gears with different outputs. The box may house gears with a mechanical advantage or a velocity advantage. An input on one side may produce several outputs from other sides of the box. In some cases the direction of ro-

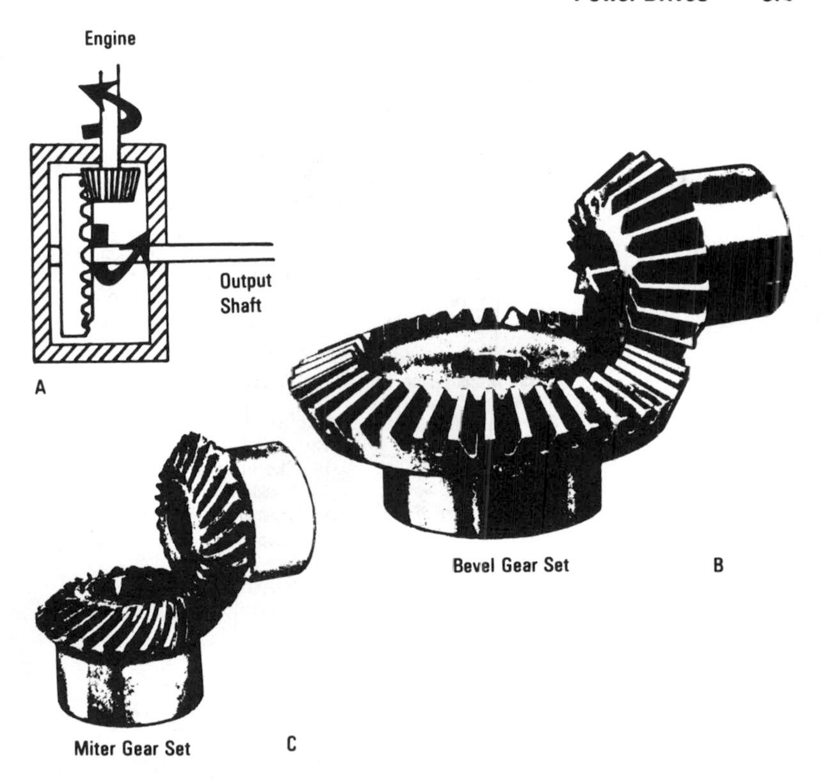

Fig. 10–19 Right-angle drive turns the input power to an output angle of 90°. Bevel gears or miter gears can be used to do the job.

tation is the only consideration. The arrangement of the gears in reference to the input shaft or gear on that shaft determines the output.

Some small engines have a gear-reduction box. Fig. 10–20A shows the O & R engine with a gear-reduction drive and a centrifugal clutch arrangement. Some engines made by O & R (Ohlsson & Rice) have a double-reduction gear unit. Fig. 10–20B illustrates Briggs & Stratton units.

Gear-Shifting Transmission—A gear-shifting transmission may be able to change direction of rotation or it may have a mechanical advantage or a velocity advantage. If three or more gears are located in the transmission with a stick shift, you can be sure it has a reverse and at least two forward gears for different speeds.

A combination of the gears may be used to cause the drive to be

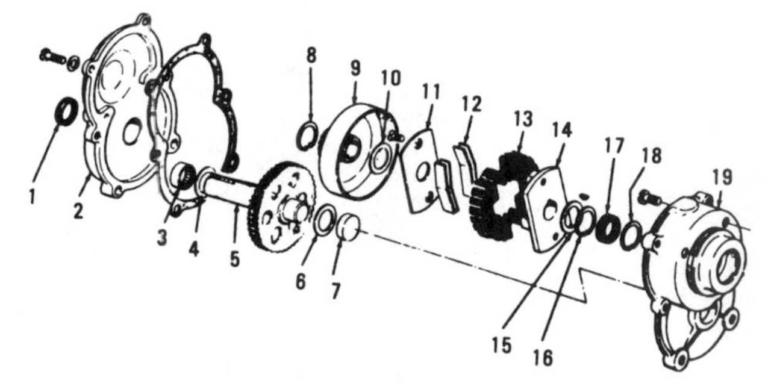

Exploded view of gear reduction unit. Clutch components may be used without gear reduction as direct drive.

1. Seal
2. Gear cover
3. Bearing
4. Thrust washer
5. Output gear & shaft
6. Thrust washer
7. Bearing
8. Snap ring
9. Clutch housing
10. Washer
11. Retainer plate
12. Clutch springs
13. Clutch shoes
14. Clutch hub
15. Washer
16. Snap ring
17. Seal
18. Snap ring
19. Crankcase cover

A

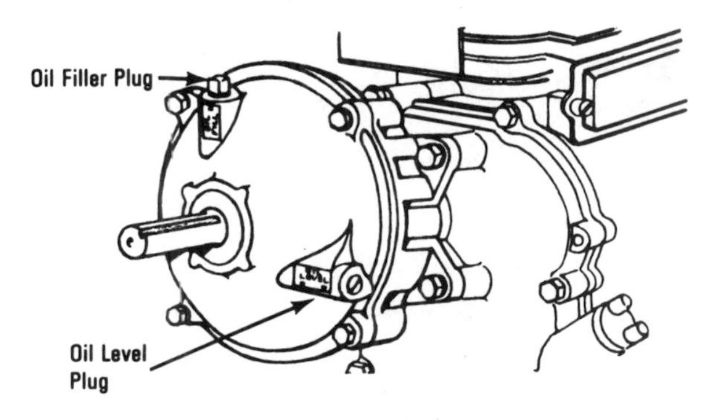

6:1 REDUCTION

B

Fig. 10–20. (A) Exploded view showing the gear-reduction unit in an O & R small engine unit. (B) The 6:1 gear reduction mounted on some models of B & S engines to slow the output for driving lawnmowers and other devices.

forward or reverse. A neutral position is also necessary for periods when no drive is needed, but it is inappropriate to turn off the engine. In neutral there is no connection between the input and output shafts.

Forward outputs are called high and low if there are only two speeds. If there are three speeds, they become low, second, and high. On some indicators, numbers are used instead of low and high. Other gears may be added and any number of speeds may be selected using the proper combination of gears.

Fig. 10–21 shows the possibilities with a gear-shifting transmis-

Gears are now in NEUTRAL as shown.
+1st Gear.† Gear 2 slides on the spline A to mesh with gear 8. The drive for the output shaft is through 5,6, the countershaft, 8,2, and 1 to the output.

+2nd Gear.† Gear 3 slides on the spline 1 to mesh with gear 7. The drive action is from the input through 5, 6 countershaft, 7,3, and 1 to the output.

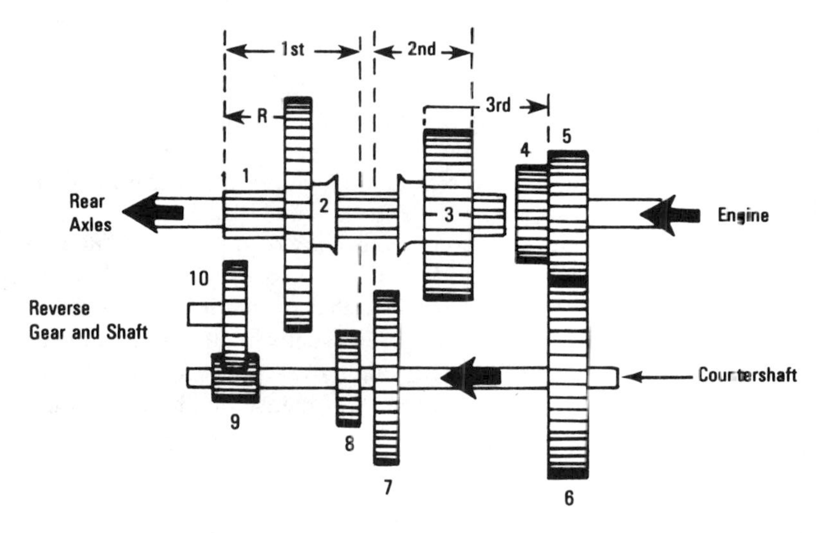

+3rd Gear.† Gear 3 slides on spline 1 to mesh with dog gear 4. The drive is from input through 4,3, and 1 to the output.

+REVERSE.† Gear 2 slides on spline 1 to mesh with gear 10. The reverse drive takes place from the input through 5,6, countershaft, 9, 10, 2, and 1 to the output.

Fig. 10–21. Transmission capable of three forward speeds and reverse.

sion capable of three forward and one reverse speeds. Some machines driven by small gasoline engines have transmissions that require shifting. That means a clutch is necessary to release the engine from the gear-shifting box during the shifting procedure. Small garden tractors and some larger lawn mowers have shifting transmissions.

Free-Wheeling Drive—This type of drive has a number of names. It may be called an overriding clutch or a "pineapple." The latter name is the result of the arrangement needed to have it operate as intended. The gears resemble a pineapple when viewed as a unit. This type of drive is usually a one-to-one type. It is a straight-through drive designed to rotate the output shaft, but the output is not able to rotate the input shaft.

Mechanisms

There are some mechanisms used in small gasoline engines that need mentioning since they will be encountered by anyone working with this type of device.

Universal Joint—The universal joint was conceived back in the days of Leonardo da Vinci. It is a device that can adjust to any misalignment of the drive source with the driven part. There are many different variations of the universal joint. However, the single and double joints are the most common types used.

The universal joint is a one-to-one driving device used to join two shafts. Check Fig. 10–22 to see how the two yokes are placed 90° to each other and joined by a centerpiece that resembles a cross. Each yoke is allowed to move up and down and usually has bearings or bushings for easier movement. Since they do move, it is necessary to have the bearings prepacked or some grease fitting installed for lubrication purposes.

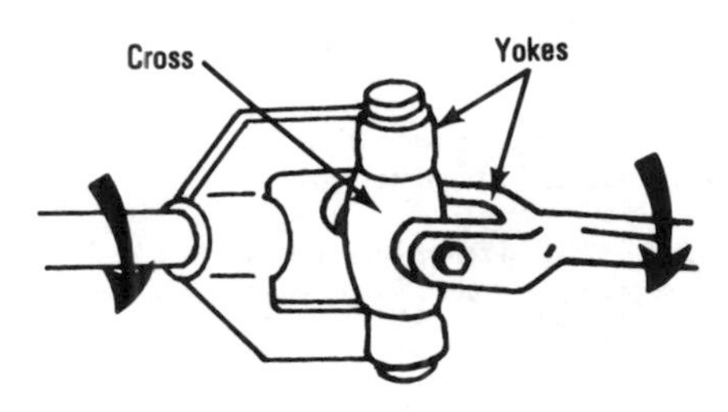

Fig. 10–22. Universal joints have many uses.

Since the two 90° sections are located so that they can move freely at the hinge point, the hinge actions will compensate for the movement in the shafts as they rotate. Automobiles use a universal joint between the transmission and the differential located at the rear of the car. Small-gasoline-engine machines, such as the Ariens snowblower, use the universal joint in the crank that rotates the snowblower's deflector (Fig. 10–23).

Flexible Shafts—Flexible shafts are used on small-gasoline-engine machines such as weed cutters. The long handle is filled with either a link-type drive or a flexible-wire-type drive. In most instances the flexible wire is used to power the cutting device (Fig. 10–24)

Differentials—A differential is usually located in a gearbox so that the proper lubrication, seals, and bearings may be attached to support the gears and shafts. In most instances the differential is used to permit one wheel to move when the other is standing still or moving in the opposite direction; at the same time it can be used to drive both wheels in the same direction. Such a device needs some special gears and special attention to details.

Fig. 10–25 shows how this is accomplished. Input is from the drive pinion to the ring gear. The ring gear is attached to and rotates the housing. The housing is not connected to either of the output shafts. There are four spider gears. The spider is rotated by the ring-gear housing. The spider gears are attached to the spider. When there is an equal load on the output shafts, spider gears serve no purpose other than as wedges between the two output shafts and their gears. The output shafts are usually the rear axles of a car, lawnmower, or tractor.

If one of the output shafts is held stationary, the spider, in order to rotate, must rotate around the output shaft that is held stationary. This means that the four spider gears will have to rotate on their axles. Rotation of the spider gears is transmitted to the opposite output gear shaft. The rotation of the spider is also transmitted to the stationary output gear shaft. This means that the moving axle of the output shaft will rotate not only with the increased velocity advantage but also with the additional velocity advantage obtained through the rotation of the spider gears. Additional velocity advantage transmitted through the spider gears will be proportional to the slowdown of the overloaded shaft. The overloaded shaft is the stationary one. This allows for an automobile to turn corners with one wheel turning faster than the other. It takes a faster-turning wheel to go around the outside of the turn

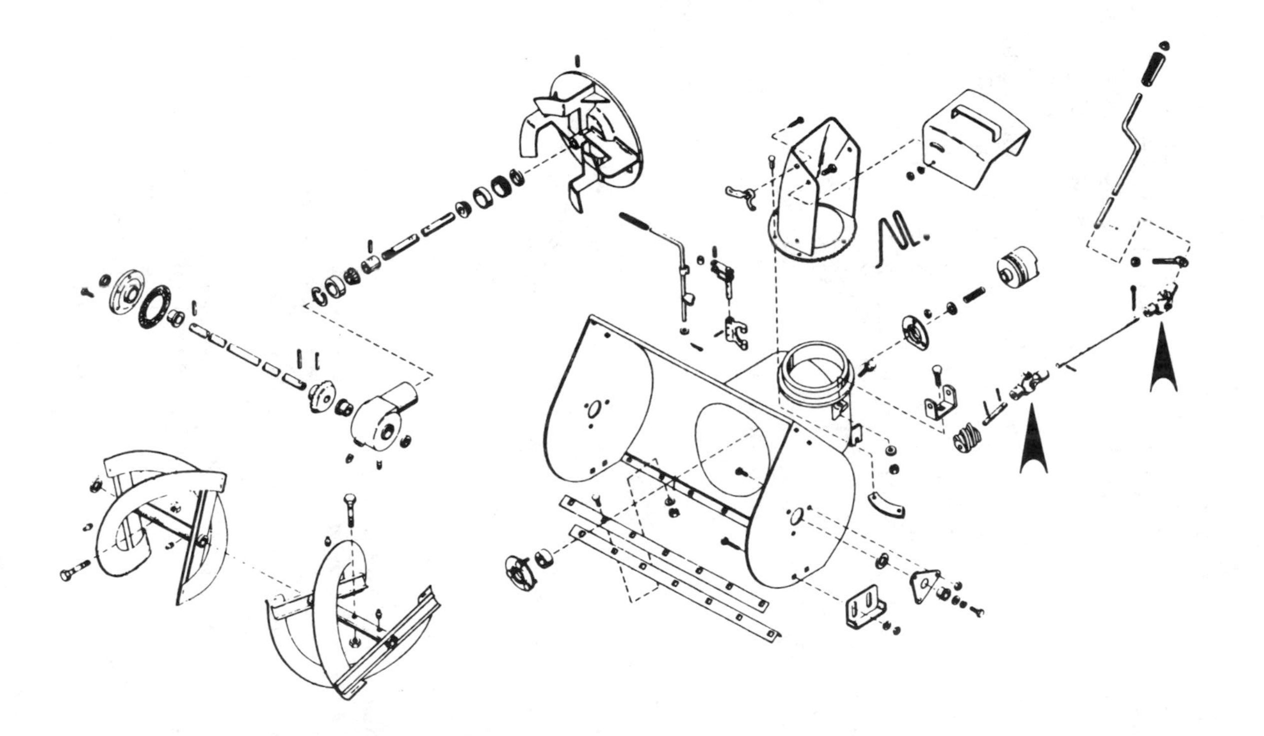

Fig. 10–23. On this snowblower, universal joints are used to adjust the snow deflector about 300°.

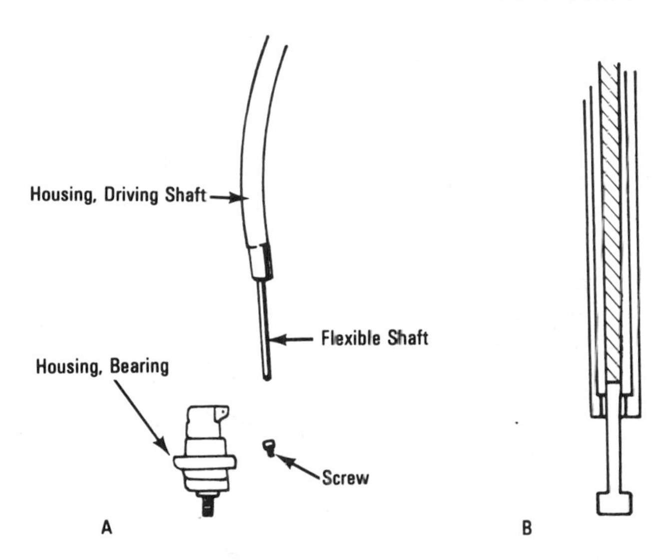

Fig. 10–24. (A) Flexible shaft enclosed in a housing. It is used to drive a housing with connections to a weed-cutter string holder. (B) Cutaway view of the wire-type drive cable or flexible shaft used on a weed cutter.

than the inside, so the outside wheel must turn faster than the inside wheel.

Most machines driven by small engines allow one axle or wheel to be driven and the other to float, or not be driven. That way the free wheel can catch up with the driven wheel or turn faster as needed on turns. However, many drives on lawnmowers drive two wheels equally so the driven wheels have to be picked up from the ground every time they are turned, or one of the wheels has to be slid across the grassy surface to catch up with the other wheel.

Sun Gear—The sun gear is much like the differential (Fig. 10–26). A planetary gear, an internal ring gear, and a sun gear are used. Four planetary gears must be used for this arrangement to operate properly.

The four planetary gears are similar to the spider gears in the differential. A bracket attached to a shaft holds the planetary gears in their respective places. The internal ring gear holds the planetary gears in mesh with the sun gear. If the ring gear is rotated when the sun gear is held stationary, the ring gear will cause the planetary gears

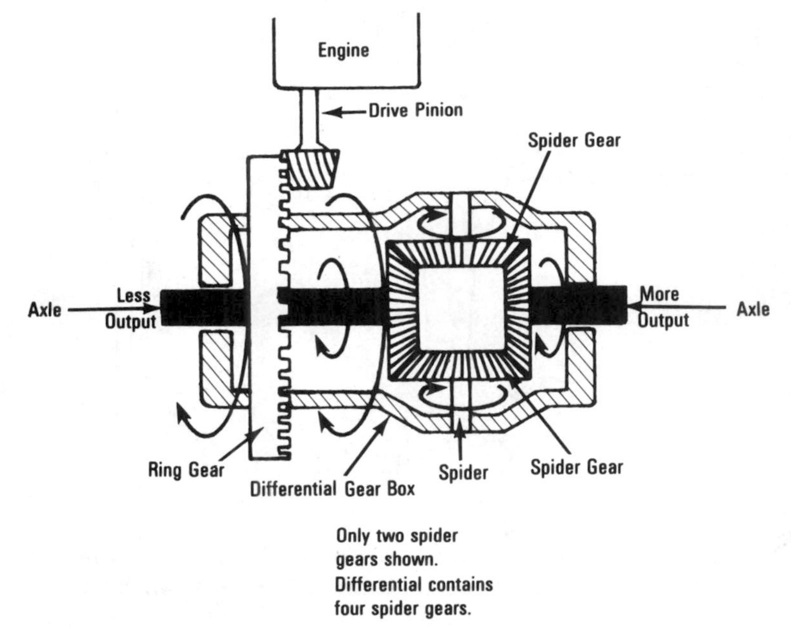

Fig. 10–25. Differential gearbox.

to travel on the outside of the sun gear without exerting to cause it to move. The energy input to the ring gear is dissipated through the rotation of the planetary gear around the stationary sun gear. If the bracket that holds the planetary gears is held steady and the ring gear is rotated in a counterclockwise (CCW) direction, the energy of the rotating ring

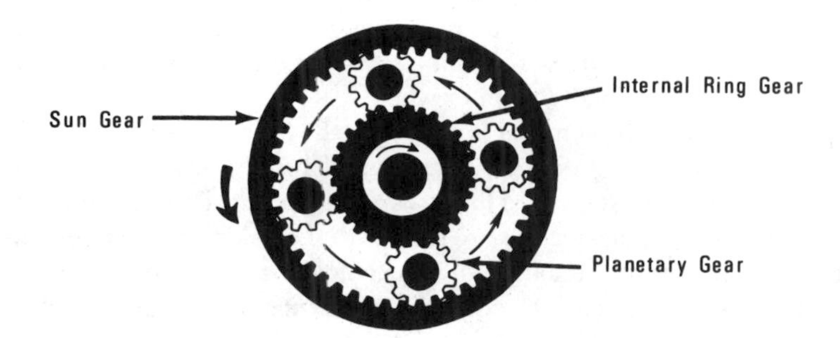

Fig. 10–26. Sun gear.

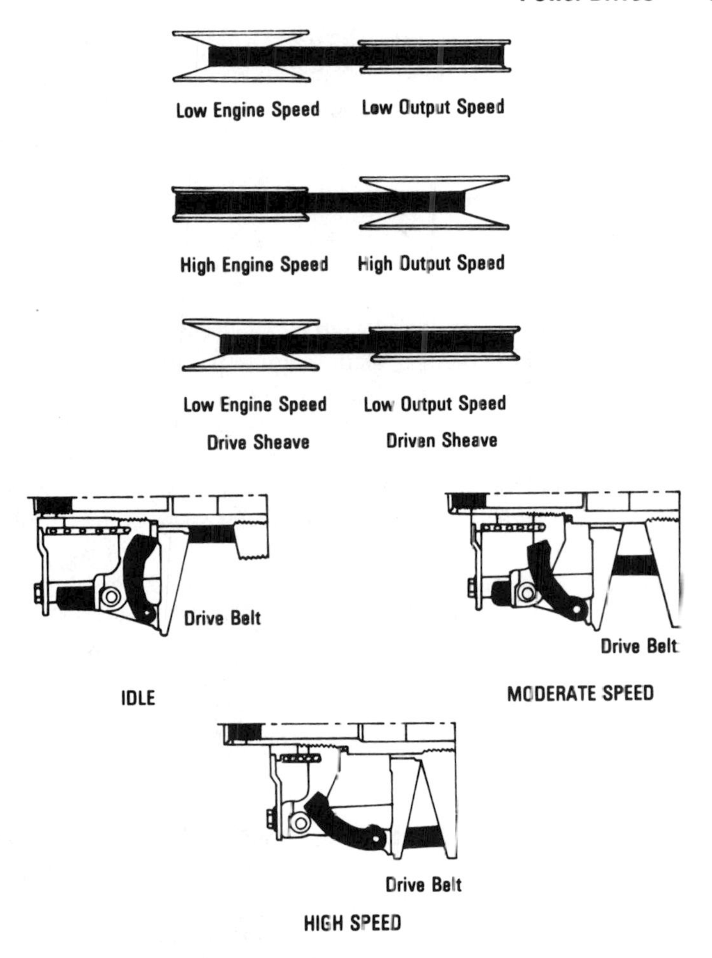

Fig. 10–27. Adjustable-speed sheaves.

gear will cause each of the planetary gears to rotate around their axles. This rotation is in the same direction (CCW) as the ring gear. However, the rotating planetary gear in a CCW direction will cause the sun gear to be driven in the opposite, or clockwise, direction (Fig. 10–26). Power input can also be at the bracket that holds the planetary gear or at the sun-gear shaft.

Adjustable-Speed Sheaves—Some small-engine-driven devices, such as snowmobiles, use the variable- or adjustable-speed qualities of the sheave. A sheave is a pulley. A V-belt is used to drive pulleys or sheaves. If the sheaves are adjustable, it is possible to increase or decrease the speed according to the ratio of the diameters of the pulleys or sheaves (Fig. 10–27). As the sheave separates, it allows the belt to slip down nearer the shaft. This effectively decreases the pulley diameter. By varying both sheaves it is possible to obtain a variety of speed characteristics for a snowmobile or any other type of driven device.

Power Take Off—Power take-off and reduction units are available for most horizontal engines since they more easily fit the needs of powered devices. The vertical shaft engine is primarily used for lawnmowers with a blade fitted directly to the shaft.

Good examples of small engine power take-off variations in shafts are shown in Fig. 10–28. This shows the standard PTO shaft-type variations found in the Honda line of small engines. Also note the reduction unit shown. The PTO shaft type varies according to the coupling mode of the attachment.

Specifications for Honda engines ranging from 2.2 horsepower to 13.0 horsepower are shown in Fig. 10–29. Fig. 10–30 shows the horsepower, engine speed and torque characteristics for each type of engine. Practical applications for each engine are also given.

Outboard Engines

There are a number of outboard engines made today. They are small and lightweight. Most of them use two-cycle engines. They may have one, two, or three cylinders. Some are made with six and others with eight. All sizes are available. But to remain within the scope of this book, we will limit our discussion to the smaller-horsepower type that in most instances uses only one cylinder.

Fig. 10–31 shows one of the smaller outboard engines made by Chrysler. This is a 4-horsepower engine. The 6- and 8-horsepower engines look larger with the gears and housing being enclosed in larger shrouds (Figs. 10–32 and 10–33).

So far we have looked at engines that are capable of being used as outboard engines on boats. If we take a closer look at the lower unit of the outboard engine, we will find the gearing needed to make the propeller move and even reverse its direction of rotation.

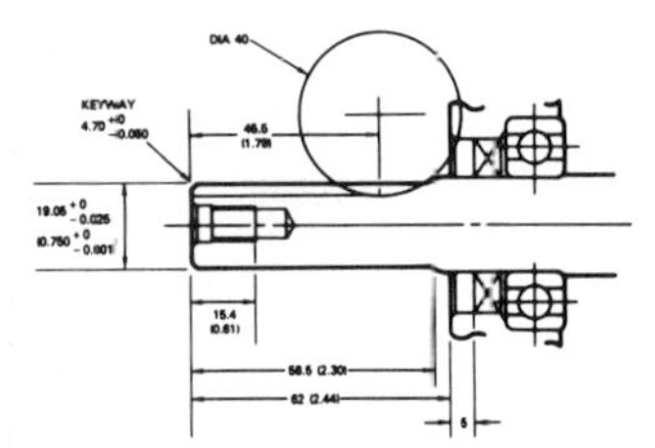

Q-Type Shaft-Flat Key For General Purpose*—See each model for specifications.*

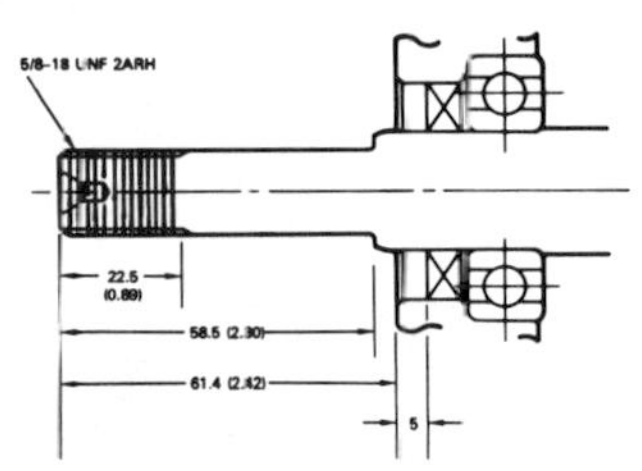

P-Type And T-Type Threaded Crankshaft*—See each model for specifications.*

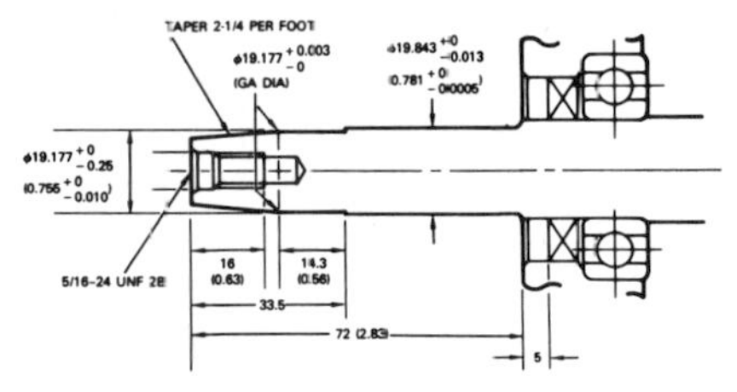

V-Type/Taper*—See each model for specifications.*

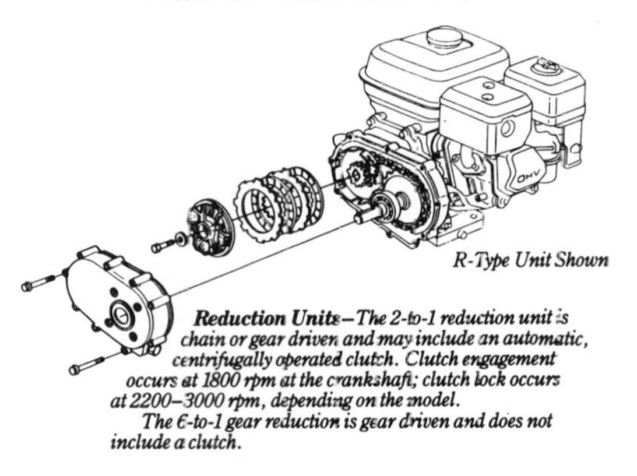

R-Type Unit Shown

Reduction Units*—The 2-to-1 reduction unit is chain or gear driven and may include an automatic, centrifugally operated clutch. Clutch engagement occurs at 1800 rpm at the crankshaft; clutch lock occurs at 2200–3000 rpm, depending on the model.*

The 6-to-1 gear reduction is gear driven and does not include a clutch.

Fig. 10–28. Standard PTO (Power Take Off) shaft-type variations in Honda horizontal engines. *(Courtesy Honda)*

SPECIFICATIONS

HORSEPOWER	MODEL	TYPE	VARIATION	OIL ALERT	HORIZONTAL SHAFT	VERTICAL SHAFT	CRANKSHAFT P.T.O.	6 to 1 REDUCTION	2 to 1 REDUCTION	TOP GOVERNED SPEED — NO LOAD	RECOIL STARTER	12V STARTER	TRANSISTOR IGNITION	C.D.I. (W/ELECTRONIC ADVANCER GX340, GX390	AIR CLEANER	FUEL TANK (QT)	WEIGHT
2.2	G100	K1	QA		•		2 19/64 x 5/8 dia. tapped 1/4 28 UNF			4200	•		•		SD	1.48	19
4	GX120	K1	QA	•	•		2 7/16 x 3/4 dia. tapped 5/16 24 UNF			3900			•		OB	2.64	34
	GX120	K1	QX	•	•					3900	•		•		DE	2.64	30
	GX120	K1	QXC	•	•					3900	•		•		CY	2.64	34
	GX120	K1	QXS(1)	•	•					3900	•		•		DE	2.64	30
	GX120	K1	HX	•	•		2 3/64 x 3/4 dia.	•		3900	•		•		DE	2.64	45
	GX120	K1	TX	•	•		2 7/16 x 5/8 threaded			3900	•		•		DE	2.64	34
	GXV120		A1			•	3 5/32 x 7/8 dia. tapped 3/8 24 UNF			3600	•		•		DE	1.06	36
	GXV120		D1			•				3600	•		•		DE	1.06	36
5.5	GX160	K1	QA		•		2 7/16 x 3/4 dia. tapped 5/16 24 UNF			3900	•		•		DE	3.88	36
	GX160	K1	QX	•	•					3900	•		•		DE	3.88	34
	GX160	K1	QXS(1)	•	•					3900	•		•		DE	3.88	34
	GX160	K1	QXC	•	•					3600	•		•		CY	3.88	34
	GX160	K1	VX	•	•		2 53/64 x 3/4 dia. taper 2 1/4" per ft.			3900	•		•		DE	3.88	38
	GX160	K1	TX	•	•		2 7/16 x 5/8 threaded			3900	•		•		DE	2.01	36
	GX160	K1	RH		•		2 3/32 x 22 mm tapped M8 x 1.25		•	3900	•		•		DE	3.88	48
	GX160	K1	QXE		•		2 7/16 x 3/4 dia. tapped 5/16 24 UNF			3900	•	•	•		DE	3.88	34
	GX160	K1	HX		•		2 3/64 x 3/4 dia.	•		3900	•		•		DE	3.88	48
	GXV160		N1			•	3 5/32 x 7/8-1 dia. tapped 3/8 24 UNF			3600	•		•		DE	2.01	40
8	GX240	K1	QA	•	•		3 31/64 x 1 dia. tapped 7/16 20 UNF			3900	•		•		DE	6.4	66
	GX240	K1	QAE	•	•		3 31/64 x 1 dia. tapped 7/16 20 UNF			3900	•	•	•		DE	6.4	62
	GX240	K1	QAS	•	•					3900	N/A		•		CY	N/A	56
	GX240	K1	QXC	•	•		3 31/64 x 1 dia. tapped 7/16 20 UNF			3900	•		•		CY	6.4	76
	GX240	K1	HA	•	•		3 15/64 x 1 dia.	•		3900	•		•		DE	6.4	56
	GX240	K1	LA(3)	•	•		2 23/64 x 25 mm tapped M8 x 1.25		•	3900			•		DE	6.4	59
	GX240	K1	PA	•	•		3 1/2 x 1 dia. 14 NF threaded			3900	•		•		DE	6.4	62
	GX240	K1	RA	•	•		2 3/64 x 22 mm tapped M8 x 1.25		•	3900	•		•		DE	6.4	67
	GX240	K1	VA	•	•		4 11/64 x 22.2 mm taper 2 1/4" per ft.			3900	•		•		DE	6.4	56
9	GX270		QA	•	•		3 31/64 x 1 dia. tapped 3/8 24 UNF			3900	•		•		DE	6.4	66
	GX270		QAE	•	•					3900	•	•	•		DE	6.4	73
	GX270		QXC	•	•					3900	•		•		CY	6.4	66
	GX270		HA	•	•		3 15/64 x 1 dia.	•		3900	•		•		DE	6.4	56
	GX270		RA	•	•		2 3/64 x 22 mm tapped M8 x 1.25		•	3900	•		•		DE	6.4	56
	GX270		PA	•	•		3 1/2 x 1 dia. 14 NF threaded			3900	•		•		DE	6.4	56
	GX270		VA	•	•		4 11/64 x 22.2 mm taper 2 1/4" per ft.			3900	•		•		DE	6.4	56
11	GX340	K1	QA	•	•		3 31/64 x 1 dia. tapped 3/8 24 UNF			3900	•			•	DE	7.4	73
	GX340	K1	QAE	•	•					3900	•	•		•	DE	7.4	79
	GX340	K1	QXC	•	•					3900	•			•	CY	7.4	74
	GX340	K1	Q1J0(4)	•	•					3900	•	•		•	DE	7.4	79
	GX340	K1	VA	•	•		4 11/64 x 7/8 taper 2 1/4" per ft.			3900	•			•	DE	7.4	74
	GX340	K1	VXE	•	•					3900	•	•		•	DE	7.4	79
	GX340	K1	HA	•	•		3 15/16 x 1 dia.	•		3900	•		•	•	DE	7.4	78
	GXV340	K1	DA2	•		•	3 5/32 x 1 dia. tapped 7/16 20 UNF			3600	•		•			W/O	86
	GXV340	K1	DAET			•				3600	•	•	•		DE	2.2	89
	GXV340	K1	DAP			•				3600	•	(5)	•		DE	W/O	86
13	GX390	K1	QA	•	•		3 31/64 x 1 dia. tapped 3/8 24 UNF			3900	•			•			73
	GX390	K1	QXC	•	•					3900	•			•			74
	GX390	K1	QAE	•	•					3900	•		•	•			79
	GX390	K1	Q1J0(4)	•	•					3900	•	•		•			79
	GX390	K1	VA	•	•		4 11/64 x 22.2 mm taper 2 1/4" per ft.			3900	•		•		DE	7.4	74
	GX390	K1	VXE	•	•					3900	•	•		•	DE	7.4	79
	GX390	K1	HA	•	•		3 15/16 x 1 dia.	•		3900	•		•		DE	7.4	78
	GXV390	K1	DA2	•		•	3 5/32 x 1 dia. tapped 7/16 20 UNF			3600	•					W/O	86
	GXV390	K1	DAET			•				3600	•	•				2.2	86
	GX360	K1	DD	•	•		3 x 1 tapped 7/16 20 UNF			3600		•				W/O	154

NOTES:(1) 12V(DC)-50 w. lamp coil (2)Fuel pump; fill tube (3)2:1 reduction no clutch (4)10 amp charging (5)Provision for electric starter SD-Semi dry OB-Oil bath DE-Dual element W/O-Without CY-Cyclone

Fig. 10–29. Specifications for Honda engines 2.2 to 13 horsepower. Note the PTO shaft sizes. *(Courtesy Honda)*

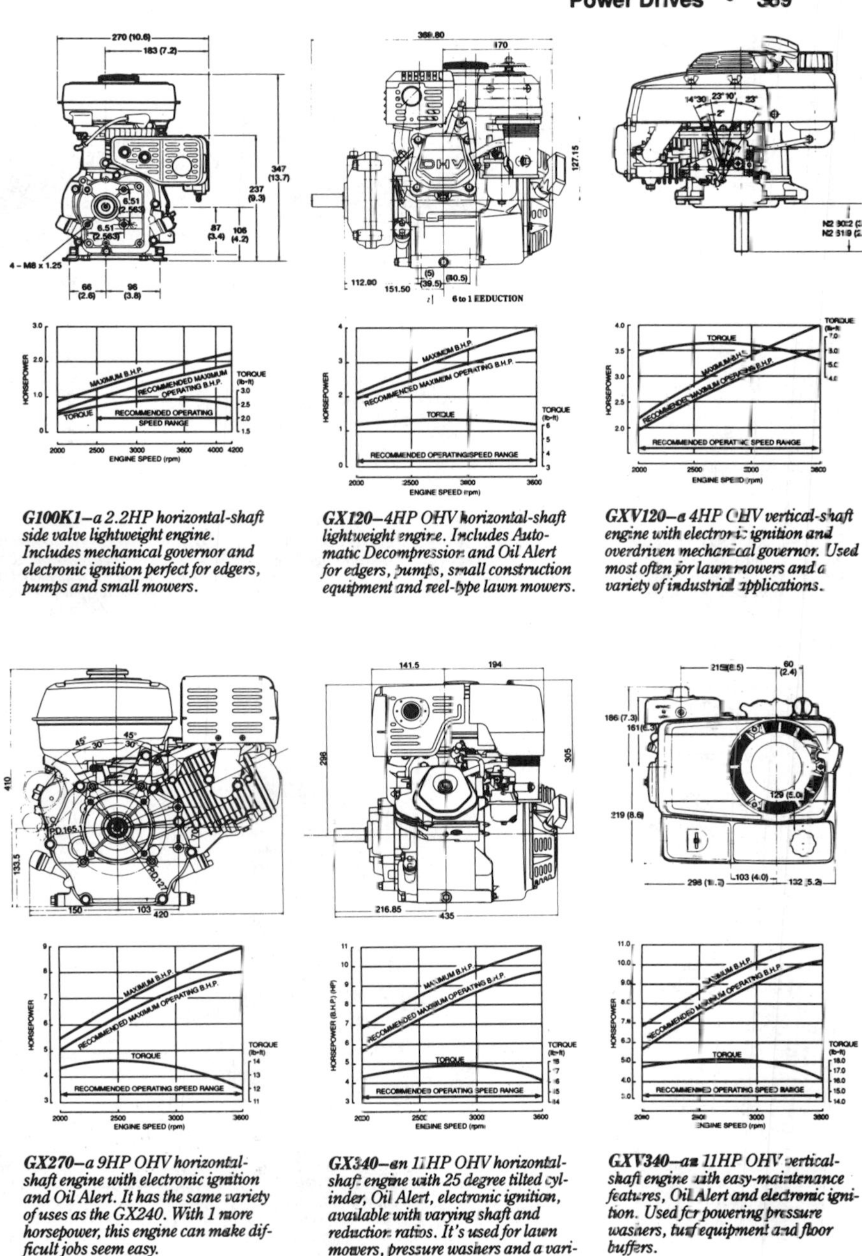

G100K1–a 2.2HP horizontal-shaft side valve lightweight engine. Includes mechanical governor and electronic ignition perfect for edgers, pumps and small mowers.

GX120–4HP OHV horizontal-shaft lightweight engine. Includes Automatic Decompression and Oil Alert for edgers, pumps, small construction equipment and reel-type lawn mowers.

GXV120–a 4HP OHV vertical-shaft engine with electronic ignition and overdriven mechanical governor. Used most often for lawn mowers and a variety of industrial applications.

GX270–a 9HP OHV horizontal-shaft engine with electronic ignition and Oil Alert. It has the same variety of uses as the GX240. With 1 more horsepower, this engine can make difficult jobs seem easy.

GX340–an 11HP OHV horizontal-shaft engine with 25 degree tilted cylinder, Oil Alert, electronic ignition, available with varying shaft and reduction ratios. It's used for lawn mowers, pressure washers and a variety of construction equipment.

GXV340–an 11HP OHV vertical-shaft engine with easy-maintenance features, Oil Alert and electronic ignition. Used for powering pressure washers, turf equipment and floor buffers.

Fig. 10–30. Horsepower, engine speed, torque characteristics of Honda engines. *(Courtesy Honda)*

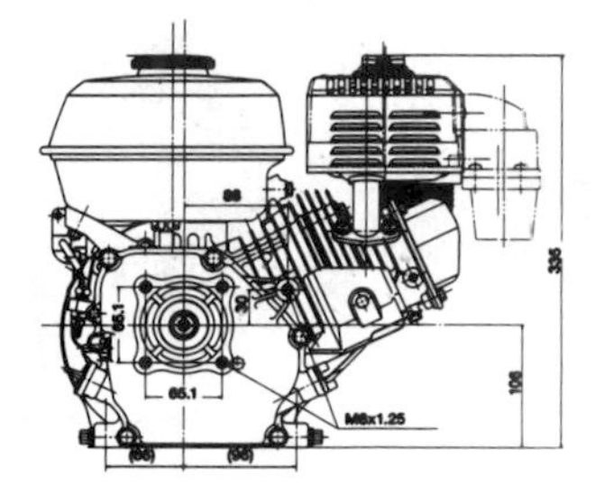

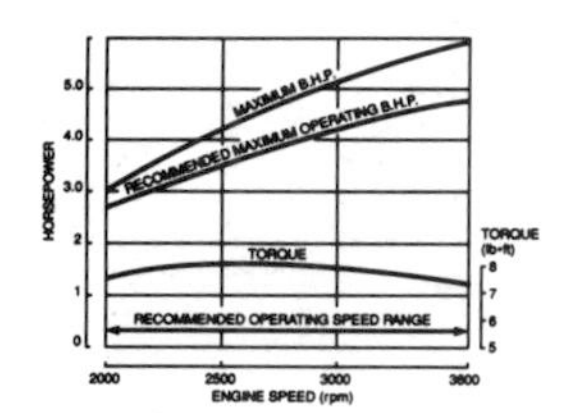

GX160–5.5HP OHV horizontal-shaft engine with electronic ignition and Oil Alert. Usage includes powering air compressors, generators, pumps, pressure washers, reel-type lawn mowers, cement trowels, and construction equipment.

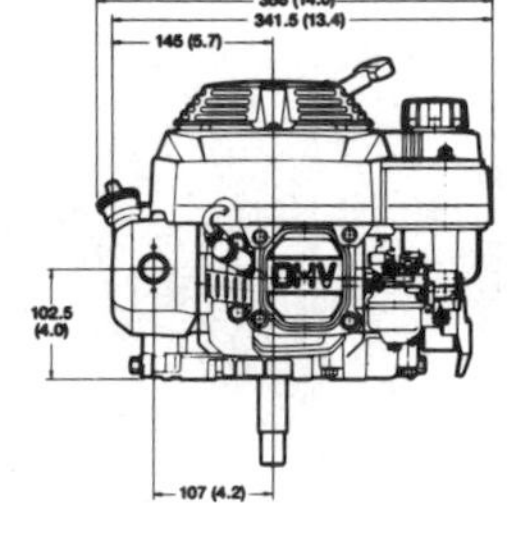

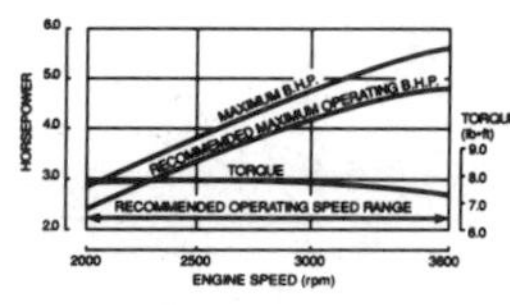

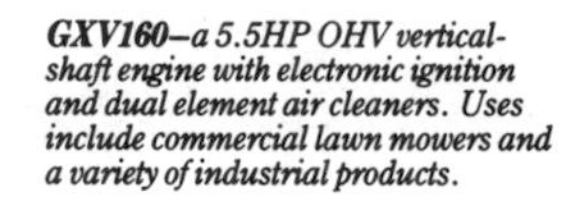

GXV160–a 5.5HP OHV vertical-shaft engine with electronic ignition and dual element air cleaners. Uses include commercial lawn mowers and a variety of industrial products.

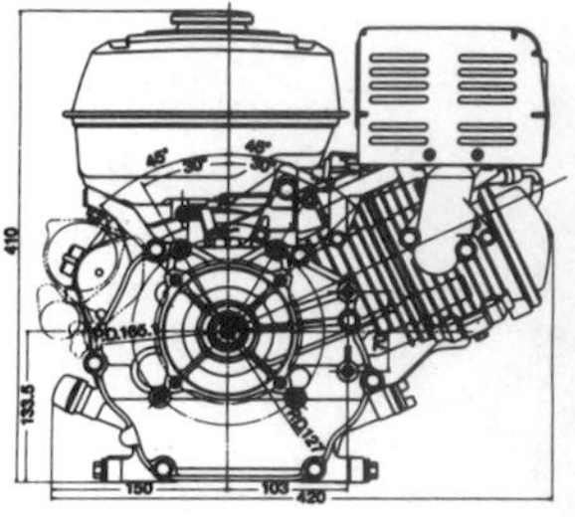

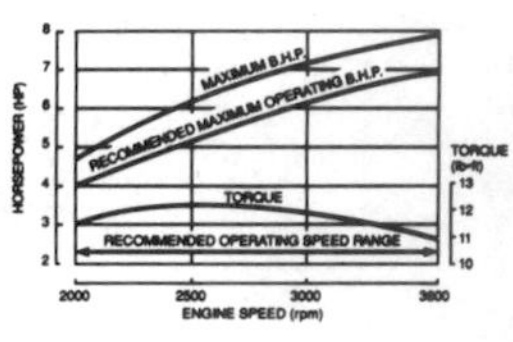

GX240–an 8HP OHV horizontal-shaft engine with electronic ignition and Oil Alert. Its uses include powering cement mixers, air compressors, and water pumps, as well as many other construction applications.

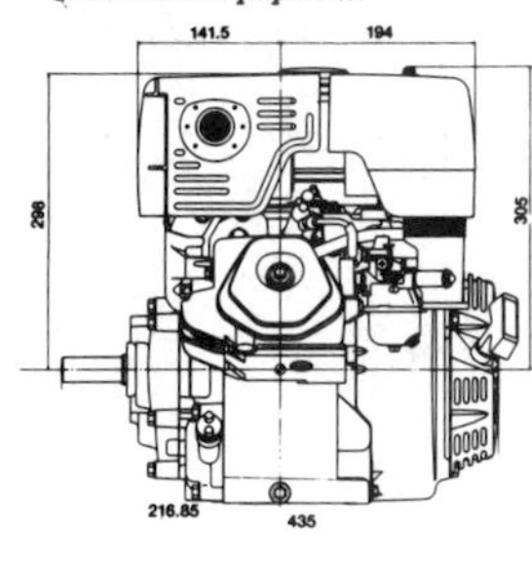

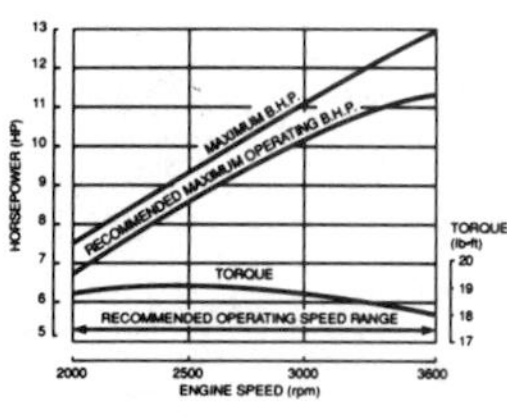

GX390–a 13HP OHV horizontal-shaft engine with automatic decompression for easy starting. The 13HP is the largest work-horse in our line of industrial engines.

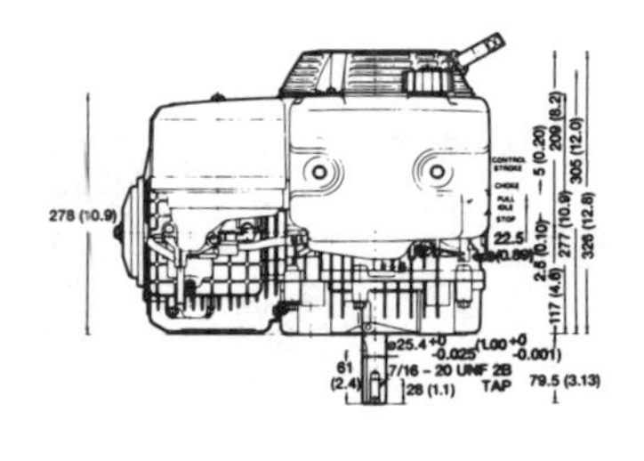

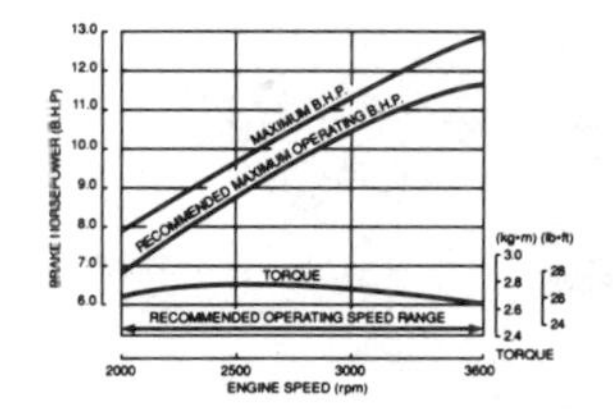

GXV390–a 13HP OHV vertical-shaft engine with features like Oil Alert, and electric start capabilities. Designed with commercial lawn & garden, and floor care equipment in mind.

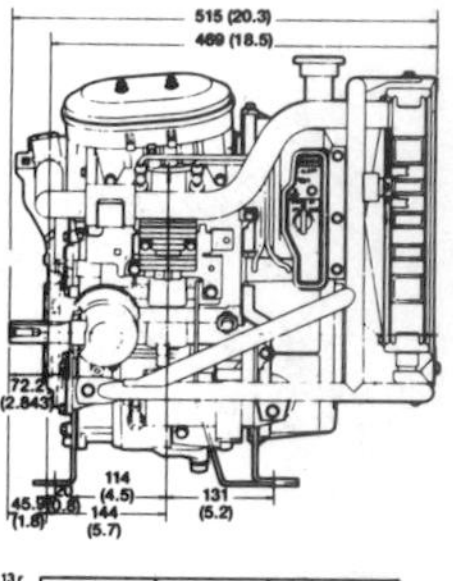

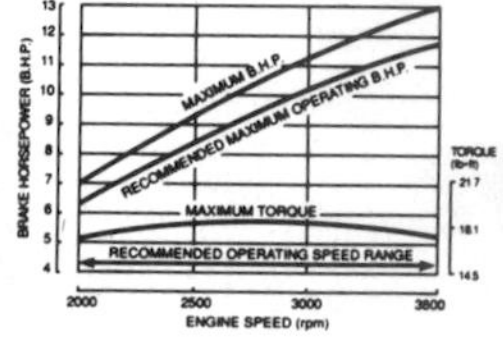

GX360–a 13HP horizontal-shaft liquid-cooled engine with twin-cylinders, Oil Alert, water temperature alert, electronic ignition and electric starter. Possible uses include lawn tractors, power washers and a variety of construction equipment.

Fig. 10–30. (Cont'd)

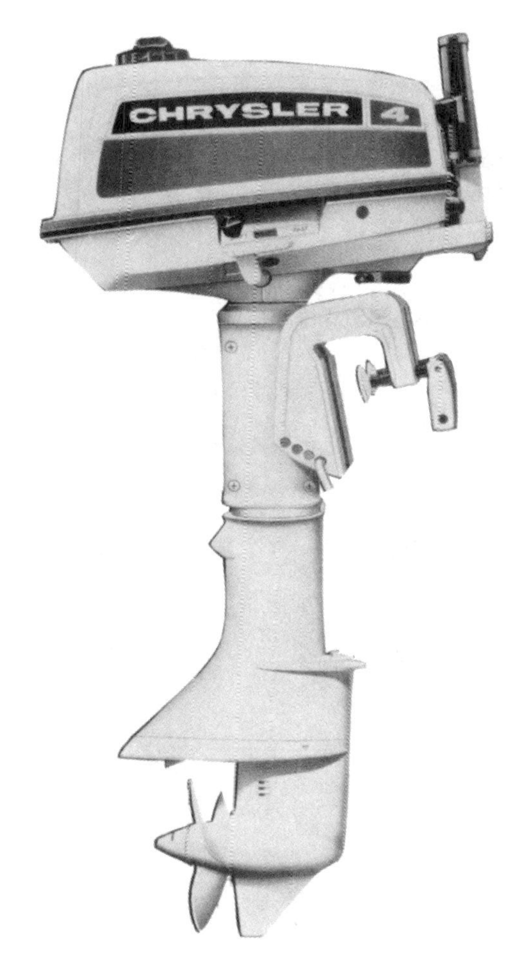

Fig. 10–31. Chrysler outboard engine. *(Courtesy Chrysler Corp.)*

Lower-Unit Construction

Fig. 10–34 shows the parts of the lower unit of an outboard engine. Compare this with Figs. 10–32 and 10–33. The lower unit contains the stern bracket so it can be fastened to the boat, the drive shaft, casing, and all the necessary shafting and gearing required to deliver the power from the engine to the propeller.

Fig. 10–35 shows how the reversing gear unit works. The shift yoke causes either the reverse or forward gear to be engaged with the drive shaft. Since the engine turns in the same direction at all times, the gearing has to be relied on to change the direction of the propeller rotation. The shift rod extends up to the control section of the engine.

The exploded view in Fig. 10–36 show how the propeller shaft assembly is arranged and how it comes apart for service. Each manufacturer has a manual for each engine that tells the do's and don'ts of disassembly and the proper procedure for reassembly. These technical manuals should be consulted before disassembly.

Some engines have a one-piece casting for the lower unit. Other

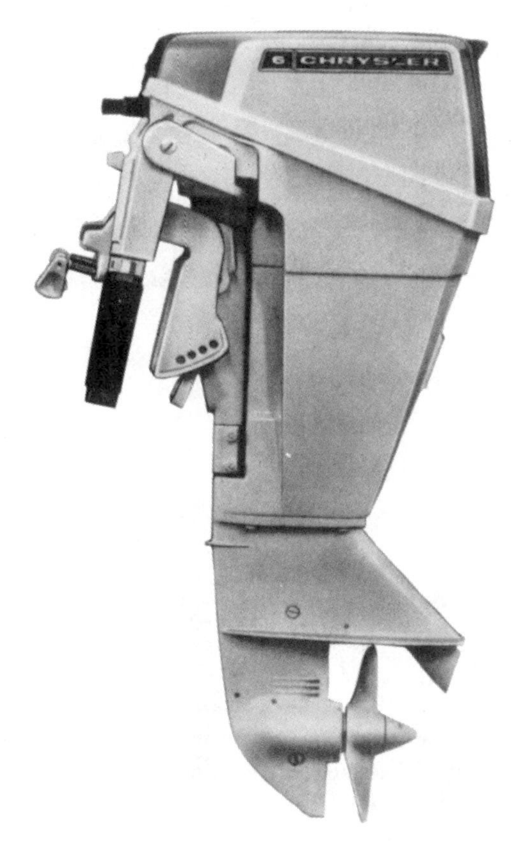

Fig. 10–32. Chrysler outboard engine. *(Courtesy Chrysler Corp.)*

designs may have two or three pieces bolted together to form the housing. If you work with the outboard lower section, make sure you place the skeg in a copper-jawed vise. Keep in mind that unless you have the manual for repairs you should not attempt to disassemble the unit. Fig. 10–37 shows a stretch-out of the lower unit's gears located in the housing.

Propeller Assembly

The propeller is subjected to much abuse from floating debris. If it hits a log or some other solid object, it may stop the engine or damage

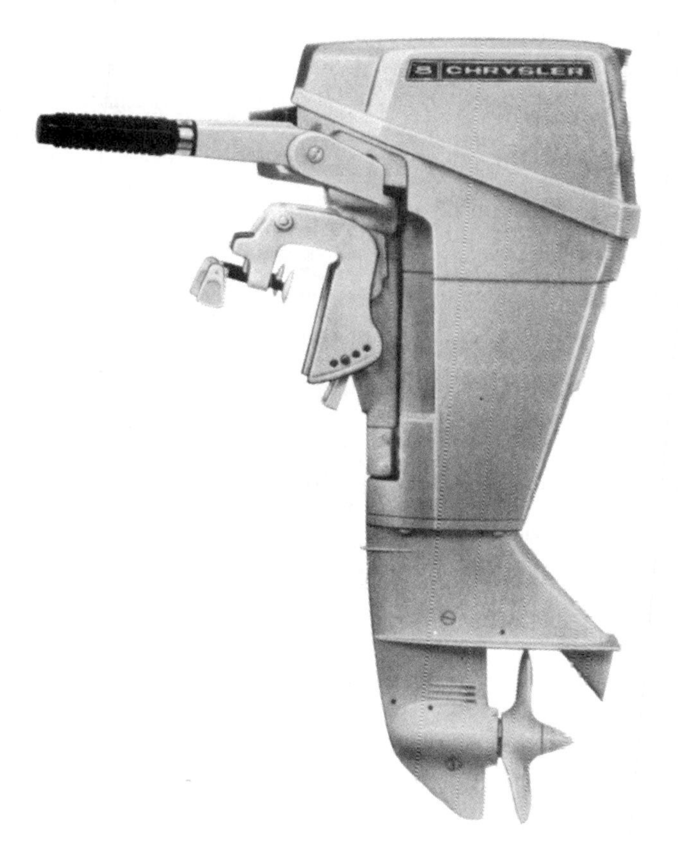

Fig. 10–33. Chrysler outboard engine. *(Courtesy Chrysler Corp.)*

the gears, shafts, or the engine itself. Therefore, a shear pin is placed in the drive shaft to the propeller (Fig. 10–38).

The propeller is driven by the drive pin through the serrated rubber insert. This insert comes to rest in a similarly serrated section of the propeller. When a submerged object is struck, the rubber insert absorbs the shock, which might otherwise shear the drive pin. Once the object has been cleared of the propeller, the propeller continues to rotate normally. This prevents the breaking of the shear pin every time an object is encountered in the water.

These lower units are in the water most of the time. They should have a sealant such as Permatex® placed on the new gaskets to make sure they do not leak. Some of the parts have a tight fit to prevent leakage. Be sure to check the entire area around the lower section to make sure there are no nicks or dents to cause trouble later with water leakage into the gear case and contamination of the lubricants.

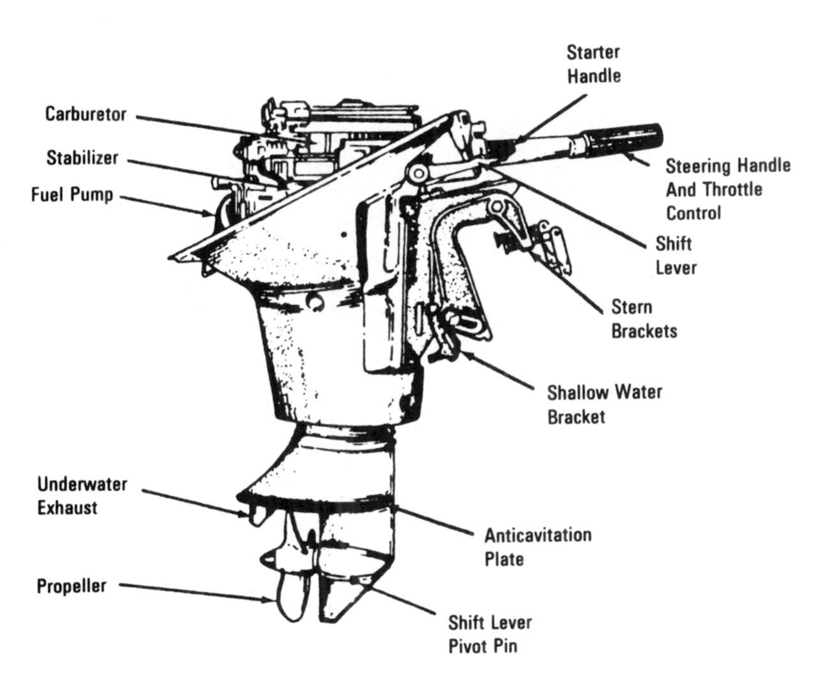

Fig. 10–34. Lower gearbox and housing for an outboard engine. *(Courtesy Johnson Motors, Outboard Marine Division)*

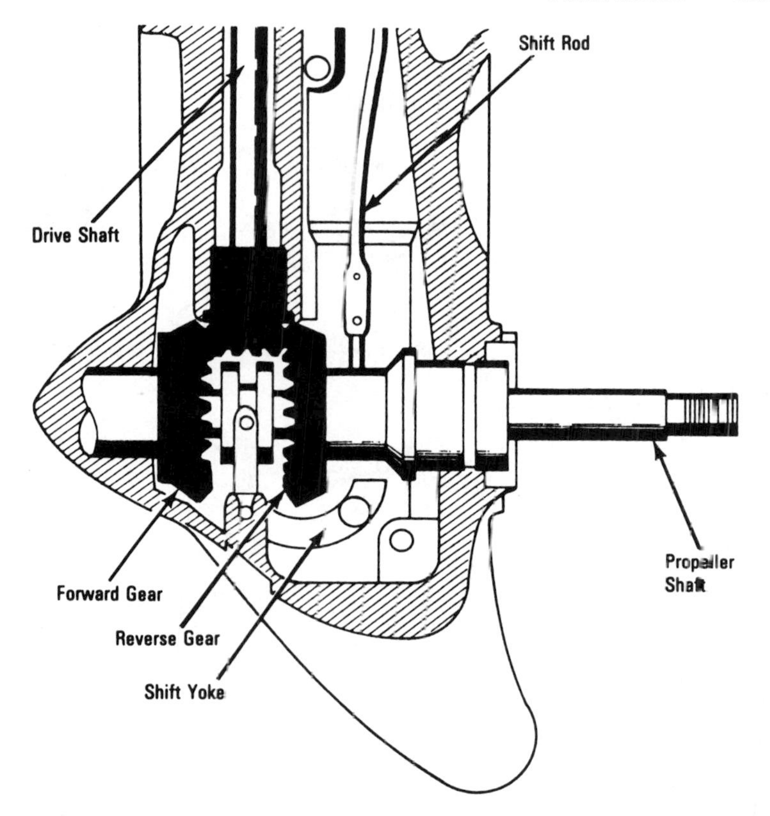

Fig. 10–35. Reverse and forward gear-shifting arrangement. *(Courtesy Johnson Motors, Outboard Marine Division)*

Chapter Highlights

1. Simple devices can be used to vary speed and increase the ability of machines to do work.
2. Friction is very important in the workings of a clutch.
3. Friction is a force that opposes motion.
4. Sliding friction causes difficulty in moving a book over a tabletop.
5. The coefficient of friction is the ratio of the force needed to overcome friction to the normal force pressing the surfaces together.

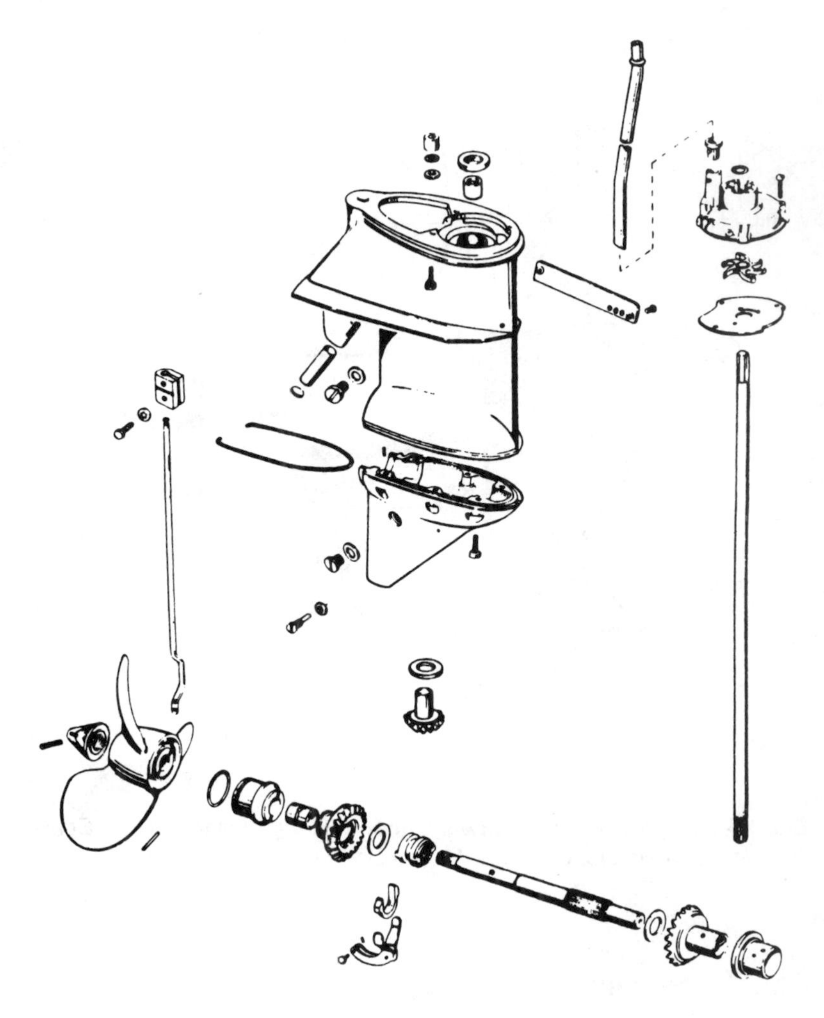

Fig. 10–36. Propeller and shaft assembly exploded. *(Courtesy Johnson Motors, Outboard Marine Division)*

6. Fluid friction is substituted for solid friction when lubricants are used between sliding surfaces.
7. There are six basic machines: screw, wedge, pulley, inclined plane, wheel and axle, and lever.

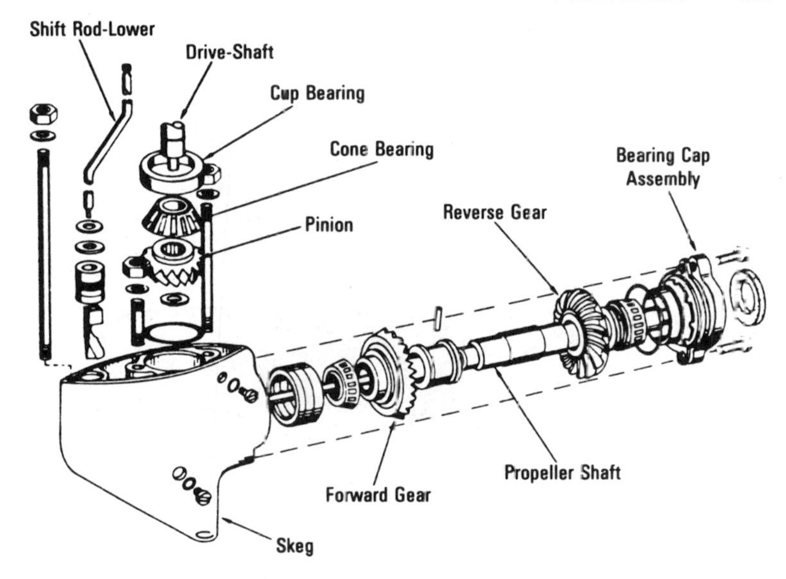

Fig. 10–37. Lower gear housing assembly exploded. *(Courtesy Johnson Motors, Outboard Marine Division)*

8. There are three classes of levers.
9. A series of pulleys with rope or chain attached in combination with one another results in what is called a block and tackle.
10. The wheel and axle is a wheel or crank rigidly attached to an axle.
11. The wedge is a double-inclined plane.
12. The screw is derived from the inclined plane.
13. It is impossible to make any machine 100 percent efficient.
14. Gear wheels are used to vary the speed or direction of a twisting force.
15. A worm gear is nothing more than a screw that is meshed with a gear wheel.
16. Clutches are used in an apparatus when you want the drive power source to be separated from the driven element.
17. Automatic friction clutches use centrifugal force to engage the mechanism.
18. Brakes use friction to slow down or stop a car.
19. Cones can be used in a clutch arrangement.

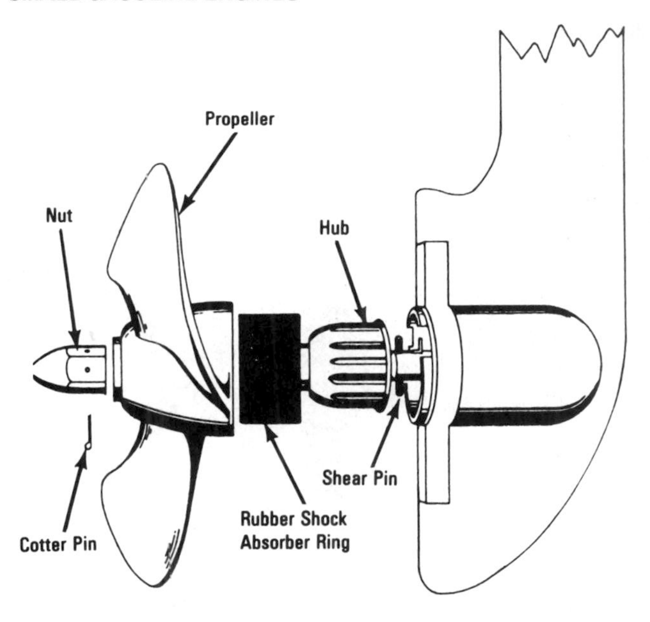

Fig. 10–38. Propeller showing shear pin and rubber shock absorber. *(Courtesy Johnson Motors, Outboard Marine Division)*

20. Centrifugal clutches use centrifugal force and friction to operate.
21. The right-angle drive consists of two gears mounted so that the two mesh at 90°.
22. A gearbox may house gears with mechanical advantage or velocity advantage.
23. The universal joint is a device that allows for any misalignment of the drive source with the driven part.
24. A differential allows for a driven wheel to make a turn without the other wheel being dragged along.
25. The sheave is a pulley. Its diameter varies with the adjustment of the width of the pulley.
26. In the outboard engine the shift yoke causes either the reverse or forward gear to be engaged with the drive shaft.
27. The propeller on an outboard engine is driven by the drive pin through a serrated rubber insert.

28. Static friction is the resistance to movement between two bodies that are in contact with one another and at rest.
29. New materials are now used for brake pads.
30. Breathing in asbestos fibers can cause cancer of the lung.

Vocabulary Checklist

asbestos
asbestosis
block
centrifugal clutch
circumference
compound machine
cone-type clutch
differential
disc clutch
flexible shaft
free-wheeling drive
gearbox
inclined plane
jackscrew
Kevlar
machine efficiency
magnetic clutch
mechanical advantage
mesothelioma
overrunning clutch
planetary gear
positive-drive clutch
power take-off
PTO
reduction units
right-angle drive
screw
serrated rubber
sheave
spider gear
sun gear
universal joint
velocity
vertical shaft machines
wedge
worm gear

Review Questions

1. What is friction?
2. What is sliding friction?
3. What is the coefficient of friction?
4. How do you reduce friction?
5. There are six basic machines. List them.
6. What is mechanical advantage?
7. What is ideal mechanical advantage?
8. What's the difference between a Class 1 and a Class 2 lever?

9. How do you get a mechanical advantage with a pulley?
10. Describe a wedge.
11. Where does the screw obtain its mechanical advantage?
12. What is the purpose of a gear?
13. What is a velocity advantage?
14. What is a worm gear?
15. What is a clutch?
16. List at least six types of clutches.
17. Where is the centrifugal clutch used?
18. What is a right-angle drive?
19. What is a universal joint?
20. What is a differential? Where is it used?
21. What is a sun gear?
22. How do adjustable-speed sheaves work to vary speed?
23. Why are outboard engines made to withstand sudden load changes?
24. How do the propellers on an outboard engine keep from stopping the engine when they hit a log or solid object?
25. What is a power take-off unit?
26. What is asbestosis?
27. What is now used for brake linings?
28. What do Formula One car racers now use for brake linings?
29. Where did S45 friction material come from?
30. What is the name of the cancer produced by asbestos fibers?

SECTION THREE

Servicing the Small Gasoline Engine

CHAPTER 11

Preventive Maintenance

Preventive maintenance is just what it says. It means you can prevent trouble with any engine if you maintain it according to the manufacturer's specifications. It is therefore important to be able to follow the instructions furnished by the manufacturer. In order to do so, it is best to keep all the materials furnished with the engine when it was purchased. If these are not available, there are technical manuals located in technical-school libraries, public libraries, and in the library of a local repairperson. There is usually a distributor for Briggs & Stratton and Tecumseh engines in every town of moderate size. If you can't find the information needed there and want your own set of manuals, you may be able to obtain them from the company or from Sears, Roebuck and Company. Sears equipment uses both Tecumseh and Briggs & Stratton engines, so they carry the full line of tech manuals for those who want to do their own repairs.

Proper care and maintenance will contribute to a long and trouble-free operation of any engine.

Care and Maintenance

Care of the engine begins before it is started up. Unpack or uncrate it carefully. Follow the manufacturer's recommendations for putting the engine in operation. Some of the procedures are just plain common sense, but we mention them here since the manufacturer recommends them. Fig. 11–1 illustrates the proper *before-starting* procedures for the Briggs & Stratton engine.

To start the engine, three steps should be followed. Fig. 11–2 shows how starting the engine is accomplished. Note that each engine has its own small differences. However, basically the engines shown here will be representative of most small gasoline engines.

Maintenance

There are some basic differences for the different series of engines. Two sets of instructions are shown in Fig. 11–3, typical of the majority of small gasoline engines. As you can see from the maintenance instructions provided by the manufacturer, generally commonsense procedures are mentioned. Such things as checking the oil level, changing the oil, cleaning and reoiling the air cleaner, cleaning the cooling system, checking the spark plugs, and removing carbon deposits are standard procedure.

Adjustments

There are some adjustments that must be made by the operator so that the engine will operate at its most efficient level. These include carburetor adjustments and, in some instances, the adjustment of the governor after the engine has run for a time. These adjustments will be found in the appendix since they vary according to the type of engine being used. Fig. 11–4 shows some of the adjustments made to various carburetors as recommended by the manufacturer.

Service Procedures

Each manufacturer recommends that certain things be done to the engine before it is started for the first time. Each manufacturer also lists some service procedures and makes available certain manuals, booklets, and sheets of specifications for their engines. These recommendations have been made after considerable testing of the engines under various conditions. Each time a person writes to the manufac-

BEFORE STARTING

1. **FILL CRANKCASE WITH OIL**—Use a high-quality detergent oil classified "For Service SC, SD, SE or MS." Nothing should be added to the recommended oil.

SUMMER
(Above 40° F.)
Use SAE 30

If not available
Use SAE 10W-30
or SAE 10W-40

WINTER
(Under 40° F.)
Use SAE 5W-20 or SAE 5W-30

If not available
Use SAE 10W or SAE 10W-30

Below 0° F.
Use SAE 10W or SAE 10W-30
Diluted 10% with Kerosene

DIRECTIONS: Place engine level. Use screwdriver or bar to remove oil filler plug. Fill crankcase to point of overflowing. POUR SLOWLY. Capacity 1-1/4 pints.

2. **FILL FUEL TANK**—Use clean, fresh, lead-free or leaded regular-grade automotive gasoline. **Fill tank completely!**

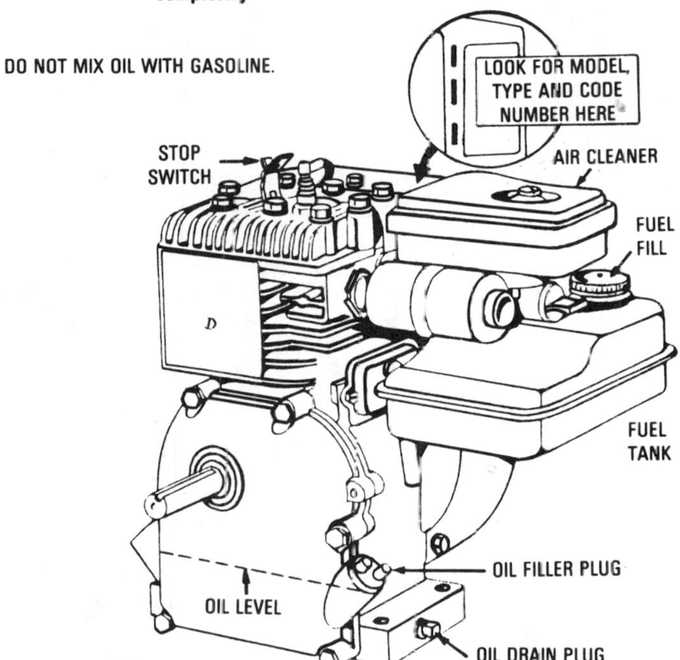

Fig. 11–1. Procedures to follow before starting the engine. *(Courtesy B & S)*

STARTING

1. **CHOKE ENGINE**—Engine may be equipped with either manual or Choke-A-Matic controls.

a. **Manual Choke and Stop**—Be sure stop switch is away from spark plug. Pull choke as illustrated.

b. **Choke-A-Matic Control**—Move control on equipment as far as possible toward "Choke" or "Start" position.

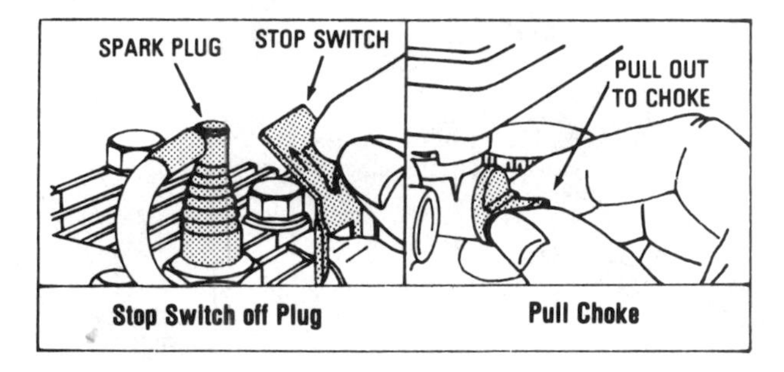

2. **START ENGINE**—Engine may be equipped with rewind or rope starter.

CAUTION: ALWAYS KEEP HANDS AND FEET CLEAR OF MOWER BLADE OR OTHER ROTATING MACHINERY.

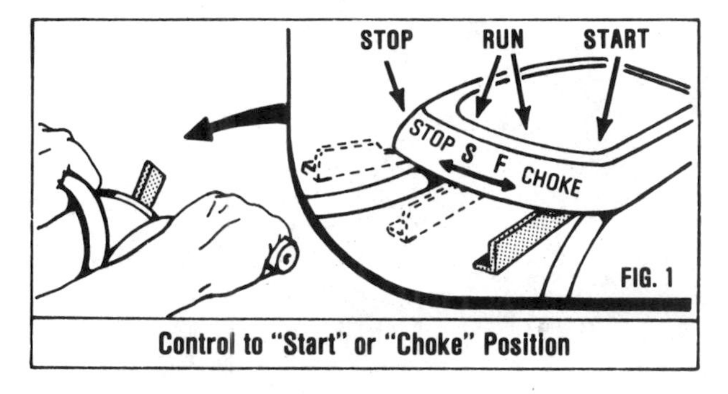

Fig. 11–2. Procedures to follow to start the engine. *(Courtesy B & S)*

a. Rewind Starter—Grasp starter as illustrated and pull out cord rapidly. Repeat if necessary with choke opened slightly. When engine starts, open choke gradually.

b. Rope Starter—Wind rope around pulley in direction shown by arrow.

Pull the rope with a quick full arm stroke. Repeat if necessary with choke open slightly. When engine starts, open choke gradually.

NOTE: ENGINE MAY NOT START if controls on a powered equipment do not close choke fully.

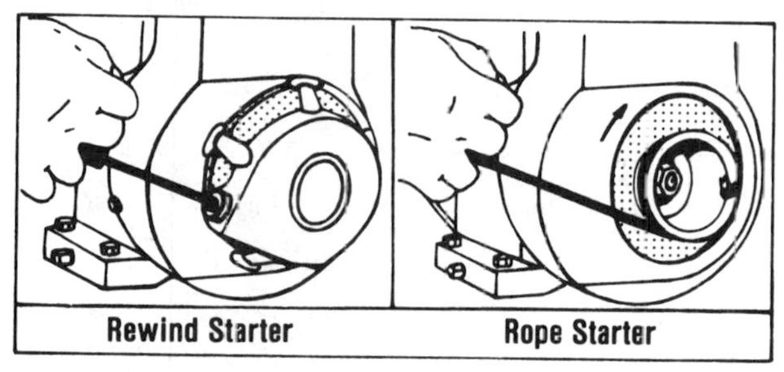

Rewind Starter | Rope Starter

3. STOP ENGINE

a. Manual Control—Push stop switch against end of spark plug.

b. Choke-A-Matic Control—Move control lever to "Stop" position.

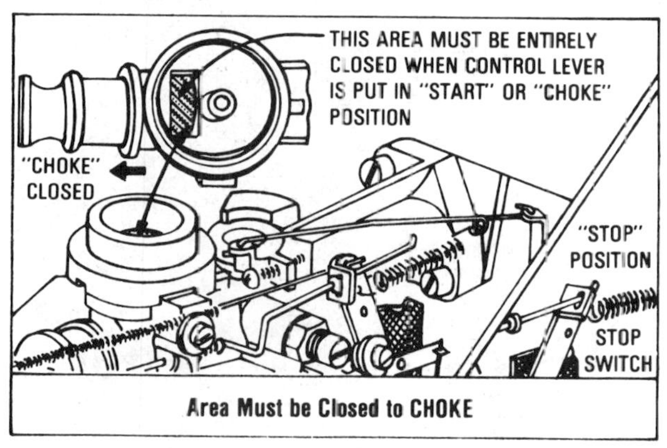

Area Must be Closed to CHOKE

Fig. 11–2. (Continued)

MAINTENANCE

1. CHECK OIL LEVEL regularly, at least after each five hours of operation. (Take care to remove dirt around filler plug.) Be sure oil level is maintained FULL TO POINT OF OVERFLOWING.

2. CHANGE OIL after first five hours of operation. Thereafter change oil every twenty-five hours of operation. Remove drain plug and drain oil while engine is warm. Replace drain plug. Remove oil filler cap and refill with new oil of proper grade. Replace filler cap.

CHECK OIL (6 to 1 Gear-Reduction Models Optional) every hundred hours by removing the oil plug in lower half of gear cover. Add SAE 10W-30 oil at upper oil filler plug until oil runs out of lower hole. Replace both plugs.

NOTE: Filler plug has vent hole and must be placed in top opening.

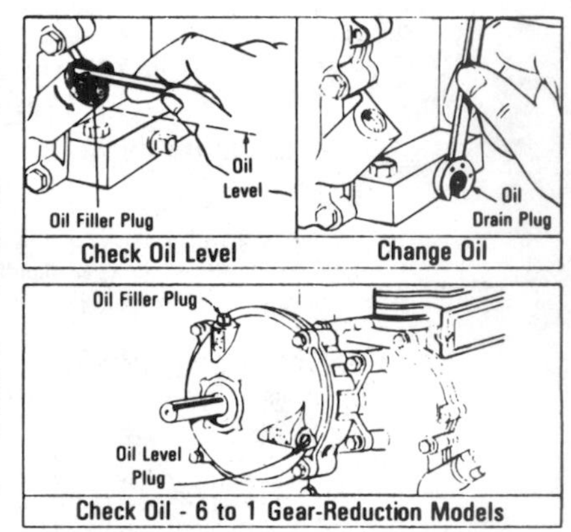

3. CLEAN AND REOIL AIR CLEANER and element every twenty-five hours under normal conditions. The capacity of the "Oil-Foam" air cleaner is adequate for a full season's use without cleaning in average homeowner lawn mower service. (Clean every few hours under extremely dusty conditions).

1. Remove screw.
2. Remove air cleaner carefully to prevent dirt from entering carburetor.
3. Take air cleaner apart.
4. A—Wash foam element in kerosene or liquid detergent and water to remove dirt.
   B—Wrap foam in cloth and squeeze dry
   C—Saturate foam in engine oil. Squeeze to remove excess oil.
   D—Assemble parts—fasten to carburetor with screw.

4. CLEAN COOLING SYSTEM—Grass or chaff may clog cooling system after prolonged service in cutting tall dry grasses or hay. Continued operation with a clogged cooling system causes severe overheating and possible engine damage. Remove blower housing and clean regularly.

5. SPARK PLUG—Clean and reset gap at 0.030" every 100 hours of operation.

CAUTION: Blast cleaning of spark plugs in machines that use abrasive grit is not recommended. Spark plugs should be cleaned by scraping or wire brushing and washing with a commercial solvent or gasoline.

6. REMOVE CARBON DEPOSITS—Clean combustion chamber, top of piston and around both valves every 100- 300 hours of operation.

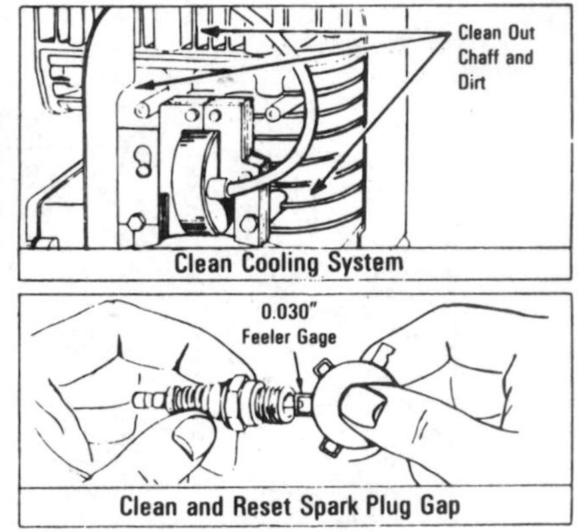

Fig. 11–3. Maintenance checks for keeping the engine operating trouble-free. *(Courtesy B & S)*

1. CHECK OIL LEVEL before starting engine and after every five hours of operation.

ADD oil as necessary to keep level FULL TO POINT OF OVERFLOWING or to FULL mark on dipstick.

Before removing oil fill plug, clean area around plug to prevent dirt from entering oil fill hole.

Engine should be in a level position when checking oil.

OIL MINDER (Optional)—Press and release bellows. If oil fills clear plastic tube, level is OK. If oil does not fill tube, add oil.

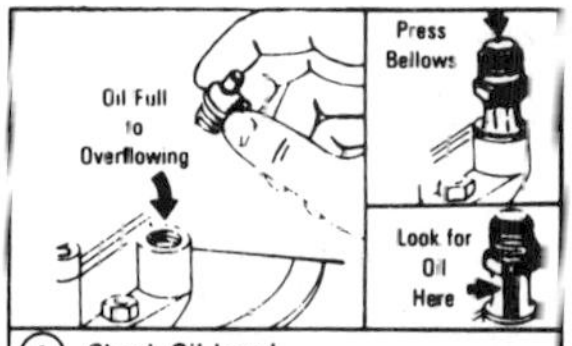

(1) Check Oil Level

2. CHANGE OIL after first five hours of operation. Thereafter change every twenty-five hours. Change oil while engine is warm. Oil may be drained through oil drain on bottom of engine. To drain completely, always place engine level when draining through the bottom. Oil may also be drained through oil fill hole or extended oil fill tube as shown.

When tipping, empty fuel tank or keep engine spark plug or muffler side up.

Oil capacity 1 ¼ pints.

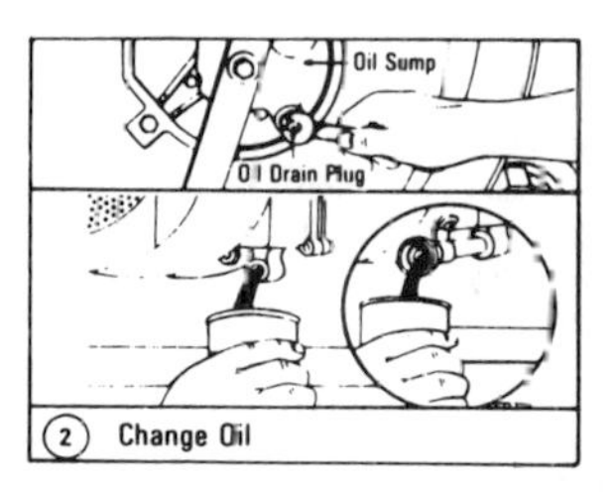

(2) Change Oil

3. CLEAN AIR CLEANER and reoil element every twenty-five hours under normal conditions. Clean every few hours under extremely dusty conditions. Poor engine performance and flooding usually indicates that the air cleaner should be serviced.

To Service:

1. Remove screw.
2. Remove air cleaner carefully to prevent dirt from entering carburetor.
3. Take air cleaner apart and clean.
   a. WASH foam element in kerosene or a liquid detergent and water to remove dirt.
   b. DRY foam completely by wrapping and squeezing in a cloth.
   c. SOAK foam with engine oil. Squeeze to distribute and remove excess oil.
4. Reassemble parts and fasten to carburetor.

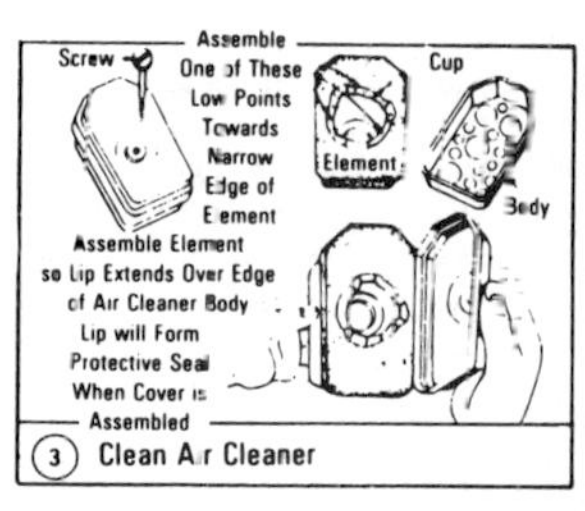

(3) Clean Air Cleaner

4. CLEAN COOLING SYSTEM—Grass, chaf or dirt may clog the rotating screen and the air cooling system, especially after prolonged service cutting dry grasses. To avoid overspeeding, overheating and engine damage, remove the blower housing and clean the area shown. This should be a regular maintenance operation.

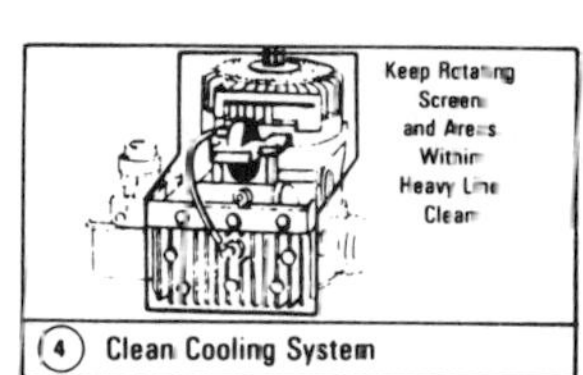

(4) Clean Cooling System

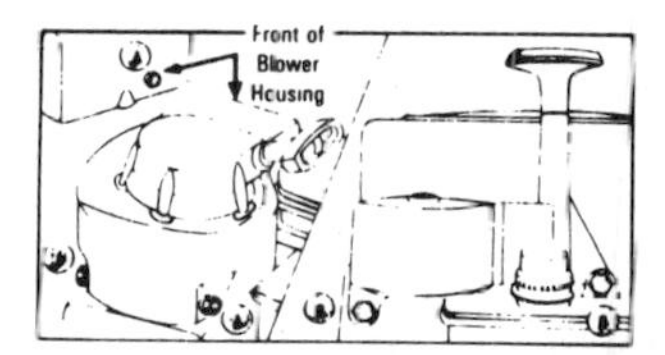

Fig. 11–3. (Continued)

ADJUSTMENTS

CARBURETOR ADJUSTMENTS

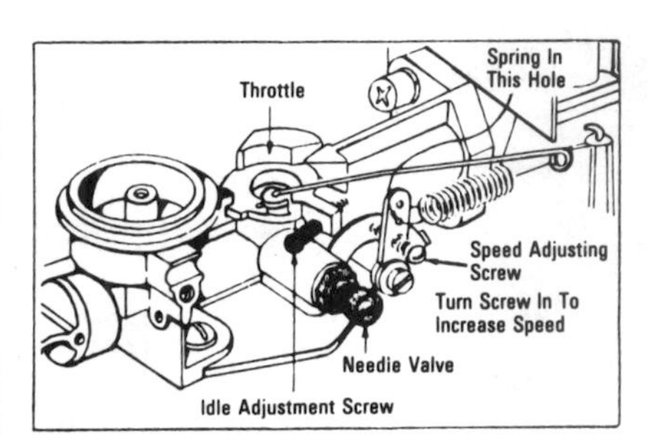

Minor carburetor adjustment may be required to compensate for differences in fuel, temperature, altitude and load.

NOTE: Adjust carburetor with fuel tank half full of regular grade gasoline.

Initial Adjustment:
Close needle valve (turn clockwise) then open 1-1/2 turns (turn counterclockwise). This initial adjustment will permit the engine to be started and warmed up before making final adjustment.

Final Adjustment: With engine running at normal operating speed (approximately 3000 rpm without load) close the needle valve (turn clockwise) until engine starts to lose speed (lean mixture). Then slowly open needle valve (turn counterclockwise) past the point of smoothest operation, until engine just begins to run unevenly. This mixture should be rich enough for best performance under load. Hold throttle in idling position. Turn idle speed adjusting screw until fast idle is obtained. (1750 rpm). Test the engine under full load. If engine tends to stall or die out, it usually indicates that the mixture is slightly lean and it may be necessary to open the needle valve slightly to provide a richer mixture. This richer mixture may cause a slight unevenness in idling.

CHOKE-A-MATIC CONTROL ADJUSTMENTS

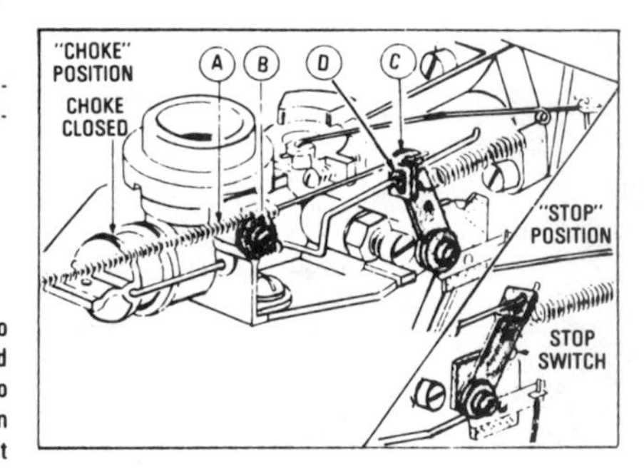

Proper choke and stop switch operation is dependent upon proper adjustment of remote controls on the powered equipment.

To Check Operation:

Remove air cleaner. Move remote-control lever to "Choke" position. The carburetor choke should then be closed. Move the remote control lever to "Stop". Speed lever on carburetor should then make good contact with stop switch to short out ignition.

Fig. 11–4. Operator adjustments for keeping the carburetor operating properly. *(Courtesy B & S)*

To Adjust:

Place remote control lever on equipment in "Fast" (high speed) position. Loosen control casing clamp screw (B) on carburetor. Move control casing (A) and wire forward or backward until speed lever (C) just touches the choke operating link at (D). Tighten casing clamp screw (B) on carburetor. Recheck operation of controls after adjustment. Replace air cleaner.

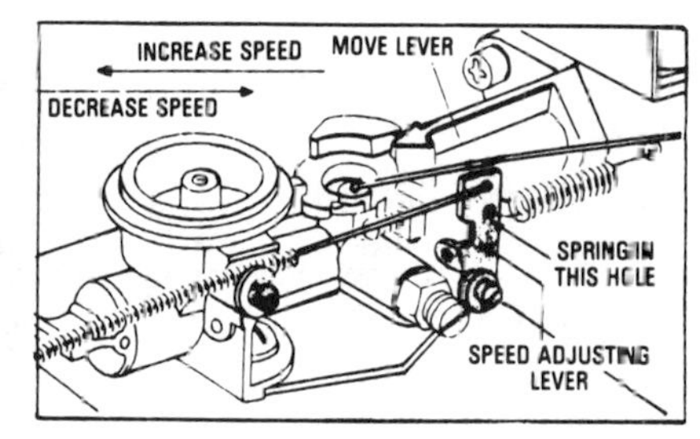

SPEED CONTROL ADJUSTMENTS

Remote Control

Controls on powered equipment should move speed lever in a direction that will elongate governor spring to increase speed.

To Adjust:

Loosen clamp screw on carburetor or fuel tank bracket and move casing in or out to obtain proper speed. Maximum recommended speed is 3600 rpm.

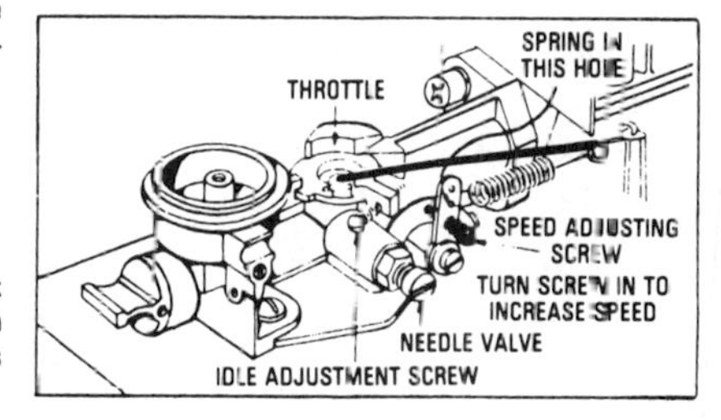

Standard Control

To increase engine speed, turn speed adjusting screw clockwise.

To decrease engine speed, turn speed adjusting screw counterclockwise.

Manual Friction Control

To increase or decrease engine speed, move speed adjusting rod as shown.

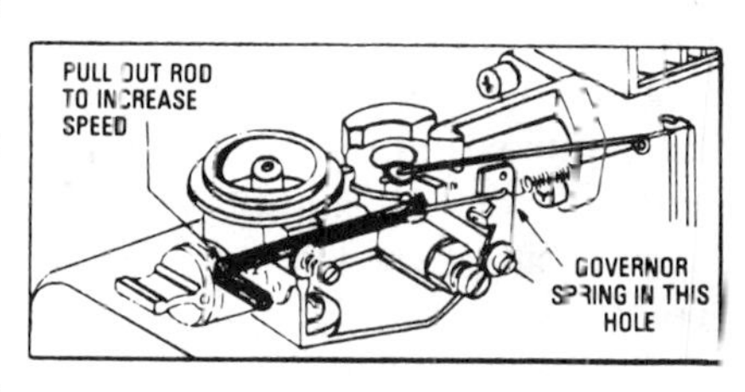

Fig. 11–4. (Continued)

turer to mention a defect or difficulty, this is taken into consideration in the next revision of the manuals and handbooks.

Operators' Manuals

Manufacturers place an operator's manual with each engine. This may be a small booklet eight to sixteen pages in length which includes all the adjustments, care and maintenance procedures as well as an exploded view of the engine or device. In most instances the device that the engine is powering is also included in the parts list with instruction on where to obtain parts if needed. Fig. 11–5 shows a weed and grass trimmer that has a two-cycle engine mounted on the end of a shaft to drive the plastic string that cuts weeds or grass. Included with this booklet is a warranty registration card that must be returned to the manufacturer. In most cases this produces marketing information for the manufacturer of the product that uses the small gasoline engine.

Technical Manuals

Technical manuals have more detail in them than do the operators' manuals (Fig. 11–6A), and in some instances provide step-by-step methods for making repairs or removing a part so it can be replaced. This type of manual is available from the service department of the engine's manufacturer. In most cases you have to pay for the information. Write or call the manufacturer—get the address off the label affixed to the engine—to obtain the price of this publication.

In some cases the small-engine manuals are available from Sears, Roebuck and Company at their parts and service department in stores in larger cities. They may also be obtained through the Sears catalog department in other locations.

If you intend to work with the engines and repair them, you should have a library of tech manuals so you can readily refer to the specifications for bearings, pistons, and other parts. No one can remember all the specifications for the many types of engines made by various manufacturers.

In some repair manuals there are *flow charts* that help you analyze the trouble and find its cause (Fig. 11–6B). Other charts or tables of specifications are available from the previously mentioned sources for

OPERATOR'S MANUAL

CAUTION
Read Rules for Safe Operation
and Instructions Carefully

Fig. 11–5. Operator's manual. *(Courtesy Kioritz)*

tech manuals. These tables or charts can help with locating information needed to obtain a part or replace one.

Engine Specifications and Identification

Engine identification refers to the manufacturer's model number. In most instances this can be found on the engine in plain view of a

Fig. 11–6. (A) Technical manual for a four-cycle engine. *(Courtesy Tecumseh)*

casual observer. The engine specifications start with the engine identification. This can be a code or just a number. In Fig. 11–7, a Tecumseh engine code is shown, and each letter and number has a meaning.

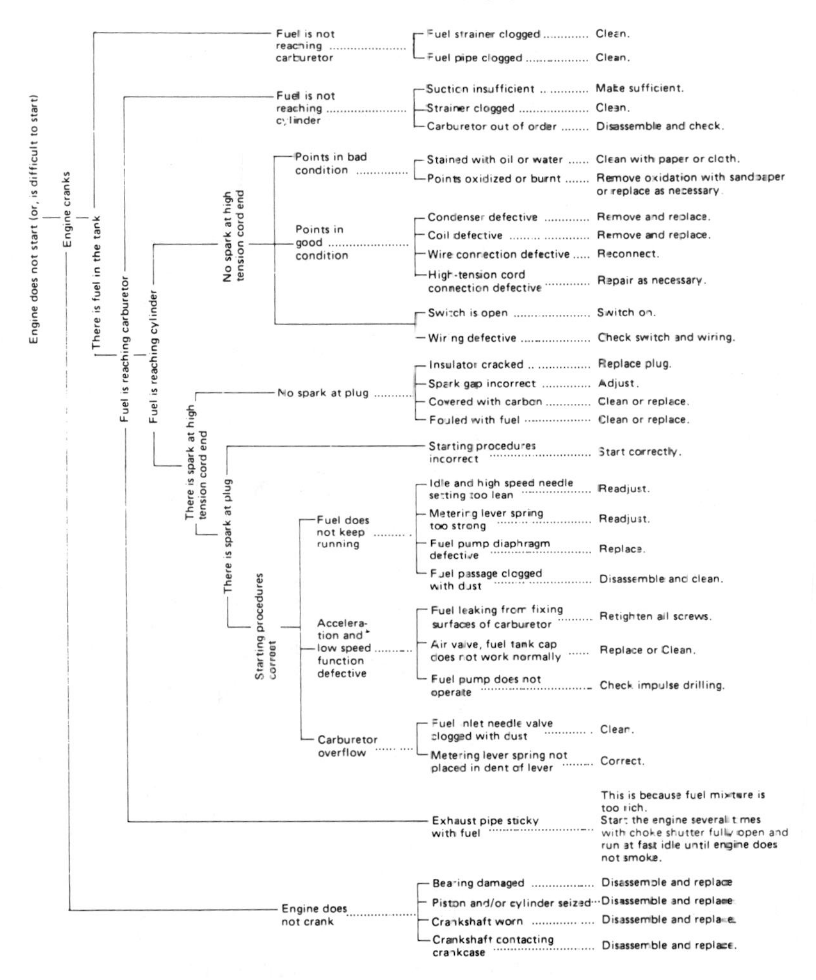

Fig. 11–6. (B) Troubleshooting chart. *(Courtesy Kioritz)*

Engine Identification

The engine identification plate is mounted on the carburetor side of the blower housing. A typical model and specification number is stamped into this plate as shown in Fig. 11–7. Both the complete model and specification numbers are needed for parts replacement.

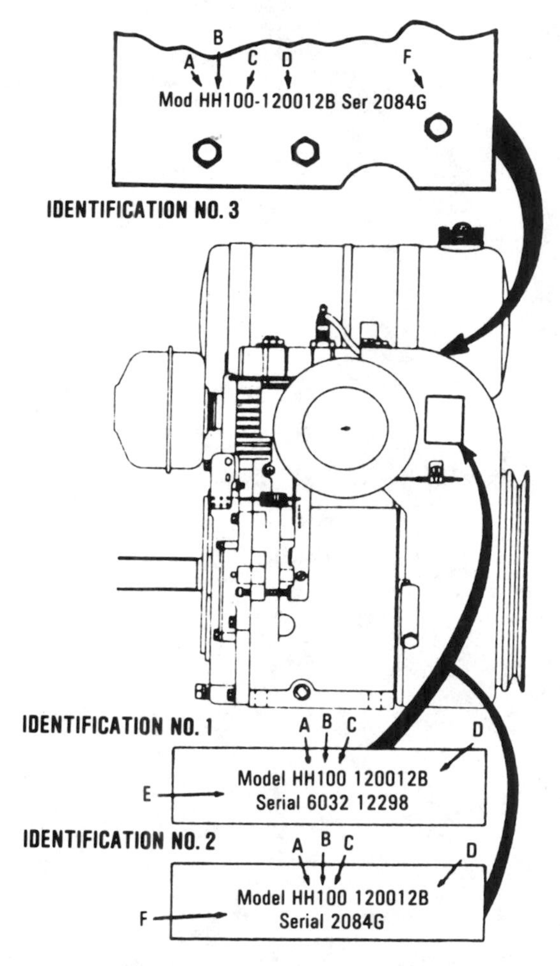

Fig. 11–7. Reading the engine code. *(Courtesy Tecumseh)*

Model and specification numbers are explained in Fig. 11–7. Match up the letters *A* through *F* with the following:

A. *H* stands for horizontal crankshaft engine.
B. *H* stands for cast-iron engine.
C. The last zero is dropped, indicating the rated horsepower of the engine. (In this illustration the horsepower is 10.)

D. *120012B* is the specification number.
E. Series number is *6032 12298.*
  1. *6032* indicates the date the engine was assembled: *6* indicates the year—1966;—*032* shows the day of the year it was assembled—in this case it was the thirty-second day of 1966, or February 2, 1966.
  2. *12298* is the factory sequence number.

F. Identification tag No. 2 is the same as Identification tag No. 1, except the factory number is left off and the *G* on the serial stands for assembly line and shift of production.
G. Identification No. 3 is stamped on the blower housing near the head bolts as indicated.

If the model is a VH80 or VH100, the letters *VH* represent the following:

A. *V* stands for vertical-crankshaft engine.
B. *H* stands for horizontal-crankshaft engine.

Briggs & Stratton engines have model numbers that tell you the type of engine and its specifications. Table 11–1 shows the B & S method of identifying engines.

Tolerances and Classes of Fit

Any time that metal-to-metal contact is desired, there is some reference to the type of fit and the tolerance of the parts.

Tolerance

Engines with very close tolerances are made today by automatic machines. These parts must also be very close to what is specified as the size in both plus and minus directions. A tolerance is the difference between the largest and the lowest limits of a dimension. Tolerance comes from the verb *to tolerate,* which means to endure, to put up with, to allow a certain amount of oversize or undersize.

Tolerance is mentioned after a dimension in terms of + or −. That is, an engine may have a cylinder that has a 2.125″ (53.975 mm) bore,

Table 11–1. Briggs & Stratton Numerical Model Number System

This handy chart explains the unique Briggs & Stratton numerical model designation system. It is possible to determine most of the important mechanical features of the engine by merely knowing the model number. Here is how it works:

A. The first one or two digits indicate the *cubic inch displacement.*
B. The first digit after the displacement indicates *basic design series* relating to cylinder construction, ignition, general configuration, etc.
C. The second digit after the displacement indicates *position of crankshaft and type of carburetor*.
D. The third digit after the displacement indicates *type of bearings* and whether or not the engine is equipped with *reduction gear* or *auxiliary drive*.
E. The last digit indicates the *type of starter*.

	First Digit After Displacement	Second Digit After Displacement	Third Digit After Displacement	Fourth Digit After Displacement
Cubic Inch Displacement	Basic Design Series	Crankshaft, Carburetor, Governor	Bearings, Reduction Gears, & Auxiliary Drives	Type of Starter
6	0	0-	0-Plain Bearing	0-Without Starter
8 9	1 2	1-Horizontal Vacu-Jet	1-Flange Mounting Plain Bearing	1-Rope Starter
10 11	3 4	2-Horizontal Pulsa-Jet	2-Ball Bearing	2-Rewind Starter
13 14	5 6	3-Horizontal Flo-Jet (Pneumatic Governor)	3-Flange Mounting Ball Bearing	3-Electric-110 Volt, Gear Drive
17 19 20	7 8 9	4-Horizontal Flo-Jet (Mechanical Governor)	4-	4-Elec. Starter-Generator-12 Volt, Belt Drive
23 24 25		5-Vertical Vacu-Jet	5-Gear Reduction (6 to 1)	5-Electric Starter Only-12 Volt, Gear Drive
30 32		6-	6-Gear Reduction (6 to 1) Reverse Rotation	6-Alternator Only*
		7-Vertical Flo-Jet	7-	7-Electric Starter, 12 Volt Gear Drive, with Alternator
		8-	8-Auxiliary Drive Perpendicular to Crankshaft	8-Vertical-pull Starter
		9-Vertical Pulsa-Jet	9-Auxiliary Drive Parallel to Crankshaft	

*Digit 6 formerly used for "Wind-Up" Starter on 60000, 80000, and 92000 Series

EXAMPLES

To identify Model 100202:

10	0	2	0	2
10 Cubic Inch	Design Series 0	Horizontal Shaft-Pulsa-Jet Carburetor	Plain Bearing	Rewind Starter

Similarly, a Model 92998 is described as follows:

9	2	9	9	8
9 Cubic Inch	Design Series 2	Vertical Shaft-Pulsa-Jet Carburetor	Auxiliary Drive Parallel to Crankshaft	Vertical Pull Starter

(Courtesy B & S)

but it can be + 0.006″ (0.1524 mm), or 2.131″ (54.1274 mm), and still be the proper size. If the under tolerance was –0.006″ (–0.1524 mm), then the cylinder could be 2.119″ (53.8226 mm) and still be within limits. If the bore read 2.125″ (53.975 mm) ±0.006″ (0.1524 mm), then the 2.119″ (53.8226 mm) to 2.131″ (54.1274 mm) are the high and low limits of the bore. Anywhere between the high and the low limits would be acceptable. Of course, the piston would have to be the correct size to match the bore of the cylinder. Pistons are not exact fits. There would be too much friction if they were, so rings are placed on the pistons to expand to the proper size and make a good seal between the cylinder wall and the piston.

Tolerances are specified for piston rings, pistons, cylinder bores, valve seats, faces, and stems. Shaft journals, bushings, and bearings are also specified as to tolerances allowed. In Table 11–2, the upper and lower limits are given for a number of engine parts. Five engine models are shown with their specific clearances and tolerances.

Manufacturers cannot make parts exactly to dimension. Therefore, they make them within a certain tolerance both + or –. Pistons and cylinders have to be mated accordingly. Bearings also have to be matched up to fit crankshafts and other places where they are used. This means the tolerance becomes very important in the remanufacturing or repairing of engines.

Allowances

An allowance is an intentional difference between mating parts. It may be either positive or negative. A positive allowance is the minimum clearance between mating parts. A negative allowance is the maximum clearance between mating parts. A positive allowance provides clearance for a running or sliding fit. A negative allowance provides interference between mating parts. This produces a force fit. There are types of fits and they are classified according to the amount of clearance.

Classes of Fit—There are basically four categories which describe how parts fit together. These categories are called classes of fit. A *running fit* means the parts must slide or rotate without hindrance. The male part is smaller than the female part by at least 0.001″ (0.0254 mm) for each inch of diameter. This indicates that there is a clearance between the two parts.

Table 11–2. Engine Specifications

Model		HH80	HH100	HH120	VH80	VH100
Displacement	Early	19.44	23.70	27.66	23.75	23.75
	Late	23.70				
Stroke		2¾"	2¾"	2⅞"	2¾"	2¾"
Bore	Early	3.0000 3.0010	3.3120 3.3130	3.500 3.501		
	Late	3.3120 3.3130				
Timing Dimension		TDC-Start 0.095 Run	TDC-Start 0.095 Run	TDC-Start 0.095 Run	Solid State	Solid State
Point Gap		0.020	0.020	0.020	Solid State	Solid State
Spark Plug Gap		0.028 0.033	0.028 0.033	0.028 0.033	0.035	0.035
Valve Clearance	Intake	0.010	0.010	0.010	0.010	0.020
	Exhaust	0.020	0.020	0.020	0.020	0.020
Valve Seat Angle		46°	46°	46°	46°	46°
Valve Seat Width		0.042 0.052	0.042 0.052	0.042 0.052	0.042 0.052	0.042 0.052
Valve Face Angle		45°	45°	45°	45°	45°
Valve Face Width		0.089 0.099	0.089 0.099	0.089 0.099	0.089 0.099	0.089 0.099
Valve Lip Width		0.06	0.06	0.06	0.06	0.06
Valve Spring Free Length	Early	2.125	2.125	2.125	1.885	1.885
	Late	1.885	1.885	1.885		
Valve Spring Compressed Tension Early Prod. Models		19-21 lbs. at 1.65 Length	19-21 lbs. at 1.65 Length	19-21 lbs. at 1.65 Length	23 lbs. ± 2.5 lbs. at 1.55 Lenght	23 lbs. ± 2.5 lbs. at 1.55 Length
Valve Spring Compressed Tension Late Prod. Models		23 lbs. ± 2.5 lbs. at 1.55 Length	23 lbs. ± 2.5 lbs. at 1.55 Length	23 lbs. ± 2.5 lbs. at 1.55 Length		
Valve Guides STD Diameter		0.312 0.313	0.312 0.313	0.312 0.313	0.312 0.313	0.312 0.313
Valve Guides Over-Size Dimensions		0.343 0.344	0.343 0.344	0.343 0.344	0.344 0.345	0.344 0.345

Table 11–2. (*Cont'd*)

Model		HH80	HH100	HH120	VH80	VH100
Dia. Crankshaft Conn. Rod Journal		1.3750 1.3755	1.3750 1.3755	1.3750 1.3755	1.3755 1.3750	1.3755 1.3750
Maximum Conn. Rod Dia. Crank Bearing		1.3765	1.3765	1.3765	1.3761	1.3761
Shaft Seat Dia. For Roller Bearings		1.1865 1.1870	1.1865 1.1870	1.1865 1.1870	1.1865 1.1870	1.1865 1.1870
Crankshaft End Play		None*	None*	None*	None*	None*
Piston Diameter	Early	2.993 2.994	3.3080 3.3100	3.4950 3.4970	3.308 3.310	3.308 3.310
	Late	3.3080 3.3100				
Piston Pin Diameter		0.6873 0.6875	0.6873 0.6875	0.6873 0.6875	0.6873 0.6875	0.6873 0.6875
Width Comp. Ring Groove		0.0950 0.0960	0.0950 0.0960	0.0950 0.0960	0.0950 0.0960	0.0950 0.0960
Width Oil Ring Groove		0.1880 0.1900	0.1880 0.1900	0.1880 0.1890	0.1800 0.1900	0.1800 0.1900
Side Clearance Ring Groove		0.0020 0.0035	0.0020 0.0035	0.0020 0.0035	0.0025 0.0030	0.0025 0.0030
Ring End Gap		0.010 0.020	0.010 0.020	0.010 0.020	0.010 0.020	0.010 0.020
Top Piston Land Clearance		0.0305 0.0335	0.030 0.035	0.031 0.036	0.0305 0.0335	0.0305 0.0335
Piston Skirt Clearance	Early	0.005 0.007	0.002 0.005	0.003 0.006	0.003	0.003
	Late	0.002 0.005				
Camshaft Bearing Dia.		0.6235 0.6240	0.6235 0.6240	0.6235 0.6240	0.6235 0.6240	0.6235 0.6240
Cam Lobe Dia. Nose to Heel		1.3045 1.3085	1.3045 1.3085	1.3045 1.3085	1.3045 1.3085	1.3045 1.3085
Magneto Air Gap		0.006 0.010	0.006 0.010	0.006 0.010	0.006 0.010	Solid .006 State .010

*Preloaded in cylinder cover 0.000 to 0.007.

(*Courtesy Tecumseh*)

Fig. 11–8. (A) Briggs & Stratton model series 60100 to 60152. *(Courtesy B & S)*

Fig. 11–8 (B) Vertical-shaft engine with pull-up starter. *(Courtesy B & S)*

A *push fit* calls for forcing, but can be assembled by hand. The parts are not quite able to move past one another. There is a slight clearance because the male part should be 0.001″ to 0.0001″ (0.0254 to 0.00254 mm) smaller. This classifies as a clearance.

A *driving fit* means the assembly must be done with some type of force such as an arbor press. The male and female parts have an interference. That means the male part must be the same size or 0.001″ (0.0254 mm) larger per inch of diameter than the female part (Fig. 11–9).

A *force fit* requires a hydraulic press to put the two parts together. There is interference of 0.002″ (0.0508 mm) for each inch of diameter.

Other types of permanent fits involve welding, brazing, and soldering. The parts are not easily separated when needed. The previous classes of fit are separable when the proper equipment is used, and this is the case for many of the problems you will encounter with mating parts in rebuilding, or reassembling, or disassembling a small gasoline engine. In the troubleshooting and repair section of this book, you will observe some of these tolerances and fits and become more aware of their importance.

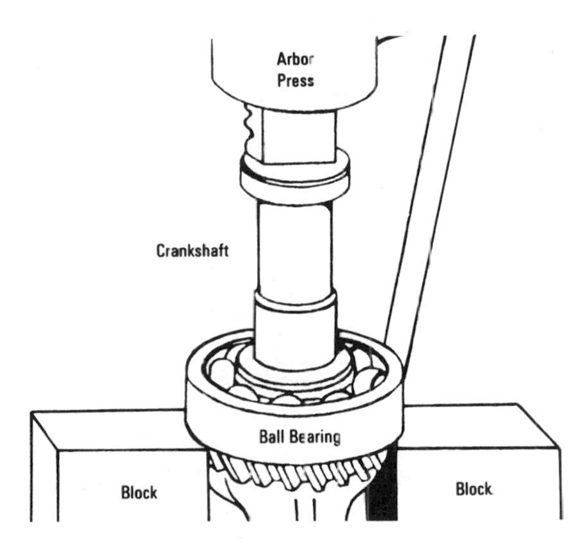

Fig. 11–9. Arbor press used to press-fit a ball bearing on a crankshaft of a small gasoline engine. *(Courtesy B & S)*

In some parts that must remain stationary on a shaft or in a hole, such as bearings, there is the shrink fit and the spun fit.

Shrink fit means that the female part is heated or the male part is chilled. This means that when normal operating temperature is reached, the female will be stretched around the male part. This allows for a more permanent fit without fear of the parts moving at the point of contact (Fig. 11–10).

Spun fit means the female part is rotated very rapidly. The friction between the stationary male part and the rotating female part produces heat. Areas of the surfaces of both the parts become very hot and reach a semimolten state. This produces what is referred to as resistance welding, or a spun fit.

Inspections

In any service procedures, the best thing to do first is to check to see that all parts of the engine are present and in the correct locations. Then check the gas tank to make sure there is fuel in the tank. Another step to take when the gasoline supply is checked is to check the oil level in the engine. If it is a two-cycle engine, there will of course be no oil in the crankcase. Check to see that the gas gets to the carburetor, and then check to see if the spark plug is being furnished a high enough voltage to create a spark.

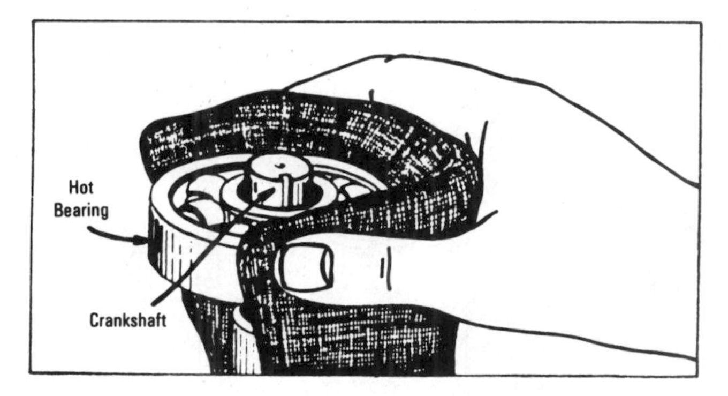

Fig. 11–10. Heating the bearing to expand it so it will fit over the cooler crankshaft. *(Courtesy B & S)*

Casual inspection of the cooling system also would be in order. Material may have gotten into the fins and caused overheating. The grass clippings, straw, dirt, and dust may have hampered the air-vane governor. All these visual inspections should be made before starting or trying to start the engine.

When inspecting an engine for reassembly, the reverse steps from disassembly should be followed. Check the tolerances and classes of fit. Make sure each part is in its proper location and in operating condition. Manufacturer's specifications should be referred to for exact tolerances and fits.

General inspection while making repairs will aid in the engine's operating the first time it is started.

Storage of Small Gasoline Engines

Most small engines are used on lawnmowers or equipment made for summer use. This means they should be properly stored during the winter. Winter equipment such as snowmobiles or snowblowers needs to be stored during the summer. The procedure for each season's storage is slightly different. One thing to keep in mind is not to run the winter engines during the summer for too long a period. They usually have a shroud that retains the engine heat for winter operation. Operation during the summer can result in overheating and the warping of some engine parts.

Storage Instructions

Storage instructions come with the operator's manual. They usually include these steps or suggestions:

1. All fuel should be removed from the fuel tank. Run the engine until it stops from lack of fuel. The small amount of fuel that remains in the sump of the tank should be removed by absorbing it with a clean, dry cloth.
2. Remove the spark plug, pour 1 ounce (0.2957 dl) (two or three tablespoons) of SAE-30 oil into the cylinder and crank slowly to distribute the oil. Replace the spark plug.

3. Clean the dirt and chaff from the cylinder, cylinder head, fins, and blower housing.
4. Store in a suitable place for the season.

Putting the Engine Back in Service

Putting the engine back in service after storage is the same as putting it into service for the first time, with one possible exception. In most storage instructions, the oil is not drained before storing and the engine has oil when removed from storage. It is therefore a good idea to change the oil before starting the engine after it has been stored.

Putting the engine in service requires that the sump be filled with oil. For summer, use SAE 30 if the temperature is over 40°F (4.4°C). If SAE 30 is not available, you can use SAE 10W-30 or 10W-40. In winter, use SAE 5W-20 or SAE 5W-30, if the temperature gets below 40°F (4.4°C). If the 5-weight oil is not available, use 10W or SAE 10W-30. Do not overfill the engine with oil. Follow the manufacturer's recommendations for filling. Most engines have a fill level marked on them. Some have a dipstick that will tell you how full the engine is.

Next fill the fuel tank. Use clean, fresh, lead-free or leaded gasoline of the grade automobile engines use. Fill the tank completely. Do not mix the oil with the gasoline unless you are putting a two-cycle engine into operation for the first time or removing it from storage.

Move the engine control or throttle setting to *FULL* or *FAST* position. *Caution:* Make sure your body is clear of any obstructions that may cause damage when the engine starts.

Rewind starters are usually employed on small gasoline engines. If so, grasp the starter handle and pull out on the cord rapidly. Return it slowly to the engine. Repeat if necessary.

Important: Some engines have a unique automatic choke. In case of flooding, move the control to stop and pull the starter six times. If the engine continues to flood, rotate the carburetor needle valve ⅛ of a turn clockwise to obtain a leaner mixture.

In most cases the carburetor needle valve has been adjusted so that the mixture is proper for average operating conditions. Normally, no changes are necessary. However, if the engine does not start easily, a minor adjustment will compensate for the difference in fuel, temperature, or altitude.

If the engine does not start easily when cold, rotate the needle valve ⅛ of a turn counterclockwise for a richer mixture.

If the engine does not start easily when warm, rotate the needle valve ⅛ of a turn clockwise for a leaner mixture.

To stop the engine, move the control lever to *STOP* position.

In-Service Adjustments

While the engine is in service, a few adjustments may be made in the carburetor controls to make sure maximum efficiency is obtained for the engine.

Remote-Control Adjustments

The remote control must be properly adjusted to stop and to operate the engine at maximum speed (Fig. 11–11A).

To check operation, move the remote-control lever to *STOP* position. The carburetor lever should contact the stop switch as shown in Fig. 11–11A. With the remote-control lever in the *FAST* position, the carburetor lever should just touch the boss as shown in Fig. 11–11B.

Carburetor Adjustments

Minor carburetor adjustments may be required to compensate for differences in fuel, temperature, altitude, and load.

All carburetor adjustments should be made with the air cleaner on the engine. The air-cleaner mounting screw *must* be in the carburetor when the engine is run. Best adjustment is made with a fuel tank *half full* of gasoline.

To adjust the carburetor:

1. Start the engine and run it long enough to warm it to operating temperature. If the engine is out of adjustment so that it will not start, close the needle valve by turning it clockwise. Then open the needle valve 1½ turns counterclockwise (Fig. 11–12).
2. Move the engine control to run the engine at normal operating speed.

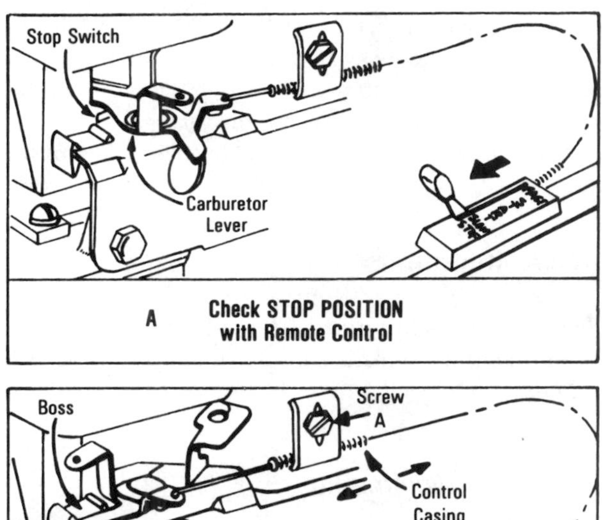

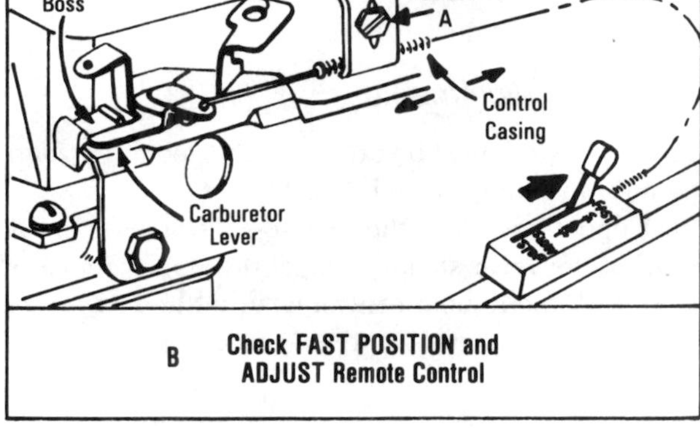

Fig. 11–11. Adjustment of the *FAST* and *STOP* positions of the remote controls of an engine. *(Courtesy B & S)*

3. Turn the needle valve in a clockwise direction until the engine starts to lose speed. This produces a lean mixture.
4. Then slowly turn the needle valve out or counterclockwise past the point of smoothest operation until the engine just begins to run unevenly. This produces a rich mixture.
5. Turn the needle valve clockwise very slowly until the engine runs evenly (Fig. 11–13). Final adjustment of the needle valve should be slightly to the rich side (counterclockwise of the midpoint).
6. Move the engine control to *SLOW*. Turn the idle-speed adjusting screw until a fast idle speed is obtained (1750 rpm).

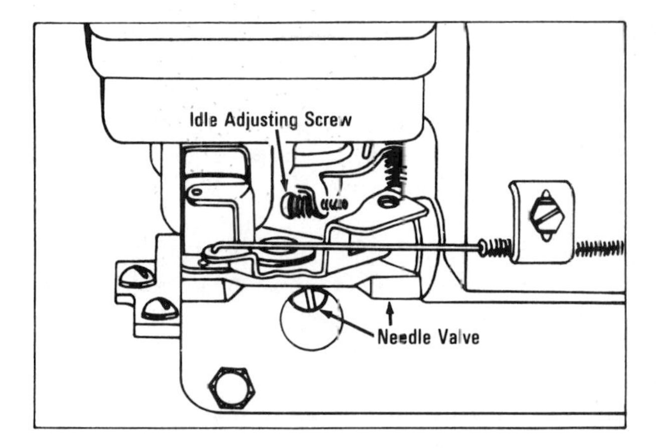

Fig. 11–12. Carburetor adjustment — engine will not start. *(Courtesy B & S)*

7. To check adjustment, move the engine control from *SLOW* to *FAST* speed. Engine should accelerate smoothly. If the engine tends to stall or die out, increase the idle speed or readjust the carburetor, usually to a slightly richer mixture.

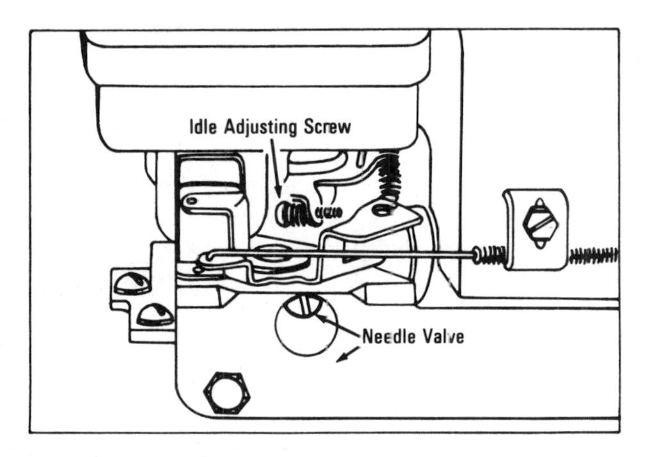

Fig. 11–13. Carburetor adjustment — uneven operation. *(Courtesy B & S)*

Tune-Up Techniques

This procedure will show what major repairs have to be made; in the case of a new engine that has been run for a season, it is more or less normal maintenance. The steps that follow are also performed as part of the complete overhaul that will be discussed in another chapter.

1. Remove air cleaner. Check for proper servicing.
2. Check oil level and drain. Clean the fuel tank and lines if separate from the carburetor.
3. Remove the blower housing. Inspect the rope and rewind assembly and starter clutch.
4. Clean the cooling fins and entire engine. Rock the flywheel to check the compression.
5. Remove the carburetor. Disassemble and inspect for wear or damage. Wash in solvent, replace parts as necessary, and reassemble. Set the initial adjustment.
6. Inspect the crossover tube or intake elbow for damaged gaskets.
7. Check the governor blade, linkage, and spring for damage or wear. If the mechanical governor is used, also check its adjustment.
8. Remove the flywheel. Check for seal leakage at both the flywheel and power take-off sides. Check flywheel key.
9. Remove the breaker cover and check for proper sealing.
10. Inspect the breaker points and condenser. Replace or clean and adjust. Check the plunger.
11. Check the coil; inspect all wires for breaks and damaged insulation. Be sure the lead wires do not touch the flywheel. Check the stop switch and lead.
12. Replace the breaker cover. Use a sealer where the wires enter.
13. Install the flywheel. Time the engine if necessary. Set the air gap. Check the spark using the spark plug tester.
14. Remove the cylinder head. Check the gasket. Remove the spark plug. Clean off the carbon and inspect the valves for seating.
15. Replace the cylinder head. Torque the bolts to the proper or specified torque. Set the spark plug gap or, if necessary, replace the plug.
16. Replace the oil and fuel. check the muffler for restrictions or damage.
17. Adjust the remote-control linkage and cable, if used, for correct operation.

18. Service the air cleaner. Check the gaskets and element for damage.
19. Run and adjust the mixture for top speed.

Preventive Maintenance Procedures

The main aim of preventive maintenance is to keep the engine operating at top efficiency for its intended or designed lifetime with few or no problems. In order to do so, an established routine should be used to maintain the engine.

Gasoline

Make sure clean and *fresh* gasoline is used in the engine. Regular or unleaded may be used in most small engines. Mix the oil with the gasoline for two-cycle engine operation. Make sure the correct ratio of oil to gasoline is observed for good maintenance-free operation of the engine. Do not purchase more than a thirty-day supply of gasoline. Fresh gasoline minimizes gum deposits and ensures a fuel with volatility tailored for the season.

Oil

For four-cycle engines, use the correct SAE viscosity recommended for the engine by the manufacturer. For the two-cycle engine, make sure the correct oil is purchased for two-cycle operation.

Use oil on the parts of the machine that require oiling. The correct weight is usually given in an operator's manual.

A good-grade detergent oil is recommended by most engine manufacturers. Summer calls for SAE 30 and winter for SAE 10W. The W stands for winter use. Multiple-viscosity oils may be used if the single-viscosity ratings are not available. That means 10W-30 or 10W-40 in summer and 5W-30 or 5W-20 in winter. No special additives are called for by the manufacturers.

Oil should be changed after each twenty-five hours of engine operation. Under dusty or dirty operating conditions the oil should be changed more frequently. In the normal running of any engine, small particles from the cylinder walls, pistons, and bearings will gradually work into the oil. Dust particles from the air also get into the oil. If the oil is not changed regularly, these foreign particles cause increased friction and a grinding action, which shortens the life of the engine. Fresh oil also assists in cooling, for old oil gradually becomes thick and loses its cooling effect as well as its lubricating qualities.

The *air cleaner* should be serviced every twenty-five hours of engine operation. Dirty operating conditions require more frequent servicing.

Cast-Iron Engine Break-in

The breaking in of a new cast-iron engine needs to be done according to the manufacturer's recommendations. Change the oil every twenty-five operating hours, but during the break-in period, change it after the first two hours.

The oil level is to be checked every five operating hours or sooner. *Do not overfill.* Screw the oil dipstick (if one is on the engine) into the filler opening as far as possible when checking the oil level. Overfilling may cause the oil seals to leak and result in excessive oil consumption.

Oil may turn black after only a few hours of operation. This is only the lead coating on the connecting rod coming off in the oil. The oil can still be used until the twenty-five-hour oil change. In some cases, such as very dusty or dirty surroundings, the oil will have to be changed more frequently .

Cooling Area

Use a small brush or compressed air to clean the cooling fin area. Wipe the entire engine clean for better cooling and appearance. Tighten all bolts. This will keep damaging vibration to a minimum. All checks, cleaning, and tightening should be done just before starting the engine, as it is cool and easier to work on.

Other Maintenance Items

If the engine uses a dry element for an air filter, make sure it is serviced often. Dry-type air cleaners are treated paper elements with rubberlike sealing edges. It is important that the edges of these elements seat properly to prevent dirt leakage. Compressed air may be used to clean the element (Fig. 11–14).

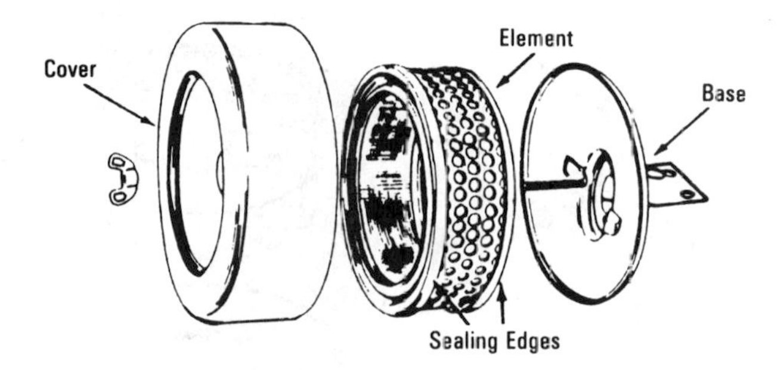

Fig. 11–14. Dry air-filter element. *(Courtesy Tecumseh)*

Compressed air should be directed from the inside of the element blowing toward the outside to dislodge the accumulated dirt. Tapping the element on a block of wood will also dislodge accumulated dirt. Wash the element in soap and water and thoroughly flush or rinse from the inside until the water is free of soap.

Allow the element to dry completely. Use low-pressure compressed air to aid in quick drying of the element. Blow air through the element from the inside.

Inspect the element for cracks and holes. If there is any doubt concerning the filter's condition, replace it.

Clean the air filter frequently to make sure full engine power and performance are available. Never operate the engine without the air filter properly assembled. Do not puncture the element. Carefully inspect the element for cracks and holes. Use a light bulb to shine light through the element. This will indicate any holes, cracks, or too much dirt.

Oil-bath air cleaners will give adequate engine protection. However, they have a tendency to spill the oil when the engine is operating on a tilt or incline. Clean all parts in a solvent and dry thoroughly (Fig. 11–15). Fill the filter with oil and assemble in the sequence shown in the exploded view. Gaskets must be used at the points shown in Fig. 11–15. Replace the gaskets if damaged. Maintain the oil level at the point indicated on the air-cleaner body. Use the same grade of oil for the air cleaner as is used in the engine crankcase.

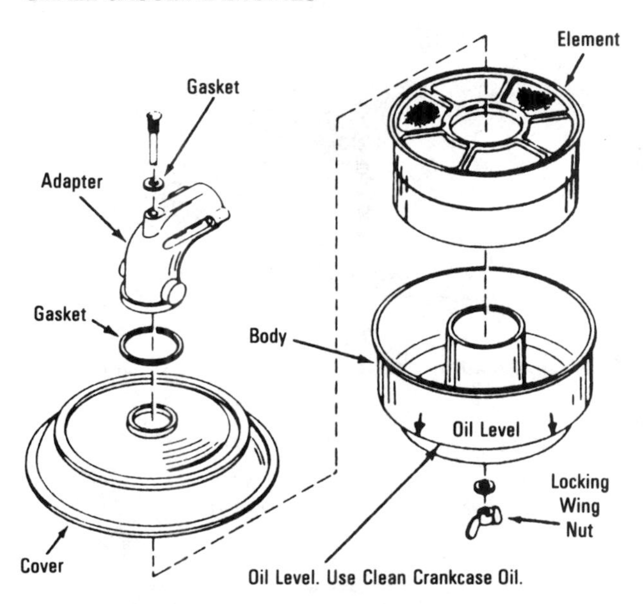

Fig. 11–15. Reassembling the element into an oil-bath air cleaner. *(Courtesy Tecumseh)*

Warranty

The warranty for the engine should be read closely. There are limitations on any warranty. Some states even have laws that govern the limits of liability and other claims that may be made by the purchaser of an engine or a product with a mounted engine as part of its complete package.

Some warranties are for ninety days and others are for one year. The limitations are usually listed (Fig. 11–16).

All warranties require you to return the defective merchandise to the manufacturer or to a representative in your area. You, the owner, will have to assume the expense of transportation to and from the repair facility or the factory.

It is best to check the exact terms and coverage provided by the warranty and the dealer when you purchase the engine. These things should be in writing so that there will be no arguments later.

ECHO PRODUCTS NINETY DAYS LIMITED WARRANTY

The Kioritz Corporation of America, referred to hereafter as the KCa, warrants to the original purchaser at resale that each new Echo product shall be free from defects in material and workmanship for a period of ninety (90) days from the date of the original purchase. In the event of a defect in an Echo product, KCA will repair and/or replace the defective part or parts free of charge.

In the event of a malfunctin or failure in an Echo product simply deliver or send, postage prepaid, within the warranty period of ninety (90) days, the complete Echo product to your nearest authorized KCA dealer. KCA reserves the right to inspect the claimed defective part or parts to determine if the defect or malfunction complained of is covered by this warranty. Upon confirmation that the malfunction or failure is the result of a defect covered by this warranty, KCA shall, within ninety (90) days after receipt of the product, at its option, repair and/or replace the defective part or parts free of charge.

Authorized KCA dealers are located throughout the United States and Canada to assure fast and efficient service of your Echo product. If you do not know the location of an authorized KCA dealer, you can obtain such information by contacting KCA at the following address. KCA assumes no responsibility for any damage, defects, or costs resulting from repairs and/or modifications of an Echo product performed by sad person who is not an authorized Echo servicing dealer or distributor.

This warranty shall only cover defects arising under normal usage.
KCA assumes no responsibility whatsoever if the Echo product shall fail during the warranty period by reason of

1. Misuse, negligence, physical damage or accident.
2. Lack of maintenance as prescribed in the Owner's Manual.
3. Improper adjustments of ignition, timing and carburetors.
4. The use of fuels and lubricating oils not approved by KCA.
5. The use of parts or accessories not approved by KCA.
6. Operation of the unit with the air filter removed.
7. Operation at engine speeds in excess of KCA recommendations.
8. Repair by any unauthorized party during the warranty period.
9. Normal wear and tear.
10. Operation of the unit with an incorrect fuel/oil ratio.

KCA MAKES NO FURTHER EXPRESS WARRANTIES OR REPRESENTATIONS EXCEPT THOSE CONTAINED HEREIN. NO REPRESENTATIVE OR DEALER IS AUTHORIZED TO ASSUME ANY OTHER LIABILITY REGARDING ECHO PRODUCTS. THE DURATION OF IMPLIED WARRANTIES GRANTED UNDER STATE LAW, INCLUDING WARRANTIES OF MERCHANTABILITY AND FITNESS FOR A PARTICULAR PURPOSE ARE LIMITED IN DURATION TO THE DURATION OF THE EXPRESS WARRANTY GRANTED HEREUNDER. KCA SHALL IN NO EVENT BE LIABLE FOR DIRECT, INDIRECT, SPECIAL OR CONSEQUENTIAL DAMAGES. SOME STATES DO NOT ALLOW LIMITATIONS ON HOW LONG AN IMPLIED WARRANTY LASTS, AND/OR DO NOT ALLOW THE EXCLUSION OR LIMITATION OF INCIDENTAL OR CONSEQUENTIAL DAMAGES, SO THE ABOVE LIMITATIONS AND EXCLUSIONS MAY NOT APPLY TO YOU. THIS WARRANTY GIVES YOU SPECIFIC LEGAL RIGHTS, AND YOU MAY ALSO HAVE OTHER RIGHTS WHICH MAY VARY FROM STATE TO STATE.

To validate this warranty, the original purchaser must complete and return within ten (10) days of the date of purchase the warranty registration card supplied with each unit.

For additional information regarding this warranty or service of Echo products, you may contact Kioritz Corporation of America at 350 Wainwright Avenue, Northbrook, Illinois 60062.

KIORITZ CORP. OF AMERICA
350 WAINWRIGHT AVE. NORTHBROOK, ILL 60062 U S A
PHONE (312) 498 1390

FILL IN FOR YOUR OWN RECORD

MODEL NO. ____________________

SERIAL NO. ____________________

DATE OF PURCHASE ____________________

Do not return parts directly to ECHO DIVISION. Any part claimed to be defective must be returned to ECHO DIVISION by delivering it to an authorized ECHO dealer or distributor.

Fig. 11–16. (A) A ninety-day warranty. *(Courtesy Kioritz)*

BRIGGS & STRATTON ENGINE WARRANTY

For ONE YEAR from purchase date, Briggs & Stratton Corp. will replace for the original purchaser, FREE OF CHARGE, any part, or parts, found upon examination by any Factory Authorized Service Center, or by the Factory at Milwaukee, Wisconsin, to be DEFECTIVE IN MATERIAL AND/OR WORKMANSHIP.

All transportation charges on parts submitted for replacement under this Warranty must be borne by purchaser.

There is no other Warranty express or implied. Briggs & Stratton Corp. shall in no event be liable for consequential damages.

BRIGGS & STRATTON CORP.

V. R. Shiely

V. R. SHIELY – PRESIDENT

NOTE: The Briggs & Stratton Engine Warranty does not cover breakage of parts or damage to parts due to abuse or failure to follow the recommended maintenance procedures. The warranty also excludes any accessories, controls or equipment which are not manufactured by Briggs & Stratton Corporation.

If warranty service is needed contact your nearest Authorized Service Center. For Prompt Attention your center will need to know the engine model, type and code number, the trouble experienced and the total number of hours the engine has run. If you differ with the decision of a Service Center on a warranty claim, ask the Service Center to submit all supporting facts to the Factory for review. If the Factory decides that your claim is justified, you will be fully reimbursed for those items accepted as defective.

FILL IN THE REQUIRED INFORMATION FOR YOUR RECORD:

(See Decal on Blower Housing for Model, Type and Code Number)

Engine Model No. ____________ Type No. ____________ Code No. ____________

Dealer Purchased From ________________________ Date ____________

Type of Equipment ________________________________

Name or Trademark of Equipment Manufacturer ________________________

BRIGGS & STRATTON ENGINES ARE MADE UNDER ONE OR MORE OF THE FOLLOWING PATENTS: 3,252,499

2,669,322	2,796,453	3,114,851	3,149,618	3,194,224	3,276,439
2,693,789	2,999,491	3,118,433	3,165,094	3,236,937	3,378,099
2,693,791	2,999,562	3,144,097	3,168,936	3,242,741	3,415,237

DESIGN

D-191,806 D-196,017 D-197,175 D-213,476

OTHER PATENTS PENDING

Fig. 11–16. **(B) A one-year warranty.** *(Courtesy B & S)*

One thing should be kept in mind. The engine should be operated according to the manufacturer's recommendations, and the recommended fuel and oil should be used and changed at the proper intervals. Most companies stand behind their engines. They want you to be a satisfied customer and make others aware of your satisfaction with the operation of their product.

Chapter Highlights

1. Proper care and maintenance will contribute to a long and trouble-free operation of any engine.
2. Care of the engine begins before it starts.
3. There are some adjustments that have to be made by the operator so that the engine will operate at its top efficiency.
4. Each engine comes with an operator's manual.
5. The engine identification plate is mounted on the carburetor side of the blower housing on most engines.
6. Engines are made today with very close tolerances by automatic machines.
7. An allowance is an intentional difference between mating parts.
8. There are four categories of classes of fit—how parts fit together.
9. In any inspection procedure, the best thing to do first is to check to see that all parts of the engine are present and in the correct locations.
10. Engines with a shroud designed to retain the heat during the winter should have it removed for summer operation.
11. Minor adjustments are made in the carburetor to compensate for differences in fuel, temperature, altitude, and load.
12. Make sure clean and fresh gasoline is used in engines.
13. Use the correct SAE rating of oil in four-cycle engines.
14. Use the correct fuel-and-oil mixture for two-cycle engines.
15. The air cleaner should be serviced every twenty-five hours of engine operation.
16. Do not overfill the oil sump or crankcase with oil.
17. Keep the fins on the air-cooled engine free of debris or any clogging materials.

Vocabulary Checklist

allowances
casual inspection
classes of fit
driving fit
force fit
identification
multiple viscosity
oil filter element
operator's manual
preventive maintenance
push fit
richer mixture
shrink fit
specifications
spun fit
technical manual
tolerance
torque
warranty

Review Questions

1. When does the care of an engine begin?
2. What does the term *preventive maintenance* mean?
3. Where do you get operator's manuals?
4. Where do you get technical manuals for engines?
5. How do manufacturers identify their engines?
6. What is tolerance?
7. What is an allowance?
8. What is a running fit?
9. Identify the following terms:
   - **a.** push fit
   - **b.** force fit
   - **c.** driving fit
   - **d.** shrink fit
   - **e.** spun fit
10. How do you check the oil level on a two-cycle engine?
11. Why do you drain all the gasoline from an engine's tank before storage?
12. Why do you remove the spark plug and put in about 1 ounce (0.2957 dl) of oil before storage of an engine?
13. Do you drain the oil from an engine's crankcase before storing?
14. After a period of storage, why do you change the oil in an engine before it is run?

15. Why should you make carburetor adjustments with half a tank of gasoline?
16. What are the first five things you do in a tune-up procedure?
17. Why do you use only fresh gasoline in an engine?
18. Do you use detergent oil in a small gasoline engine?
19. How often do you check the engine's oil level?
20. Does black oil mean it is dirty?
21. How do you clean dry air-cleaner elements?
22. What is a warranty?

CHAPTER 12

Troubleshooting

Troubleshooting is slightly different for the two-cycle engine than for the four-cycle engine. The differences should be apparent by this time. Keep in mind that the two-cycle engine uses air and fuel just like the four-cycle. However, the two-cycle has gasoline and oil mixed and no oil reserve in the crankcase. Therefore, the two-cycle does not require the oil level in the crankcase to be checked as part of the checkup procedures. The two-cycle also has reeds to be checked in some instances, which are not used on the four-cycle. The carburetors are slightly different also. There are variations in carburetors even on the four-cycle, and there are variations in the ignition systems within a category of engines. This means that knowing both the model number and manufacturer is very important.

Keep these small differences in mind as you go about your troubleshooting routine. In this chapter we have taken the four-cycle and the two-cycle engines and provided separate sections since they do have slight differences that demand individual attention. Notice, however, as you read over the troubleshooting procedures, that there are some very close resemblances in the general steps that apply to both types of engines.

Two-cycle engines are used as outboard engines and on mopeds (Fig. 12–1). They find use on weed cutters and trimmers, lawnmowers, snowmobiles, chain saws, and small electrical generators. Many other applications have also been devised.

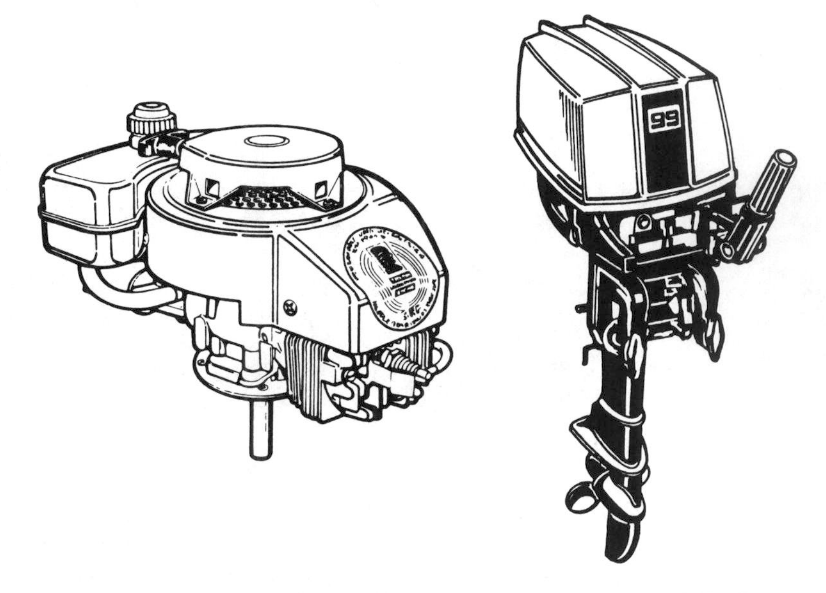

Fig. 12–1. Two-cycle gasoline engine with a vertical crankshaft and a two-cycle outboard motor. *(Courtesy Tecumseh)*

Sears, Roebuck and Company markets a line of Gas Bushwackers® with bladed trimmers. They are an extension of the nylon-cord grass and weed cutter. They have three types or sizes. Each has a different size engine. The largest engine is 37.7 cc. The other two engines are 28.0 cc and 26.2 cc. As you can see, these smaller engines are not rated as to horsepower but according to their cubic centimeter (cc) rating. The newer metric standards abbreviate cubic centimeter as cm^3, but buyers are not used to it in an engine rating specification so "cc" will probably be around for some time.

A cubic centimeter can be converted to cubic inches by multiplying cc by 0.061025. This means that the 28-cc engine mentioned above equals 1.7087 cubic inches. Since 28 is a larger number and would sound better when you are trying to sell something, you can see how easy it is to cling to the cc rating. If you wish to convert cubic inches to cubic centimeters, you multiply the cubic inches by 16.3867265. Take a look at the three sizes of 37.7, 28.0, and 26.2 cc. If these were given in cubic inches they would read, respectively, 2.3, 1.7, and 1.6 cubic inches.

To Check Compression—Spin the flywheel in reverse rotation (counterclockwise) to obtain accurate compression check. The flywheel should rebound sharply, indicating satisfactory compression. If the compression is poor, look for:

1. Loose spark plug.
2. Loose cylinder head bolts.
3. Blown head gasket.
4. Warped cylinder head.
5. Worn bore and/or rings.
6. Broken connecting rod.

To Check Ignition—Remove the spark plug. Spin the flywheel rapidly with one end of the ignition cable clipped to the spark plug tester (Fig. 12–2). The other end of the tester is grounded on the cylinder head. If the spark jumps the 0.166″ (4.2164 mm) tester gap, you may assume the ignition system is functioning satisfactorily. Try a new spark plug. If the spark does not occur, look for:

1. Incorrect armature air gap.
2. Worn bearings and/or shaft on flywheel side.
3. Sheared flywheel key.
4. Incorrect breaker point gap.
5. Dirty or burned breaker points.
6. Shorted ground wire (when so equipped).
7. Shorted stop switch (when so equipped).
8. Condenser failure.
9. Armature failure.

Two-Cycle Engine Troubleshooting Chart

Cause	Remedy
Engine Fails to Start or Starts with Difficulty	
No fuel in tank.	Fill the tank with clean, fresh fuel.
Fuel shutoff valve closed.	Open the valve.
Obstruction in the fuel line.	Clean the fuel screen and line. If necessary, remove and clean the carburetor.

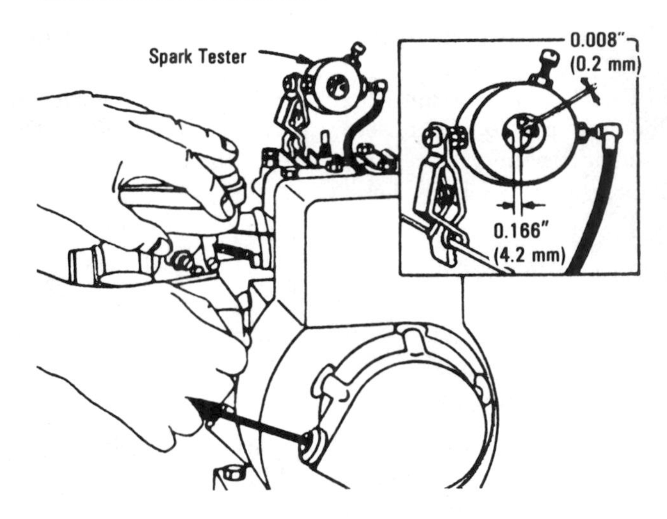

Fig. 12–2. Spark tester. *(Courtesy B & S)*

Two-Cycle Engine Troubleshooting Chart, cont'd

Cause	Remedy
Engine Fails to Start or Starts with Difficulty, cont.	
Tank cap vent obstructed.	Open vent in fuel tank cap, or replace the cap.
Water in fuel.	Drain the tank. Clean the carburetor and fuel lines. Dry the spark plug and points. Fill the tank with clean, fresh fuel.
Engine overchoked.	Close the fuel shutoff and pull the starter until the engine starts. Reopen the fuel shutoff for normal fuel flow immediately after the engine starts.
Improper carburetor adjustment.	Adjust the carburetor.
Loose or defective magneto wiring.	Check the magneto wiring for shorts or grounds; repair if necessary.
Faulty magneto.	Check timing, point gap; if necessary, overhaul the magneto.
Spark plug fouled.	Clean and regap the spark plug.
Spark plug porcelain cracked.	Replace the spark plug.
Poor compression.	Overhaul the engine.

Two-Cycle Engine Troubleshooting Chart, cont'd

Cause	Remedy
Engine Knocks	
Carbon in combustion chamber.	Remove the cylinder head or cylinder and clean the carbon from the head and piston.
Loose or worn connecting rod.	Replace connecting rod.
Loose flywheel.	Check the flywheel key and keyway. Replace parts if necessary. Tighten the flywheel nut to proper torque.
Worn cylinder.	Replace cylinder.
Improper magneto timing.	Time magneto.
Engine Lacks Power	
Choke partially closed.	Open the choke.
Improper carburetor adjustment.	Adjust carburetor.
Magneto improperly timed.	Time magneto.
Worn piston or rings.	Replace piston or rings.
Air cleaner fouled.	Clean air cleaner.
Reed fouled or sluggish.	Clean or replace reed.
Improper amount of oil in fuel mixture.	Drain tank. Fill tank with correct mixture.
Crankcase seals leaking.	Replace worn crankcase seals. Some engines have no lower seal. Check the bearing surface of the crankshaft.

Clean or Replace Reeds—There are a number of reed designs used in two-cycle engines. Fig. 12–3A indicates five types. Fig. 12–3B shows how to check for the ten-thousandths clearance. Reeds should not bend away from the plate more than 0.010″ (0.254 mm). When replacing reeds in units where reeds can be removed, be sure the sealing surface of the reed faces the sealing surface of the adapter or body (Fig. 12–3B).

In servicing the reeds, clean them and the stops and adapters of oil or other foreign matter. All serviceable reeds plates have locating "smudge" marks on the smooth side of the reed as shown in Fig. 12–3C. The torque requirement on screws securing the reed stop and reeds to the reed plate on Type 640 engines is 6 to 9 inch-pounds (0.678 to 1.017 Nm). Snowblower engines with reeds on the carburetor adapter require 10 to 15 inch-pounds (1.13 to 1.695 Nm) of torque.

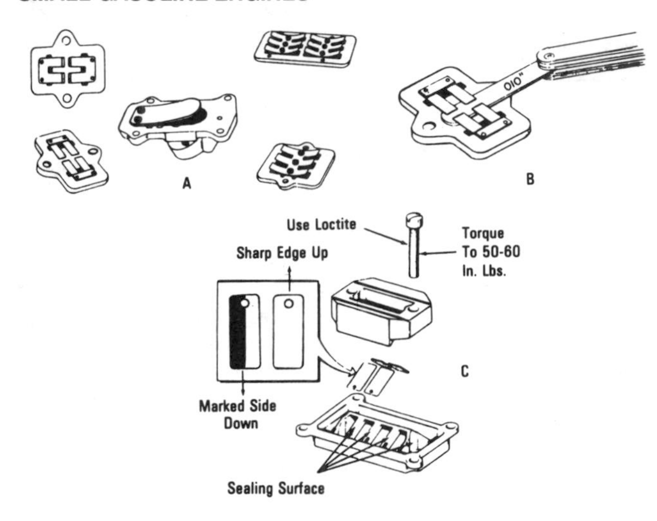

Fig. 12–3. Reeds used in two-cycle engine carburetors. *(Courtesy Tecumseh)*

Use Loctite® on the screws. The Type 639 engine reed assembly is shown in Fig. 12–3C. If the smudge marks are gone, feel for a rough edge on the reed. This edge must face away from the sealing surface. Torque the screw to 50 to 60 inch-pounds (5.650 to 6.780 Nm). Use Loctite on the screw threads.

Two-Cycle Engine Troubleshooting Chart, cont'd

Cause	Remedy
Engine Overheats	
Engine improperly timed.	Time engine.
Carburetor improperly adjusted.	Adjust carburetor.
Air flow obstructed.	Remove any obstructions from the air passages in shrouds.
Cooling fins clogged.	Clean cooling fins.
Excessive load on engine.	Check operation of driven equipment. Reduce excessive load.
Carbon in combustion chamber.	Remove cylinder head or cylinder and clean carbon from the head and piston.
Improper amount of oil in fuel mixture.	Drain the tank. Fill with correct mixture.

Two-Cycle Engine Troubleshooting Chart, cont'd

Cause	Remedy
Engine Runs Unevenly or Surges	
Fuel tank cap vent hole clogged.	Open the vent hole.
Governor parts sticking or binding.	Clean; if necessary, repair governor parts.
Carburetor throttle linkage or throttle shaft and /or butterfly binding or sticking.	Clean, lubricate, or adjust linkage and deburr the throttle shaft or butterfly.
Engine Vibrates Excessively	
Engine not securely mounted.	Tighten loose mounting bolts.
Bent crankshaft.	Replace crankshaft.
Driven equipment out of balance.	Recheck driven equipment.

Carburetor Adjustments

Since most of the trouble with two-cycle engines will be the carburetor out of adjustment, it is best to take a closer look at two types of carburetors used on these engines. Fig. 12–4 gives some service hints for the *float-feed carburetor.* These are general hints and do not give specific adjustments for individual carburetors.

Fig. 12–5 gives service hints for the *diaphragm carburetors.* These adjustments will in many cases aid in the quick return of the engine to service.

Four-Cycle Engine Troubleshooting

Most complaints concerning engine operation can be classified as one, or a combination, of the following:

1. Will not start.
2. Hard starting.
3. Kicks back when starting.
4. Lack of power.
5. Vibration.
6. Erratic operation.
7. Overheating.
8. High oil consumption.

When the cause of the malfunction is not readily apparent, you should perform a check of the compression, ignition, and carburetor

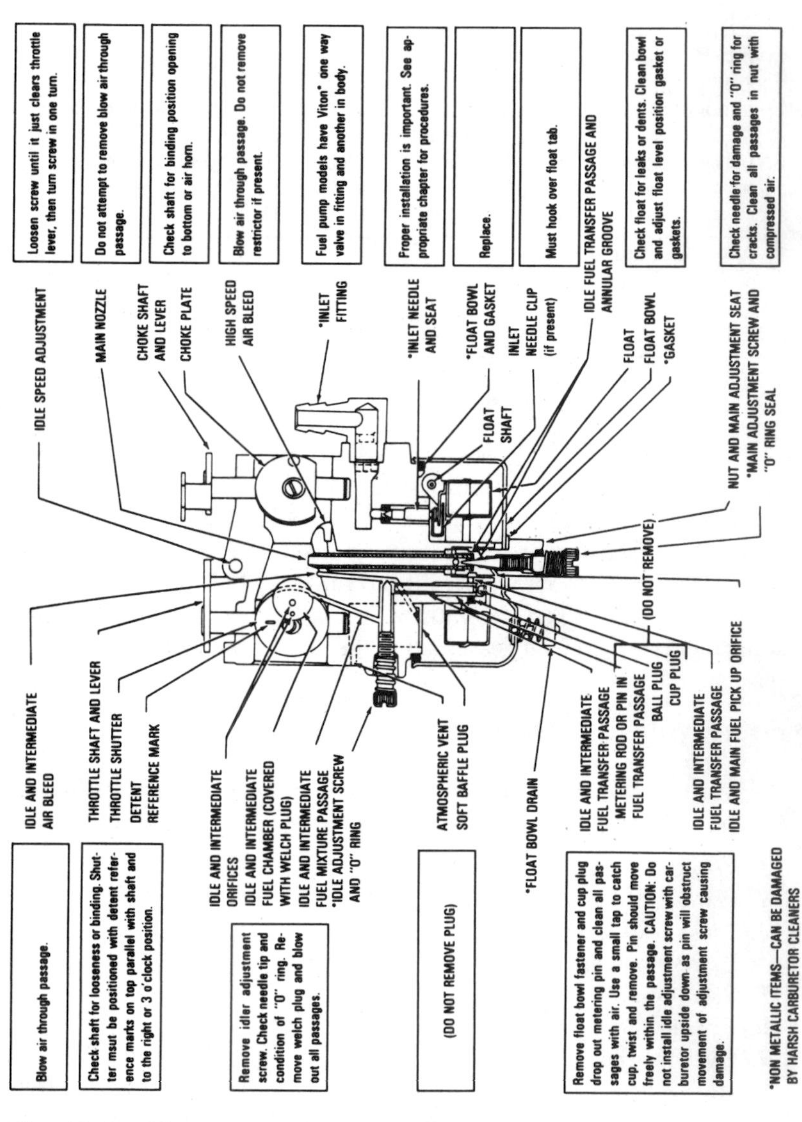

Fig. 12–4. Float-type carburetor adjustments for a two-cycle engine. *(Courtesy Tecumseh)*

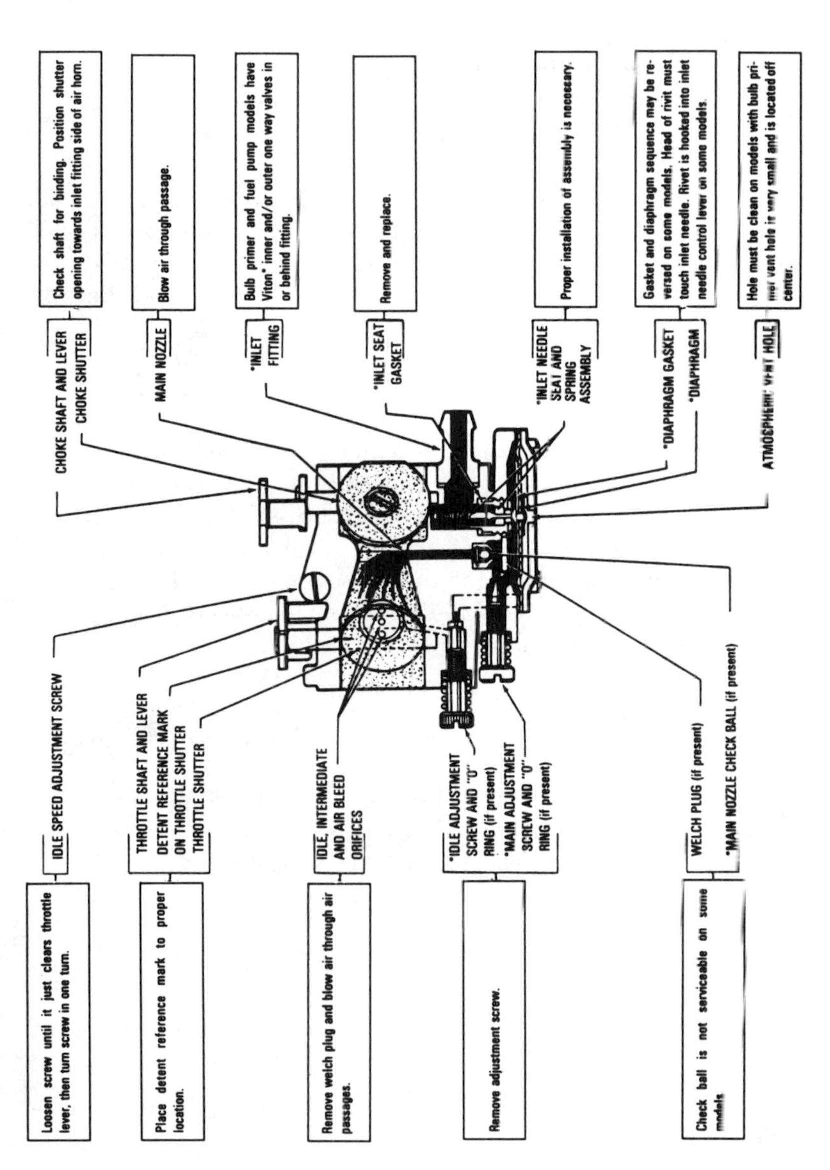

Fig. 12–5. Service hints for diaphragm-type carburetors used in two-cycle engines. *(Courtesy Tecumseh)*

systems. The checkup can be performed in a matter of minutes when done systematically. This is the surest and quickest way to locate a problem source. The basic checkup procedure is the same for all engine models; any variation by model will be shown.

Keep in mind that four-cycle engines may be used on any number of different pieces of power equipment. They can be found on lawnmowers, snowblowers, plows, and small electric generators. They can also be found powering water pumps and any number of other agricultural-type equipment. Therefore, what appears to be an engine malfunction may be a fault of the powered equipment rather than the engine. If the equipment is suspect, see the next several paragraphs.

Equipment Problems

Frequently, what appears to be a problem with engine operations, such as hard starting, vibration, and noise, may be the fault of the equipment powered rather than the engine itself. Since many various types of equipment are powered by these small four-cycle gasoline engines, it is not possible to list all the various conditions that may exist. Listed, however, are some of the most common effects of equipment problems and what to look for as the most common cause.

Hard Starting, Kickback, or Will Not Start

1. Loose blade (lawnmower). Blade must be tight to the shaft or adapter.
2. Loose belt. A loose belt, like a loose blade, can cause a backlash effect, which will counteract the engine cranking effort.
3. Starting under load. The unit should be disengaged when the engine is started; or, if the equipment is engaged, it should not have a heavy starting load.
4. Check the remote Choke-A-Matic control assembly for proper adjustment.
5. Check the interlock system for shorted wires, loose or corroded connections, or defective modules or switches.

Vibration—This can be caused by a number of things, some of which are listed below:

1. Cutter blade is bent or out of balance (lawnmower). Remove the blade and balance.

2. Crankshaft is bent. Replace. This happens when a rotary mower hits a large object when used as a weed or grass cutter with the blade attached to the crankshaft.
3. Worn blade coupling (lawnmower). Replace if the coupling allows the blade to shift, causing unbalance.
4. Mounting bolts are loose. Tighten.
5. Mounting deck or plate is cracked. Repair or replace.

Power Loss—This can be indicated by a stalled engine or one that will not do the job it is supposed to do even though it is running. Usually the engine doesn't come up to full speed.

1. Bind or drag in the unit. If possible, disengage the engine from the equipment and operate the unit manually to feel for any binding action.
2. Grass cuttings build up under the deck of the lawnmower.
3. No lubrication in the transmission or gear box.
4. Excessive drive belt tension may cause seizure.

Noise—You have to know the normal noise generated by the engine and the equipment it powers to be able to detect anything that sounds different. Noises will differ according to the equipment being powered. However, some general descriptions follow:

1. Cutter blade coupling or pulley—an oversize or worn coupling can result in knocking, usually under acceleration. Check for fit or tightness.
2. No lubrication in the transmission or gearbox.

Compression Checks

Compression checks are done the same way as on the two-cycle engine (see section above). However, there are a few other things to be checked in the slightly more complicated four-cycle engine. To check the compression of the engine, spin the flywheel in reverse rotation (counterclockwise) to obtain an accurate compression check. The flywheel should rebound sharply. This indicates satisfactory compression. A hand-held compression checker may be inserted into the spark plug hole to check more accurately, if required. If the compression is poor, look for:

1. Loose spark plug.
2. Loose cyclinder head bolts.
3. Blown head gasket.
4. Burnt valves and/or seats.
5. Insufficient tappet clearance.
6. Warped cylinder head.
7. Warped valve stems.
8. Worn bore and/or rings.
9. Broken connecting rod.

Ignition Checks

Ignition and carburetion should be checked first if the engine does not start. To check the ignition, remove the spark plug. Spin the flywheel rapidly with one end of the ignition cable clipped to the ignition tester (Fig. 12–2) and with the other end of the tester grounded on the cylinder head. If the spark jumps the 0.166″ (4.20 mm) gap, you may assume the ignition system is functioning satisfactorily. That means you should try a new spark plug or clean the one that is clogged or damaged.

If the spark does not occur, look for:

1. Incorrect armature air gap.
2. Worn bearings and/or shaft on the flywheel side.
3. Sheared flywheel key.
4. Incorrect breaker point gap.
5. Dirty or burned breaker points.
6. Breaker plunger stuck or worn.
7. Shorted ground wire (when so equipped).
8. Shorted stop wire (when so equipped).
9. Condenser failure.
10. Armature failure.
11. Improperly operating interlock system.

If the engine runs but misses during operation, a quick check to determine if ignition is or is not at fault can be made by inserting the spark plug tester (Fig. 12–2) between the ignition cable and the spark plug. A spark miss will be readily apparent. While conducting this test on Magna-Matic-equipped engines, Models 9, 14, 19, and 23, set the tester gap at 0.060″ (1.524 mm).

Honda's G100 series of engines utilize the breaker point ignition system. To adjust the timing or to check the timing refer to Fig. 12–6. Note the location of the index mark. Fig. 12–7 shows how the breaker points are disassembled. On reassembly, be sure the plate is not pinching the wire. After installing, adjust the ignition timing (Fig. 12–8).

Carburetion Checks

Before making a carburetion check, be sure the fuel tank has an ample supply of fresh, clean gasoline. On gravity feed (Flo-Jet) models,

a. To check timing:

(1) Remove the fan cover. Connect timing tester terminals to the engine stop switch and the cylinder.

(2) Watch the tester and rotate the flywheel to align the "F" mark on the flywheel with the index mark on the cylinder barrel.
The timing is correct if the tester operates when these marks align.

Ignition timing	20° BTDC

b. To adjust the timing:

(1) Remove the flywheel, woodruff key and point cover. Install the special tool on the crankshaft and reinstall the woodruff key to index the tool on the shaft.
Tighten the 12 mm nut to secure the tool.

(2) Rotate the tool clockwise to see if the points start to open when the tool pointer aligns with the index mark on the cylinder. If not—

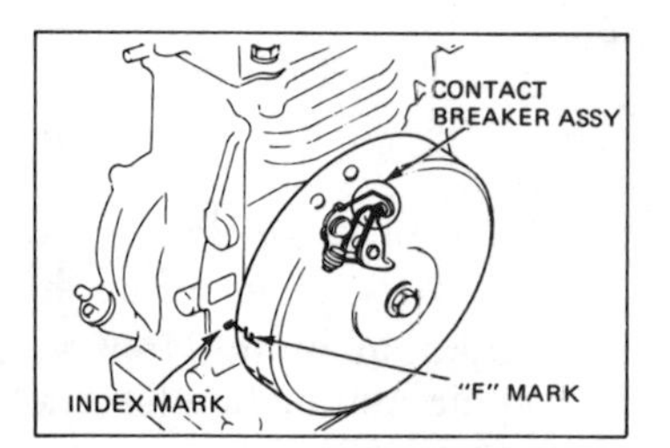

(3) Loosen the breaker plate attaching screw and move the plate in either direction to adjust the point gap.

(4) After adjusting, tighten the screw and recheck timing.

(5) When the timing is correct measure the point gap with a wire-type gauge. If the gap does not fall within range, replace the points. Recheck and adjust as necessary.

Breaker points gap	0.3–0.4 mm (0.012–0.016 in)

Fig. 12–6. Ignition timing for the Honda G100 series of engines. *(Courtesy Honda)*

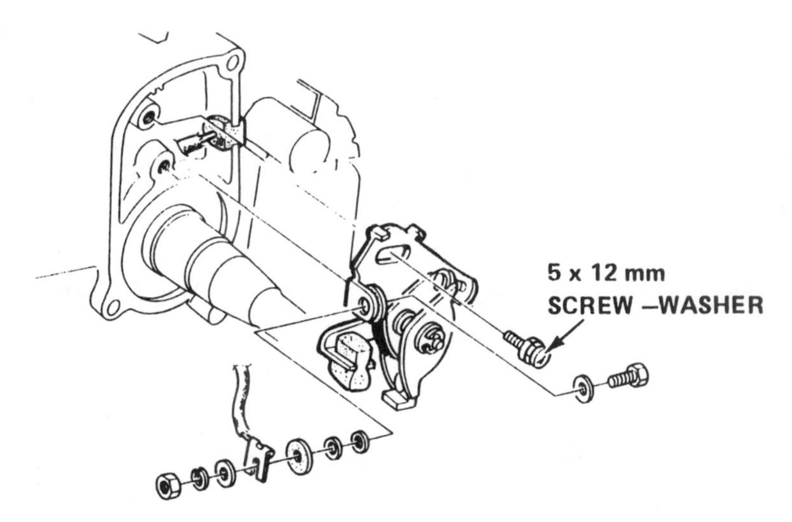

Fig. 12–7. **Breaker points — exploded view.** *(Courtesy Honda)*

see that the shutoff valve is open and the fuel flows freely through the fuel line. On all models, inspect and adjust the needle valves. Check to see that the choke closes completely. If the engine will not start, remove and inspect the spark plug. If the plug is *wet*, look for:

1. Overchoking.
2. Excessively rich fuel mixture.

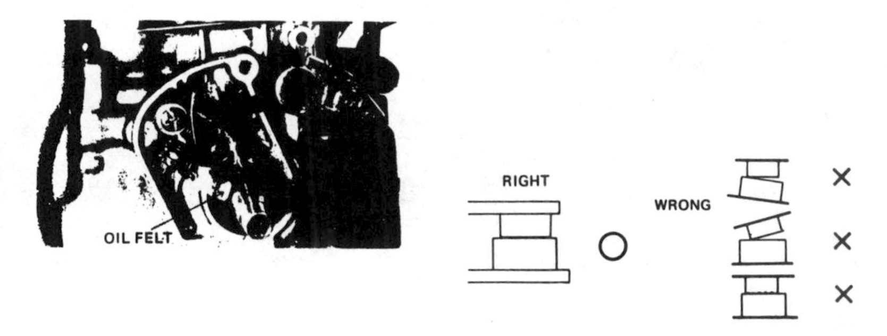

Fig. 12–8. Checking the breaker points for burning, pitting, fouling or abnormal wear. Check the oil felt for a "wet condition." Polish the point faces with a point file or fine-grade emery cloth, if necessary. Apply oil to the felt if dry. *(Courtesy Honda)*

3. Water in the fuel.
4. Inlet valve stuck open (Flo-Jet carburetor).

If the spark plug is *dry,* look for:

1. Leaking carburetor mounting gaskets.
2. Gummy or dirty screen or check valve (Pulsa-Jet and Vacu-Jet carburetors).
3. Inlet valve stuck shut (Flo-Jet carburetors).
4. Inoperative pump (Pulsa-Jet carburetors).

Fuel Test—A simple check to determine if the fuel is getting to the combustion chamber through the carburetor is to remove the spark plug and pour a small quantity of gasoline through the spark plug hole. Replace the plug. If the engine fires a few times and then quits, look for the same condition as for a *dry* spark plug.

Chapter Highlights

1. The four- and two-cycle engines have different troubleshooting procedures.
2. A cubic centimeter can be converted to cubic inches by multiplying cc by 0.061025.
3. There are checkoff charts and lists of troubles and what is causing them or how to remedy them for both two- and four-cycle engines.
4. Most of the trouble with two-cycle engines is the carburetor adjustment.
5. Frequently, what appears to be a problem with engine operation, such as hard starting, vibration, and noise, may be the fault of the equipment powered rather than the engine itself.
6. You have to know the normal noise generated by the engine and the equipment it powers to be able to detect anything that sounds different.
7. Compression checks on two- and four-cycle engines are done the same way.
8. Before making a carburetor check, be sure the fuel tank has an ample supply of fresh, clean gasoline.
9. Check the oil felt on the ignition system for a wet condition.
10. Be sure the plate is not pinching the wire when reassembling the ignition system.

Vocabulary Checklist

belt tension
breaker point ignition
compression
condenser
cubic centimeter
cubic inch
fouling
ignition break point
kickback
Loctite
obstruction
overchoking
reeds
sheared
smudge marks
spark plug porcelain
tappet clearance
troubleshooting
valve stems
warped cylinder head
wet condition

Review Questions

1. Why is troubleshooting different for the two-cycle engine than for the four-cycle engine?
2. What are some uses for the two-cycle engine?
3. Why are some applications better for the two-cycle engine than for the four-cycle engine?
4. How many cubic inches of displacement is represented by 28 cc?
5. List three troubles that may cause an engine to fail to start or to start with difficulty.
6. If compression of an engine is poor, what is the possible cause?
7. How do you check for ignition operation?
8. List at least nine ways to check for ignition problems.
9. If an engine knocks, what could be causing it?
10. If an engine lacks power, what could cause it?
11. Why do the reeds in a two-cycle engine need replacing?
12. How do you clean the reeds in a two-cycle engine?
13. What causes an engine to overheat?
14. What can cause an engine to run unevenly or in surges?
15. What can cause excess engine vibration?
16. List eight of the complaints concerning engine operation for a four-cycle engine.

17. What can cause hard starting, kickback, or no starting in equipment operated by a small engine?
18. What causes power loss to become apparent?
19. What nine things do you look for if the compression of an engine is low or poor?
20. If a spark does not occur at the spark plug, what eleven things do you look for as possible causes?
21. What type of ignition system does the Honda G100 have?

CHAPTER 13

Engine Block Servicing

In some cases the cylinder block needs servicing other than a complete overhaul. In this chapter we take a look at the cylinders, valves, piston, and other parts associated with the engine block itself. Some of these are easily repaired and others require more complete overhaul procedures. These overhaul procedures will be discussed in the next chapter.

So far you have been troubleshooting the engine and have found a number of things that may require a disassembly of the engine in order to get to the part that needs service, repair, or replacement. In this chapter we will take a look at some of these problems and how to solve them as quickly and easily as possible, using some of the manufacturer's recommended procedures.

Disassembling, Testing, and Inspecting the Four-Cycle Engine

In order to work on an engine block, you must first disassemble the engine. There are some suggested procedures that will make it easier to put the engine back together once it is overhauled (Fig. 13–1). The overhaul procedure for disassembly and inspection is as follows (repairs and reassembly will follow in the next chapter):

- Drain the oil (Fig. 13–2). This becomes self-explanatory when you forget and then turn the engine to any position other than upright and find yourself with oil running out of openings you never imagined. Remove the plug and allow the oil to drain until most of it has been removed. The rest can be absorbed with rags later as you work with the engine.
- Remove the air cleaner and stud, the fuel pipe and tank assembly, the air-cleaner elbow or pipe, the carburetor and linkage, the carburetor intake elbow, and the muffler (Fig. 13–3).

  Keep in mind that some of these parts will not be on all engines. These are listed in order to make you aware of the procedure for many types of engines. You will have to adjust your work according to the engine you are working on. The muffler, of course, should not be removed when hot because you may get burned. It is also a good idea to drain the carburetor before you put it aside, because the gas may drain on your working surface and create a fire hazard.
- Check the space between the upper and lower carburetor body or carburetor-to-tank fit. Check the carburetor throttle shaft and bushings for wear (Fig. 13–4). This type of inspection is useful when you decide to reassemble the engine. You can either repair it now or tag it, naming the defect that you spotted, and then repair it before reassembly of the engine.
- Disassemble the carburetor. Inspect all parts and label those that need repair or replacement.
- If the engine has an electric starter, remove and inspect the brushes and armature condition. Also check the commutator segments and the whole surface area (Fig. 13–5).
- Remove the blower housing. Place it aside for closer inspection later when you reassemble the engine.
- Check the flywheel compression by spinning the flywheel counterclockwise. If it snaps back with some degree of force, the compression is all right.
- Remove the spark plug. Adjust the gap to 0.030″ (7.62 mm). Clean and wash the plug or replace it if the porcelain is cracked or chipped. The electrodes should also be in good shape. Inspect closely to make sure any cleaning materials have been removed.
- Disassemble the fuel tank and bracket assembly or carburetor.

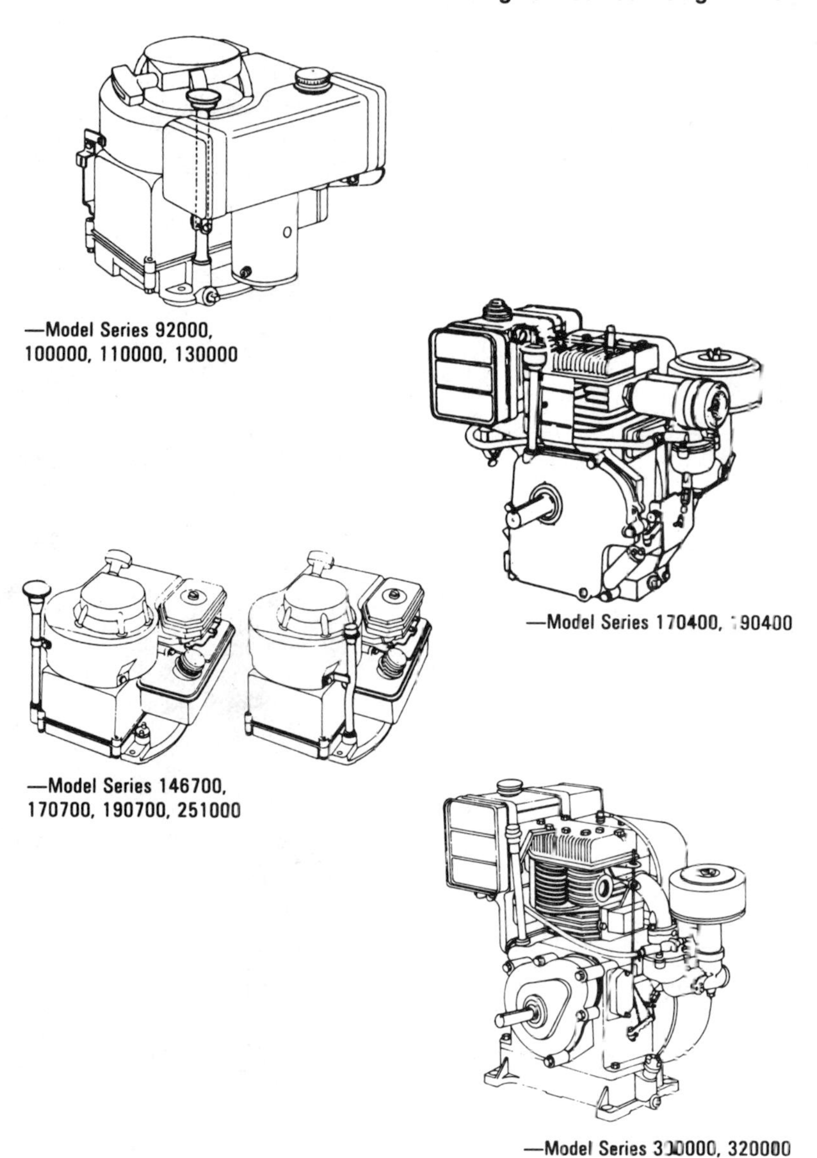

Fig. 13–1. Five models of Briggs & Stratton engines most often encountered. *(Courtesy B & S)*

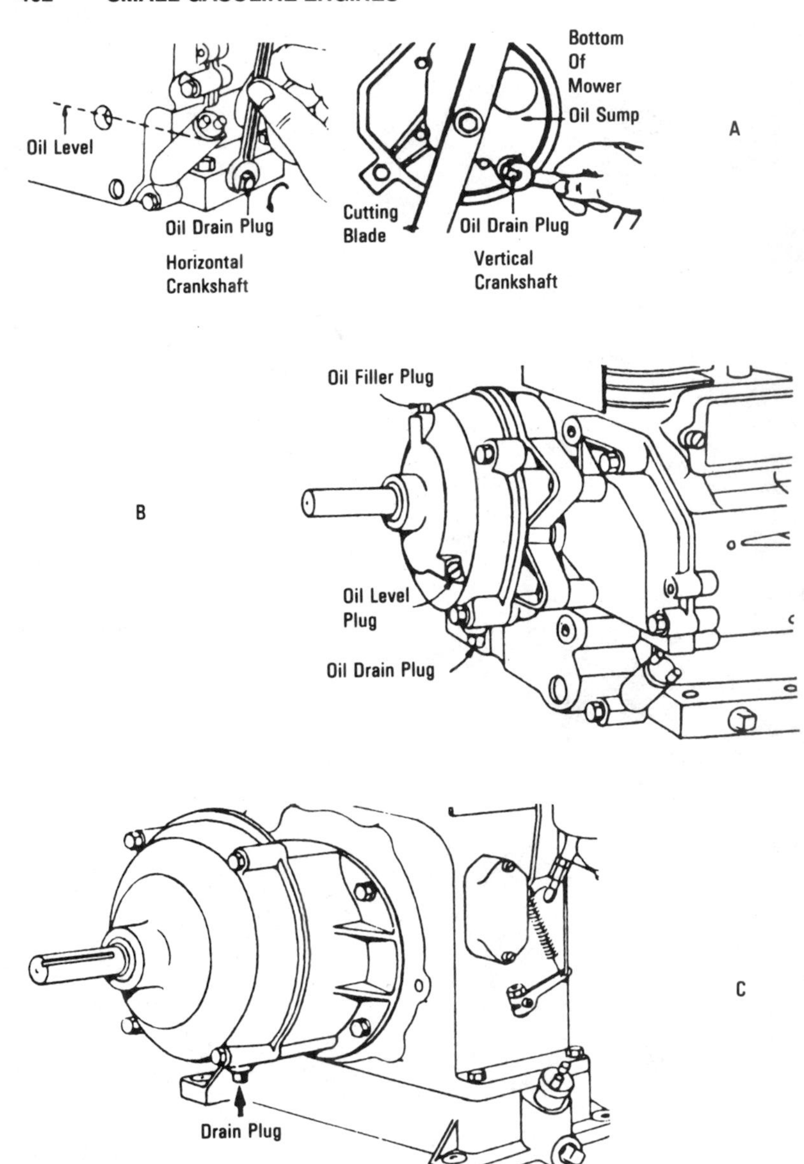

Fig. 13–2. Draining the oil means removing the drain plugs and allowing the oil to flow out. *(Courtesy B & S)*

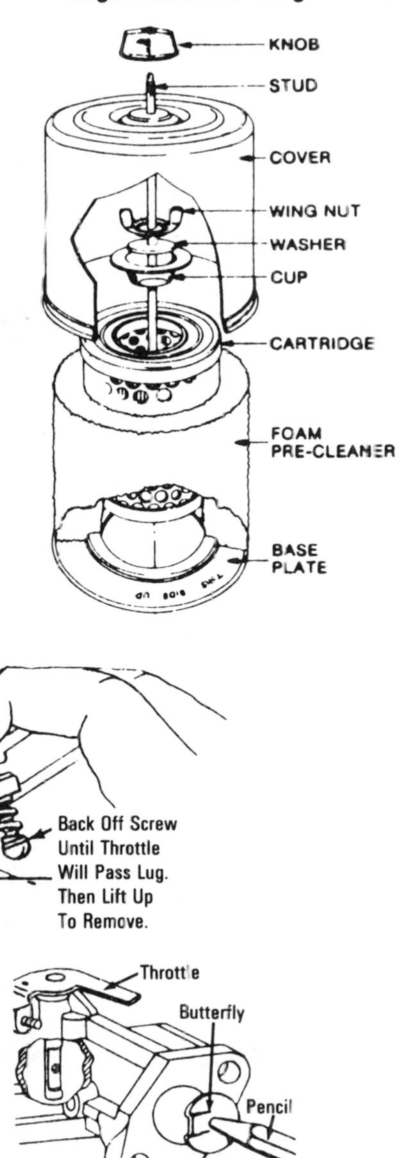

Fig. 13–3. Exploded view of the foam-type air cleaner. *(Courtesy B & S)*

Fig. 13–4. Removing the throttle from the various types of carburetors. *(Courtesy B & S)*

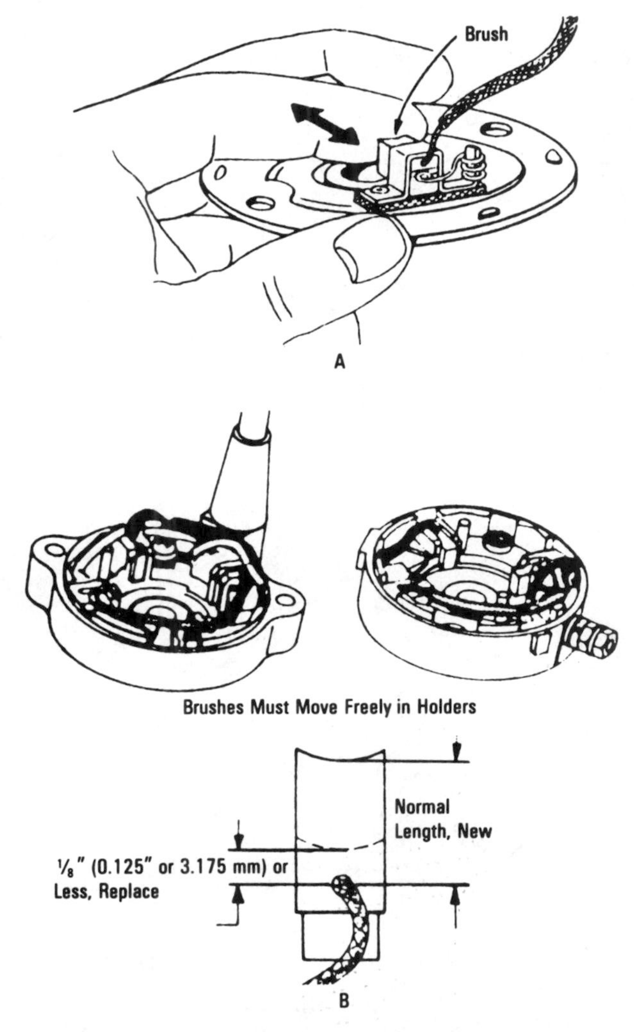

Fig. 13–5. Checking the brushes in the electric starter. *(Courtesy B & S)*

Some carburetors have hoses that are cracked and leak at the connection points. Also check the snap action of the hose clamps. The tank filter (usually a piece of screen) should be cleaned. The tank should have grass clippings or dirt and sludge removed. If there is a float to check the gas tank fuel level, make sure it is working properly.

- Disassemble the rope starter pulley by removing it with a pulley puller if necessary. In most engines the whole starter unit is lifted off without too much effort. However, in some older models you may have to use a puller to remove the hand-wound rope pulley that is used to start the engine (Fig. 13–6).
- Remove the governor blade (Fig. 13–7). Place to one side and check visually that all parts are clean and operating properly.
- Remove the breather or valve cover (Fig. 13–8). Do this carefully so as not to lose the screws or gasket. The hose should be carefully removed at this time and inspected for fit and aging.
- Remove the cylinder head and shield (Fig. 13–9). Always note the position of the different head screws so that they may be properly reassembled. If a screw is used in the wrong position, it may be too short and not engage enough threads, or it may be too long and bottom on a fin, either breaking the fin or leaving the cylinder head loose.
- Check tappet clearance. After the valves have been removed and refaced, insert the valves in their respective positions in the cylinder. Replace a valve if the margin is $\frac{1}{64}$″ (0.4 mm) or less after refacing. See Table 13–1 for clearances.
- After inspecting the tappets for clearance and the condition of the valve seat, remove the valve and spring if necessary (Fig. 13–10). To remove the valves, use retainers. Slip the upper jaw of the valve compressor tool over the top of the valve chamber,

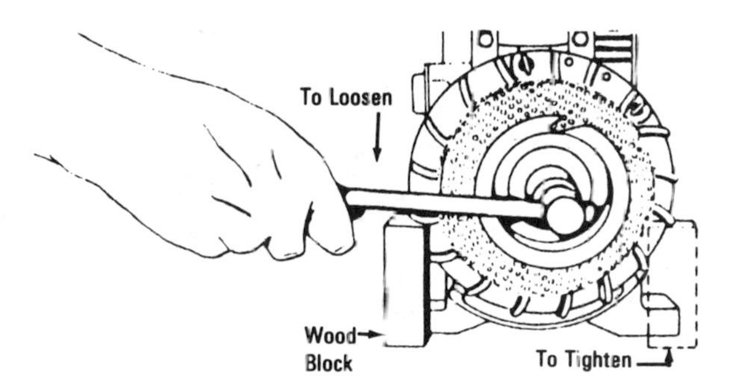

Fig. 13–6. Removing the flywheel. *(Courtesy B & S)*

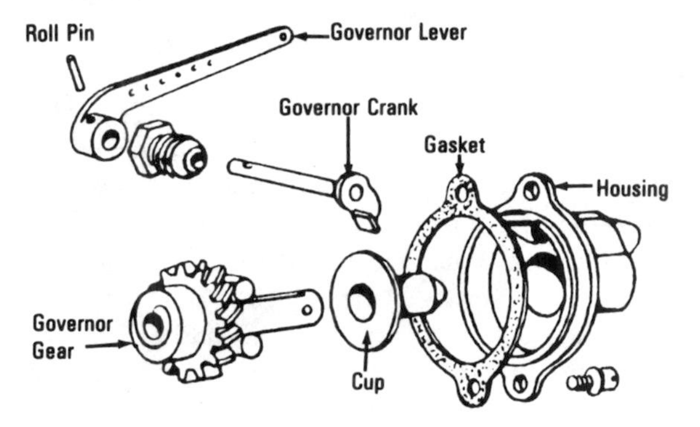

Fig. 13–7. Exploded view of the mechanical governor. *(Courtesy B & S)*

and the lower jaw between the spring and retainer. Compress the spring. Remove the retainer. Pull out the valve (Fig. 13–10C and D).

- Remove the rope starter, rewind or windup starter. Then remove the flywheel. See Chapter 2 for proper tools and procedure for pulling flywheel.
- Carefully remove the breaker point cover. If you damage the cover, replace to ensure a proper dust seal (Fig. 13–11).
- Check the breaker point plunger hole (Fig. 13–12). Breaker point assemblies are removed by loosening the screw holding

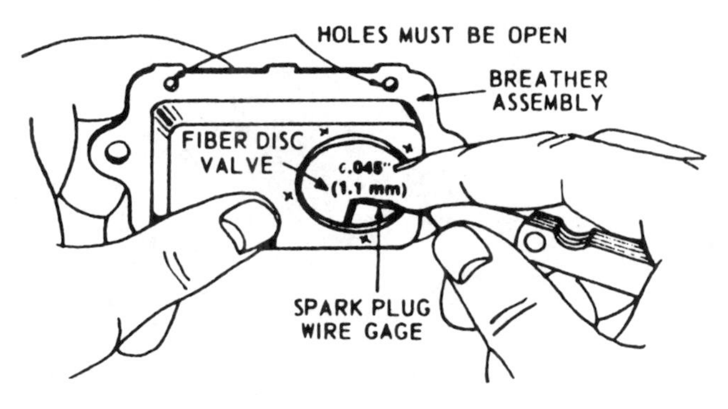

Fig. 13–8. Checking the fiber disc valve clearance. *(Courtesy B & S)*

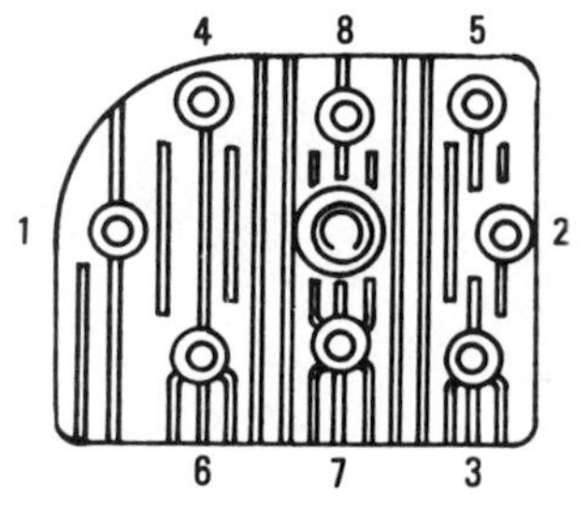

Aluminum Cylinder Engines
(14 Cu. In. and Less)
Long Screws in These
3 Holes

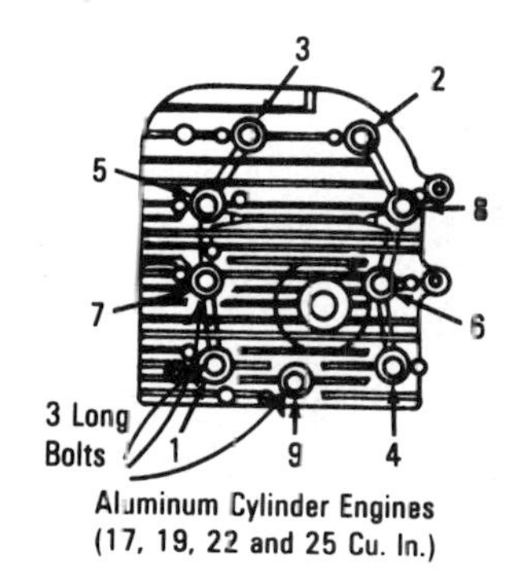

Aluminum Cylinder Engines
(17, 19, 22 and 25 Cu. In.)

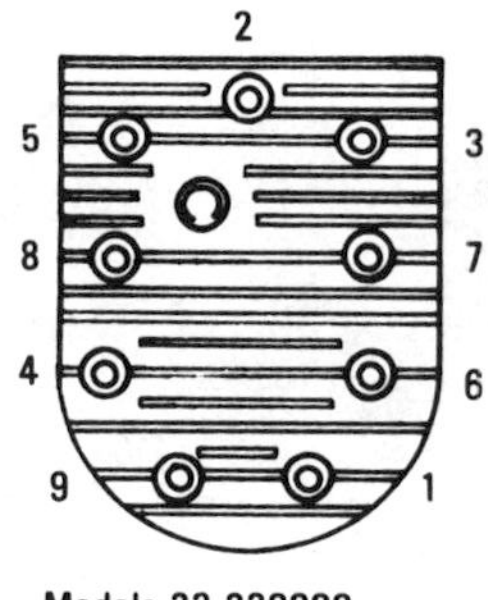

Models 23-230000
240000-300000-320000

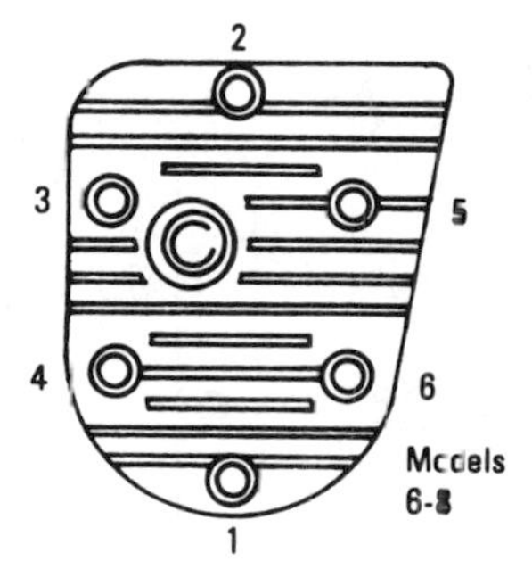

Models
6-8

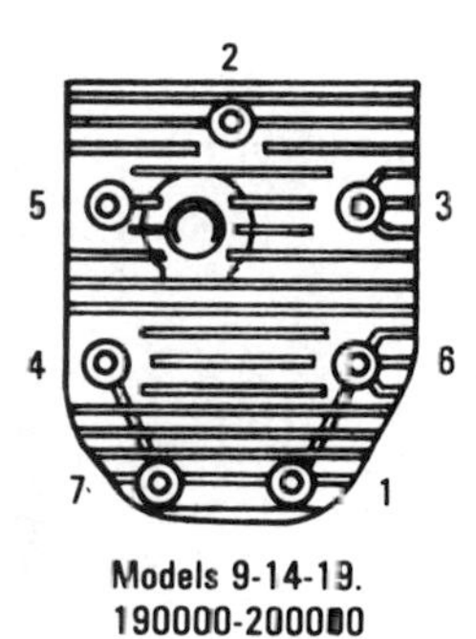

Models 9-14-19.
190000-200000

Fig. 13–9. Arrangement of bolts on cylinder heads of different models. Numbers represent the sequences for tightening the bolts. *(Courtesy B & S)*

Table 13–1. Valve Tappet Clearance

Model Series	Intake				Exhaust			
	Maximum		Minimum		Maximum		Minimum	
Aluminum Cylinder	Inches	Milli-meters	Inches	Milli-meters	Inches	Milli-meters	Inches	Milli-meters
6B, 600000, 8B, 80000, 82000, 92000, 94000, 100000, 110000, 130000, 140000, 170000, 190000, 220000, 250000	0.007	0.18	0.005	0.13	0.011	0.28	0.009	0.23
Cast-Iron Cylinder	Inches	Milli-meters	Inches	Milli-meters	Inches	Milli-meters	Inches	Milli-meters
5, 6, 8, N, 9, 14, 19, 190000, 200000	0.009	0.23	0.007	0.18	0.016	0.41	0.014	0.36
23, 230000, 240000, 300000, 320000	0.009	0.23	0.007	0.18	0.019	0.48	0.017	0.43

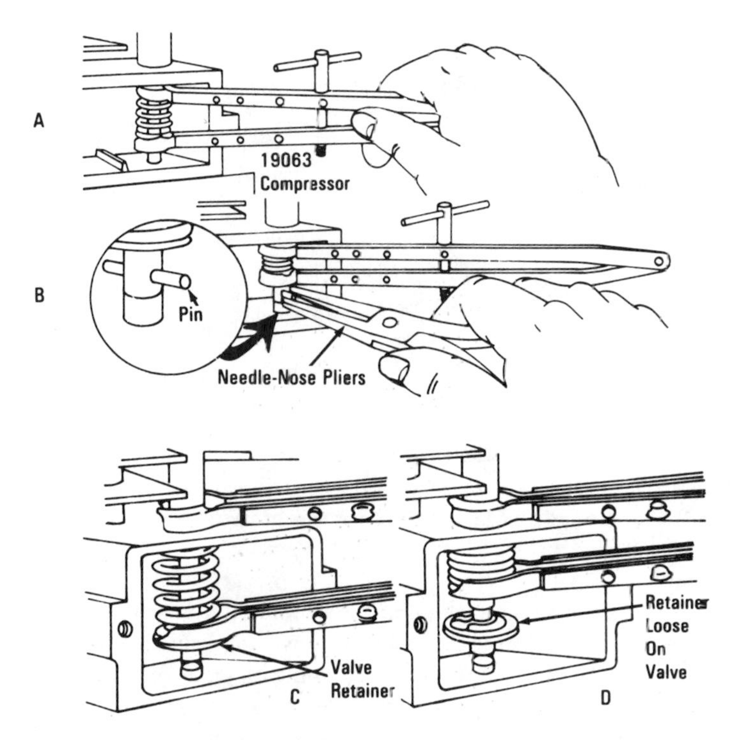

Fig. 13–10. (A & B) Using the valve spring compressor to remove the valves and the long-nose to remove the valve pin. (C & D) Using the valve compressor to aid in removing the valve retainer. *(Courtesy B & S)*

the post. Condensers on most models also include the breaker points. The condenser is removed by loosening the screw holding the condenser clamp.

- Test the coil on the armature. If the coil is open or shorted, replace. Check the high-voltage lead from the coil to the spark plug for signs of deterioration or breaks in the insulation.
- Remove the crankshaft (Fig. 13–13). To remove the crankshaft from engines made of aluminum alloy, remove the rust or burrs from the power take-off end of the crankshaft. Remove the crankcase cover or sump. If the sump or cover sticks, tap lightly with a soft hammer on alternate sides near the dowel. Turn the

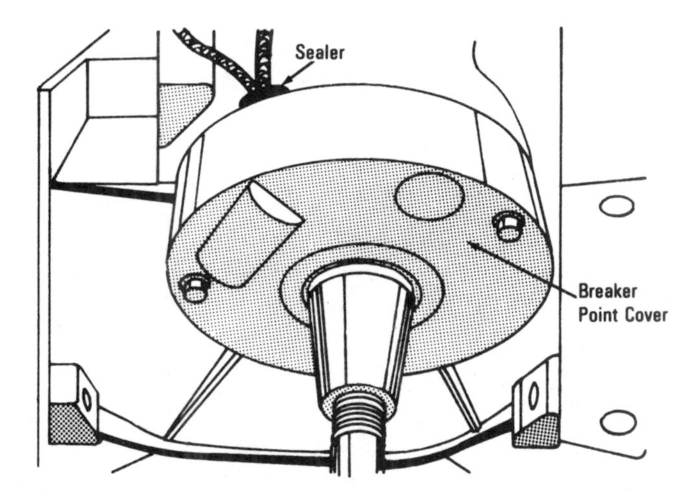

Fig. 13–11. Location of the sealer for the breaker point cover. *(Courtesy B & S)*

crankshaft to align the crankshaft and cam gear timing marks. Lift out the cam gear. Then remove the crankshaft. On ball-bearing models, the crankshaft and cam gear must be removed together. On cast-iron engines, remove the magneto. Then remove the burrs and rust from the power take-off end of the crankshaft. Remove the crankshaft.

- Remove the auxiliary power take-off, if the engine has one (Fig. 13–14). This auxiliary power take-off shaft is perpendicular to the crankshaft. It rotates at the rate of one revolution for every

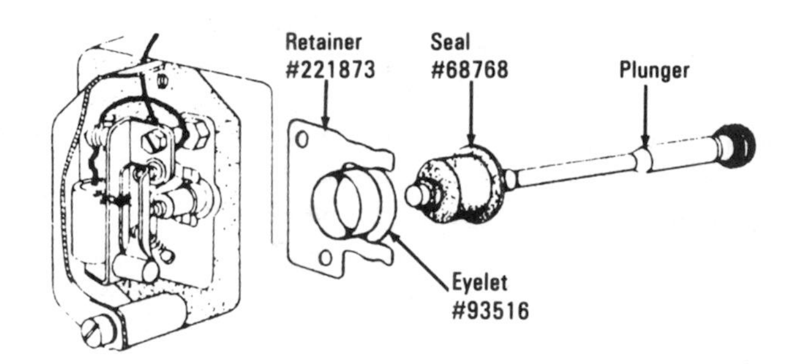

Fig. 13–12. Seal assembly for the external breaker. *(Courtesy B & S)*

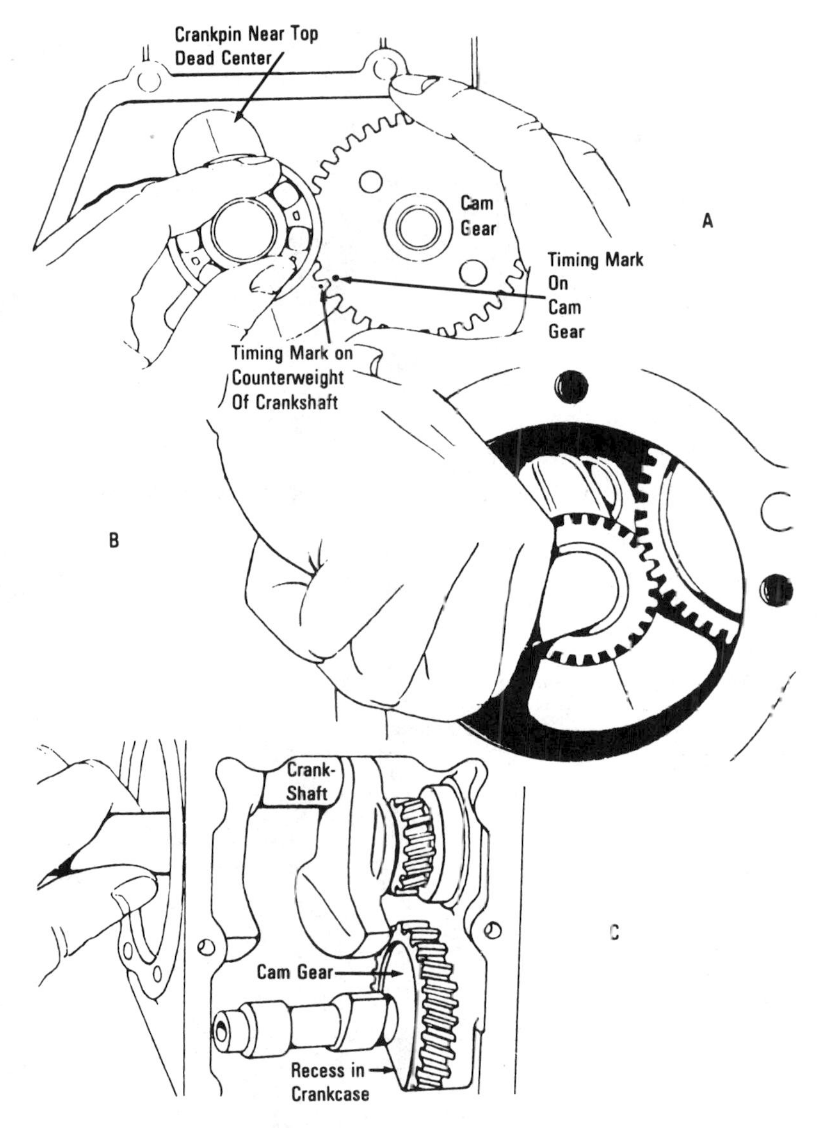

Fig. 13–13. (A) In ball-bearing engines alignment of marks is needed to remove the crankshaft. (B) Removing the crankshaft in the cast-iron engines. (C) Location of the crankshaft in relation to the cam gear. *(Courtesy B & S)*

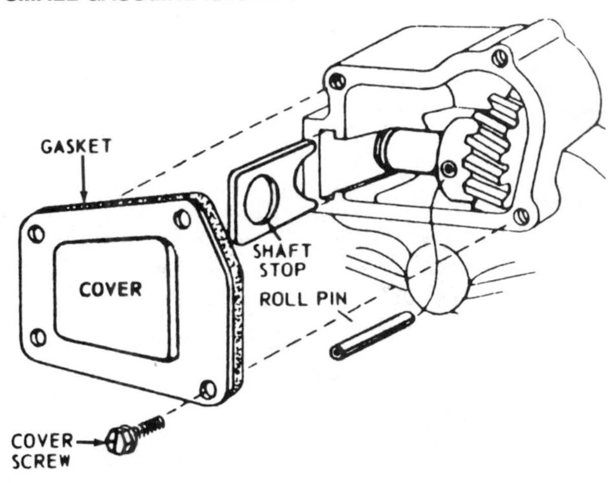

Fig. 13–14. Removing the cover on the auxiliary power take-off box. *(Courtesy B & S)*

Table 13–2. Engine Identification for Briggs & Stratton Engines

Series	Cubic Inches	Horsepower
60000	6	2.0
80000	8	2.5
92000	9	3.0 or 3.5
94000	9	3.5
100000	10	4.0
110000	11	4.0
130000	13	5.0
140000	14	6.0
170000	17	7.0
190000	19	8.0
200000	20	8.0
220000	22	10.0
230000	23	10.0
240000	24	12.0
250000	25	11.0
300000	30	12.0
320000	32	16.0

$8\frac{1}{2}$ revolutions of the crankshaft. On these models, the cam gear, the worm gear, and oil slinger are a factory assembly and not available as separate pieces (Fig. 13–15).

- Inspect seals for damage. Remove and replace if necessary (Fig. 13–16).
- Check the mechanical governor parts. Worn linkage or damaged governor springs should be replaced to ensure proper governor operation. If the spring or linkage is changed, change

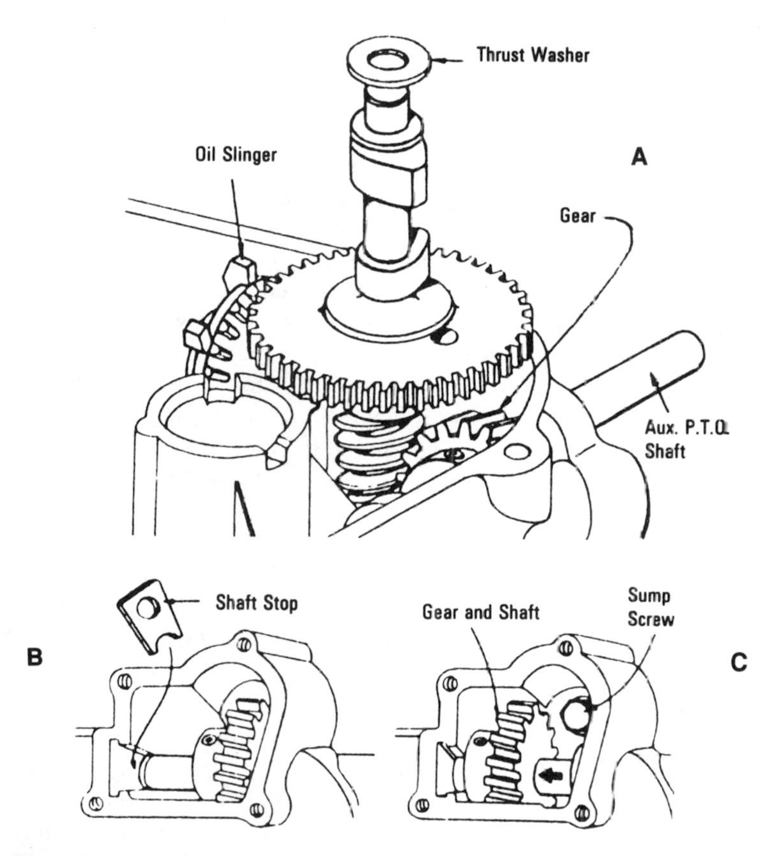

Fig. 13–15. (A) Auxiliary power take-off engine and location of the cam gear and oil slinger. (B) Shaft stop for the auxiliary power take-off. (C) Sump screw location on the auxiliary power take-off models. *(Courtesy B & S)*

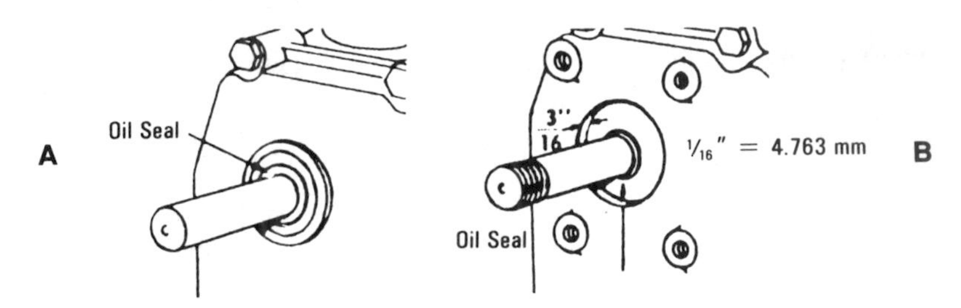

Fig. 13–16. Replacing oil seals. (A) Most oil seals are pressed in, flush with the hub. (B) However, ball-bearing models use a seal pressed 3⁄16″ (4.76 mm) below the crankcase mounting flange. *(Courtesy B & S)*

and adjust to no-load rpm or check top no-load rpm with the engine assembled. If a governor spring must be replaced, consult the engine parts list. Choose the proper governor spring by engine-type number. After a new governor spring is installed, check the engine's top governed speed with an accurate tachometer. Run the engine at half throttle to allow the engine to reach normal operating temperature before measuring the speed with a tachometer. To account for tolerances, which may be required by tachometer manufacturers, we suggest that the top governed speed of the engine be adjusted to at least 200 rpm lower than the maximum speeds in Table 13–3.

- Inspect the oil slinger. The oil slinger is driven by the cam gear. Old-style slingers using a die-cast bracket assembly have a steel bushing between the slinger and the bracket. Replace the bracket on which the oil slinger rides if worn to a diameter of 0.49″ (12.4 mm) or less. Replace the steel bushing if it is worn. Newer-style oil slingers have a stamped-steel bracket. A spring washer is used only on models 100900–130900. Inspect the gear teeth, old or new style. Replace if worn.
- Inspect the cam gear. Remove the crankshaft cover. If the cover sticks, tap it lightly with a soft hammer on alternate sides near the dowel. Turn the crankshaft to align the crankshaft and the cam gear timing marks. Lift out the cam gear; then remove the crankshaft. On ball-bearing models, the crankshaft and cam gear must be removed together (Fig. 13–13A). This works well with the aluminum-cylinder engines. However, with the cast-

Table 13–3. Lawnmower-Blade Rotation Speeds (Rotary lawnmower blades—maximum rotational speeds, which will produce blade-tip speeds of 19,000 feet per minute)

Blade Length	Maximum Rotational rpm
18″—45.72 cm	4032
19″—48.26 cm	3820
20″—50.80 cm	3629
21″—53.34 cm	3456
22″—55.88 cm	3299
23″—58.42 cm	3155
24″—60.96 cm	3024
25″—63.50 cm	2903
26″—66.04 cm	2791

Check the manufacturer's recommended speeds to be sure. Each model may vary slightly.

(Courtesy B & S)

iron-cylinder engines (Model series 5, 6, 8, and N with plain bearings) remove the magneto. Remove burrs and rust from the power take-off end of the crankshaft. Remove the crankshaft as shown in Fig. 13–13A. Engines with ball bearings are slightly different. Remove the magneto. Drive out the cam gear shaft while holding the cam gear to prevent dropping (Fig. 13–17). Push the cam gear into the recess. Pull the crankshaft out from the magneto side. The models with plain bearings are slightly different once again. Remove the crankshaft cover. Rotate the crankshaft to the approximate position shown in Fig. 13–13B. Pull out the crankshaft from the power take-off side, twisting slightly, if necessary, to clear the cam gear.

Model series 9, 14, 19, 23, 200000, 230000, 240000, 300000, and 320000 with ball bearings call for the removal of the crankshaft cover and bearing support first. On models 240000, 300000, and 320000 the piston and rod must be removed from the engine. Rotate the crankshaft to the position shown in Fig. 13–13B. On some models it may be necessary to position the crankshaft approximately 180° from the position shown in Fig. 13–13B. Pull the crankshaft out, turning as needed (Fig. 13–13C).

To remove the cam gear from all cast-iron models except the model series 300000 and 320000, use a long punch to drive the cam

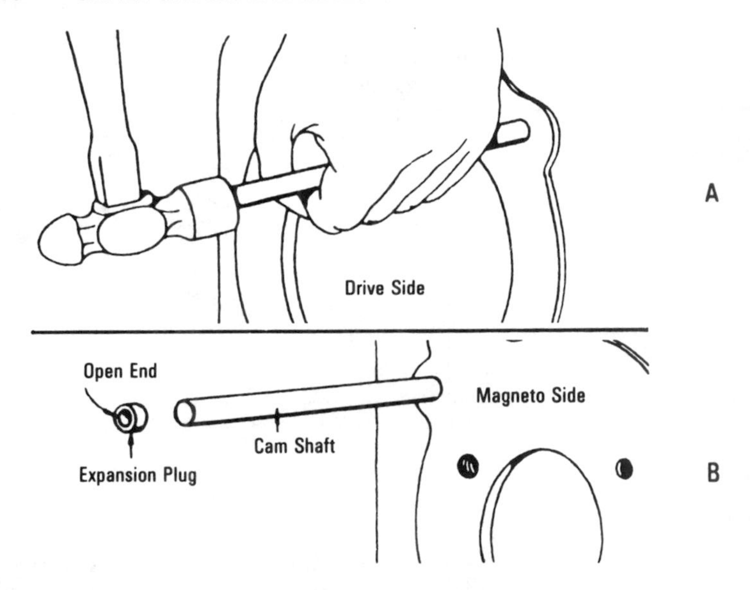

Fig. 13–17. Drive out the cam gear shaft while holding the cam gear to prevent dropping. Push the cam gear into the recess. Pull the crankshaft out from the magneto side. *(Courtesy B & S)*

gear shaft out toward the magneto side. Save the plug (Fig. 13–17). Do not burr or peen the end of the shaft while driving the shaft out. Hold the cam gear while removing the punch so the gear does not drop and nick.

On model series 300400 and 320400, remove the short bolt and Belleville washer from the power take-off drive gear (Fig. 13–18). Loosen the long bolt and Belleville washer two turns on the magneto side and tap the head of the bolt with a hammer to loosen the cam gear shaft. Turn the bolt out while pushing out the cam gear shaft. Remove the bolts from the cam gear bearing (Fig. 13–19). While holding the cam gear, remove the cam gear bearing and remove the cam gear (Fig. 13–20).

Model series 301400, 302400, 325400, and 326400 call for loosening the long bolt two turns. Use the hammer to drive out the cam gear shaft and cam gear plug. Loosen the bolt while pushing out the cam gear shaft and plug. Remove the bolts and cam gear bearing (Fig. 13–19). Remove the cam gear as shown in Fig. 13–20.

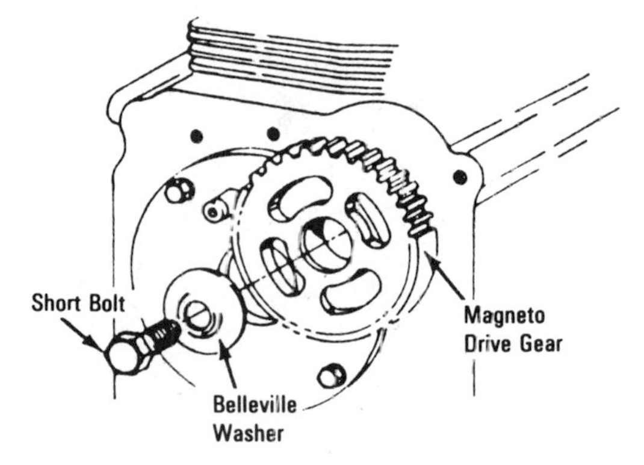

Fig. 13–18. Removing the short bolt. *(Courtesy B & S)*

- When checking the tappets also check the cam for visible signs of wear.
- Check the connecting rod and piston. To remove the piston and connecting rod from the engine, bend the connecting-rod lock down (Fig. 13–21). Remove the connecting-rod cap. Remove any carbon or ridge at the top of the cylinder bore. This will prevent breaking of the rings. Push the piston and rod out through the top of the cylinder.

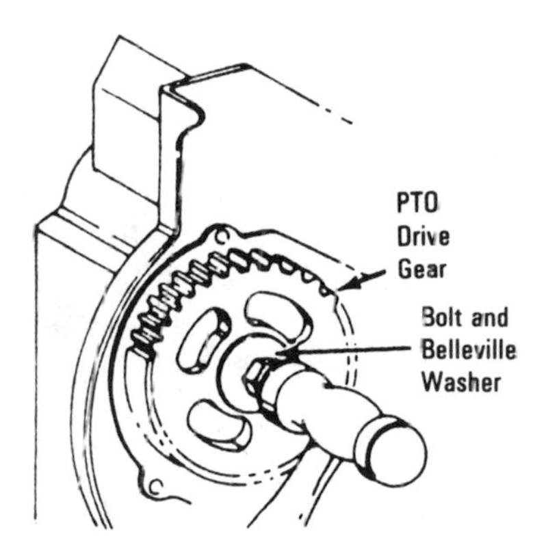

Fig. 13–19. Removing the long bolt using the hammer to loosen the cam gear shaft. *(Courtesy B & S)*

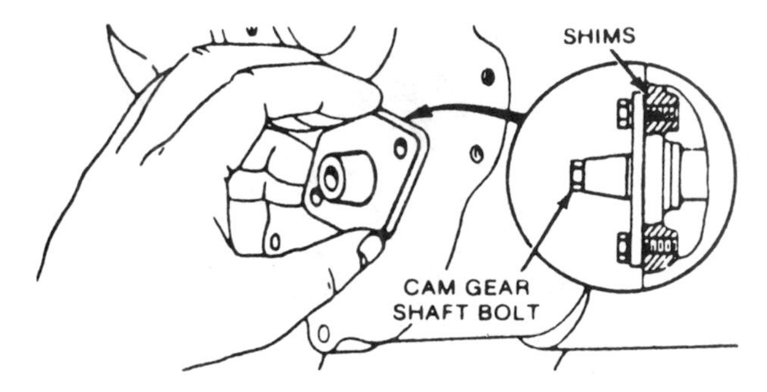

Fig. 13–20. Removing the cam gear bearing. *(Courtesy B & S)*

Pistons used in sleeve-bore aluminum alloy engines are marked with an *L* on top of the piston (Fig. 13–22). These pistons are tin-plated. This piston assembly is *not* interchangeable with the piston used in aluminum-bore (Kool Bore®) engines. Pistons used in the aluminum-bore engines are *not* marked on top of the piston. The piston is chrome-plated and is not to be used in a sleeve-bore engine.

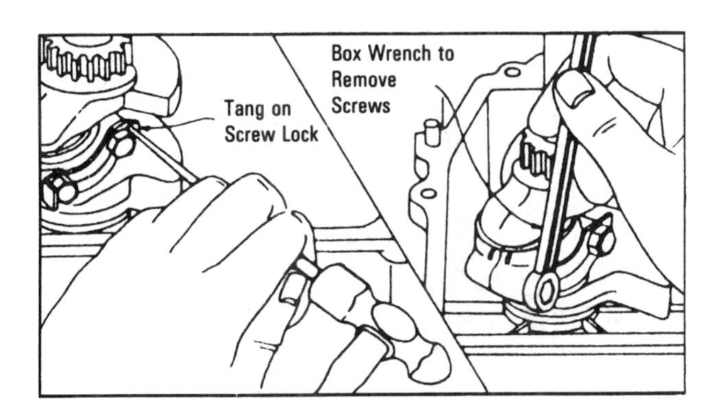

Fig. 13–21. Bending the rod lock. *(Courtesy B & S)*

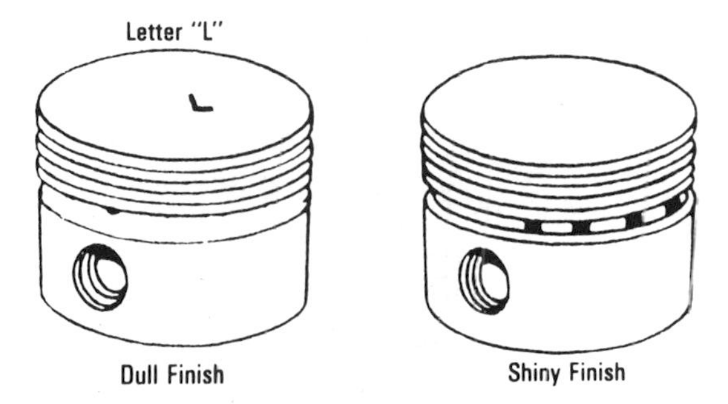

Fig. 13–22. Piston variations. *(Courtesy B & S)*

Remove the connecting rod from the piston. Remove the piston pin lock with long-nose pliers. One end of the pin is drilled to facilitate removal of the lock (Fig. 13–23). Remove the rings one at a time, as shown in Fig. 13–24, slipping them over the ring lands. Use a ring expander to prevent damage to the rings and piston.

Check the Piston—If the cylinder is to be resized, there is no reason to check the piston since a new oversized piston assembly will be used. If, however, the cylinder is not to be resized and the piston shows no signs of wear or scoring, the piston should be checked.

In order to check the piston you may have to clean the carbon from the top ring groove. Place a new ring on the groove. Check the remain-

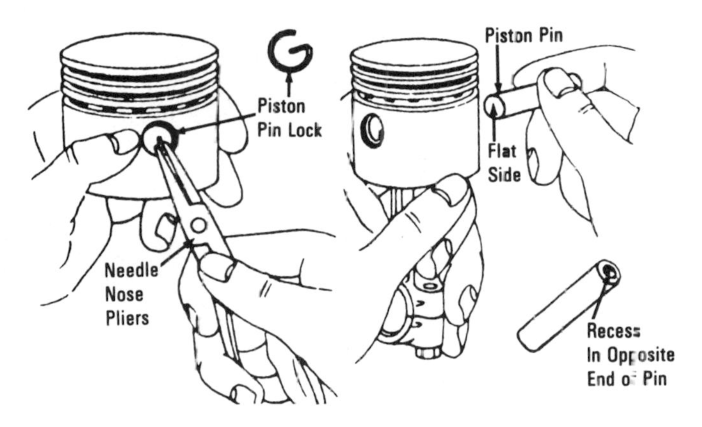

Fig. 13–23. Removing the connecting rod. *(Courtesy B & S)*

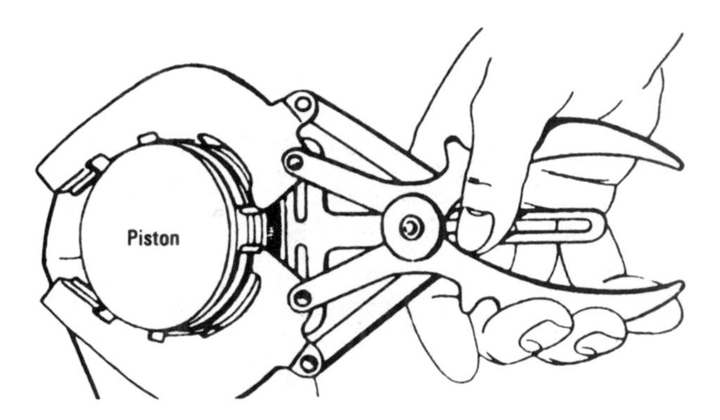

Fig. 13–24. Removing piston rings. *(Courtesy B & S)*

ing space in the groove with a feeler gage (Fig. 13–25). If a 0.007″ (0.18-mm) feeler gage can be inserted (all models), the piston is worn and should be replaced.

Check the Rings—To check the rings, first clean all the carbon from the ends of the rings and from the cylinder bore. Insert the old rings one at a time 1″ (25.4 mm) down into the cylinder. Check the gap with a feeler gage (Fig. 13–26). If the ring gap is greater than shown in Table 13–4, the ring should be rejected. Do not deglaze cylinder walls when installing piston rings in the aluminum-cylinder engines.

Chrome ring sets are available for all current aluminum- and cast-iron-cyclinder models. No honing or deglazing is required. The cylinder bore can be a maximum of 0.005″ (0.13 mm) oversize when using chrome rings.

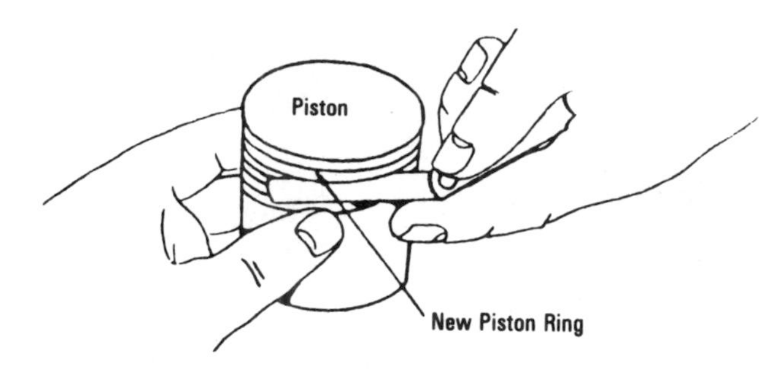

Fig. 13–25. Checking the ring grooves. *(Courtesy B & S)*

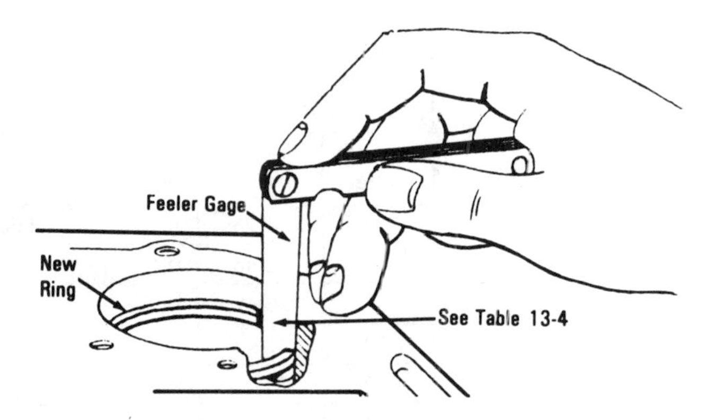

Fig. 13–26. Checking the ring gap. *(Courtesy B & S)*

Check the Connecting Rod—If the crankpin bearing in the rod is scored, the rod must be replaced. Rejection sizes of the crankpin bearing hole and the piston pin bearing hole are shown in Table 13–5. Piston pins 0.005″ (0.13 mm) oversize are available in case the connecting rod and piston are worn at the piston pin bearing. If, however, the

Table 13–4. Ring-Gap Rejection Sizes

Basic Model Series	Compression Ring		Oil Ring	
Aluminum Cylinder	**Inches**	**Millimeters**	**Inches**	**Millimeters**
6B, 60000, 8B, 80000 82000, 92000, 110000, 111000 100000, 130000 140000, 170000, 190000, 250000	0.035	0.80	0.045	1.14
Cast-Iron Cylinder	**Inches**	**Millimeters**	**Inches**	**Millimeters**
5, 6, 8, N, 9 14, 19, 190000 200000, 23 230000, 240000 300000, 320000	0.030	0.75	0.035	0.90

(Courtesy B & S)

Table 13–5. Connecting-Rod Reject Sizes

Basic Model Series	Crankpin Bearing		Piston Pin Bearing	
Aluminum Cylinder	**Inches**	**Millimeters**	**Inches**	**Millimeters**
6B, 60000	0.876	22.25	0.492	12.50
8B, 80000	1.001	25.43	0.492	12.50
82000, 92000, 110000	1.001	25.43	0.492	12.50
100000	1.001	25.43	0.555	14.10
130000	1.001	25.43	0.492	12.50
140000, 170000	1.095	27.81	0.674	17.12
190000	1.127	28.63	0.674	17.12
220000, 250000	1.252	31.80	0.802	20.37
Cast-Iron Cylinder	**Inches**	**Millimeters**	**Inches**	**Millimeters**
5	0.752	19.10	0.492	12.50
6, 8, N	0.751	19.08	0.492	12.50
9	0.876	22.25	0.563	14.30
14, 19, 190000	1.001	25.43	0.674	17.12
200000	1.127	28.63	0.674	17.12
23, 230000	1.189	30.20	0.736	18.69
240000	1.314	33.38	0.674	17.12
300000, 320000	1.314	33.38	0.802	20.37

(Courtesy B & S)

crankpin bearing in the connecting rod is worn, the rod should be replaced. Do not attempt to "file" or "fit" the rod.

Check the Piston Pin—If the piston pin is worn 0.005″ (0.01 mm) out of round or below the rejection sizes listed in Table 13–6, it should be replaced.

- Inspect and check the crankshaft. Table 13–7 shows the rejection sizes of the various wear points of the crankshaft. Discard crankshaft if it is worn smaller than the size shown. Keyways should be checked to be sure they are not worn or spread. Re-

Table 13–6. Piston-Pin Rejection Sizes

Basic Model Series	Piston Pin		Piston Bore	
Aluminum Cylinder	**Inches**	**Millimeters**	**Inches**	**Millimeters**
6B, 60000	0.489	12.42	0.491	12.47
8B, 80000	0.489	12.42	0.491	12.47
82000, 92000, 110000	0.489	12.42	0.491	12.47
100000	0.552	14.02	0.554	14.07
130000	0.489	12.42	0.491	12.47
140000, 170000, 190000	0.671	17.04	0.673	17.09
220000, 250000	0.799	20.29	0.801	20.35
Cast-Iron Cylinder	**Inches**	**Millimeters**	**Inches**	**Millimeters**
5, 6, 8, N	0.489	12.42	0.491	12.47
9	0.561	14.25	0.563	14.30
14, 19, 190000	0.671	17.04	0.673	17.09
200000	0.671	17.04	0.673	17.09
23, 230000	0.734	18.64	0.736	18.69
240000	0.671	17.04	0.673	17.09
300000, 320000	0.799	20.29	0.801	20.35

(Courtesy B & S)

move the burrs from the keyway edges to prevent scratching the bearing. Fig. 13–27 shows the various points to be checked on the crankshaft.

Note: Undersize (0.20″) (0.508 mm) connecting rods may be obtained for use on reground crankpin bearings. Complete instructions are included in the undersize rod package.

- Check the armature assembly and back plate on the Magna-Matic series. Usually the coil and armature are not separated but are left assembled for convenience. However, if one or both need replacement, you can further disassemble them by unfastening the coil primary wire and the coil ground wire. Pry out the clips that hold the coil and coil core to the armature (Fig.

Table 13–7. Crankshaft Reject Sizes

Model Series	Power Take-Off Journal		Magneto Journal		Crankpin Journal	
Aluminum Cylinder	**Inches**	**Millimeters**	**Inches**	**Millimeters**	**Inches**	**Millimeters**
6B, 60000	0.873	22.17	0.873	22.17	0.870	22.10
8B, 80000*	0.873	22.17	0.873	22.17	0.996	25.30
82000, 92000*, 94000, 110900*, 111200, 111900*	0.873	22.17	0.873	22.17	0.996	25.30
100000, 130000	0.998	25.35	0.873	22.17	0.996	25.30
140000, 170000	1.179	29.95	0.997†	25.32†	1.090	27.69
190000	1.179	29.95	0.997†	25.32†	1.122	28.50
220000, 250000	1.376	34.95	1.376	34.95	1.247	31.67
Cast-Iron Cylinder	**Inches**	**Millimeters**	**Inches**	**Millimeters**	**Inches**	**Millimeters**
5, 6, 8, N	0.873	22.17	0.873	22.17	0.743	18.87
9	0.983	24.97	0.983	24.97	0.873	22.17
14, 19, 190000	1.179	29.95	1.179	29.95	0.996	25.30
200000	1.197	29.95	1.179	29.95	1.122	28.50
23, 230000‡	1.376	34.95	1.376	34.95	1.184	30.07
240000	Ball	Ball	Ball	Ball	1.309	33.25
300000, 320000	Ball	Ball	Ball	Ball	1.309	33.25

*Auxiliary-drive models PTO bearing reject size—1.003″ (25.48 mm).
†Gear-reduction PTO—1.179″ (29.95 mm).
‡Synchro-balance magento bearing reject size—1.179″ (29.95 mm).

(Courtesy B & S)

13–28). The coil core is a slip fit in the coil and can be pushed out of the coil. It may be a good idea to repair this assembly and then put it aside for reassembly into the engine after repairs are made on the cylinder block. In order to do so, reassemble by pushing the coil core into the coil with rounded side toward the ignition cable. Place the coil and core on the armature with the coil retainer between the coil and the armature and with

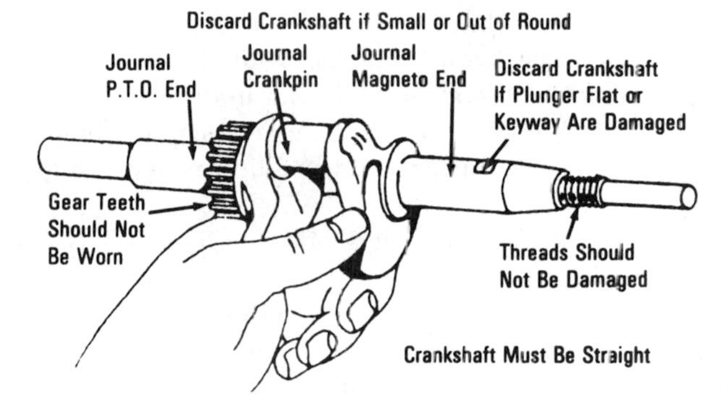

Fig. 13–27. Check points on the crankshaft. *(Courtesy B & S)*

the rounded side toward the coil. Hook the lower end of the clip into the armature, then press the upper end onto the coil core (Fig. 13–28).

Fasten the coil ground wire (bare double wires) to the armature support. Once you have it back together as a unit place the assembly against the cylinder around the rotor and bearing sup-

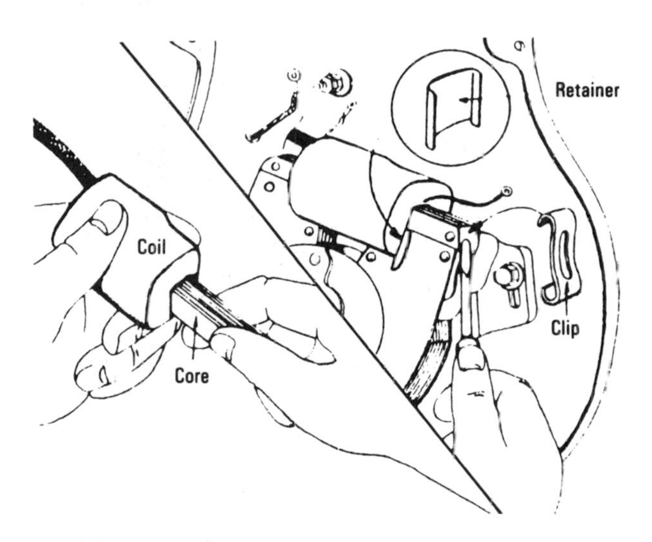

Fig. 13–28. Removing the coil from the armature. *(Courtesy B & S)*

port. When it is time to reassemble the engine, insert the three mounting screws together with the washer and lock washer into the three long oval holes in the armature. Tighten them enough to hold the armature in place, but loose enough that the armature can be moved for adjustment of the rotor timing.

- Check the rotor on the Magna-Matic engine models 9, 14, 19, 23, 191000, and 231000. The flywheel has already been removed. The armature gap on engines equipped with this type of ignition system is fixed and can change only if wear occurs on the crankshaft journal and/or main bearing. Check for wear by inserting a feeler gage ½″ (12.7 mm) in width at the points between the rotor and the armature. Minimum feeler gage thickness is 0.004″ (0.1 mm). Keep the feeler gage away from the magnets on the rotor or you will have a false reading (Fig. 13–29).

Remove the Rotor—The rotor is held in place by means of a Woodruff key and a clamp on newer engines, and a Woodruff key and a setscrew on older engines (Fig. 13–30). The rotor clamp must always remain on the rotor unless the rotor is in place on the crankshaft and within the armature or a loss of magnetism will occur.

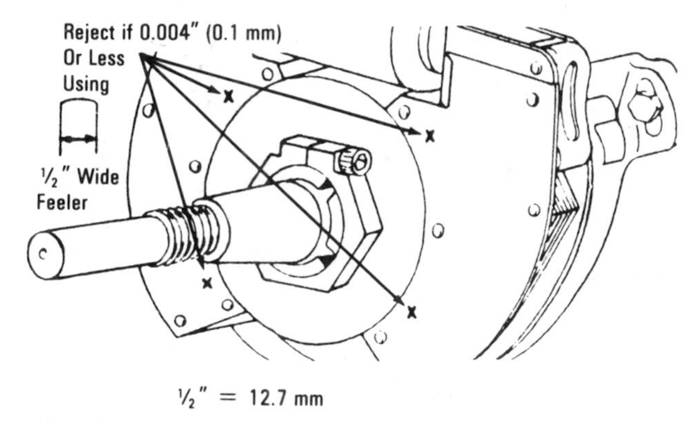

Fig. 13–29. Checking the armature gap on a Magna-Matic ignition model. *(Courtesy B & S)*

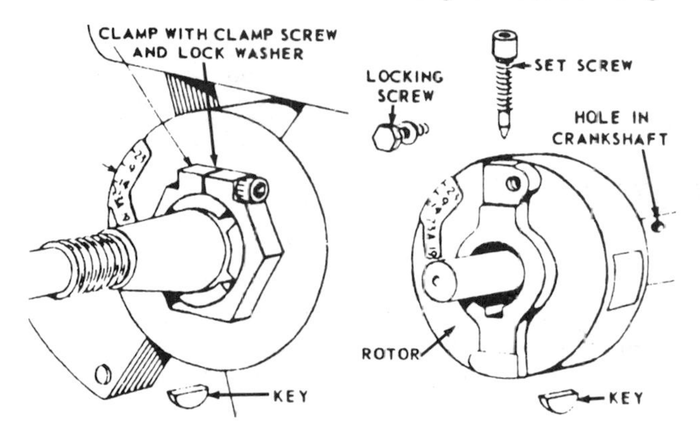

Fig. 13–30. The rotor clamp must always remain on the rotor (unless the rotor is in place on the crankshaft and within the armature) or a loss of magnetism will occur. *(Courtesy B & S)*

- Loosen the socket head screw in the rotor clamp, allowing the clamp to loosen. It may be necessary to use a puller to remove the rotor from the crankshaft. On older models, loosen the small lock screw, then the setscrew.
- Test the coil or armature and leads. Use an ohmmeter to check the continuity of the primary of the coil. You should read a small resistance for the primary. The ground lead and the lead that goes to the points are the primary leads. The secondary leads are the ground lead once again and the high-voltage lead that goes to the spark plug. The secondary of the coil should read a higher resistance than the primary. A reading of zero on either represents a short. A reading of infinity (the meter doesn't move on any scale) indicates an open. In either case, the coil should be replaced.
- Check automatic spark advance. Inspect the gear teeth for wear and nicks. Camshaft and cam gear journals and lobe rejection sizes are shown in Table 13–8. Also check the automatic spark advance on models equipped with Magna-Matic (Fig. 13–31). Place the cam gear in the normal operating position with the movable weight down. Press the weight down, then release. The spring should lift the weight. If not, the spring is stretched or the weight is binding. Repair or replace.

Table 13–8. Cam Gear Reject Sizes

Model Series	Cam Gear or Shaft Journal		Cam Lobe	
Aluminum Cylinder	**Inches**	**Millimeters**	**Inches**	**Millimeters**
6B, 60000	0.498	12.65	0.883	22.43
8B, 80000*	0.498	12.65	0.883	22.43
82000, 92000, 94000	0.498	12.65	0.883	22.43
110900, 111200, 111900	0.436 Magneto 0.498 PTO	11.07 Magneto 12.65 PTO	0.870	22.10
100000, 130000	0.498	12.65	0.950	24.13
140000, 170000, 190000	0.498	12.65	0.977	24.82
220000, 250000	0.498	12.65	1.184	30.07
Cast-Iron Cylinder	**Inches**	**Millimeters**	**Inches**	**Millimeters**
5, 6, 8, N	0.372	9.45	0.875	22.23
9	0.372	9.45	1.124	28.55
14, 19, 190000	0.497	12.62	1.115	28.32
200000	0.497	12.62	1.115	28.32
23, 230000	0.497	12.62	1.184	30.07
240000	0.497	12.62	1.184	30.07
300000	†	†	1.184	30.07
320000	†	†	1.215	30.86

*Auxiliary-drive models PTO—0.751″ (19.08 mm).
†Magneto side—0.8105″ (20.59 mm); PTO—0.6145″ (15.61 mm).

(Courtesy B & S)

Model series 111200 and 111900 have a cam gear with Easy-Spin® plus a compression release on the exhaust cam. In the starting position, the actuator cam moves the rocker cam so it will open the exhaust valve at the same time as the Easy-Spin lobe. When the engine starts, the actuator cam moves out and lets the rocker cam move down and the exhaust valve operates normally. To check, move the actuator cam to the running position (Fig. 13–32). Push the rocker cam against the actuator cam. Release the actuator cam. The actuator cam spring should pull the actuator cam against the shoulder pin, causing the rocker cam to raise up to starting position (Fig. 13–33). There should be no binding. Replace if binding exists.

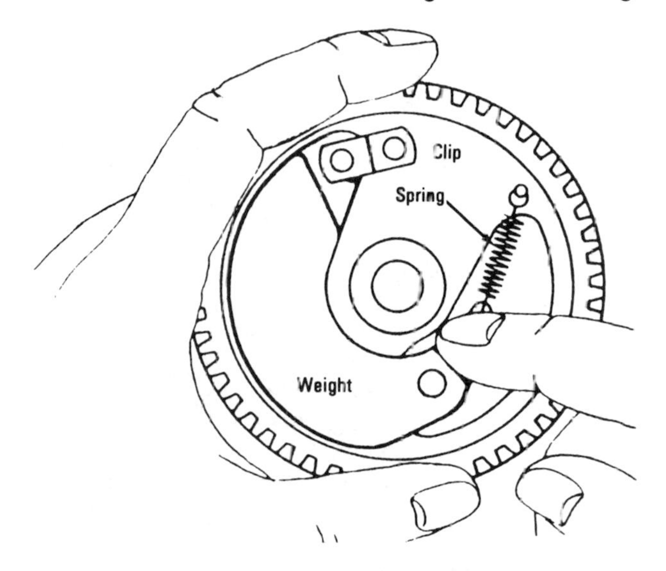

Fig. 13–31. Checking the automatic spark advance. *(Courtesy B & S)*

- Inspect the cylinder after the engine has been disassembled. Visual inspection will show if there are any cracks, stripped bolt holes, broken fins, or if the cylinder wall is damaged. Use a tele-

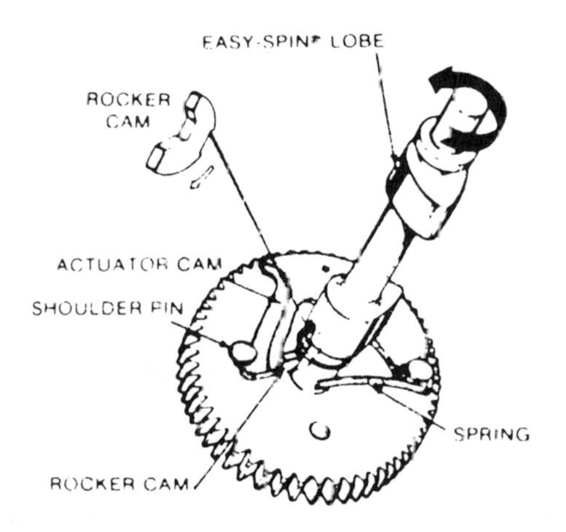

Fig. 13–32. Running position of the camshaft. *(Courtesy B & S)*

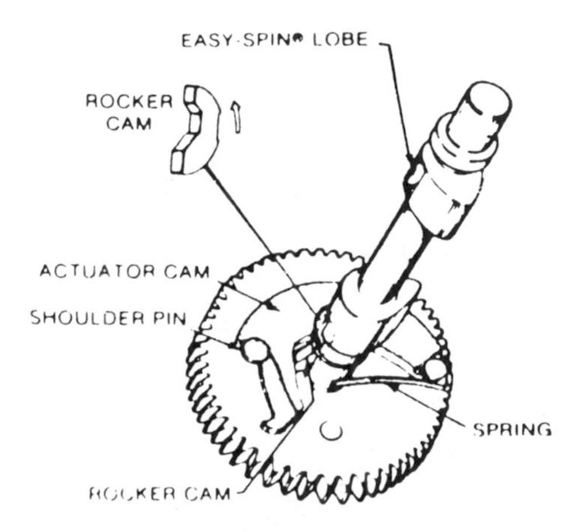

Fig. 13–33. Start position of the camshaft. *(Courtesy B & S)*

scoping gage and dial indicator or inside micrometer to determine the size of the cylinder bore. Measure at right angles (Fig. 13–34).

If the cylinder bore is more than 0.003″ (0.08 mm) oversize or 0.0015″ (0.04 mm) out of round on cast-iron cylinders, or 0.0025″ (0.06 mm) out of round on lightweight cylinders, it must be resized.

The Honda 2.2 hp engine (Fig. 13–35) is shown as an example of a slightly different arrangement for tightening the cylinder head bolts.

So far, you have disassembled the engine. All parts have been inspected and you are ready to repair those parts which need attention. Chapter 14 covers the repair procedures and contains some suggested practices to make the job go as smoothly as the manufacturer suggests.

Disassembling, Testing, and Inspecting the Two-Cycle Engine

The operation of the two-cycle engine is different from the four-cycle. For one thing, unlike the four-cycle, it does not use oil. Therefore, you do not have to drain the oil before disassembling the engine.

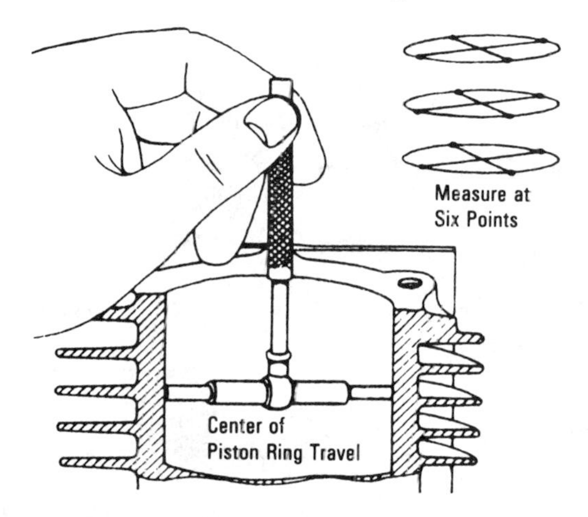

Fig. 13–34. Checking cylinder bore. Check at least six points. *(Courtesy B & S)*

Disassembly of the entire engine means that you have reason to believe there are troubles that cannot be found or corrected by using the troubleshooting techniques already suggested. Or in some cases, you already have the trouble located and must disassemble the engine to get to the damaged part.

The two-cycle engine is smaller, in most cases, than the four-cycle. The two-cycle has fewer parts, so it will probably be easier to work with than the four-cycle. The engine used here for an example is made by Tecumseh. This is a popular type and will be encountered by most people when they attempt to learn about this type of engine.

Disassembly of a Split-Crankcase Engine

- Remove the shroud and fuel tank if present (Fig. 13–36). Remove the flywheel and inspect for stator striking, magnet strength, and other visual signs of wear (Fig. 13–37).
- Remove the stator and test (Fig. 13–38). Inspect the coil assembly for cracks in the insulation or for evidence of overheating or any mechanical damage. Make sure the electrical leads are intact, especially where they enter the coil.

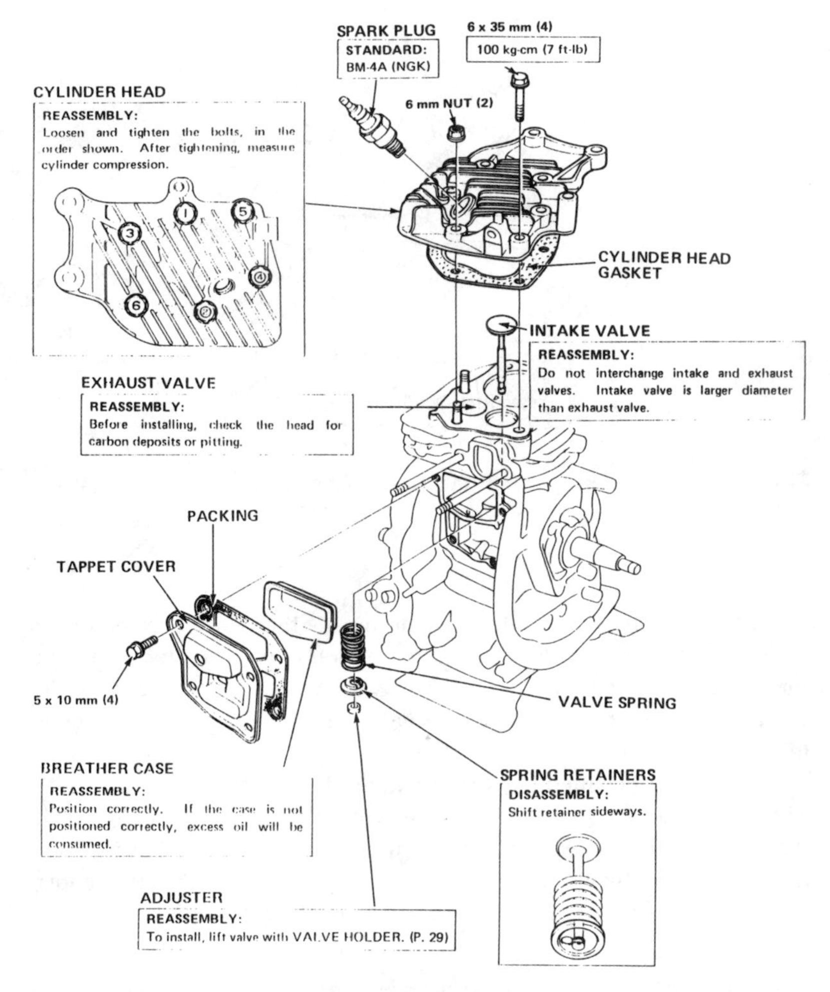

Fig. 13–35. Honda model G100 engine servicing of cylinder head and valves. *(Courtesy Honda)*

The coil can be tested with an ohmmeter as explained previously with the four-cycle engine. Inspect the condenser for visible damage. Look especially for damaged terminal lead, dents or gouges in the can, or a broken mounting clip. Use a condenser (capacitor) tester to check for high-voltage break-

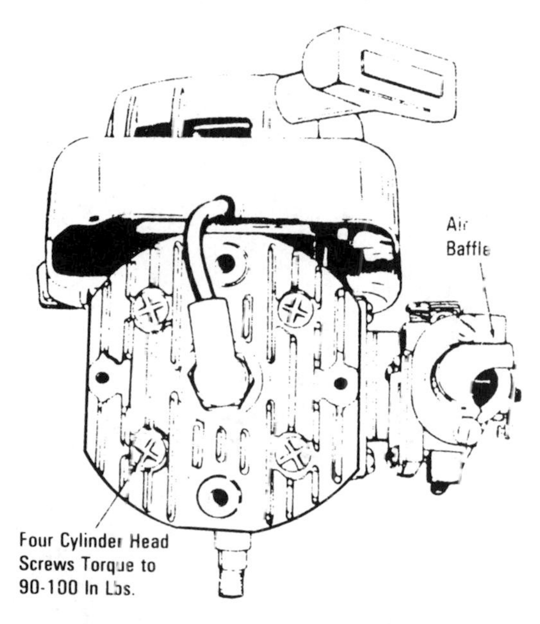

Fig. 13–36. A split-crankcase engine. *(Courtesy Tecumseh)*

down or shorts. Also check for an open condenser. Replace if in doubt.

- Remove the carburetor and governor linkage. Make a note of the *exact positions* of wire links and springs or proper carburetion adjustments cannot be made.
- Lift off the reel plate and gasket if they are present and inspect. Clean the reeds, stops, and adapter of oil, or other foreign matter (Fig. 13–39). Check reeds for seal against the adapter. Reeds should not bend away from the plate more than 0.010″ (0.254 mm).
- Remove the spark plug and inspect (Fig. 13–40). Grasp the high-tension wire (spark plug wire) by the insulation and hold the terminal end about $\frac{1}{8}$″ (0.3175 mm) from the metal body of the spark plug. Turn the flywheel (which has been replaced temporarily for this check). If you get a bright spark, then the magneto is working. Remove the spark plug and reconnect the high-voltage lead. Ground the plug and crank the engine once

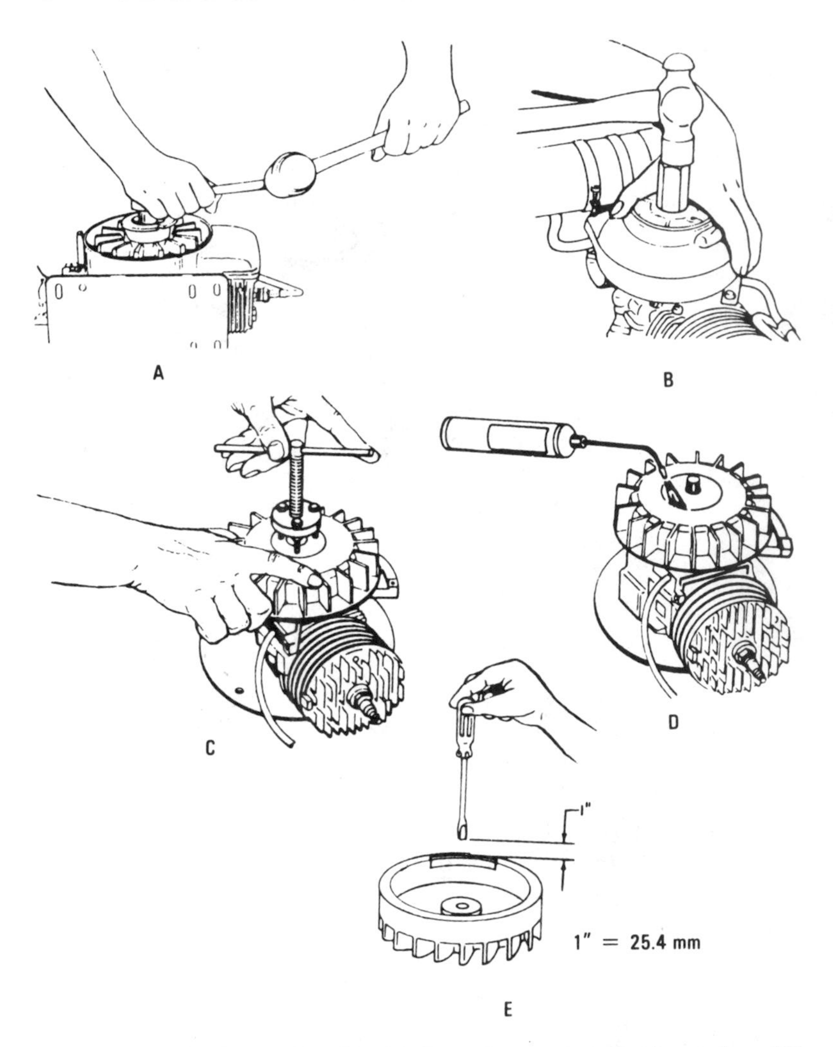

Fig. 13–37. (A) Removing flywheel nut from a two-cycle engine. (B) Using a knockoff tool to remove the flywheel. (C) Using a flywheel puller. (D) Using a torch to apply heat to remove the flywheel. (E) Test of the magnets on the flywheel. Hold screwdriver 1″ (25.4 mm) above the magnet; it should be pulled to the magnet. Place the flywheel on a wooden benchtop. *(Courtesy Tecumseh)*

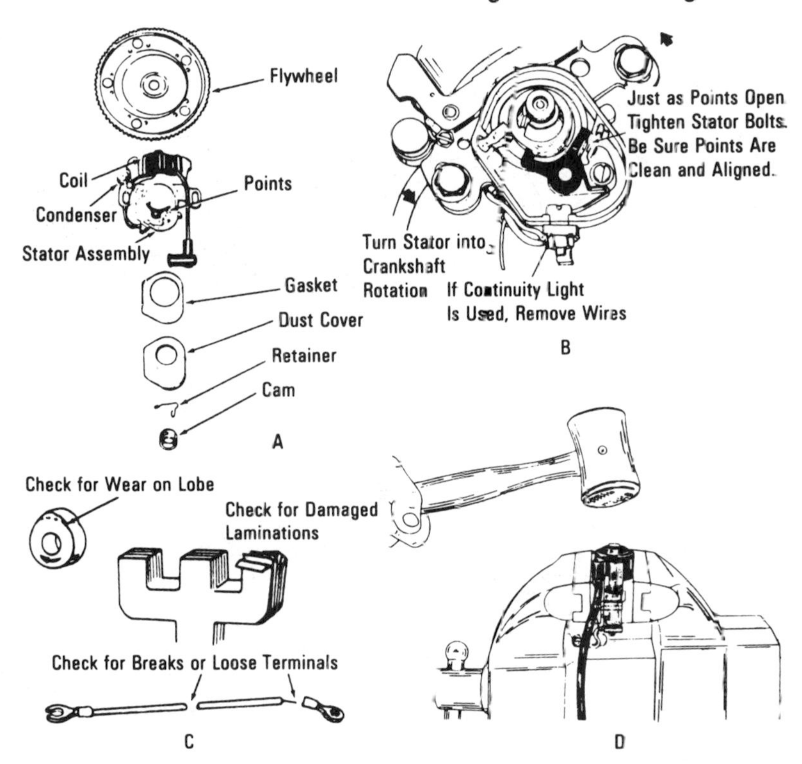

Fig. 13–38. (A) Coil assembly located under the flywheel. (B) Adjustable timed unit (stator can be rotated). (C) Checking the coil and armature. (D) Removing the coil from the armature. *(Courtesy Tecumseh)*

again (Fig. 13–40B). If a spark jumps the spark gap, the ignition system is operating satisfactorily.

Spark plugs should be removed, cleaned, and adjusted periodically. Check the point gap with a wire feeler gage (Fig. 13–40C). Replace if the points are pitted and burned or the porcelain is cracked. Be sure the plug is cleaned of all foreign matter before reinstalling it.

- Remove the muffler. Be sure the muffler or ports are not clogged with carbon. The lock tabs on the mufflers used on production engines do not have the ends bent over the muffler

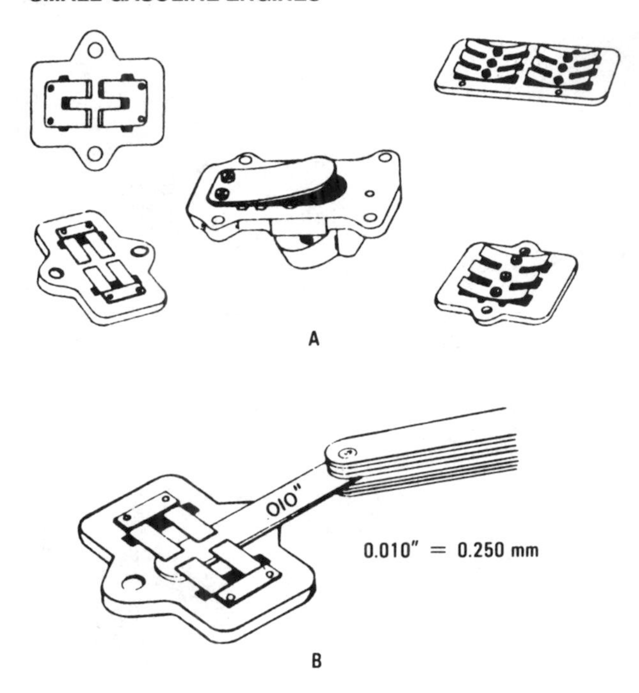

Fig. 13–39. (A) Different types of reeds. (B) Checking the reed clearance. *(Courtesy Tecumseh)*

screws (Fig. 13–41). Whenever replacing the muffler for whatever purpose (or the muffler screws have been removed) the lock tabs must be bent over the muffler screw heads.

- Remove the transfer port cover. Check for seal. See Fig. 13–42 for the location of the transfer port cover.
- Remove the cylinder head, if present. Some engines use Loctite on the screws holding the cylinder head to the cylinder. Removing the screws on the engines treated with Loctite can be done, but it creates a problem. This is true especially with screws that have a slotted head for a straight screwdriver blade. The screws can be removed if heat is applied to the head of the screws with the tip of a soldering gun.

Engines with a governor mounted on the power take-off end of the crankshaft can be disassembled by removing screws that hold the outboard bearing housing to the crankcase (Fig.

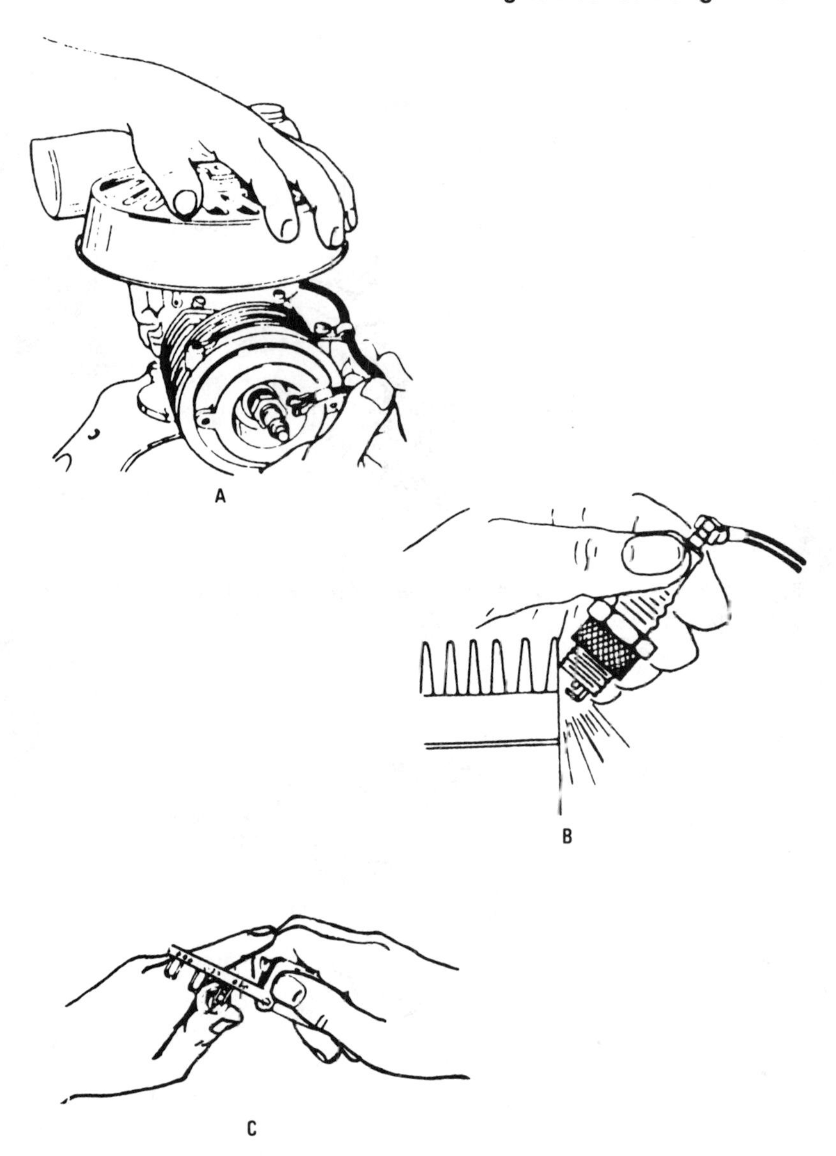

Fig. 13–40. (A) Checking the spark from the wire to the plug. (B) Checking the spark plug. (C) Checking the gap with wire gage. *(Courtesy Tecumseh)*

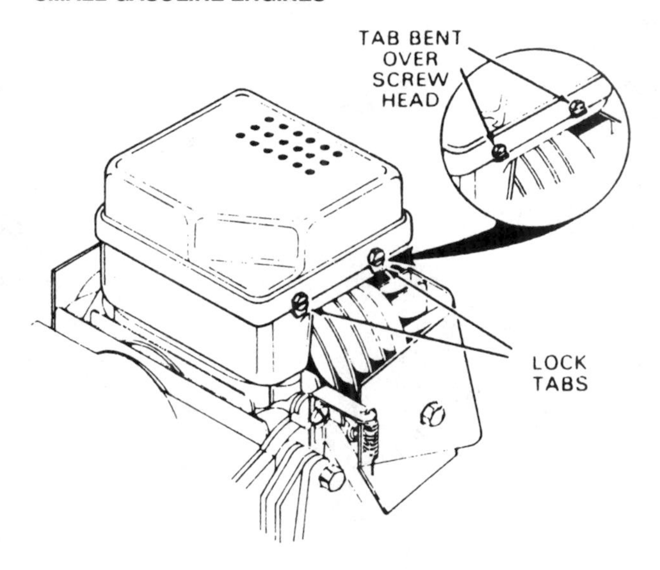

Fig. 13–41. Location of the lock tabs on the muffler. *(Courtesy Tecumseh)*

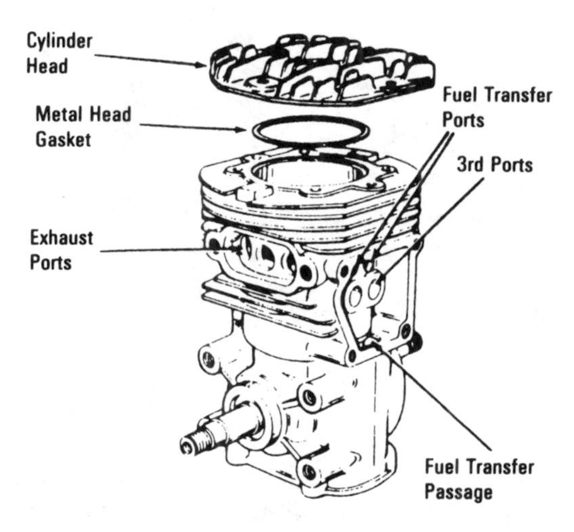

Fig. 13–42. Location of transfer port covers. *(Courtesy Tecumseh)*

13–43). Clean the power take-off end of the crankshaft. Remove the outboard bearing housing and bearing. Loosen the setscrew that holds the governor assembly to the crankshaft. Slide the entire governor assembly from the crankcase. Remove the screw that holds the governor bell crank bracket to the crankcase. Remove the governor bell crank and bracket.

- Make the matching marks on the cylinder and crankcase as shown in Fig. 13–44. Remove four nuts and lock washers that hold the cylinder to the crankcase. Remove the cylinder by pulling straight out from the crankcase.
- To separate the crankcase halves, remove all of the screws that hold the crankcase halves together. Hold the crankshaft vertically. Grasp the top half of the crankcase and hold firmly. Strike the top end of the crankshaft firmly with a rawhide mallet (Fig. 13–45) while holding the assembly over a bench to prevent damage to the parts when they fall. The top half of the crankcase should separate from the remaining assembly. Invert the assembly and repeat the procedure to remove the other casting half from the crankshaft on the ball-bearing units.

Each time the crankshaft is removed from the crankcase,

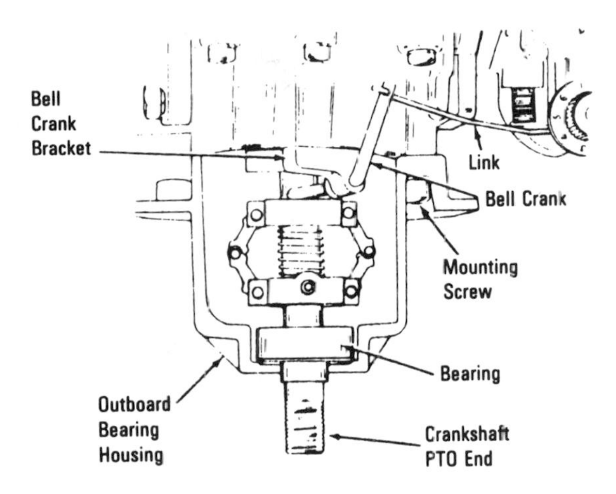

Fig. 13–43. Mechanical governor on PTO end of engine. *(Courtesy Tecumseh)*

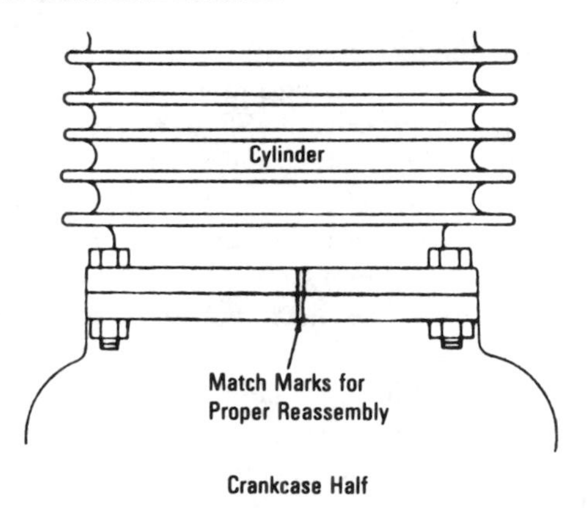

Fig. 13–44. Matching marks on the cylinder and crankcase. *(Courtesy Tecumseh)*

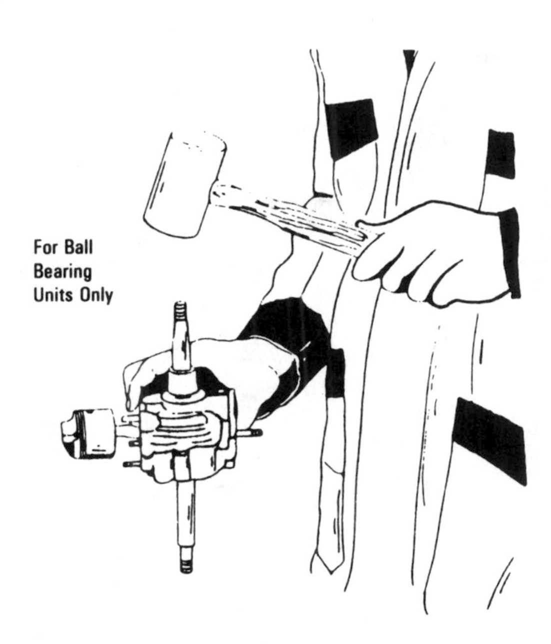

Fig. 13–45. Separating crankcase halves. *(Courtesy Tecumseh)*

seals at the ends of the crankcase should be replaced (Fig. 13–46).

Use seal protectors where necessary when sliding the crankshaft through the new seals.

The gasket surface where the crankcase halves join must be thoroughly cleaned before reassembly. Clean each half with a cloth saturated with a lacquer thinner. Do not buff. Do not use a file or an abrasive on these surfaces. If one half is in need of replacement, both halves must be replaced together (Fig. 13–47).

- Remove the seals. Discard all seals removed from the crankcase. Check the condition of the seal retainers and retainer springs (Fig. 13–48).

To remove the seals, use an ice pick, an old carburetor main adjustment needle, or a screwdriver. Insert the tool between the retainer spring and crankcase retainer spring groove so that the point of the tool is near the gap in the spring. Carefully pry the retainer spring out of the spring groove, then remove the retainer and seal. Seal removal can be done with the crankshaft in place, but it is easier with it removed.

Caution: Use care when you remove the retainer spring from the crankcase groove. Excessive pressure could damage the crankcase. This is especially true on the magneto side, which has a thin-walled hub.

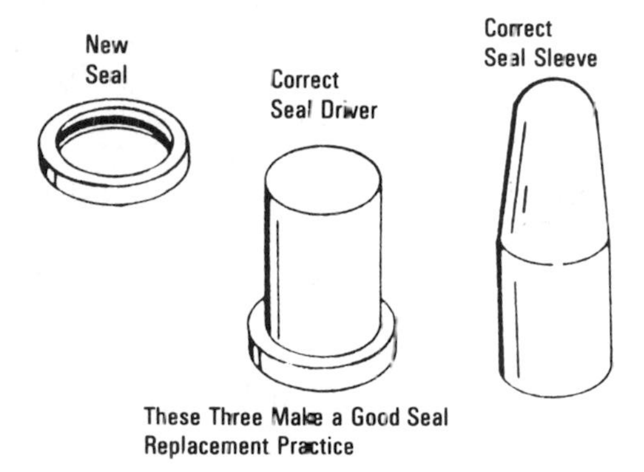

Fig. 13–46. Crankcase seals. *(Courtesy Tecumseh)*

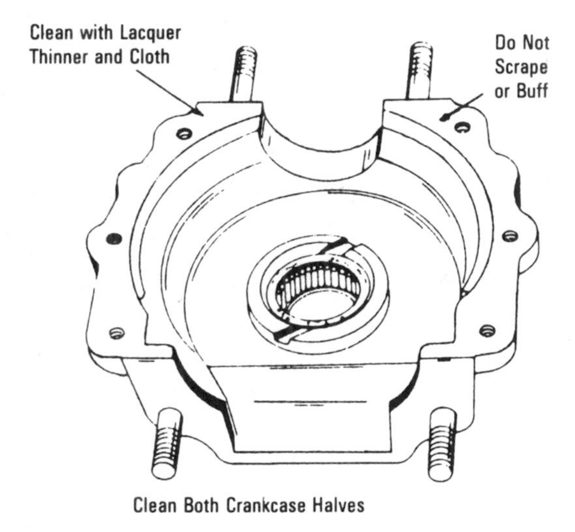

Fig. 13–47. Clean gasket surfaces thoroughly where crankcase halves join. *(Courtesy Tecumseh)*

Do not scratch or nick the crankshaft where the seal makes contact. Newer engine models have a small recess cast into the seal receptacle. This allows easy access to the seal and the retainers (Fig. 13–49).

Outboard engine oil seals (the power take-off end oil seal) will seal down or outward from the crankcase (Fig. 13–50).

- Disassembly of the connecting rod for engines that use a solid bronze or aluminum connecting rod is done by removing two self-locking capscrews holding the connecting rod to the crankshaft. Remove the rod cap (Fig. 13–51). Make a note of the match marks on the rod and cap. They have to be aligned when reassembled.

Engines using steel connecting rods are equipped with needle bearings at both crankshaft and piston pin end. Remove the two screws that hold the connecting rod and cap to the crankshaft. Take care not to lose needle bearings during removal. Needle bearings at the piston pin end of the steel rods are caged and can be pressed out as an assembly if damaged (Fig. 13–52).

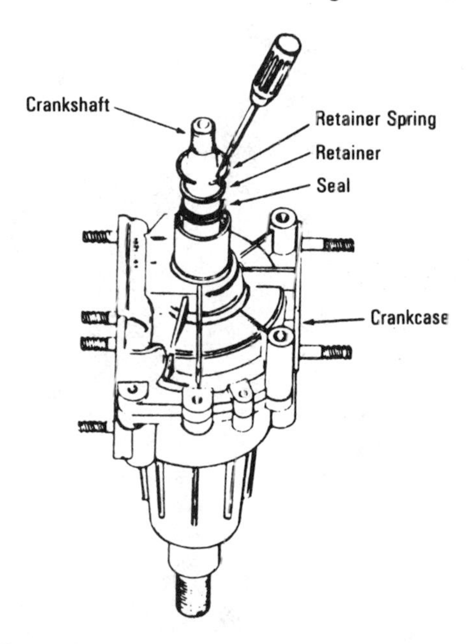

Fig. 13–48. Removing a seal. *(Courtesy Tecumseh)*

- Check the connecting rod for cracks or distortion (Fig. 13–52). Check the bearing surfaces for scoring or wear. Bearing diameters should be within the limits indicated by the manufacturer.

Two arrangements of needle bearings are shown in Fig. 13–53. The split-row and the single-row needles are shown. These are the two basic arrangements for needle bearings.

Service needles are supplied with a beeswax coating. This coating tends to hold the needles in position.

Checking the crankshaft for wear can be done with the aid of a micrometer (Fig. 13–54). A micrometer can be used to check the bearing journals for out-of-round condition. The main bearing journals should not be more than 0.0005″ (0.0127 mm) out of round. Connecting rod journal should not be more than 0.001″ (0.0254 mm) out of round. Replace a crankshaft that is not within these limits. Do not attempt to regrind the crankshaft since undersized parts are not available.

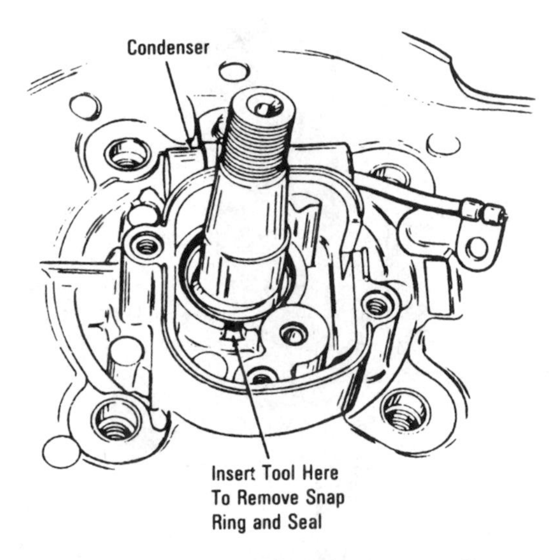

Fig 13–49. Newer models of engines have a small recess cast into the seal receptacle. This allows easy access to the seal and retainers for removal. *(Courtesy Tecumseh)*

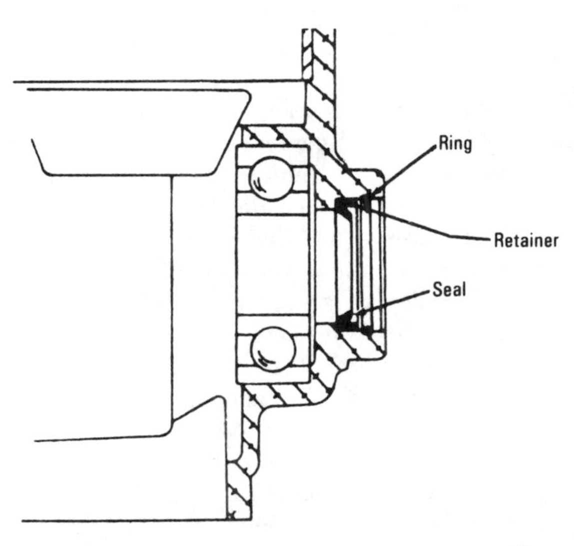

Fig. 13–50. Outboard engine oil seals on power take-off end. *(Courtesy Tecumseh)*

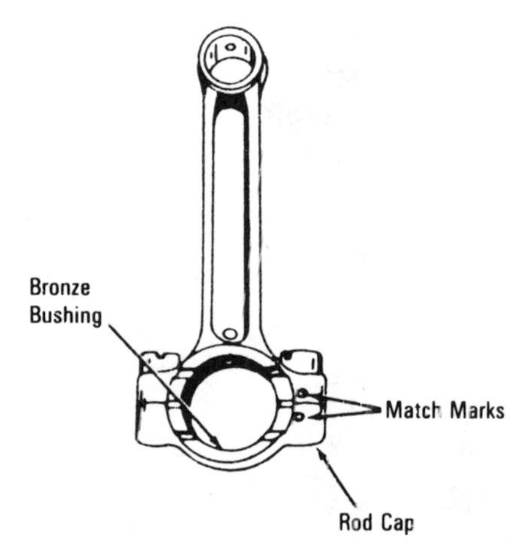

Fig. 13–51. Parts of the connecting rod. Note the match marks so they can be reassembled correctly. *(Courtesy Tecumseh)*

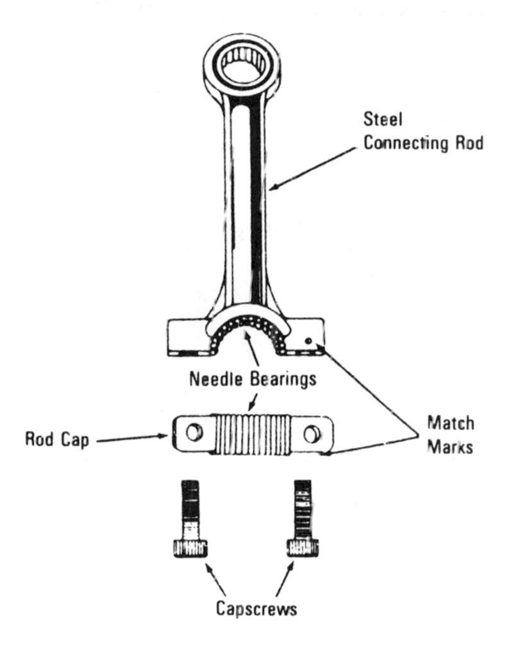

Fig. 13–52. Check connecting rod for cracks or distortion. Check the bearings for scoring or wear. *(Courtesy Tecumseh)*

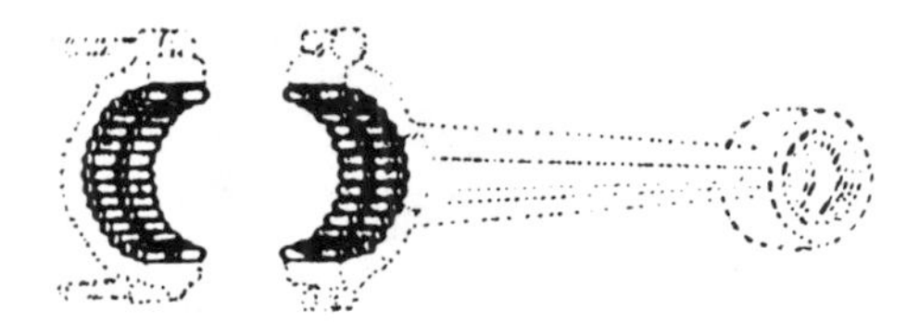

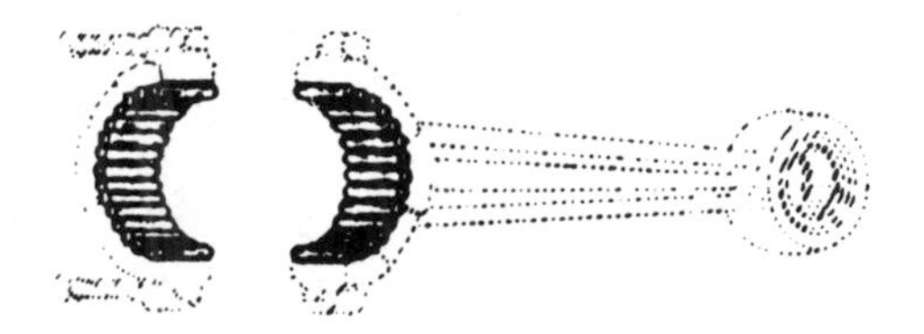

Fig. 13–53. Two basic arrangements of needles in rods. *(Courtesy Tecumseh)*

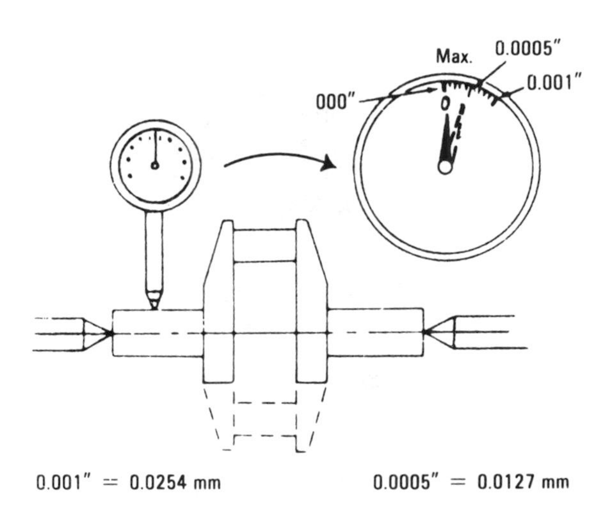

Fig. 13–54. Using a micrometer to check the bearing journals for out-of-round condition. *(Courtesy Tecumseh)*

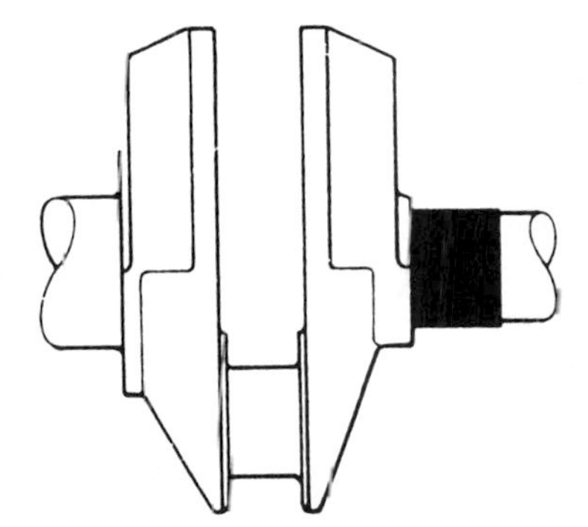

Fig. 13–55. Check the bearing surface for wear or scoring. *(Courtesy Tecumseh)*

- Check tapered portion of the crankshaft (magneto end) keyways, and threads. Damaged threads may be restored with a thread die. If the taper of the shaft is rusty, it indicates that the engine has been operating with a loose flywheel. Clean the rust off the taper and check for wear. If the taper or the keyway is worn, replace the crankshaft (Fig. 13–55).
- Check all journal diameters. They should be within the limits established by the manufacturer.
- Check the oil seal contact surfaces for any micro scratches that could cause premature wear to the oil seals. Check the crankshaft for bend by placing it between two pivot points or dead centers. Position the dial indicator sensor onto the crankshaft bearing surface and rotate the shaft. A significant variance in the indicator readings is indicative of the amount of bend in the crankshaft. General maximum limits are 0.002″ to 0.004″ (0.0508 to 0.1016 mm) true indicator reading (T.I.R.) (Fig. 13–54).

Cylinders, Reeds, and Compression Release

Two-cycle engines have reed valve assemblies that need attention from time to time. The cylinder head, gaskets, and reed valves require service to keep them operating as designed.

Cylinder and Head Service—Servicing should include checking the cylinder for bore damage or scoring. Check for broken or cracked fins and warped head or head mounting surface. If the head is warped extensively, it should be replaced. Always replace the head gasket and torque it to the proper specification.

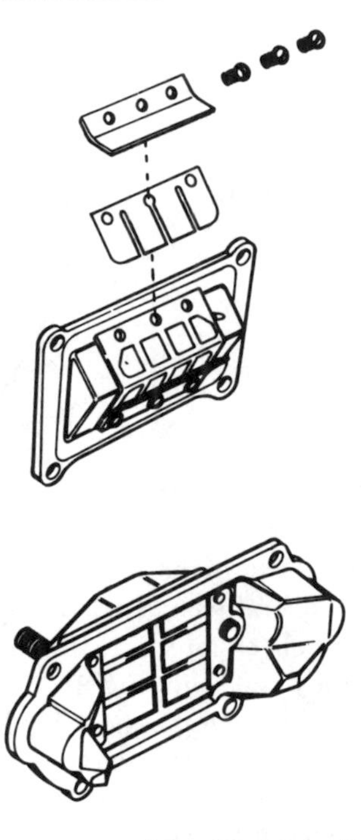

Fig. 13–56. Reed valves assembly. *(Courtesy Tecumseh)*

Gaskets—Replace all gaskets in reassembly and make sure all sealing surfaces do not leak. Leaking gaskets cause erratic engine running, hard starting and can damage the internal components by causing an imbalance in the fuel-oil-air mixture.

Reed Valves—Make sure the reeds and sealing surfaces are free of dirt and foreign matter. Check the reeds for a good seal against the sealing surface of the adapter. Reeds should not bend away from the sealing surface more than 0.010″ (0.254 mm).

If the reeds are serviceable, the smooth side of the reed must be located against the sealing surface. Service reeds have "smudge" marks on the smooth side. If these marks are gone, feel for a rough edge and disassemble away from the sealing surface (Fig. 13–56).

Cranking compression pressures bleed past the reeds into the muffler (Fig. 13–57). Once the engine is started, a high-pressure build-up between the reeds forces the reeds against their seats, stopping

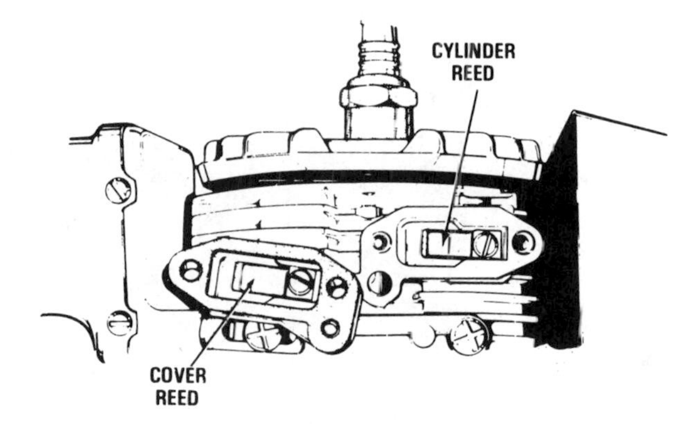

Fig. 13–57. Reed cover and location on engine. *(Courtesy Tecumseh)*

compression bleed-off and allowing the engine to run at full compression.

Replace both reeds if either is defective. When installing the reeds, be sure the colored side of the reed faces its seating surface. If in doubt, feel for a rough edge on the reed. The rough edge must be installed away from the seating surface. Assemble reed stops and tighten the self-tapping screws. The double-reed type has automatic compression release. See Fig. 13–57 for the *cover reed* and the *cylinder reed* locations. A poor running engine or lack of power may be caused by a leaking reed or cover gasket.

Automatic Compression Release.—Cranking compression pressures bleed past reed valves, through a port, into the piston pin and out the exhaust port (Fig. 13–58). As the engine starts and compression increases, the reed will be forced against the bottom port, sealing it. The engine will then run under full compression (Fig. 13–59).

Cylinder Exhaust Ports.—The muffler and cylinder exhaust ports should be cleaned after each seventy-five (75) to one hundred (100) hours of operation. It is recommended that the cylinder head be removed and carbon cleaned from the ports, the cylinder head, and the top of the piston (Fig. 13–60).

Do not scratch the metal surfaces. With the cylinder head removed, remove any carbon deposits from the cylinder wall, head, and the top of the piston. Using a pointed 3/8″ wooden dowel or similar tool,

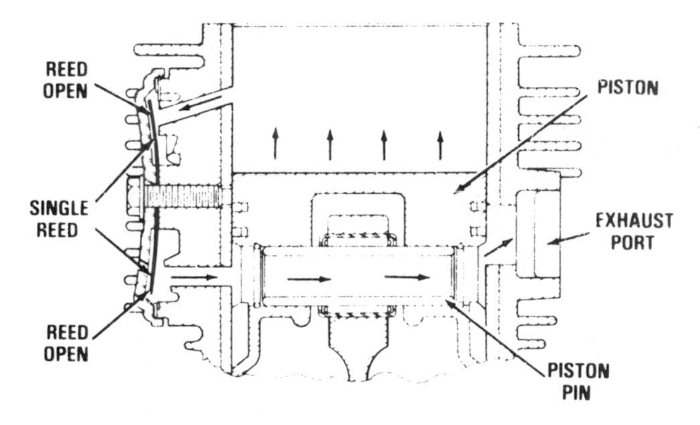

Fig. 13–58. Single-reed-type automatic compression release. *(Courtesy Tecumseh)*

remove the carbon from the exhaust ports. Be sure to remove all loose carbon particles from the engine.

Now that you have disassembled the two-cycle engine, identified any problems related to its performance, and replaced the defective parts, you will find the next chapter, on repair, interesting. It examines the cylinder block overhaul procedure. This procedure will aid you in repairing the engine block the putting it back into operation.

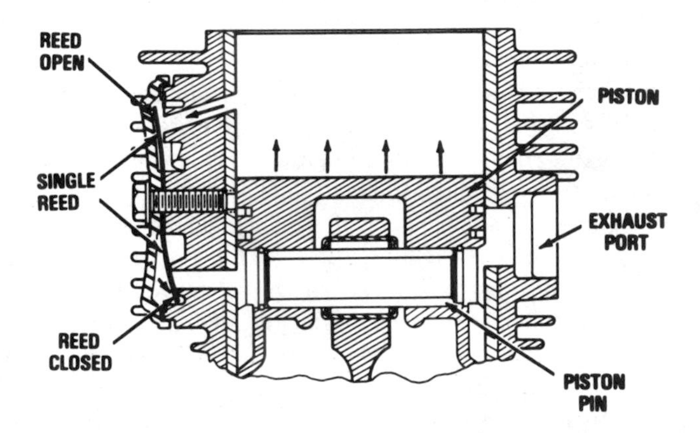

Fig. 13–59. Seating of the reed as the engine runs. *(Courtesy Tecumseh)*

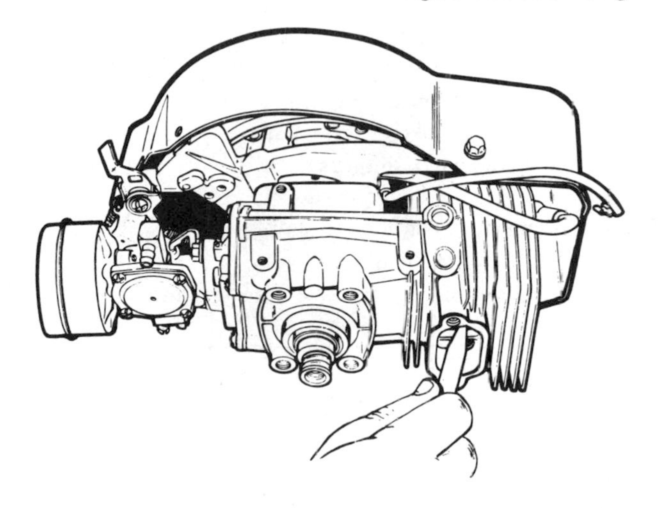

Fig. 13–60. Removing carbon deposits with a scraper. *(Courtesy Tecumseh)*

Chapter Highlights

1. In order to work on the engine block you have first to disassemble the engine.
   This chapter describes:
2. How to check an engine while disassembling it.
3. How to remove the crankshaft.
4. How to remove the auxiliary power take-off.
5. How to inspect the cam gear.
6. How to remove the cam gear from the engine.
7. How to remove the piston.
8. How to remove the connecting rod.
9. How to check the connecting rod.
10. How to check the piston, piston pin, and grooves.
11. How to remove the Magna-Matic series ignition system.
12. How to remove the rotor.
13. How to test the armature and leads.
14. How to inspect and remove the automatic spark advance.
15. How to resize cylinder bores.
16. How to disassemble a split-crankcase engine.

17. How to lift off reed plates and gasket in a two-cycle engine.
18. How to remove seals on the engine.
19. How to remove and replace needle bearings.
20. How to check the crankcase for wear.
21. How to check journal diameters.
22. How to check the reeds in a two-cycle engine.
23. How to replace reeds or head gasket.
24. How the reeds automatically seat once in use.
25. How to clean carbon from the two-cycle engine.

Vocabulary Checklist

actuator cam
armature
automatic compression release
Belleville washer
bracket
cam gear
chrome rings
coil auxiliary power take-off
commutator
compression release
cranking compression
cylinder block
cylinder exhaust ports
cylinder head
cylinder reeds
dial indicator
disassembly
gaskets
governor
journal
piston pin
reed valves
retainer
ring groove
seal protectors
secondary voltage
self-tapping screws
spark advance rocker cam
split-crankcase engine
tachometer
tappets
telescoping gage
transfer port
undersize
valve cover
Woodruff key

Review Questions

1. Before you work on an engine block, what do you have to do to it?
2. What is one of the first steps in disassembling an engine?
3. How do you remove the valves from the engine block?

4. Why do you remove the burrs and rust from a crankshaft power take-off end before taking out the crankshaft?
5. How do you remove the cam gear from all cast-iron engines?
6. When checking the tappets, what else should you check for visible signs of wear?
7. If the cylinder is resized, what do you have to do about the piston?
8. What is the first thing to check in working with the piston rings?
9. Do you have to deglaze the cylinder walls when installing piston rings in the aluminum engine?
10. What has to be done if the crankpin bearing in the rod is scored?
11. When do you look for undersize connecting rods?
12. How do you test the continuity of the primary of the ignition coil?
13. What do you use to check the size of the cylinder bore?
14. What is a split-crankcase engine?
15. Why don't you have to remove the reeds in a four-cycle engine disassembly?
16. Does the two-cycle engine have a muffler? Why?
17. How do you separate the crankcase halves of a split-crankcase engine?
18. How do you check the crankshaft for wear?
19. What part of the two-cycle engine often requires attention from time to time?
20. Why should you always replace the head gasket when the head is removed and reassembled?
21. Which side of the reeds have smudge marks?
22. What symptom is noticed when there is a leaking reed or leaking cover gasket?

CHAPTER 14

Repairing the Engine

So far you have taken the engine apart and have learned all its parts and how to make certain adjustments. Now that you have arrived at a point where you know how the engine works and why each part operates as it does, you know that some engines will need a complete overhaul. In this chapter you will learn how to overhaul an engine. Reassembly will be covered in the chapter following this one.

The four-cycle engine will be used to illustrate repair procedures. Some steps for the two-cycle are different, but generally speaking, the worn or defective parts are replaced. However, some repair procedures are unique to a particular type of engine. We attempt to present both cases here so that you will be knowledgeable in both types of engines and in items that need special attention.

As you work with engines and make comparisons on what you observe in disassembly, you will be able to discern which of these differences are important. These should be noted for future reassembly and operation of the engine.

Cleaning Parts

All parts should be inspected and cleaned before they are reassembled into an overhauled engine. Some of the cleaning consists of soaking the part in a solvent and blowing it dry with compressed air.

Other cleaning consists of removing small particles of grinding dust with soap and water. Suggested cleaning procedures will be pointed out as the parts of the engine are being repaired or worked on.

One of the most important items to consider when working on an engine is to keep the workshop clean. By keeping the workbench free of excess tools, parts, and small filings from previous jobs, you save many hours of unnecessary work. Keep the workbench free of everything except the parts and tools needed for the job at hand.

Dirty and oil-soaked rags should be placed in a metal container with a lid. Oil-soaked rags should not be used to wipe parts under repair. Keep your wrenches and tools clean. Do not put a dirty wrench back into the toolbox or on the tool board.

When a hot plate is used to heat oil to remove bearings, make sure it does not ignite an oil-soaked rag that happens to be lying on the bench. Handle the cleaning rags properly by discarding them whenever they get to a point of no further cleaning ability.

Another thing to keep in mind is clean hands. Of course, you won't be able to keep your hands spotless with the oil and grease encountered in an engine, but you can keep them clean of oil and grease. Oil and grease attract dirt that can be passed on to parts that need to be absolutely clean. Wipe your hands with a clean cloth, and place the used cloth in the proper metal container.

Resizing Cylinders

You have already inspected the disassembled engine and labeled the possible problems. Visual inspection shows if there are any cracks, stripped bolt holes, or broken fins. It may also uncover a damaged cylinder wall. This inspection reinforces your earlier inspection that resulted in the tagging of the cylinder as a possible problem. In Chapter 13, you checked the cylinder bore (Fig. 13–34). If the cylinder bore is more than 0.003″ (0.08 mm) oversize or 0.0015″ (0.04 mm) out of round on cast-iron cylinders, or 0.0025″ (0.06 mm) out of round on lightweight cylinders, it must be resized.

Next Oversize

Standard-bore cylinders are given in Table 14–1. Once you remove the piston and the engine block is bare, it is time to hone or deglaze the cylinder bore.

Table 14–1. Standard-Bore Size for Basic Engine Models

Basic Engine Model or Series	Standard-Bore Size Diameter			
Aluminum Cylinder	Maximum		Minimum	
	Inches	Millimeters	Inches	Millimeters
6B, 60000 before Ser. #5810060	2.3125	58.74	2.3115	58.71
60000 after Ser. #5810030	2.3750	60.33	2.3740	60.30
8B, 80000, 82000	2.3750	60.33	2.3740	60.30
92000, 94000	2.5625	65.09	2.5215	65.06
100000	2.5000	63.50	2.4990	63.47
110000	2.7812	70.64	2.7802	70.62
130000	2.5625	65.09	2.5615	65.06
140000	2.7500	69.85	2.7490	69.82
170000, 190000	3.0000	76.20	2.9990	76.17
220000, 250000	3.4375	87.31	3.4365	87.29

(Courtesy B & S)

Deglazing is the breaking of the smooth surface on the inside of the cylinder wall. This can be done with a very fine abrasive. It allows the rings to be properly lubricated and to seat during break-in. In some cases this deglazing is all that is necessary. To glaze means to become glassy in appearance. Cylinder walls become glazed or very smooth from the pistons rubbing against them during normal operation. Use a piece of sandpaper to roughen them up a bit and reform them to fit the new piston rings.

Note: Briggs & Stratton aluminum engines are *not* to be deglazed when installing rings in the engines.

If chrome rings are used (they are available for most models), they do not require honing or breaking the glaze to seat properly. Chrome rings are used to control oil consumption. Chrome rings are used in engines where the bores are worn to 0.005″ (0.13 mm) over standard.

Honing is the refinishing of the cylinder bores with stones especially made to do the job. If a boring bar is used to make the cylinder a larger diameter, then a hone must be used to refinish the inside of the cylinder. Honing is the process whereby a very fine finish is imparted to the cylinder wall. However, the wall is not so fine that it is glazed. Fig. 14–1 shows what the inside of the wall looks like when finished with a hone. A crosshatch pattern is applied by moving the hone up and down slightly while it is rotated by a drill press or portable electric drill.

Always resize (using the hone) to exactly 0.010″ (0.25 mm), 0.020″ (0.508 mm), or 0.30″ (0.762 mm) over standard size as shown in Table 14–1. If this is done accurately, the stock oversize rings and pistons will fit perfectly and proper clearances will be maintained. Cylinders, either cast iron or aluminum, can be quickly resized with a good hone such as shown in Fig. 14–2. Use the stones and lubrication recommended by the hone manufacturers to produce the correct cylinder wall finish.

Honing Setup

Clean the cylinder at the top and bottom to remove burrs and pieces of base and head gasket. Fasten the cylinder to a heavy iron bracket (Fig. 14–3) or use a honing plate (Fig. 14–4). Some cylinders may require shims. Use a level to align the drill press spindle with the bore (Fig. 14–5).

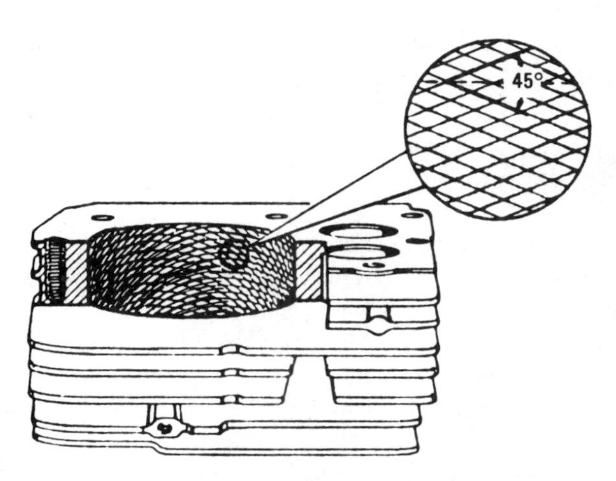

Fig. 14–1. Cylinder wall honing, or crosshatching. *(Courtesy B & S)*

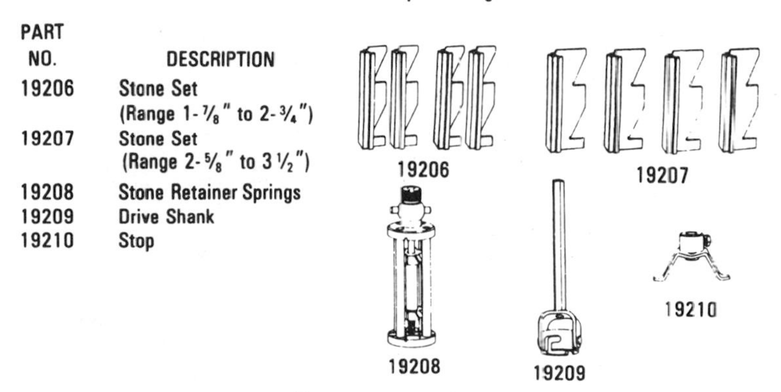

A 19205 HONE SET

All Aluminum Cylinder Engines

PART NO.	DESCRIPTION
19206	Stone Set (Range 1-⅞" to 2-¾")
19207	Stone Set (Range 2-⅝" to 3 ½")
19208	Stone Retainer Springs
19209	Drive Shank
19210	Stop

B 19211 HONE SET

All Cast Iron Cylinder Engines

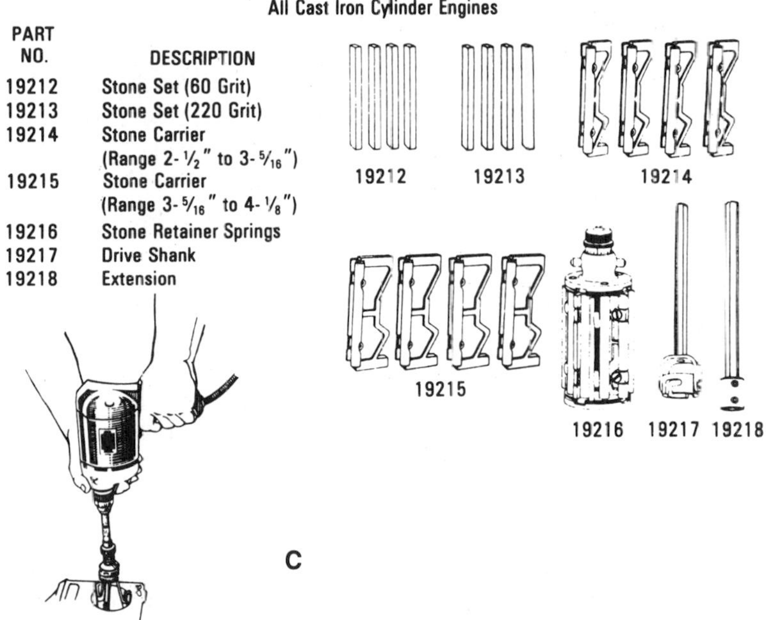

PART NO.	DESCRIPTION
19212	Stone Set (60 Grit)
19213	Stone Set (220 Grit)
19214	Stone Carrier (Range 2-½" to 3-5/16")
19215	Stone Carrier (Range 3-5/16" to 4-⅛")
19216	Stone Retainer Springs
19217	Drive Shank
19218	Extension

Fig. 14–2. (A) Hone set supplied by Briggs & Stratton for work on their engines. *(Courtesy B & S)* **(B) Hone set for cast-iron-cyclinder engines supplied by Briggs & Stratton for work on their engines.** *(Courtesy B & S)* **(C) Using a hand drill to hone a cylinder.** *(Courtesy Lisle)*

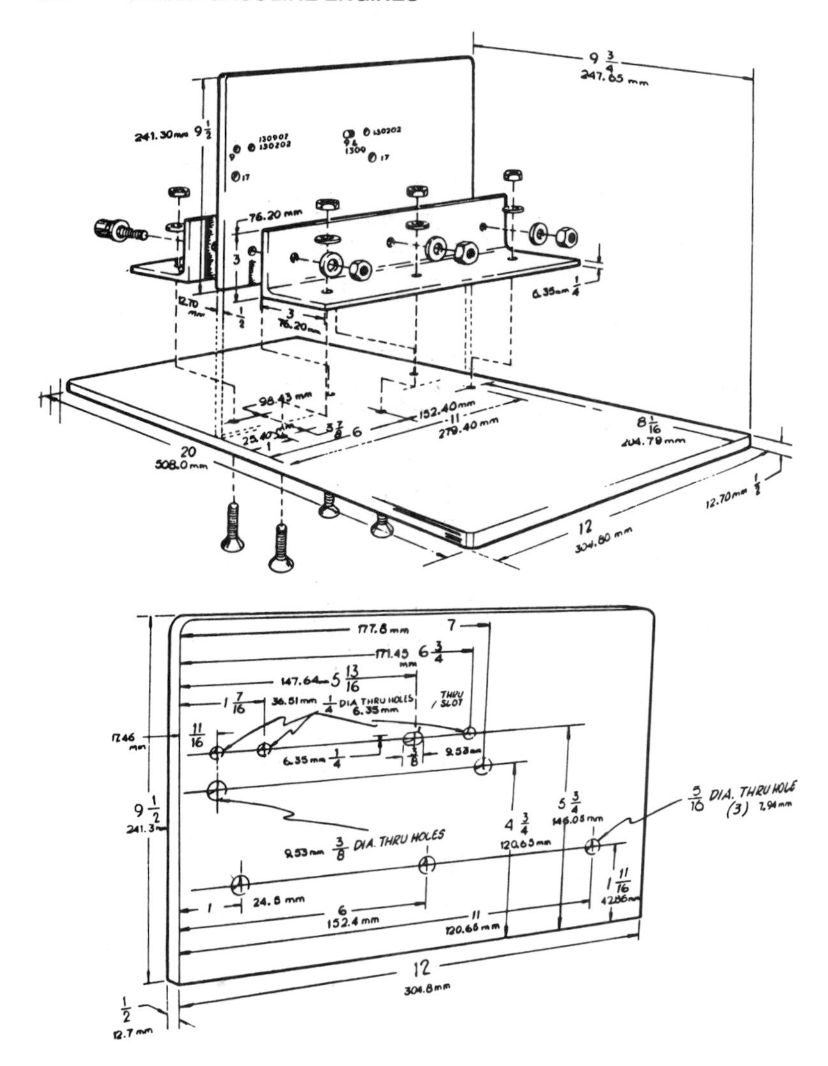

Fig. 14–3. Stand recommended by Briggs & Stratton to mount engine while honing. *(Courtesy B & S)*

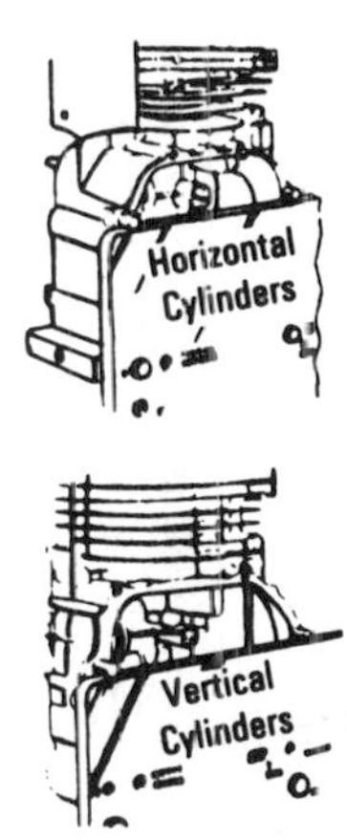

Fig. 14–4. Vertical- and horizontal-crankshaft engines mounted on the honing plate. *(Courtesy B & S)*

Liberally oil the surface of the drill press table and set the plate and the cylinder on it. Do not anchor to the drill press plate. If you are using the portable drill, however, set the plate and cylinder on the floor. Place the cylinder hone drive shaft in the chuck of the drill or portable drill.

Slip the hone into the cylinder (Fig. 14–6). Connect the drive shaft to the hone. Set the stop on the drill press so the hone can only extend ¾″ (19 mm) to 1″ (25.4 mm) from top or bottom of the cylinder. If you are using a portable drill, cut a wood block to place inside the cylinder as a stop for the hone.

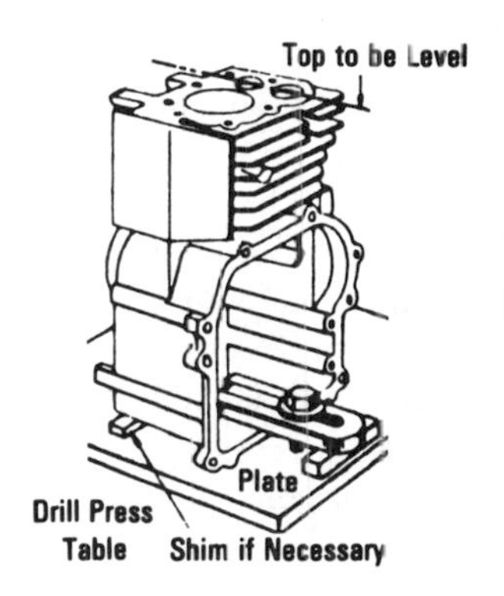

Fig. 14–5. Attaching the engine block onto the honing plate. *(Courtesy B & S)*

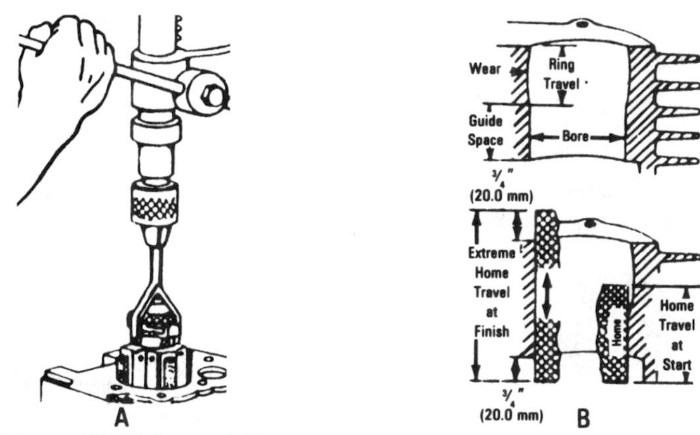

Fig. 14–6. (A) Using a drill press to hone an engine *(Courtesy B & S).* **(B) Travel of the hone.** *(Courtesy B & S).*

Honing the Cylinder

Place the hone in the middle of the cylinder bore. Tighten the adjusting knob with your finger or with a small screwdriver. Tighten until the stones fit snugly against the cylinder wall. Do not force. The hone should operate at 300 to 700 rpm. Lubricate the hone as recommended by the manufacturer.

Connect the drive shaft to the hone. Be sure that the cylinder and hone are centered and aligned with the drive shaft and drill spindle. Start the drill and, as the hone springs, move it up and down at the lower end of the cylinder (Fig. 14–6). The cylinder is not worn at the bottom but is round so it will guide the hone to straighten the cylinder bore. As the bottom of the cylinder increases the diameter, gradually increase the strokes until the hone travels the full length of the bore. Do not extend the hone more than ¾″ (19 mm) to 1″ (25.4 mm) at either end of the cylinder bore.

Cutting tension will decrease. As this happens, stop the hone and tighten the adjusting knob. Check the cylinder bore frequently with an accurate micrometer. Hone about 0.0005″ (0.01 mm) large to allow for shrinkage when the aluminum cylinder cools.

The cast-iron cylinders call for a stone change when you are within 0.0015″ (0.04 mm) of the desired size. Then you can use the finishing stones on the honing process. Always hone 0.010″ (0.254 mm), 0.020″

(0.508 mm), or 0.030″ (0.762 mm) *above* the standard dimensions given in Table 14–1.

The finished resized cylinder should have a crosshatch pattern (Fig. 14–1). The proper stones, lubrication, and spindle speed along with rapid movement of the hone within the cylinder during the last few strokes will produce this finish. Cross-hatching will allow for proper lubrication and ring break-in.

Cleaning—It is very important that the entire cylinder be thoroughly cleaned after honing. Wash the cylinder carefully in a solvent such as kerosene or commercial solvent. The cylinder bore should then be cleaned with *a brush, soap, and hot water.*

Repairing Valves, Guides, and Seats

Valves play an important part in the operation of an engine. They allow the fuel to enter at the correct time, and the combustion process takes place when both valves are tightly closed. The exhaust valve becomes very hot when it is operating normally. The temperatures inside the combustion chamber can reach as high as 5000°F (2760°C). This means the exhaust valve is exposed to extremely high temperatures. The exhaust valve opens to allow the hot gases to exit the chamber. The only cooling that a valve receives is the heat dissipated during the time it contacts the valve guides and the cylinder block when it is seated. The time that this cooling takes place is very limited, since the valve opens and closes 1000 times per minute at 2000 rpm. If the engine is rotating at 4000 rpm, as it does with an 18″ (45.7 cm) lawnmower blade attached, the valve opens and closes 2000 times per minute or 33⅓ times per second. Any carbon buildup around the valve seat can cause trouble with the proper seating of the valve and can cause it to overheat.

Refacing Valves and Seats

The faces and seats on valves should be resurfaced with a valve grinder or cutter to an angle of 45° (Fig. 14–7). Keep in mind that some engines have a 30° intake valve and seat. Valve and seat should then be lapped with a fine lapping compound to remove grinding marks and assure a good seal.

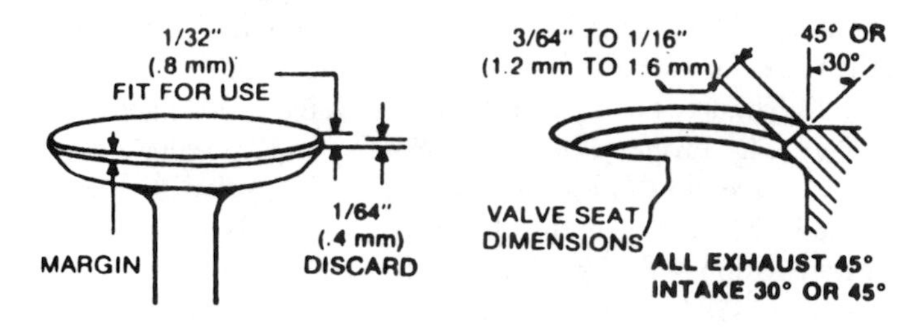

Fig. 14–7. Valve and seat dimensions. *(Courtesy B & S)*

Valve seat width should be $\frac{3}{64}''$ to $\frac{1}{16}''$ (1.91 to 1.587 mm) (Fig. 14–7). If the seat is wider, a narrowing stone or cutter should be used. If either the seat or the valve is badly burned, it should be replaced. Replace the valve if the margin is $\frac{1}{64}''$ (0.4 mm) or less after resurfacing.

Larger engines may have different values for their valves and seats. Tecumseh has a different-size valve system (Fig. 14–8). The intake valve gap is 0.010″ (0.254 mm) and the exhaust valve gap is 0.020″ (0.508 mm). They are set when the engine is cold. The engine can be

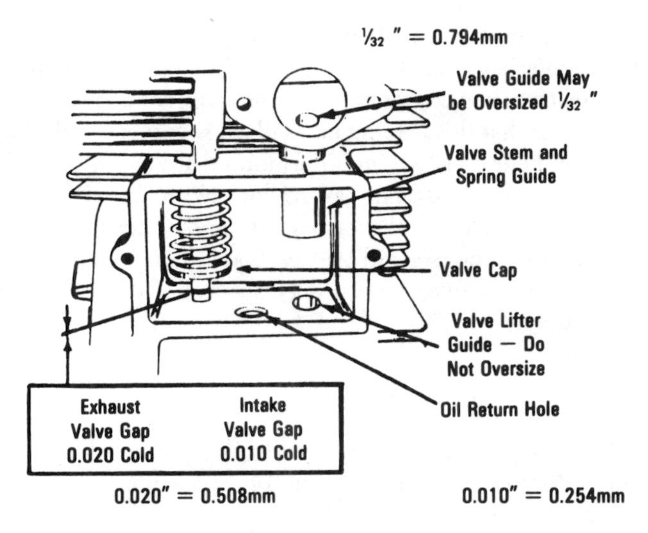

Fig. 14–8. Location of the valves in a Tecumseh engine. *(Courtesy Tecumseh)*

rotated with the piston, connecting rod, and crankshaft still in place. With the rotation, the valve lifters are at the lowest point—that's when the piston is at top dead center (TDC) on the compression stroke and both valves are closed completely. The valve stem ends should be ground flat (Fig. 14–9). An uneven or concave end can cause abnormal wear on the valve train. Fig. 14–10 shows the valve grinding specifications. Correct valve grinding procedures are necessary to obtain a satisfactory result. Face and seat angles must be accurate and the surfaces must be smooth. Lapping is recommended. A lip must be left on the valve head after resurfacing (Fig. 14–10). The valve without the lip would burn and distort very quickly. Use a "V" block and the flat side of a grindstone to grind for valve clearance. The end of the stem must be flat and true or abnormal wear will occur.

Valve Springs

The outside of the valve stem guide also guides the valve spring. A valve-spring cap is used only on the bottom of the spring. A pin through the valve stem retains the spring (Fig. 14–11). If the valve guides are worn, enlarge them and a step-type reamer. Oversize valves are ⅓₂″ (0.08 mm) larger in diameter than standard. The valve lifter guide is not to be enlarged. Lifters with oversize stems are not available.

Valve springs should be replaced when an engine is overhauled. Weak valve springs will spoil the best overhaul job. Valve-spring free length should be checked (Fig. 14–12). Comparing one spring with the other can be a quick check for any difference. If a difference is noticed, carefully measure the free lengths and compare the length and strength of each spring. Table 14–2 will give the specifications for

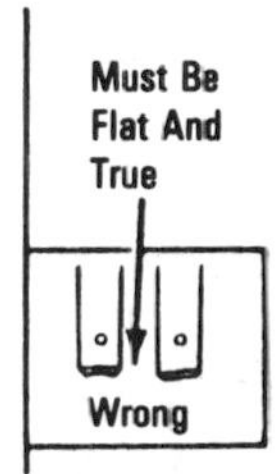

Fig. 14–9. Incorrect wear or grinding pattern for the end of the valves. *(Courtesy Tecumseh)*

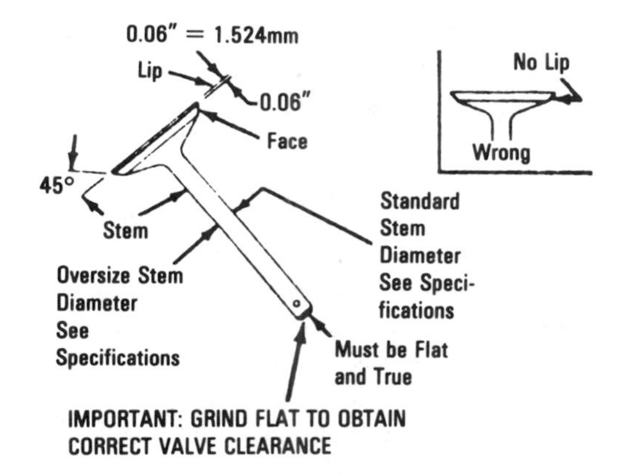

Fig. 14–10. Parts of a valve. *(Courtesy Tecumseh)*

valves, seat angles, width face angle, and lip width for five models made by Tecumseh. The valve-spring free length and compressed tension is also given. Check valve-spring free length. It should be 2.125″ (53.975 mm) long on early production engines and 1.885″ (47.879 mm) on later production models. If both ends are not parallel, replace with new springs.

Valve Guides

The information about valve guides has to be directed toward specific models of engines to be applicable in the best sense and usable. Therefore, the following will refer to Briggs & Stratton models 5, 6, 8, 6B, 60000, 8B, 80000, 82000, 92000, 94000, 100000, 110000, and 130000.

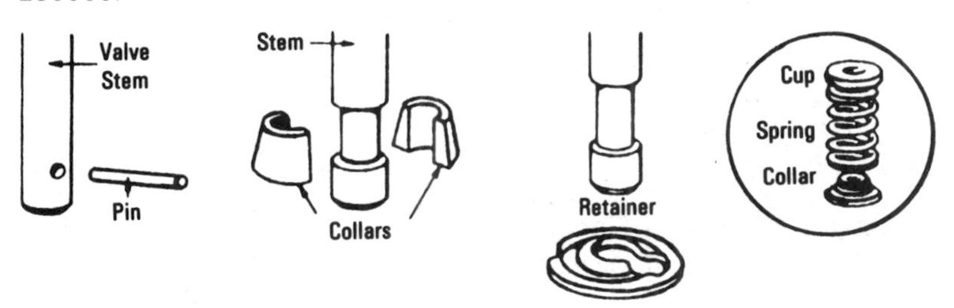

Fig. 14–11. Valve-spring retainers. *(Courtesy B & S)*

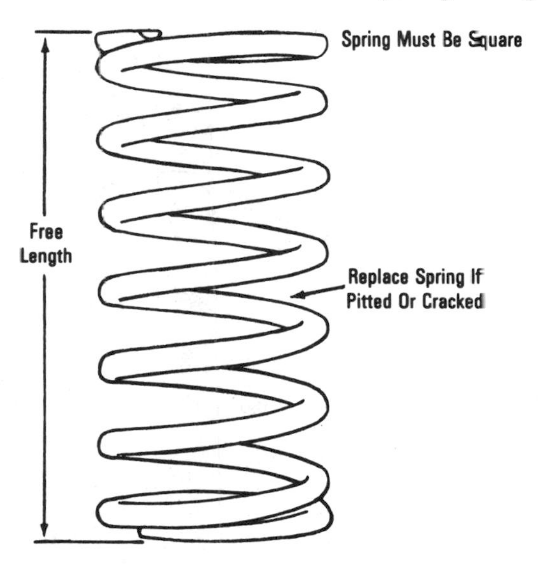

Fig. 14–12. Valve spring. *(Courtesy Tecumseh)*

If the flat end of the valve guide plug gage (Tool 19122) can be inserted into the valve guide for a distance of 5⁄16″ (7.94 mm), the valve guide is worn and should be rebushed in the following manner (Figs. 14–13 and 14–14). Place the pilot of the counterbore reamer (Tool 19064) in the valve guide (Fig. 14–13A). Install the pilot bushing (Tool 19191) over the counterbore reamer and lower the pilot bushing to rest in the valve seat. Hold the replacement valve guide bushing (Tool 63709) on top of the pilot bushing next to the reamer. Make a mark on the reamer 1⁄16″ (1.59 mm) above the top of the replacement bushing (Fig. 14–13B).

Ream out the valve guide until the mark on the counterbore reamer is level with the top of the pilot. (Fig. 14–13C). Lubricate the reamer with kerosene or equivalent lubricant.

Place the replacement bushing in the reamed-out hole (Fig. 14–14). Press the replacement bushing down until it is flush with the top of the reamed-out hole. Use the valve guide bushing driver (Tool 19065) (Fig. 14–14A).

Finish the reaming of the replacement bushing with a valve guide bushing finish reamer (Tool 19064) (Fig. 14–14B). Be sure to lubricate the reamer with kerosene or equivalent lubricant.

Table 14–2. Table of Specifications for Tecumseh Engines

Model		HH80	HH100	HH120	VH80	VH100
Displacement	Early	19.44	23.70	27.66	23.75	23.75
	Late	23.70				
Stroke		2¾″	2¾″	2⅞″	2¾″	2¾″
Bore	Early	3.0000/3.0010	3.3120/3.3130	3.500/3.501		
	Late	3.3120/3.3130				
Timing Dimension		TDC-Start 0.095 Run	TDC-Start 0.095 Run	TDC-Start 0.095 Run	Solid State	Solid State
Point Gap		0.020	0.020	0.020	Solid State	Solid State
Spark Plug Gap		0.028/0.033	0.028/0.033	0.028/0.033	0.035	0.035
Valve Clearance	Intake	0.010	0.010	0.010	0.010	0.020
	Exhaust	0.020	0.020	0.020	0.020	0.020
Valve Seat Angle		46°	46°	46°	46°	46°
Valve Seat Width		0.042/0.052	0.042/0.052	0.042/0.052	0.042/0.052	0.042/0.052
Valve Face Angle		45°	45°	45°	45°	45°
Valve Face Width		0.089/0.099	0.089/0.099	0.089/0.099	0.089/0.099	0.089/0.099
Valve Lip Width		0.06	0.06	0.06	0.06	0.06
Valve-Spring Free Length	Early	2.125	2.125	2.125	1.885	1.885
	Late	1.885	1.885	1.885		
Valve-Spring Compressed Tension Early Production Models		19–21 lb. at 1.65 Length	19–21 lb. at 1.65 Length	19–21 lb. at 1.65 Length	23 lb. ± 2.5 lb. at 1.55 Length	23 lb. ± 2.5 lb. at 1.55 Length
Valve-Spring Compressed Tension Late Production Models		23 lb.± 2.5 lb. at 1.55 Length	23 lb. ± 2.5 lb. at 1.55 Length	23 lb. ± 2.5 lb. at 1.55 Length		
Valve Guides STD Diameter		0.312/0.313	0.312/0.313	0.312/0.313	0.312/0.313	0.312/0.313
Valve Guides Oversize Dimensions		0.343/0.344	0.343/0.344	0.343/0.344	0.344/0.345	0.344/0.345

(Courtesy Tecumseh)

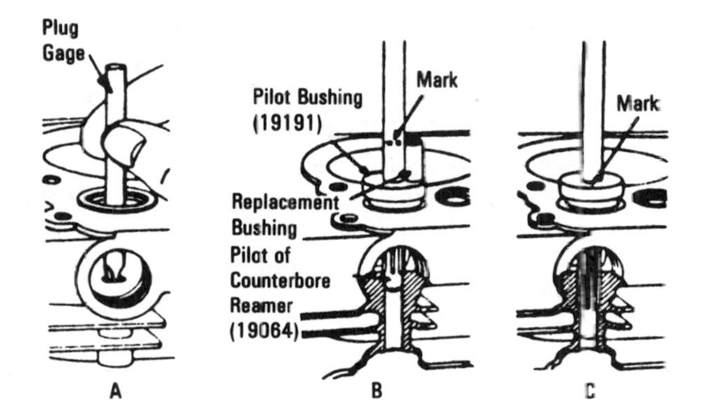

Fig. 14–13. Bushing valve guides. *(Courtesy B & S)*

It is usually not necessary to bush factory-installed brass valve guides. However, if bushing is required, *do not* remove the original bushing, but follow the standard procedure for finishing.

In models 9, 14, 19, 23, 140000, 170000, 190000, 200000, 220000, 230000, 240000, 250000, 300000, and 320000, there is a slightly different method used for checking valve guide wear and removing the valve guide bushings. If the flat end of the valve guide plug gage (Tool 19151) can be inserted into the valve guide for a distance of 5⁄16″ (7.9 mm), the guide is worn and should be rebushed in the manner shown in Fig. 14–13A.

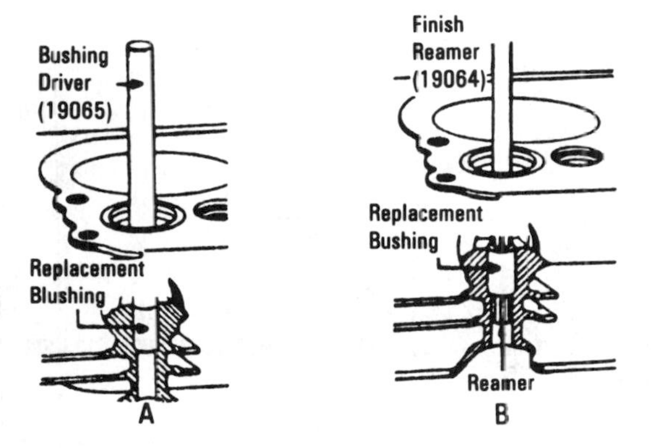

Fig. 14–14. Bushing valve guides. *(Courtesy B & S)*

Removing Valve Guide Bushings—The Briggs & Stratton kit 19232 is used to remove valve guide bushings in aluminum engines. To remove the factory- or field-installed guide bushings on aluminum engines, rotate the nut 19239 up to the head of 19238 puller screw. Center the washer on the valve seat. In some model intake seats, a larger washer may be required. Lubricate the cutting surface of the screw and inside of the guide bushings with Stanisol® or kerosene. Insert screw 19238 through washer 19240. Center the washer on the seat (Fig. 14–15).

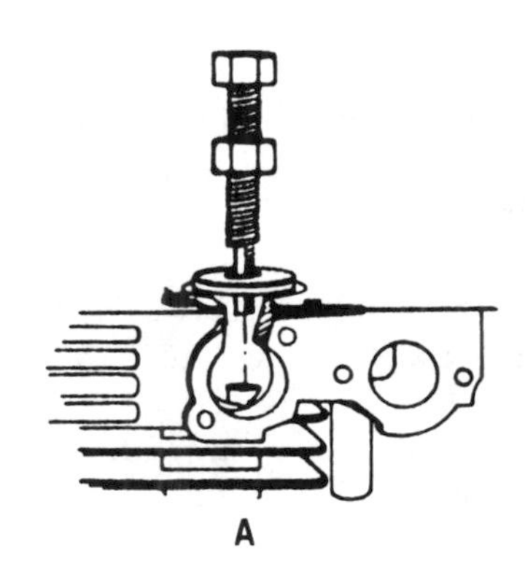

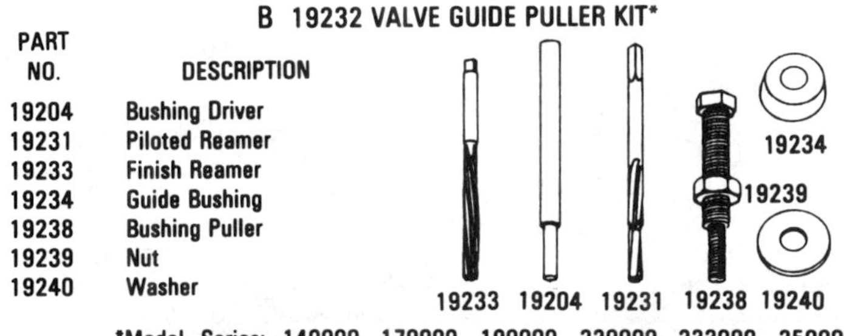

B 19232 VALVE GUIDE PULLER KIT*

PART NO.	DESCRIPTION
19204	Bushing Driver
19231	Piloted Reamer
19233	Finish Reamer
19234	Guide Bushing
19238	Bushing Puller
19239	Nut
19240	Washer

*Model Series: 140000, 170000, 190000, 220000, 233000, 250000, 300000, 326000.

Fig. 14–15. (A) Removing valve guide bushings. *(Courtesy B & S)* **(B) Valve guide puller kit** *(Courtesy B & S).*

Fig. 14–15(B) is the tool kit that is suggested by Briggs & Stratton for removing valve guide bushings. Note the numbers of the various tools used in the operations.

Next, use a 3/4" (19 mm) socket to turn the screw clockwise to a depth of 1/4" (6.5 mm), or until the bushing starts to turn, then STOP. While holding the screw stationary, turn the nut down onto the washer until the bushing is free.

Repairing Worn Guides—Aluminum or sintered iron guides can be repaired by using the tool kit already mentioned. Place the piloted counterbore reamer (Tool 19231) into the worn guide. Slide the reamer guide (Tool 19234) down the shank of 19231 and center on the valve seat (Fig. 14–13B). Place the bushing (231218) next to the reamer on the reamer guide (Fig. 14–13B). Mark the reamer 1/16" (1.6 mm) above the bushing (Fig. 14–13C).

Use Stanisol or kerosene to lubricate the reamer while it is turning clockwise. Continue reaming until the mark on the reamer is flush with the top of the reamer guide bushing (Fig. 14–13C).

Installing the Replacement Bushing—Clean out all the chips from the previous operation. Place the grooved end of the service bushing (231218) into the valve guide (Fig. 14–14A). Use bushing driver 19204 to press the bushing into the guide until it is flush with the top of the guide or until it bottoms. Place the reamer guide bushing (Tool 19234) on the valve seat and slide the finish reamer (Tool 19233) through the center of the bushing (Fig. 14–14B). Use Stanisol or kerosene as a lubricant while turning the reamer clockwise. After reaming is done, continue to turn the reamer clockwise while removing it upward. Clean out all chips before reassembling the engine.

Repairing Worn Valve Guides with Reamer and Bushing—Place the piloted counterbore reamer (19183) into the worn guide. Slide the reamer guide bushing (19192) down the shank of the reamer and center in the valve seat (Fig. 14–13B). Slide the replacement bushing (230655) next to the reamer shank on the reamer guide bushing. Mark the reamer 1/16" (1.6 mm) above the bushing as shown in Fig. 14–13C. Use Stanisol or kerosene to lubricate the reamer while turning it clockwise. Continue reaming until the mark on the reamer is flush with the top bushing. *Do not ream through the whole guide.* Continue to turn the reamer clockwise while withdrawing the reamer from the engine.

Installing Replacement Bushing—Clean out all chips. Press in the valve guide bushing (230655). Use the bushing driver (19204) until it is flush with the top of the guide or until it bottoms (Fig. 14–14A).

The bushing (230655) is finish-reamed to size at the factory. No further reaming is necessary. A standard valve can now be used.

Note: Cast-iron engines use sintered (gray) valve guide bushings. *Do not remove these bushings.*

To install brass bushings, use reaming valve guide (230655) or (231218) bushing.

Caution: Valve seating should be checked after bushing the guide, and corrected if necessary by refacing the seat.

Valve Seat Inserts

Cast-iron-cylinder engines are equipped with an exhaust-valve seat insert. This insert can be removed and a new one installed. The intake side must be counterbored to allow for installation of an intake-valve seat insert (Figs. 14–16, 14–17, and 14–18).

Aluminum-alloy-cylinder engines are equipped with inserts on exhaust and intake sides. Table 14–3 shows tools to use and valve seats for each model.

Removing the Valve Seat Insert—Use the valve seat puller. This is Tool 19138 as shown in Fig. 14–19. Select the proper puller nut as shown in Table 14–3. Be sure the puller body does not rest on the valve seat insert (Fig. 14–20). *Note:* On aluminum engines it may be necessary to grind the puller nut until the edge is 1/32″ (0.8 mm) thick in order to get the puller nut under the valve insert (Fig. 14–20).

Driving in the New Valve Seat Insert—Select the proper valve seat insert and the correct pilot and driver according to Tables 14–3 and 14–4. You will note that one side of the seat insert is chamfered at the outer edge. This side should go down into the cylinder.

Insert the pilot into the valve guide. Then drive the valve insert into place with the driver (Fig. 14–21). The seat should then be ground lightly and the valves and seats lapped lightly with grinding compound. Clean thoroughly.

Note: Aluminum-alloy-cylinder models use the old insert as a spacer between the driver and the new insert. Drive the new insert until it bottoms; the top of the insert will be slightly below the cylinder

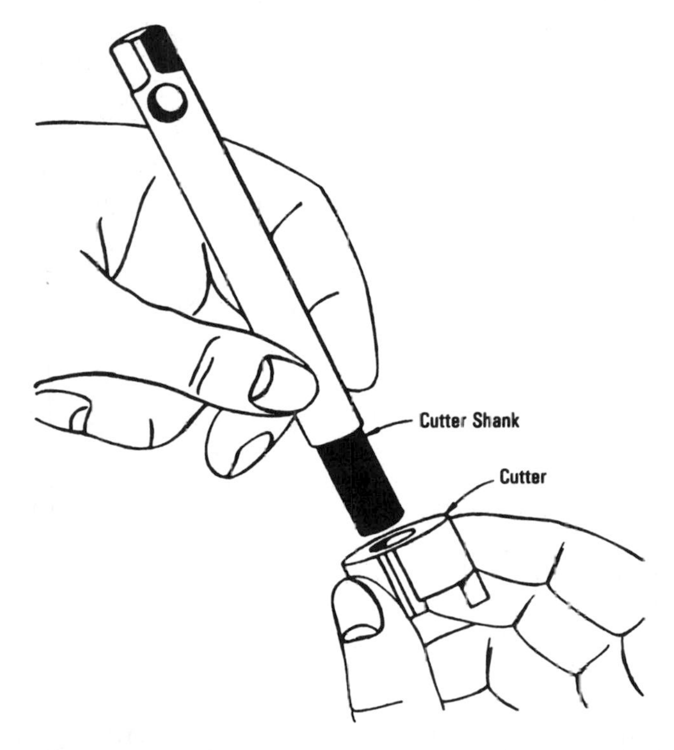

Fig. 14–16. Inserting a cutter shank into the cutter. *(Courtesy B & S)*

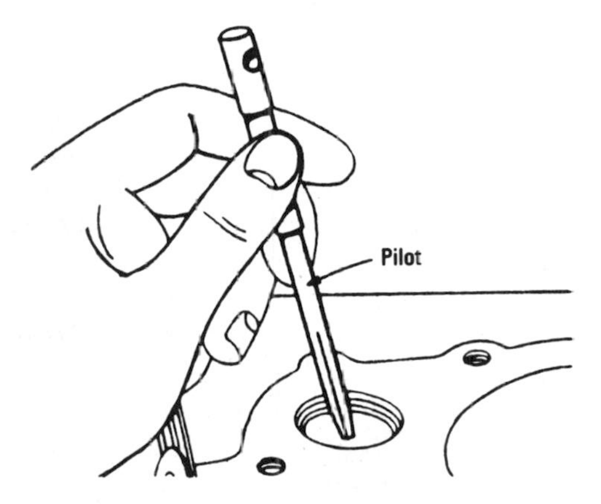

Fig. 14–17. Inserting the pilot. *(Courtesy B & S)*

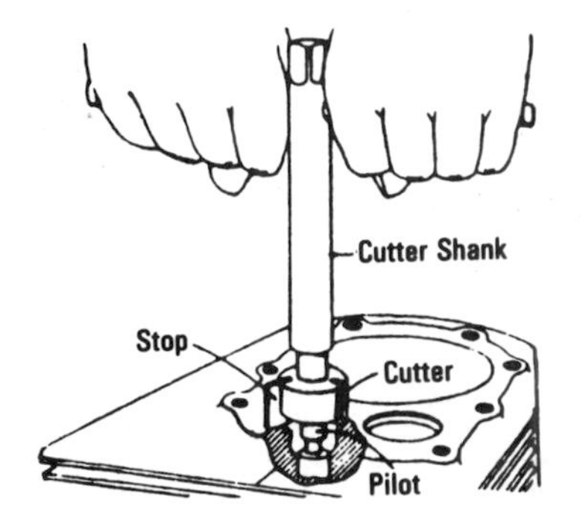

Fig. 14–18. Counterboring for a valve seat. *(Courtesy B & S)*

head gasket surface. Then peen around the insert as shown in Fig. 14–22.

On the cast-iron-cylinder models the valve seat must be counterbored to allow installation of the intake-valve seat insert. Select the proper seat insert, cutter shank, counterbore cutter, pilot, and driver according to Table 14–4.

Insert the pilot in the intake-valve guide (Fig. 14–23). Assemble the correct counterbore cutter to the cutter shank as shown in Fig. 14–24.

Counterbore the cylinder by hand until the stop on the cutter touches the top of the cylinder (Fig. 14–25). Do not force the cutter to one side or it will cut oversize. Blow out all chips. Use the knock-out pin (19135) to remove the cutter from the cutter shank.

Valve Life

Valve life is that period of time in which a valve operates without having to be repaired or replaced. Valve life can be shortened by burning. Burning can occur when pieces of combustion deposit lodge between the valve seat and valve face. This prevents the valve from closing completely. This is most common on engines that operate for long periods of time with constant speed and constant load.

However, valve life (exhaust valves in particular) can be extended by using a rotator that turns the exhaust valve a slight bit each time it is lifted. This wipes away the deposits that tend to lodge between the valve face and seat.

Table 14–3. Valve Seat Inserts

Basic Model Series	Intake Standard	Exhaust Standard	Exhause Stellite®	Insert* Puller Assembly	Puller Nut
Aluminum Cylinder					
6B, 8B	211291	211291	210452	19138	19140 Ex. 19182 In.
60000, 80000	210879†	211291	210452	19138	19140 Ex. 19182 In.
82000, 92000, 94000, 110000	210879	211291	210452	19138	19140 Ex. 19182 In.
100000, 130000, 131000	211787	211172	211436	19138	19182 Ex. 19139 In.
140000, 170000, 190000	211661	211661	210940‡	19138	19141
251000	211661	211661	210940	19138	19141
220000, 252000, 253000	261463	211661	210940		
Cast-Iron Cylinder					
5, 6, N	63838	21865		19138	19140
8	210135	21865		19138	19140
9	63007	63007		19138	19139
14, 19, 190000	21880	21880	21612	19138	19141
200000, 23, 230000	21880	21880	21612	19138	19141
240000	21880	21612	21612	19138	19141
300000, 320000		21612	21612	19138	19141

*Included puller and no. 19182, 19141, 19140, and 19139 nuts.

†211291 used before serial no. 5810060; 210808 used from serial no. 5810060 to no. 6012010.

‡Before code no. 7101260 replace cylinder.

(Courtesy B & S)

Another answer to short valve life is to use Stellite® exhaust valves. These valves resist heat and give longer life. Fig. 14–26 shows a standard valve with a Rotocap® to rotate the valve for longer life. Fig. 14–27 shows the Stellite valve and its parts number. Note the retainer and pin.

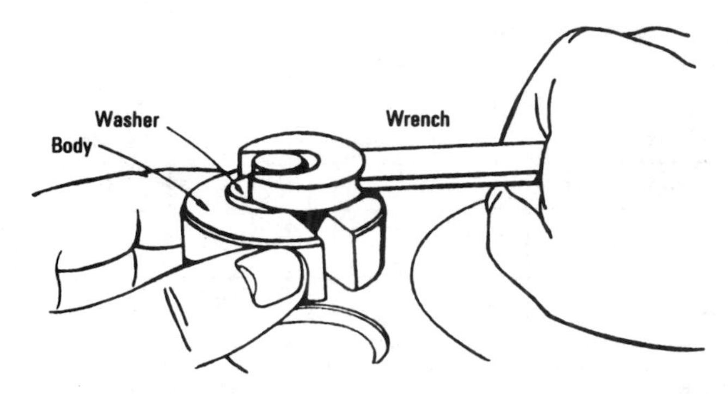

Fig. 14–19. Removing a valve seat using the washer-and-screw technique. *(Courtesy B & S)*

Fig. 14–28 shows the Stellite valve and the Rotocap, which produces even longer life for exhaust valves.

Table 14–5 lists the part numbers for basic model series and what is needed for valves, springs, Rotocaps, retainers, and pins to make the conversion to this type of valve.

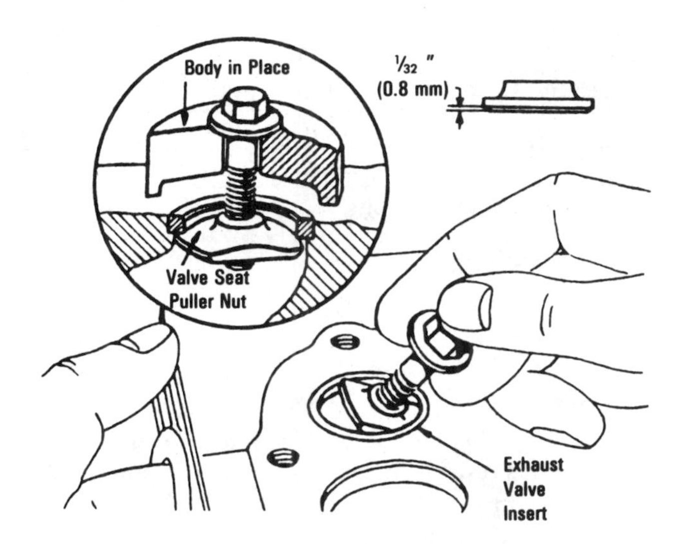

Fig. 14–20. Inserting the valve seat puller. *(Courtesy B & S)*

Table 14–4. Valve Seat Insert and Counterbore Tools

Basic Model Series	Counterbore Cutter	Shank	Cutter & Driver Pilot	Insert Driver
Aluminum Cylinder				
6B, 8B			19126	19136
60000, 80000			19126	19136
82000, 92000, 94000			19126	19136
100000, 130000			19126	19136
140000, 170000, 190000			19127	19136
Cast-Iron Cylinder				
5, 6, N	19133	19129	19126	19136
8	19132	19129	19126	19136
9	19132	19129	19127	19136
14, 19, 190000	19131	19129	19127	19136
200000, 23	19131	19129	19127	19136
230000, 240000	19131	19129	19127	19136
300000, 320000			19127	19136

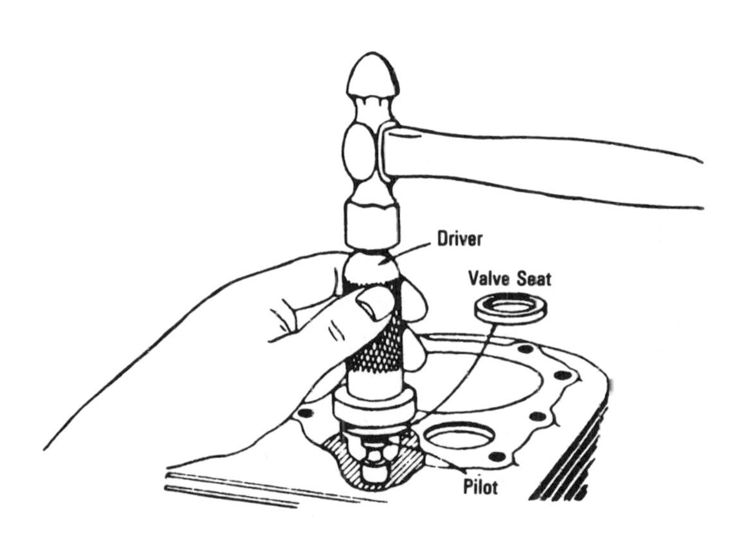

Fig. 14–21. Driving in the valve seat. *(Courtesy B & S)*

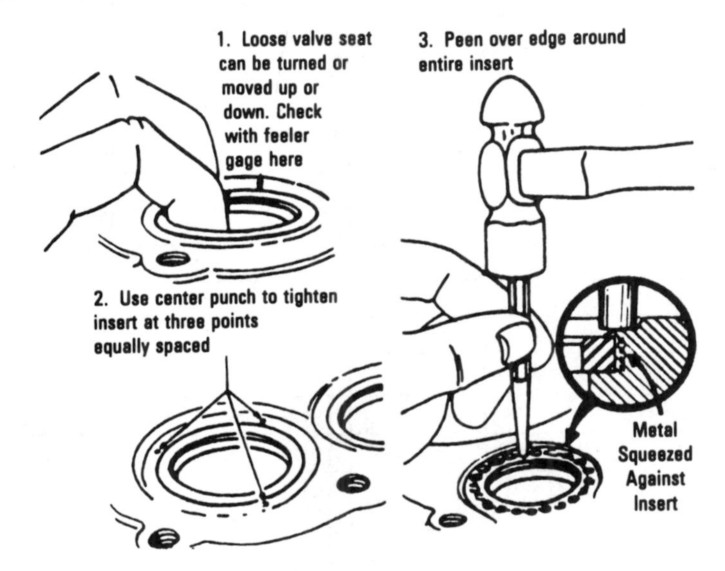

Fig. 14–22. Peening the valve seat. *(Courtesy B & S)*

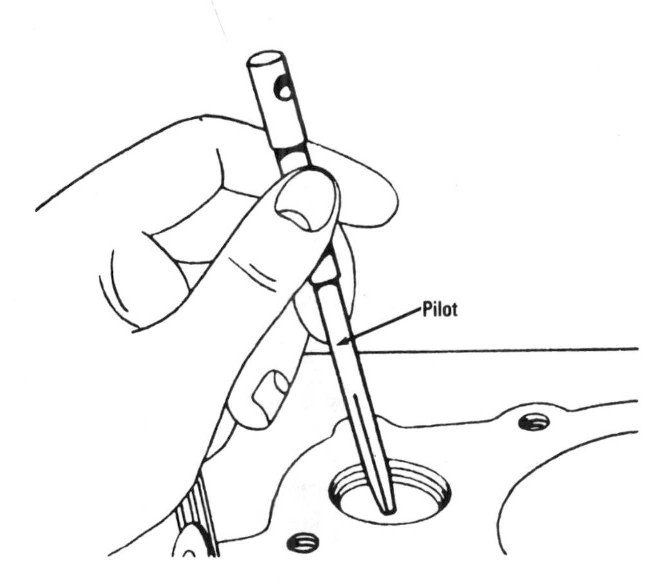

Fig. 14–23. Inserting the pilot. *(Courtesy B & S)*

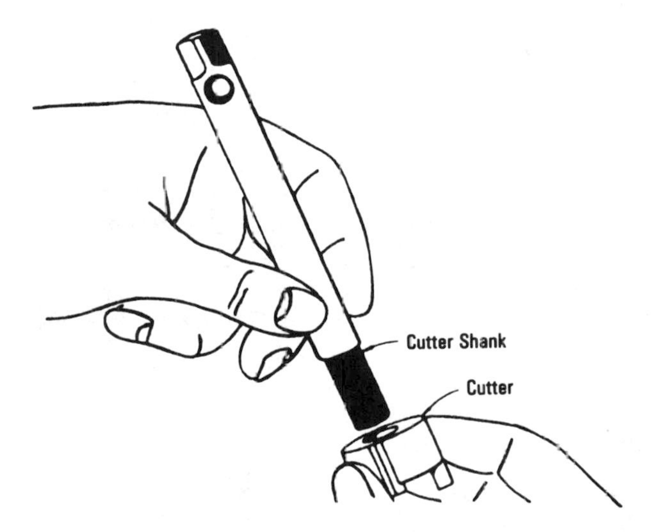

Fig. 14–24. Inserting the cutter into the shank. *(Courtesy B & S)*

Replacing Main Bearings

Ball Bearings—To install ball bearings, heat them to about 325°F (162°C). This is done by placing them in a bath of hot oil. Once the

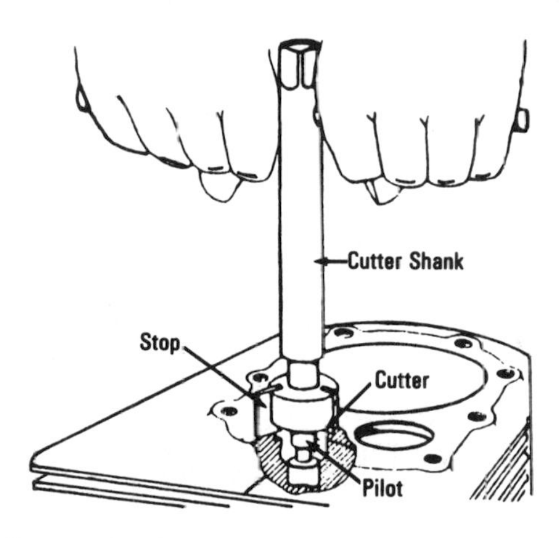

Fig. 14–25. Counterboring for a valve seat. *(Courtesy B & S)*

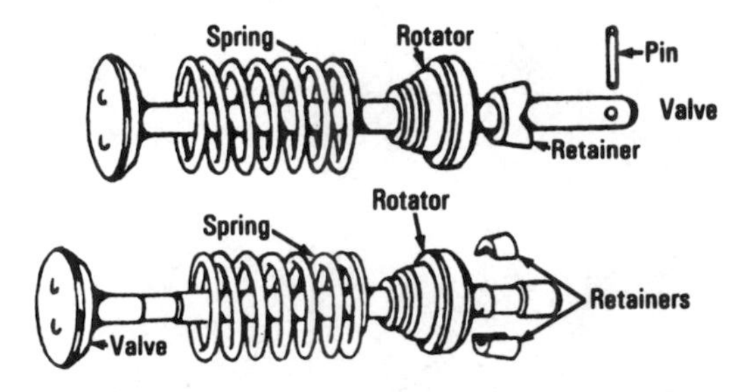

Fig. 14–26. Standard and Rotocap valves. *(Courtesy B & S)*

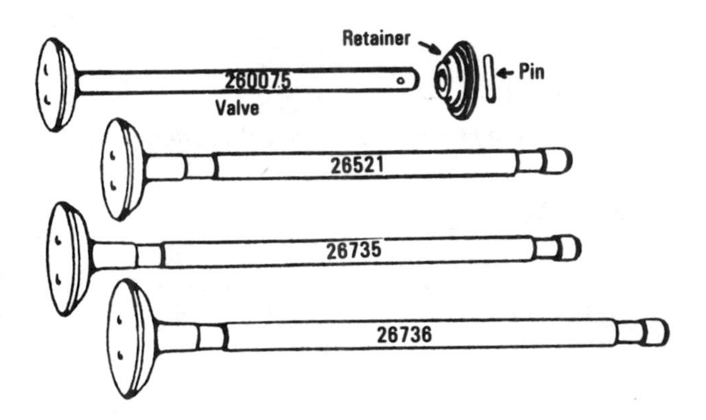

Fig. 14–27. Stellite valves. *(Courtesy B & S)*

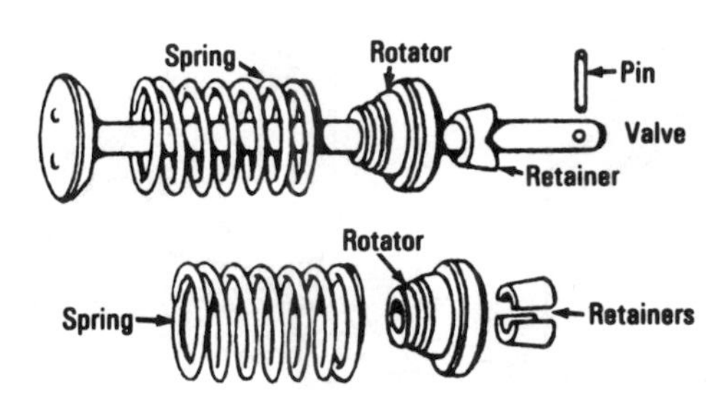

Fig. 14–28. Stellite valves and a Rotocap. *(Courtesy B & S)*

Table 14–5. **Stellite Valve and Rotocap Conversion**

Basic Model Series	Stellite Valve	Rotocap Only Conversion			
		Spring	Rotocap	Retainer	Pin
Aluminum Cylinder					
60000*, 80000*, 82000*, 92000*, 94000*	260443	26826	292259	230127	230126
100000, 130000	260860	26826	292259	230127	230126
140000, 170000, 190000, 200000, 250000	390420	26828	292260	93630	
Cast-Iron Cylinder					
14, 19, 190000, 200000	26735	26828	292260	68283	
23, 230000	261207	26828	292260	68283	
240000, 300000, 320000	261207	26828	292260	68283	(Stellite® Std.)

*To use Rotocap only 26973 standard valve must be used.

Note: Rotocap not used with LP Gas on 6, 8, and 10 cu. in. engine.

(*Courtesy B & S*)

bearings are hot, place the crankshaft in a vise with the bearing side up. When the bearing is hot, it will be a slip fit on the crankshaft journal. Grasp the bearing with the shield down and slide it on the crankshaft (Fig. 14–29). The bearing will tighten while cooling to room temperature. Do not quench. Allow the bearing to cool down normally.

Plain Bearings—Plain bearings should be checked for scoring or wear. Replace if the plug gage will enter. Try the gage at several locations in the bearing (Fig. 14–30). Also check Table 14–6 for cylinder bearing reject sizes.

The crankcase cover or bearing support should be replaced if the bearing is worn or scored. Select the correct assembly part number by using the parts list covering the engine. Then refer to Table 14–7 for the main bearing gages. Check Table 14–7 for models where it is necessary to replace the support and the cover if the bearing shows scoring or wear. Use plug gage 19117.

Replacing Magneto Bearings

In aluminum-cylinder engines (except 171700 and 191700), there is no removable bearing. The cylinder must be reamed out so a replacement bushing can be installed. Place the pilot guide bushing in the sump bearing, with the flange of the pilot guide bushing toward the inside of the sump.

Assemble the sump on the cylinder. Be careful that the pilot guide bushing does not fall out of place. Place the reamer guide bushing into the oil seal recess in the cylinder. The reamer guide bushing, along

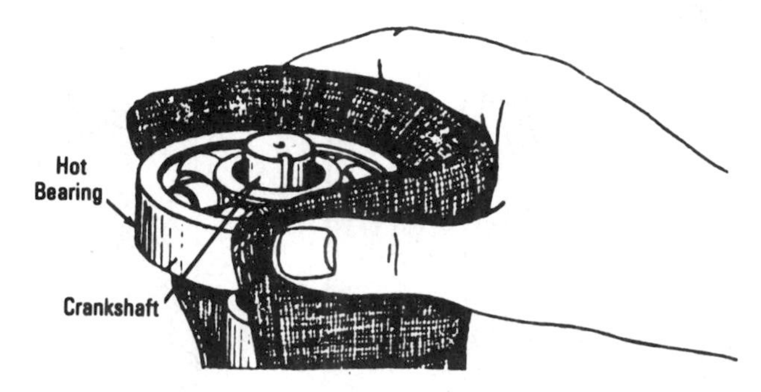

Fig. 14–29. Installing hot bearings over the crankshaft. Handle with a cloth to avoid burns. *(Courtesy B & S)*

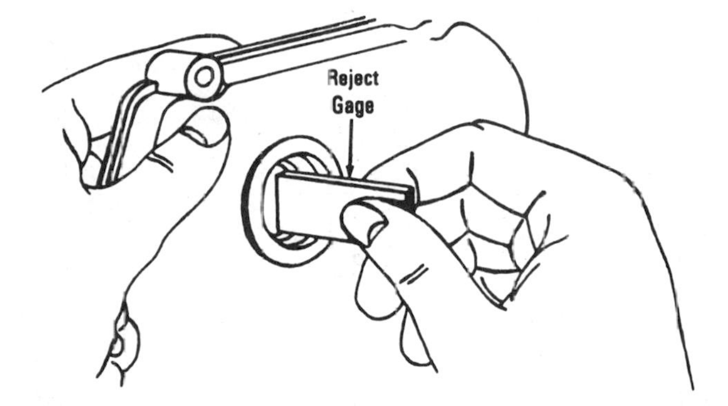

Fig. 14–30. Checking a bearing with a reject gage. *(Courtesy B & S)*

with the pilot guide bushing, will center the counterbore reamer with the opposite bearing even though the old bearing might be badly worn.

Place the counterbore reamer on the pilot and insert it into the cylinder until the tip of the pilot enters the pilot guide bushing in the sump (Fig. 14–31).

Now rotate the reamer clockwise. Keep an even, steady pressure on the reamer until it is completely through the bearing. The reamer can be lubricated with Stoddard® Solvent or with kerosene.

Counterbore reaming may be undertaken without any lubricant. However, as aluminum material builds up on the reamer flutes, there will be eventual damage to the reamer and oversize counterbores will be the result. So keep the flutes clean. The best way is to use lubrication or stop and clean them once they become clogged.

Remove the sump and pull the reamer out. Do this without backing it through the bearing. Clean out the reaming chips. Remove the reamer guide bushing from the oil seal recess.

The insert in Fig. 14–32 shows how the new bushing is held with the notch toward the cylinder and in line with the notch on the inside of the cylinder and against the reamed-out bearing. Note the position of the split in the bushing. At a point opposite to the split in the bushing, use a chisel or screwdriver and hammer to make a notch in the reamed-out cylinder bearing at a 45° angle.

Now press in the new bushing. Be careful to align the oil notches

Table 14–6. Cylinder Bearing Reject Size Chart

Basic Engine Model or Series	PTO Bearing		Bearing Magneto	
Aluminum Cylinder	**Inches**	**Millimeters**	**Inches**	**Millimeters**
6B, 8B*	0.878	22.30	0.878	22.30
60000, 80000*	0.878	22.30	0.878	22.30
82000, 92000*, 94000*	0.878	22.30	0.878	22.30
110900*, 111900*	0.878	22.30	0.878	22.30
100000, 130000	1.003	25.48	0.878	22.30
140000, 170000	1.185	30.10	1.004[†]	25.50[†]
190000	1.185	30.10	1.004[†]	25.50[†]
220000 Horiz.	Ball	Ball	Ball	Ball
220000 Vert., 250000	1.383	35.13	1.383	35.13
Cast-Iron Cylinder	**Inches**	**Millimeters**	**Inches**	**Millimeters**
5, 6, 8, N	0.878	22.30	0.878	22.30
9	0.988	25.09	0.988	25.09
14	1.185	30.10	1.185	30.10
19, 190000, 20000	1.185	30.10	1.185	30.10
23, 230000[‡]	1.382	35.10	1.382	35.10
240000, 300000	Ball	Ball	Ball	Ball
320000	Ball	Ball	Ball	Ball

*Auxiliary-drive models PTO bearing reject size 1.003″ (25/48 mm).

[†] Gear-reduction PTO—1.185″ (30.10 mm).

[‡] Synchro-balanced magneto bearing reject size 1.185″ (30.10 mm).

(Courtesy B &S)

with the driver and support until the outer end of the bushing is flush with the end of the reamed-out cylinder hub (Fig. 14–33). If the oil notches do not line up, the bushing can be pressed through into the recess in the cylinder support and then reinstalled.

With a blunt chisel or screwdriver, drive a portion of the bushing into the notch previously made in the cylinder (Fig. 14–32). This is called *staking* and is done to prevent the bushing from turning.

Basic Engine Model Series	Cylinder Support	Pilot	Counter-bore Reamer	Reamer Guide Bushing Mag.	Reamer Guide Bushing PTO	Bushing Driver	Pilot Guide Bushing Mag.	Pilot Guide Bushing PTO	Finish Reamer	Plug Gage
5, N, 6, 8	19123	19096	—	—	—	19124	19094	19095	19166	
6B, 60000, 8B, 80000, 82000, 92000 94000, 110900, 111200, 111900	19123	19096	19099	19101	19100	19124	19094	19094	19095	19166
8BHA*, 80590* 81590*, 92590* 80790*, 81709* 92990*, 110990* 111990*	19123	19096	19099	19101	†	19124	19094	†	19095	19166
100000 130000	19123	19096	19099 Mag 19172 PTO	19101	19186V 19170H	19124	19094	19168	19095 Mag 19173 PTO	19166 19178
140000 170000, 190000	19123	19096	19172 Mag 19174 PTO	19170	19171	19179	19168	19169	19173 Mag 19175 PTO	19178
171700 191700	19123	19096	19174 Mag 19174 PTO	19201	19171	19179	19169	19169	19175 PTO 19175 Mag	19178
9, 14, 19, 20, 23, 191400, 193400, 200400, 230000	Replace support and cover									19117

*Replace sump if PTO bearing is worn.

†Use sump or cover with 7/8″ (22.3 mm) diameter bearing and 19094 guide.

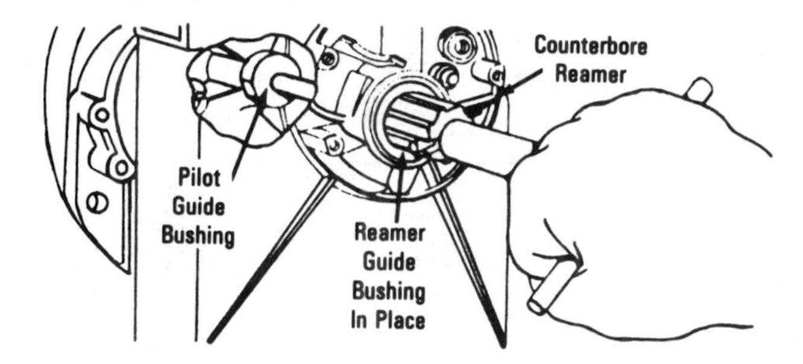

Fig. 14–31. Counterbore reaming. *(Courtesy B & S)*

Reassemble the sump to the cylinder with the pilot guide bushing in the sump bearing. Place the finish reamer on the pilot and insert the pilot into the cylinder bearing until the tip of the pilot enters the pilot guide bushings in the sump bearing (Fig. 14–34).

Use some kerosene, Stoddard Solvent, or fuel oil to lubricate the reamer. Then ream the bushing by turning the reamer clockwise with a steady, even pressure. Do this until the reamer is completely through the bearing. Improper lubricants will produce a rough bearing surface.

Remove the sump and pull the reamer out without backing it

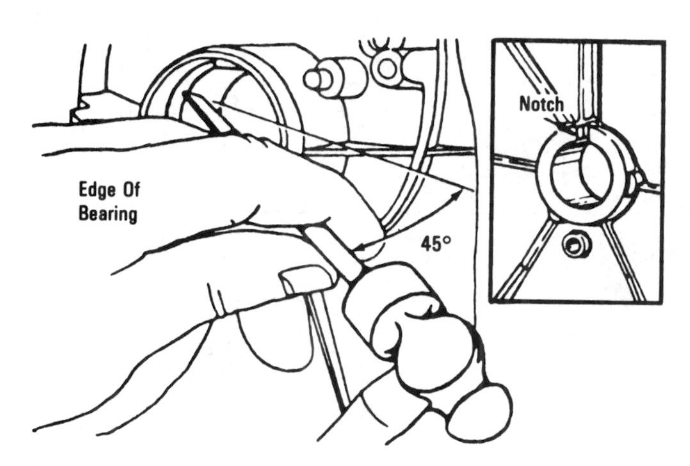

Fig. 14–32. Notching the cylinder hub. *(Courtesy B & S)*

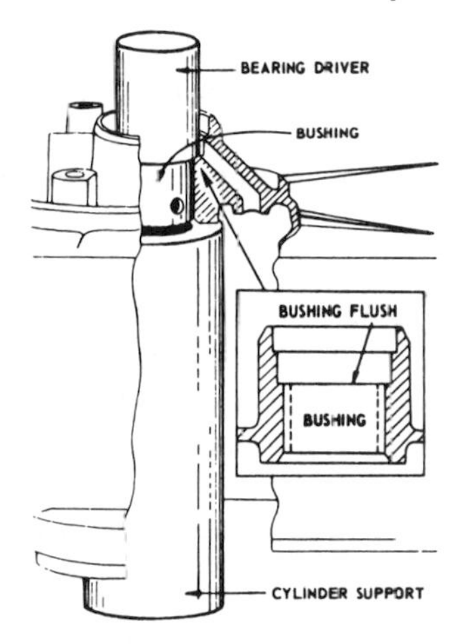

Fig. 14–33. Pressing in a new bushing. *(Courtesy B & S)*

through the bearing. Remove the pilot guide bushing. Remove all the reaming chips.

On most cylinders, the breaker point plunger hole enters the reamed-out main bearing and a burr is formed by the counterbore reaming operation. Burrs can be removed using the finish reamer (19058). Clean out the dirt and reaming chips.

Replacing the Bearing in Models 171700 and 191700—The last procedure excepted these two models. Therefore, a new type of procedure is needed for them.

Counterbore worn bearings by using the proper tools listed in

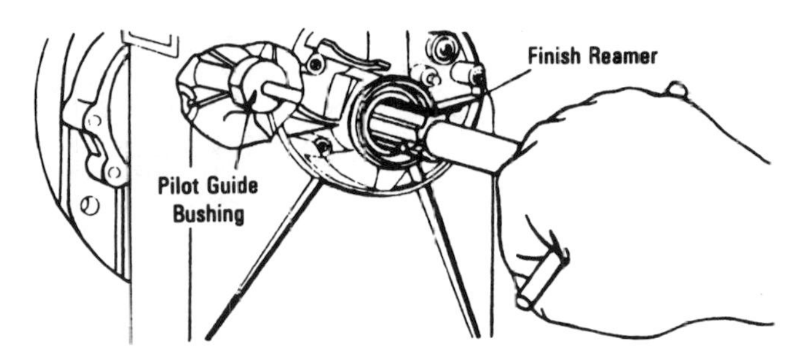

Fig. 14–34. Shell reaming. *(Courtesy B & S)*

Table 14–7. Use the procedure for a standard magneto bushing. Then place the new bushing against the reamed-out bearing on the inside of the cylinder. The bushing notch is against the cylinder and in line with the oil hole. Fig. 14–35 shows the location of the oil hole in the bushing. Note the position of the split in the bearing. At a point opposite the split in the bushing, use a chisel or screwdriver and hammer to make a notch in the reamed-out bearing at a 45° angle. Press in the new bushing from the outside cylinder (Fig. 14–36). Use care to keep the notch in line with the oil hole as seen in Fig. 14–35. This staking is done to once again prevent the bushing from turning.

Finish-ream the bushing using the same procedure as for standard magneto bushings. Clean out the breaker point plunger hole with the finish reamer (19058). Clean out the dirt and chips.

Replacing the Power Take-Off (PTO) Bearing— On aluminum-cylinder engines, the sump or crankcase cover bearing can be repaired in the same manner as the magneto bearing. However, one bearing should be completely repaired before starting the other bearing. After the bearings are finished, press in the new oil seals. This can be reviewed in Chapter 13. Fig. 13–16 shows the procedure.

Note: For models 8B-HA, 80590, 81590, 82590, 80790, 81790, 82990, 92590, 92990, 11099, and 111990, the magneto bearing can be

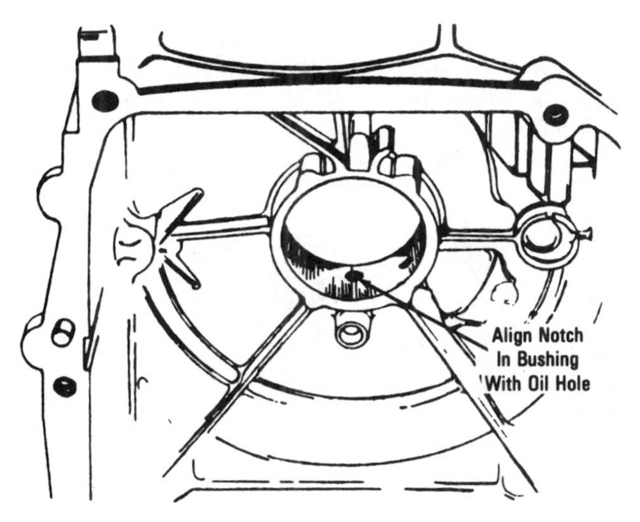

Fig. 14–35. Location of oil hole. *(Courtesy B & S)*

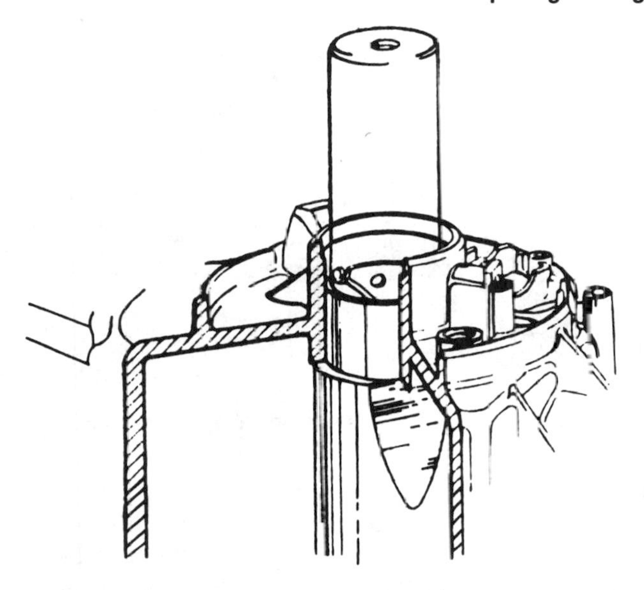

Fig. 14–36. Pressing in a new bushing. *(Courtesy B & S)*

replaced as already described. If the sump bearing is worn, the sump must be replaced. No tools are available for replacing the sump bearing.

Cam Gear Bearings—Check the cam gear bushing with plug gage 19164 (Fig. 14–37). If ¼″ (6.35 mm) or more of the gage enters the bearing bore, the bearing is worn beyond reject and the cylinder, sump, or crankcase must be replaced.

Note: On model series 111200 and 111900, the plug gage (19164) is used on the sump or crankcase cover cam gear bearing bore. Reject size of the cylinder cam bearing is 0.443″ (11.25 mm) or larger.

Installing the Breaker Point Plunger Bushing and Plunger in the Cylinder (Internal Breaker)

Flywheel-type internal breakers have the breaker point on the condenser (Fig. 14–38). The condenser is removed by loosening the screw holding the clamp. If the breaker point plunger hole becomes

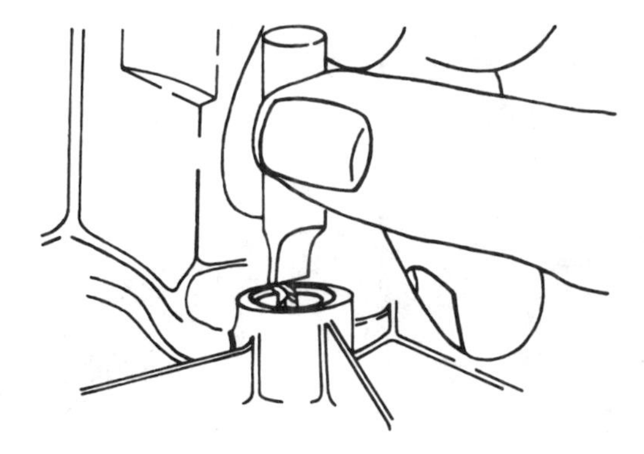

Fig. 14–37. Checking the cam gear bearing. *(Courtesy B & S)*

worn excessively, oil will leak past the plunger and may get on the points. This can cause burning. To check, loosen the breaker point mounting screw and move the breaker points out of the way. Remove the plunger. If the flat end of the plug gage (19055) will enter the plunger hole for a distance of ¼″ (6.35 mm) or more, the hole should be rebushed (Fig. 14–39).

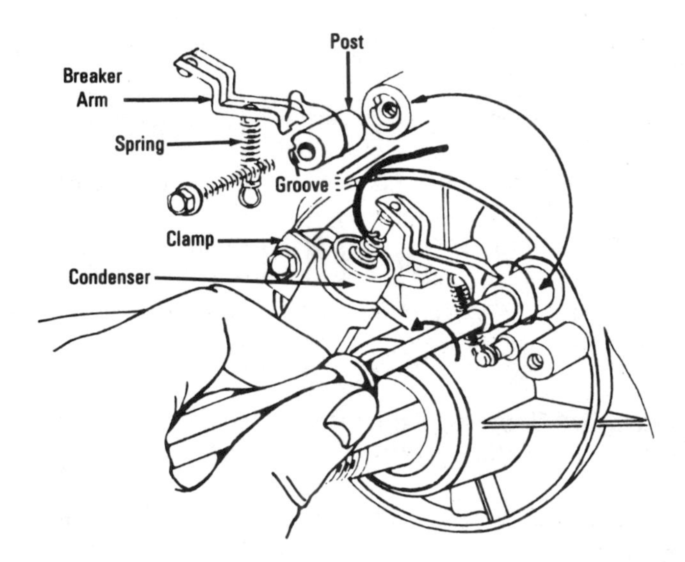

Fig. 14–38. Removing the breaker point assembly *(Courtesy B & S)*

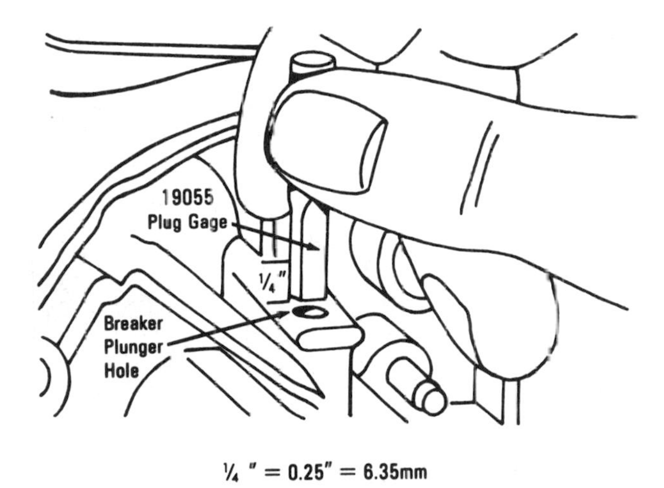

¼ " = 0.25" = 6.35mm

Fig. 14–39. Checking the breaker plunger hole with plug gage. *(Courtesy B & S).*

Breaker Point Plunger Bushing—Before you can install the breaker point bushing, you have to remove the breaker points, the armature, the crankshaft, and the starter. The old plunger hole is reamed out with a 19056 reamer (Fig. 14–40A). This should be done by hand. The reamer should be in alignment with the plunger hole. Drive the bushing (23513) with the driver (19057) until the upper end of the bushing is flush with the top of the boss (Fig. 14–40B).

Finish-ream the bushing with reamer 19058. All reaming chips or dirt must be removed (Fig. 14–40C).

Breaker Point Plunger—If the breaker point plunger is worn to a length of 0.870" (22.1 mm) or less, it should be replaced. Plungers must be inserted with the groove at the top when installed or oil will enter the breaker box (Fig. 14–41).

Replacing the Breaker Plunger and Bushing (External Breaker)

Model series 19D, 23D, 193000, 200000, 230000, 243000, 300000, and 3200000 use the external breaker. Fig. 14–42 shows the location of the ignition system in model series 19D and 23D.

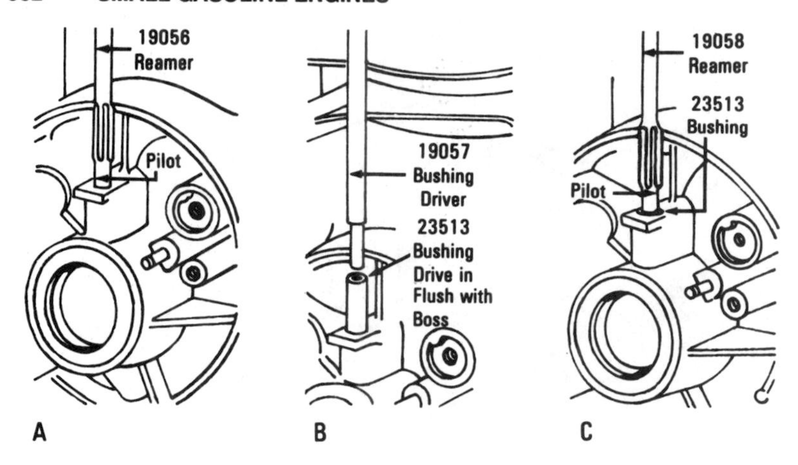

Fig. 14–40. Installing a breaker plunger bushing. *(Courtesy B & S)*

First the plunger and threaded bushing have to be removed. Fig. 14–43 shows the way the threaded bushing is removed. Fig. 14–43D shows the threaded bushing.

Place a thick washer of ⅜″ (9.525 mm) inside diameter over the end of the bushing and screw on a ⅜-24 nut [⅜-24 nut means it is ⅜″ (9.255 mm) in diameter and has 24 threads to the inch (Fig. 14–43A and B)]. Tighten the nut to pull the bushing. After the bushing has been moved about ⅛″ (3.175 mm), remove the nut and put on a second thick washer (Fig. 14–43C). A total stack of ⅜″ (9.525 mm) washers will be required to remove the bushing completely. Be sure the plunger does not fall out of the bushing as it is removed.

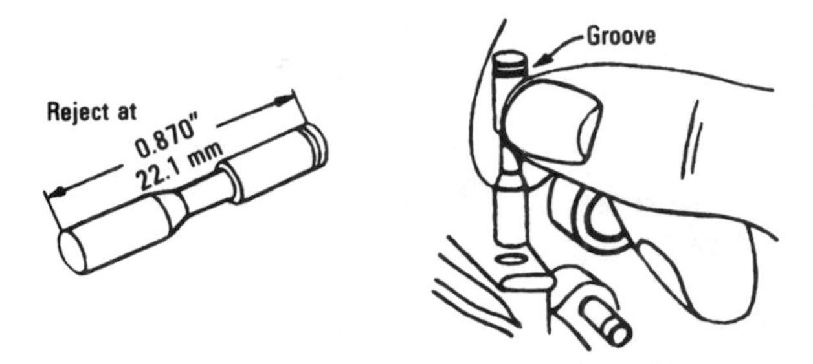

Fig. 14–41. Inserting a plunger and its reject limits. *(Courtesy B & S)*

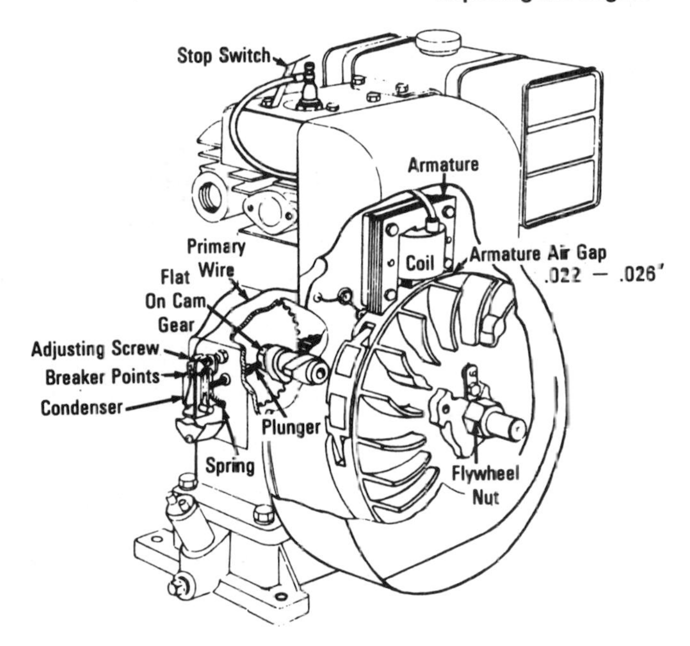

Fig. 14–42. Ignition system location on the model series 19D and 23D. *(Courtesy B & S)*

Now that the bushing and plunger have been removed, you can replace them with a new plunger and bushing. The large end of the plunger goes opposite the threads on the bushing. Screw the ⅜-24 nut onto the threads to protect them as shown in Fig. 14–44.

Insert the bushing into the cylinder. Place a piece of tubing such as 295840 (piston pin) against the nut, as shown in Fig. 14–45. Use a hammer to drive the bushing into the cylinder until the square shoulder on the bushing is flush with the face of the cylinder. Check to be sure the plunger operates freely.

Removing Bushings in Other Model Series of Engines—Pull the plunger out as far as possible. Use a pair of pliers to break the plunger off as close to the bushing as possible (Fig. 14–46A). Use a ¼-20 tap [¼″ (6.35 mm) diameter and 20 threads per inch] or a self-threading screw to thread the hole in the bushing to a depth of approximately ½″ to ⅝″ deep [12.7 to 15.875 mm (Fig. 14–46B)]. Use a ¼-20 × ½″ hexagonal head screw and two spacer washers as shown in Fig. 14–46C to pull

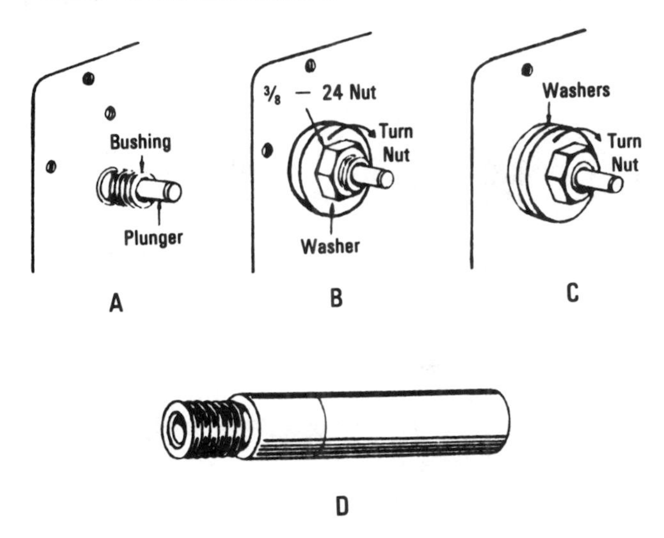

Fig. 14–43. Removing the plunger and threaded bushing. ***D*** **is the bushing with a threaded end.** *(Courtesy B & S)*

the bushing out of the cylinder. The bushing will be free when it has been extracted 5/16″ (7.938 mm). *Carefully* remove the bushing and the remainder of the broken plunger. Do not allow the plunger or chips to drop into the crankcase. Once the bushing has been removed—by snapping off the plunger and making threads in the bushing for the screw to remove it—you can install a new bushing. This old bushing may be used in driving in the new bushing. If it has been mutilated in the removal stage, use a piece of tubing of the same diameter as the bushing to be installed.

Installing the Bushing Using the Flush-Mounted Design—First locate the new bushing. Place the plunger into the bushing as shown in Fig. 14–47. Insert the plunger and bushing combination into the cylin-

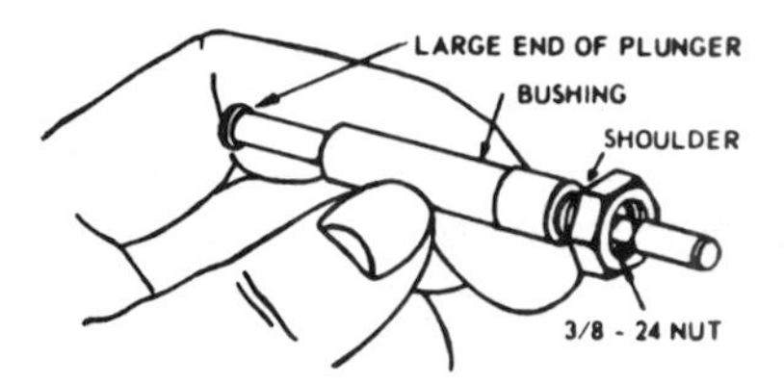

Fig. 14–44. Plunger and bushing with nut attached and plunger located in the bushing. *(Courtesy B & S)*

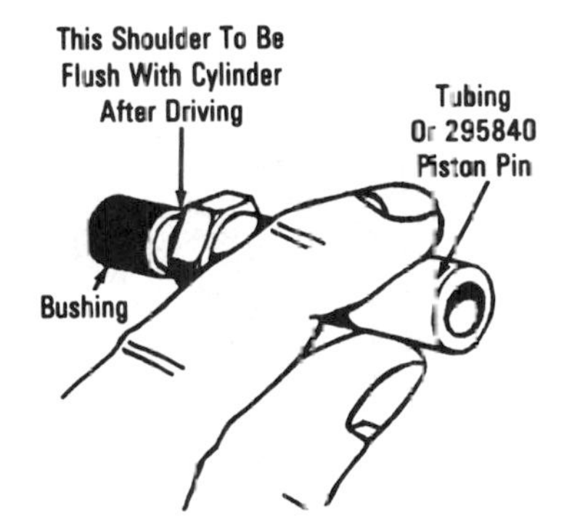

Fig. 14–45. Inserting the bushing means driving it in with a piece of tubing used to absorb the hammer blows and keep the bushing formed. *(Courtesy B & S)*

der. Use a hammer and drive the bushing into the crankcase by pounding on the old bushing or tubing end. The new bushing should be flush with the face of the cylinder. Check to make sure the plunger operates freely without restrictions.

Replacing the Armature and Governor Blade

Replacing the armature is very important inasmuch as the gap is critical. The governor blade must also be installed at the same time (Fig. 14–48). The mounting holes in the armature laminations are slot-

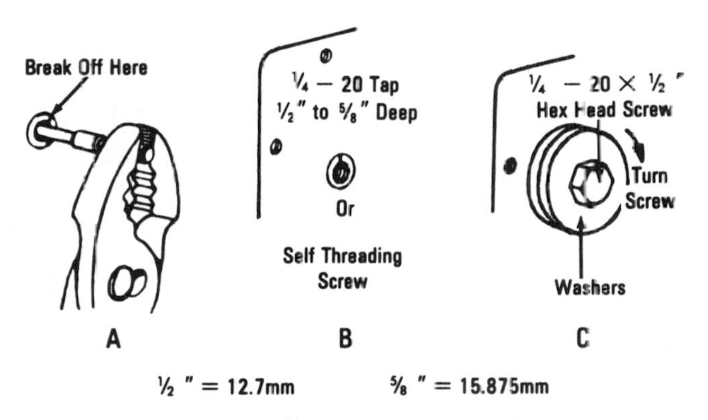

Fig. 14–46. Removing the bushing and plunger. (A) Break off the plunger. Tap the bushing with a ¼-20 or a self-tapping screw (B) and (C) put washers under the screw head and tighten. This will pull the bushing out of the crankcase. *(Courtesy B & S)*

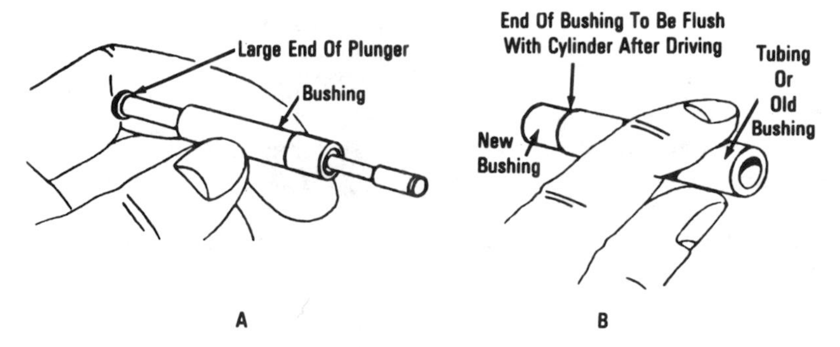

Fig. 14–47. (A) Insert the new plunger into the bushing. *(Courtesy B & S)* **(B) Insert the plunger and the bushing into the cylinder.** *(Courtesy B & S)*

ted. Push the armature up as far as possible and tighten one mounting screw to hold the armature in place.

There are three armature styles that must be considered (Fig. 14–49). Set the air gap between the flywheel and the armature as shown in Table 14–8. With the armature up as far as possible and one screw tightened, slip the proper gage between the armature and flywheel (Fig. 14–50). Turn the flywheel until the magnets are directly below the armature. Loosen the one mounting screw and the magnets should pull the armature down firmly against the thickness gage. Then tighten the mounting screws. Make sure the air vane is free to move once the screws have been rotated by hand. The armature should not touch the flywheel.

Fig. 14–51 shows the timing marks for aligning the flywheel to make sure the points open at the right time. Note the location of the armature mounting bracket. Fig. 14–52 shows the armature air gap and mounting bracket. This is the arrangement for model series 193000, 230000, 243000, 300000, and 320000.

On the Magna-Matic ignition models (9, 14, 23, etc.) the rotor and armature are correctly timed at the factory and require timing only if the armature has been removed from the engine or if the cam gear or crankshaft has been replaced.

Check Fig. 14–53. If necessary to adjust these models or if you have removed the armature from the engine, proceed as follows to realign. With the point gap set at 0.020″ (0.508 mm) turn the crankshaft

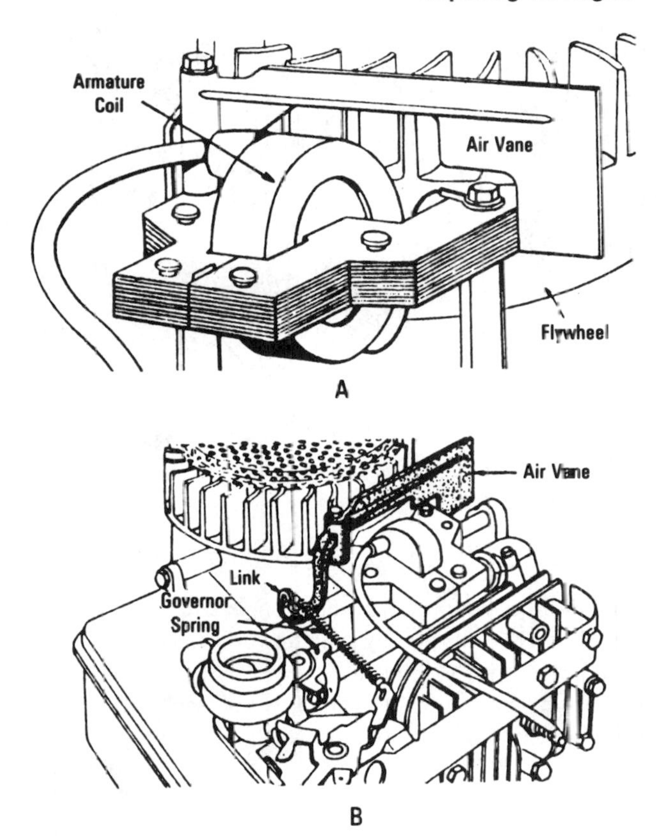

Fig. 14–48. (A) Install the armature and governor blade. *(Courtesy B & S)* **(B) Location of the air vane governor in reference to the armature.** *(Courtesy B & S)*

in the normal direction of rotation until the breaker points close and just start to open. Use a timing light or insert a piece of tissue paper between the breaker points to determine when the points begin to open. With the three armature mounting screws slightly loose, rotate the armature until the arrow on the armature lines up with the arrow on the rotor as shown in Fig. 14–53. Align with the corresponding number of the engine model. On model 9, align with 9, etc. Retighten the armature mounting screws.

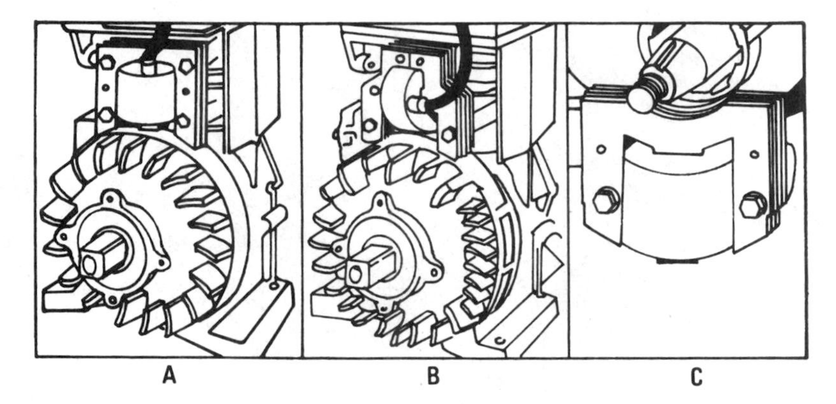

Fig. 14–49. Variations in armature styles. *(Courtesy B & S)*

Replacing the Automatic Spark Advance

On Magna-Matic models, an automatic spark advance is used. Check the automatic advance by placing the cam gear in the normal operating position with the movable weight down (Fig. 14–54). Release. The spring should lift the weight. If not, the spring is stretched or the weight is binding. Repair or replace.

Repairing or Replacing the Throttle Shaft Bushing

Before replacing the carburetor in the engine, it is a good idea to check the throttle shaft bushing for excess play. If the play is excessive, or the throttle moves too much within the bushing, replace it.

Repairing the Carburetor

Carburetor repairs have been discussed in great detail in Chapter 7. The adjustments for linkages are shown in detail in the appendix. Replace the carburetor at this point and make the necessary adjustments for proper linkage. Operation of the engine and carburetor adjustments for richness or leanness of mixture can be made later.

Table 14–8. Specifications for All Briggs & Stratton Popular Engine Models

Spark plug gap: 0.030″ (0.75 mm)

Condenser capacity: 0.18 to 0.24 μF

Contact point gap: 0.020″ (0.50 mm)

Basic Model Series	Armature				Flywheel Puller Part No.	Flywheel Nut Torque		
	Two-Leg Air Gap		Three-Leg Air Gap					
	Inches	Millimeters	Inches	Millimeters		Foot-Pounds*	Kilogram Meters*	Newton Meters*
Aluminum Cylinder								
6B, 60000, 8B	0.006 0.010	0.15 0.25	0.012 0.016	0.30 0.41	19069	55	7.6	74.6
80000, 82000, 92000, 110000	0.006 0.010	0.15 0.25	0.012 0.016	0.30 0.41	19069	55	7.6	74.6
100000, 130000	0.010 0.014	0.25 0.36	0.012 0.016	0.30 0.41	None	60	8.3	81.4
140000, 170000, 190000, 220000, 250000	0.010 0.014	0.25 0.36	0.016 0.019	0.41 0.48	19165 or 19203†	65	9.0	88.1
Cast-Iron Cylinder								
5, 6, N, 8			0.012 0.016	0.30 0.41	None	55	7.6	74.6
9					119068 or 19203	60	8.3	81.4
14					19068 or 19203	65	9.0	88.1
19, 190000, 200000	0.010 0.014	0.25 0.36	0.022 0.026	0.56 0.66	19068 or 19203	115	15.9	155.9
23, 230000	0.010 0.014	0.25 0.36	0.022 0.026	0.56 0.66	19068 or 19203	115	15.9	155.9
240000, 300000, 320000	0.010 0.014	0.25 0.36			19068 or 19203	145	20.0	196.6

*For rewind starter engines use 19161 clutch wrench.

†Use on mode l250000 built after 1975.

(Courtesy B & S)

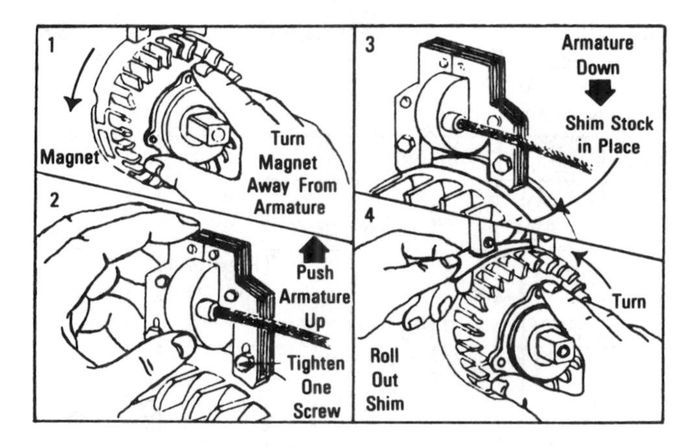

Fig. 14–50. Adjusting the air gap for an armature. *(Courtesy B & S)*

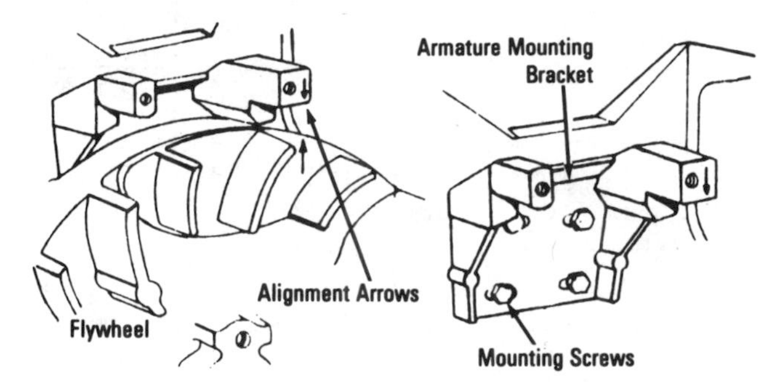

Fig. 14–51. Timing marks. Note armature mounting bracket. *(Courtesy B & S)*

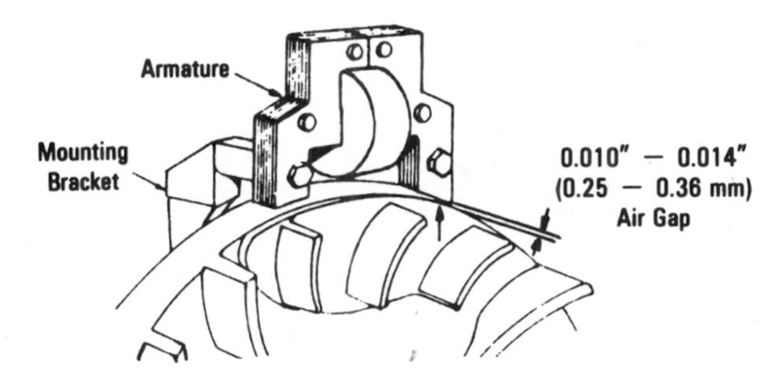

Fig. 14–52. The armature air gap. *(Courtesy B & S)*

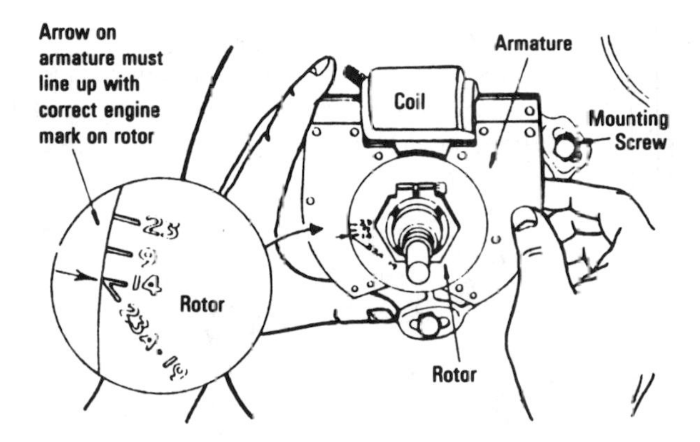

Fig. 14–53. Adjusting the rotor timing on a model series 9 engine. *(Courtesy B & S)*

Replacing the Repaired, Reroped, or Respringed Starter

If the starter has been in need of repair or replacement, it should be done at this time. Refer to Chapter 6 for details on repair and replacement of parts for starters. Also check Chapter 6 for the proper procedures of replacing the starter. In most instances it involves the placement of two or four screws into the air shroud or cowling of the engine.

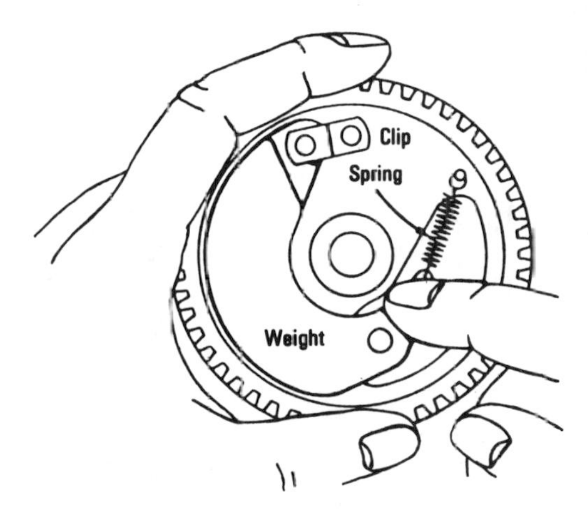

Fig. 14–54. Checking the automatic spark advance. *(Courtesy B & S)*

Crankshaft Ball Bearings

Bearings for the crankshaft have already been removed, inspected, and replaced. In some cases the crankshaft has been bent by the lawnmower blade hitting some stationary object. There are a number

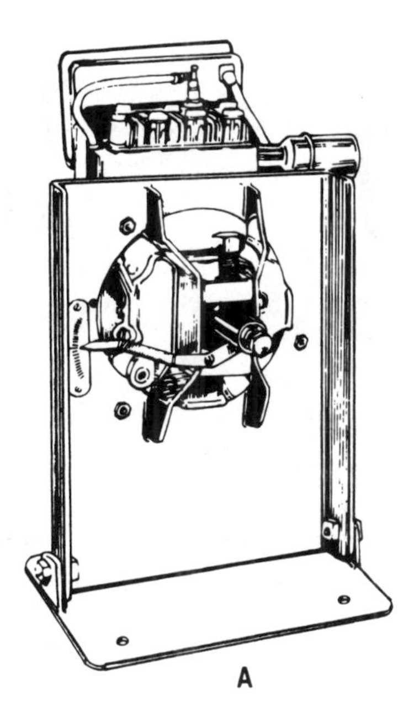

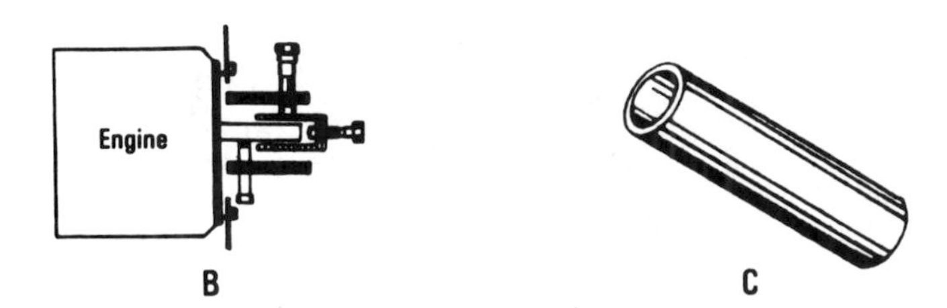

Fig. 14–55. (A) Crankshaft straightener. (B) Side view of the mount. (C) Adapter sleeves available for all sizes of power take-off shafts. *(Courtesy Frederick Manufacturing)*

of machines for straightening the crankshaft without removing the engine from the lawnmower. Fig. 14–55 is such a device.

The engine is bolted to the engine mount. The adapter sleeve (Fig. 14–55C) is slipped over the crankshaft power take-off end. The sleeve is a shaft extender. It equalizes the stress and prevents the hardened screw from marking the shaft. The end screw in the sleeve is for easy removal.

Tighten the pressure equalizer so that is is snug. Simply rotate the shaft and take a reading on the meter until it reads zero. The machine is designed so you don't bend a flange, crack the housing, or ruin a bearing.

Chapter Highlights

This chapter has described:

1. How to clean parts of the disassembled engine.
2. How to resize cylinders.
3. How to hone a cylinder bore.
4. How to clean after honing.
5. How to repair valves, guides, and seats.
6. How to remove guide bushings.
7. How to replace bushings.
8. How to remove valve seat inserts.
9. How to replace valve seat inserts.
10. How to grind valves.
11. How to replace bearings.
12. How to replace magneto bearings.
13. How to do staking.
14. How to repair cam gear bearings.
15. How to replace the breaker point plunger bushing.
16. How to repair or replace the throttle shaft bushing.

Review Questions

1. What needs to be done to all parts of an engine before it is reassembled?

Vocabulary Checklist

armature lamination
chrome rings
counterbore
cylinder walls
drill spindle
honing
hot plate
oversize
plunger
pressure equalizer
ream
refacing valves
resizing cylinders
staking
Stanisol
valve guide
valve guide bushing
valve seat inserts
valve seats
valve spring
workbench

2. Where do you put dirty rags in the workshop?
3. Why should you keep your hands clean when reassembling an engine?
4. What is meant by an oversize piston?
5. What is honing?
6. How is honing done?
7. How do you clean the cylinder bore after honing?
8. How do you reface a valve seat?
9. What is TDC?
10. What is BDC?
11. Where is the valve-spring cap used?
12. How often should valve springs be replaced?
13. What is a valve guide?
14. What is a valve guide plug gage?
15. Where is the valve guide bushing located?
16. What is the purpose of a piloted counterbore reamer?
17. What kind of valve guide bushings do cast-iron engines use?
18. What type of engines use inserts on the exhaust and intake valves?
19. How do you seat the new valve seat insert?
20. What is meant by valve life?
21. What's the difference between a ball bearing and a plain bearing?
22. Do you need a lubricant for counterbore reaming?

23. What is staking?
24. What does PTO stand for?
25. How do you check the cam gear bushing?
26. How do you straighten an engine's crankshaft without removing it from the engine?

CHAPTER 15

Reassembling, Inspecting, and Testing

In this chapter you will learn how to put the engine back together. But before you do that, you need to be able to inspect, test, and recheck the parts to make sure they are in good enough condition to warrant further use. If not, you will have to decide whether or not to replace with new parts. In some instances you will need to know machine processes, which will be explained. You also need to know the usual workbench procedures and the tools to use during reassembly. These will be explained as the steps are being taken to put the engine in working condition.

So far you have been able to inspect and analyze the extent of damage or wear on the parts as they were disassembled. Now that the inspection stage has been bridged, you can be sure you're replacing the rebuilt or new part to make the engine operate properly for a number of hours.

Before you start the reassembly process, take a look at the exploded views shown in Fig. 15–1. They will give you some idea as to how the parts come apart and how they can be reassembled. Of course, they can't be put back exactly as shown in the exploded view since there are certain processes or procedures that must be followed to get the valves back into their proper locations.

Clearances and adjustments must be made for each system in the engine. These will be described as they come up and you will be able to make the adjustments as needed.

Tappets, Cam Gear, and Camshaft

The camshaft must be replaced first so the tappets can be reassembled and other adjustments can be made. The automatic-spark models must have the cam gear in the normal operating position with the movable weight down in order to check (Fig. 15–2). Press the weight down. Release. The spring should lift the weight. If not, the spring is stretched or the weight is binding. Repair or replace.

Model series 111200 and 111900 (4.0 hp) have the Easy-Spin® lobe and a compression release on the exhaust cam (Fig. 15–3). This easy-start device allows you to pull the starter rope with very little effort. In the starting position, the actuator cam moves the rocker cam so it will open the exhaust valve at the same time as the Easy-Spin lobe. When the engine starts, the actuator cam moves out and lets the rocker cam move down and the exhaust valve operates normally. Fig. 15–14 shows the start position of the compression release cam.

To check the proper placement of the cam, move the actuator cam to the run position (Fig. 15–4). Push the rocker cam against the actuator cam. Release the actuator cam. The actuator cam spring should pull the actuator cam against the shoulder pin, causing the rocker cam to raise up to the starting position shown in Fig. 15–3. There should be no binding. Replace if binding occurs.

Crankshaft and Bearing Support

In the aluminum-alloy engine models the tappets are inserted first, the crankshaft next, and then the cam gear. When inserting the cam gear, turn the crankshaft and cam gear so that the timing marks on the gears align (Fig. 15–5).

Note: Model series 94000, 171700, 191700, 251700, and 252700 have a removable timing gear. When installing the timing gear, have the inner chamber toward the crankpin. This assures the mark will be visible.

Aluminum Engines—Ball Bearing—On crankshafts with ball

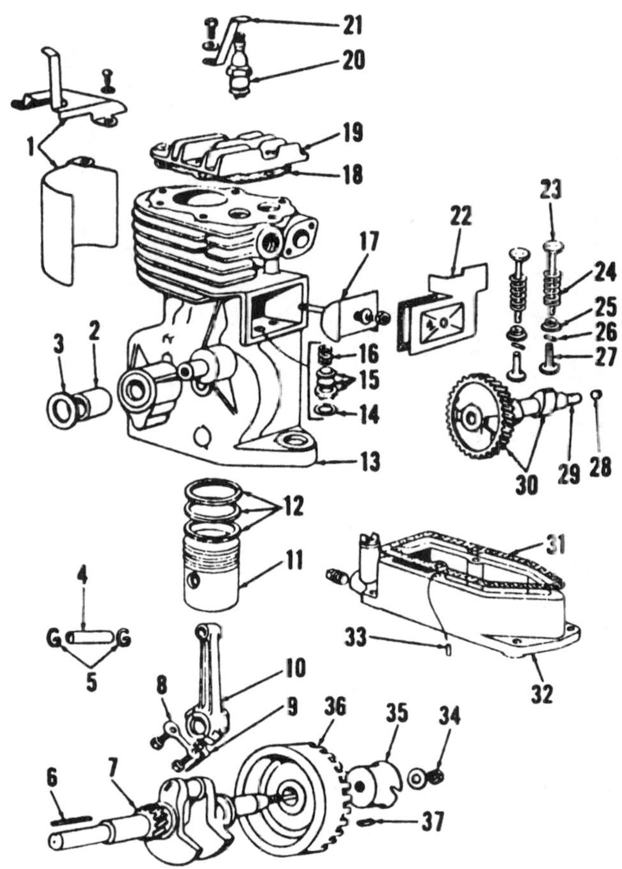

1. Shields
2. Bushing
3. Oil seal
4. Piston pin
5. Snap rings
6. Output drive key
7. Crankshaft
8. Rod bolt lock
9. Oil dipper
10. Connecting rod
11. Piston
12. Piston rings
13. Crankcase
14. Gasket
15. Breather body and valve assembly
16. Spring
17. Oil spray shield
18. Cylinder head gasket
19. Cylinder head
20. Spark plug
21. Grounding spring
22. Tappet chamber cover
23. Valves
24. Valve springs
25. Spring retainers
26. Retainer pins
27. Tappets
28. Plug
29. Camshaft
30. Cam gear
31. Gasket
32. Engine base
33. Dowel
34. Flywheel nut
35. Starter cup
36. Flywheel
37. Flywheel key

Fig. 15–1. (A) Exploded view of the model 6 engine assembly. Models N, 5, and 8 are similar. *(Courtesy B & S)*

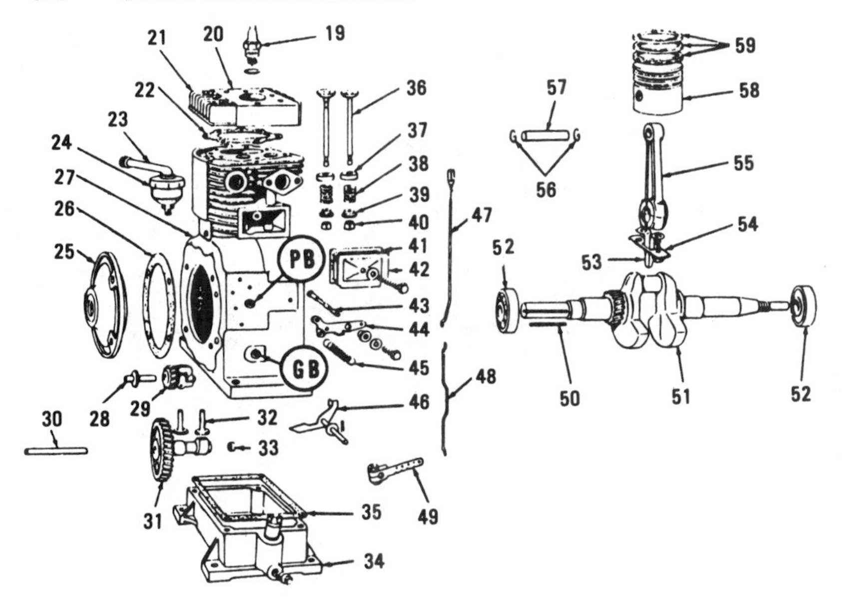

19. Spark plug
20. Air baffle
21. Cylinder head
22. Gasket
23. Breather tube
24. Breather
25. Main bearing plate
26. Gasket
27. Engine crankcase & cylinder block
28. Governor shaft
29. Governor gear & weight unit
30. Camshaft
31. Cam gear
32. Tappets
33. Camshaft plug
34. Engine base
35. Gasket
36. Valves
37. Valve spring washers
38. Valve springs
39. Spring retainers (or "Roto-caps")
40. Keepers
41. Gasket
42. Tappet chamber cover
43. Breaker point plunger
44. Governor control lever
45. Governor spring
46. Governor crank
47. Governor control rod
48. Governor link
49. Governor lever
50. Output drive key
51. Crankshaft
52. Ball bearings (on models so equipped)
53. Oil dipper
54. Rod bolt lock
55. Connecting rod
56. Piston pin retaining rings
57. Piston pin
58. Piston
59. Piston rings

Fig. 15–1. (B) (*Left*) Exploded view of model 9 or 14 engine assembly not having Magna-Matic ignition system. Breaker plunger bushing (PB) and governor crank bushing (GB) in the engine crankcase (27) are renewable. Breaker plunger (43) rides against a cam on cam gear (31). (*Right*) Exploded view of the engine crankcase and cylinder block on models with Magna-Matic ignition. Note the ignition advance weight (AW). Rod, piston, and crankshaft are shown above. *(Courtesy B & S)*

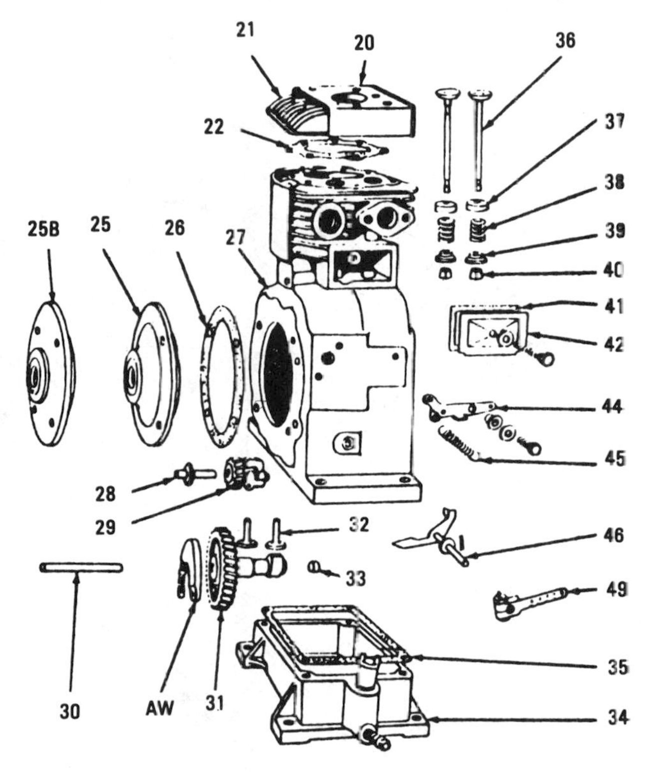

AW. Advance weight
20. Air baffle
21. Cylinder head
22. Gasket
25. Bearing plate, plain bushing
25B. Bearing plate, ball bearing
26. Gasket
27. Crankcase & cylinder
28. Governor shaft
29. Governor gear & weight unit
30. Camshaft
31. Cam gear
32. Tappets
33. Camshaft plug
34. Engine base
35. Gasket
36. Valves
37. Valve spring washers
38. Valve springs
39. Spring retainers (or "Roto-caps")
40. Keepers
41. Gasket
42. Tappet chamber cover
44. Governor control lever
45. Governor spring
46. Governor crank
49. Governor lever

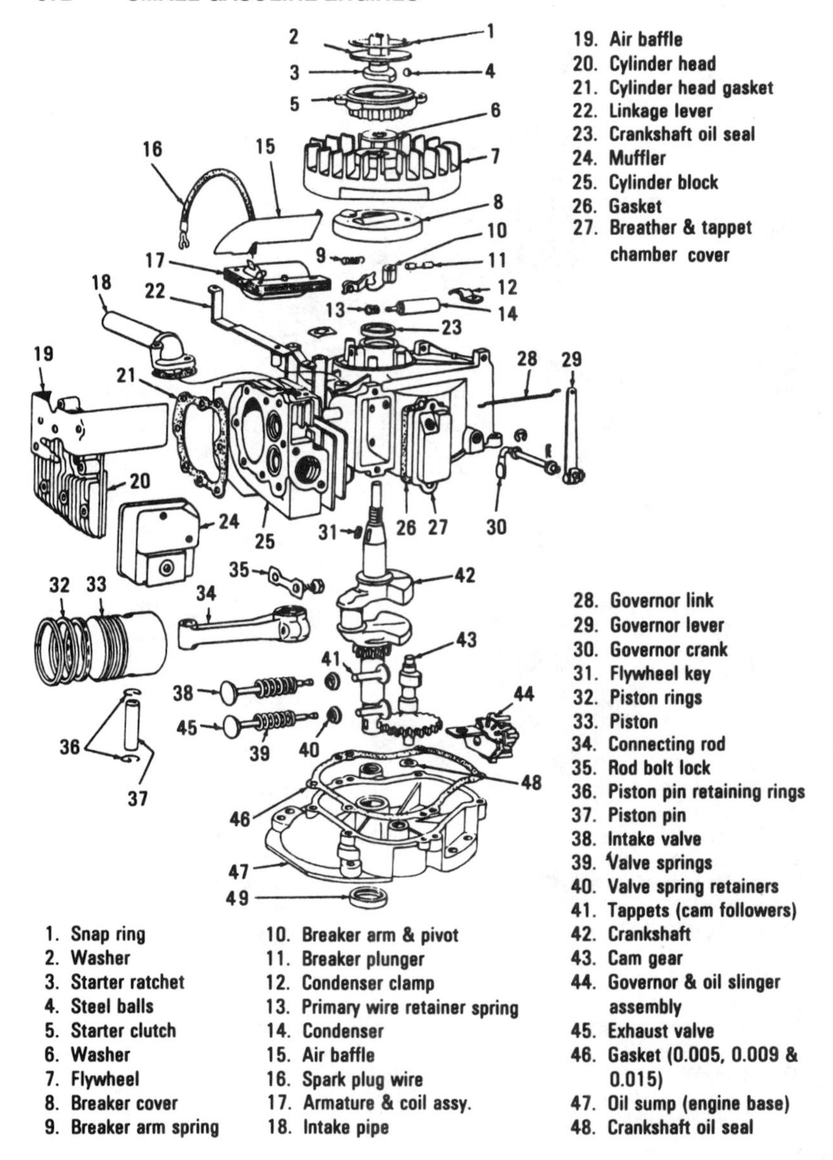

Fig. 15–1. (C) Exploded view of series 100000 vertical-crankshaft engine with a mechanical governor. Series 130000 and 140000 vertical crankshaft models with mechanical governor are similar.

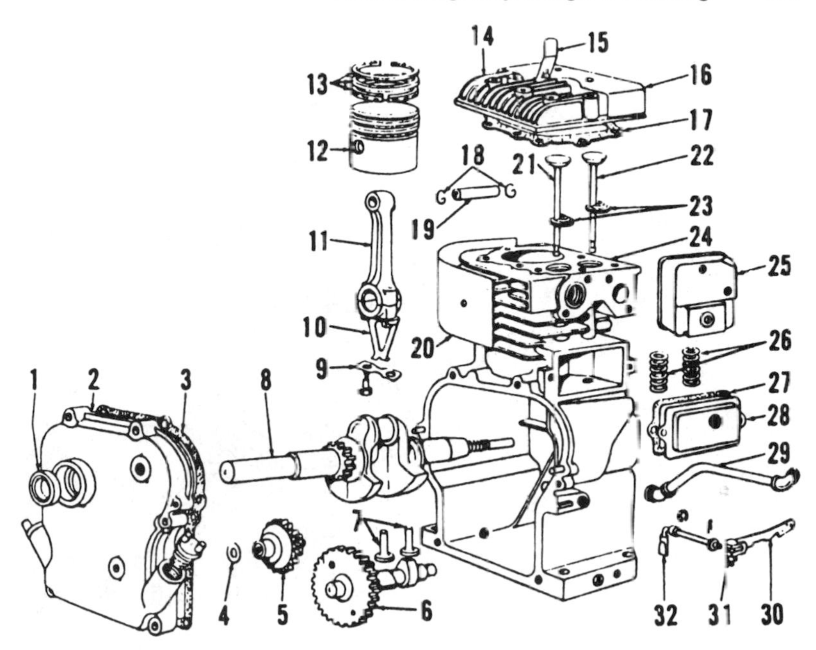

1. Crankshaft oil seal
2. Crankcase cover
3. Gasket (0.005, 0.009 or 0.015)
4. Thrust washer
5. Governor assembly
6. Cam gear and shaft
7. Tappets (cam followers)
8. Crankshaft
9. Rod bolt lock
10. Oil dipper
11. Connecting rod
12. Piston
13. Piston rings
14. Cylinder head
15. Spark plug ground switch
16. Air baffle
17. Cylinder head gasket
18. Piston pin retaining rings
19. Piston pin
20. Air baffle
21. Exhaust valve
22. Intake valve
23. Valve spring retainers
24. Cylinder block
25. Muffler
26. Valve springs
27. Gasket
28. Breather & tappet chamber cover
29. Breather pipe
30. Governor lever
31. Clamping bolt
32. Governor crank

Fig. 15–1. (D) Exploded view of series 100000 horizontal-crankshaft engine assembly, except for series 130000 and late series 140000. *(Courtesy B & S)*

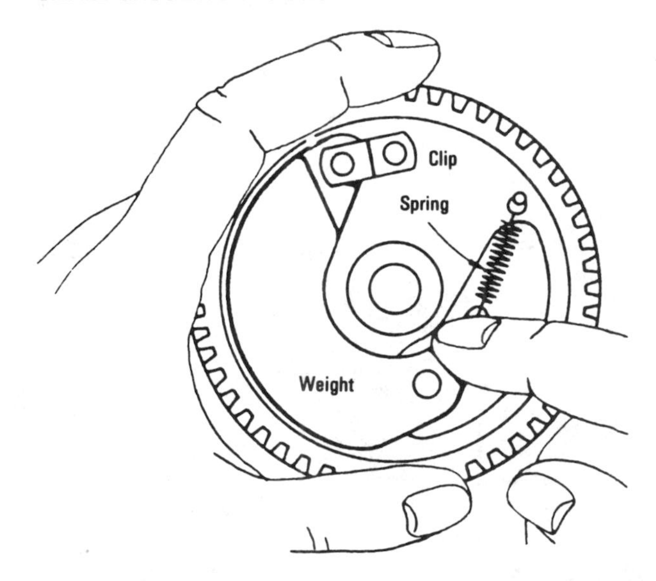

Fig. 15–2. Automatic spark advance on Magna-Matic models. *(Courtesy B & S)*

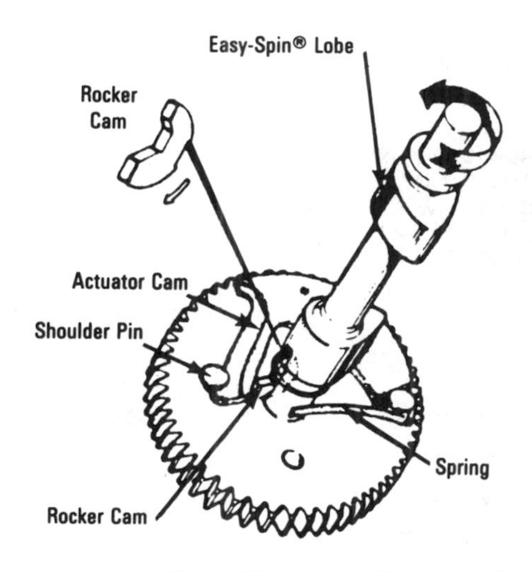

Fig. 15–3. The start position of the cam. *(Courtesy B & S)*

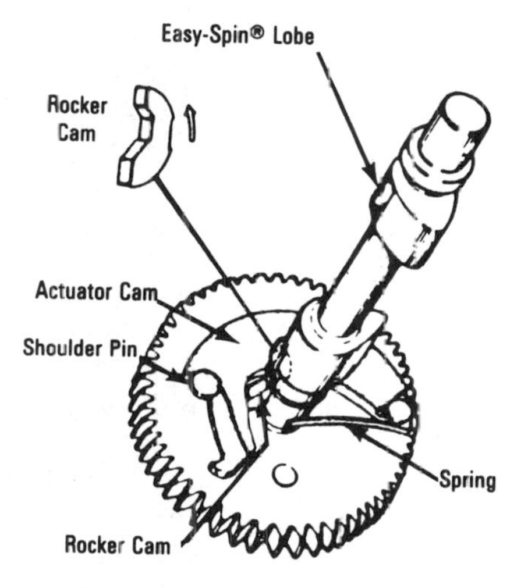

Fig. 15–4. Checking the compression release cam. *(Courtesy B & S)*

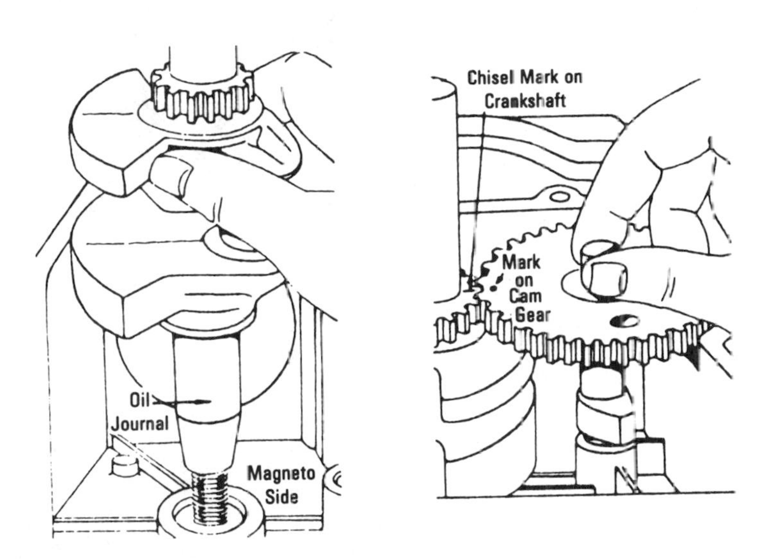

Fig. 15–5. Aligning timing marks. *(Courtesy B & S)*

bearings, the gear teeth are not visible for alignment of the timing marks. Therefore, the timing mark is on the counterweight (Fig. 15–6). On ball-bearing-equipped models, the tappets are installed first. The crankshaft and cam gear must be inserted together. Align the timing marks as shown in Fig. 15–6. Insert the crankshaft and cam gear. Thrust washers cannot be used on engines with two ball bearings. The thrust washer is added to the magneto end of the crankshaft instead of the power take-off end of the aluminum engines. This is the only difference between these engines and those with the plain bearings.

All Models—The crankshaft end play on all models, plain and ball bearing, should be 0.002″ (0.05 mm) to 0.008″ (0.20 mm). The method of obtaining the correct end play varies, however, between cast-iron, aluminum, plain- and ball-bearing models. New gasket sets include three crankcase cover or bearing support gaskets—0.005″ (0.13 mm), 0.009″ (0.23 mm), and 0.015″ (0.38 mm).

Aluminum Engines—Plain Bearing—To eliminate excess wear and engine noise, the end play in the crankshaft has to be controlled. In aluminum engines the end play should be 0.002″ (0.05 mm) to 0.008″ (0.20 mm) with one 0.015″ (0.38 mm) gasket in place (Fig. 15–7).

If the end play is less than 0.002″ (0.05 mm), which would be the case if a new crankcase or sump cover is used, additional gaskets of

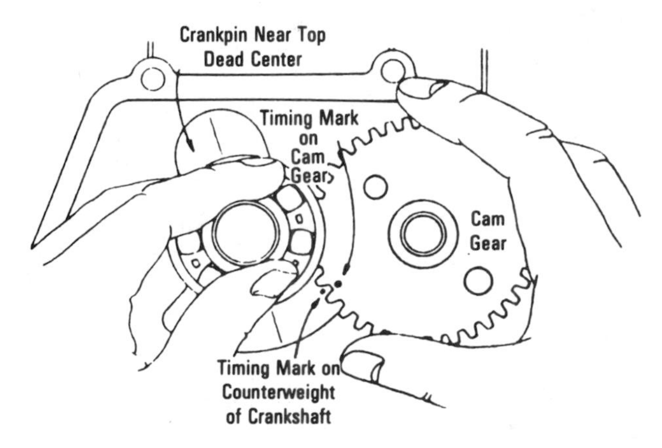

Fig. 15–6. Aligning timing marks on ball-bearing engines. *(Courtesy B & S)*

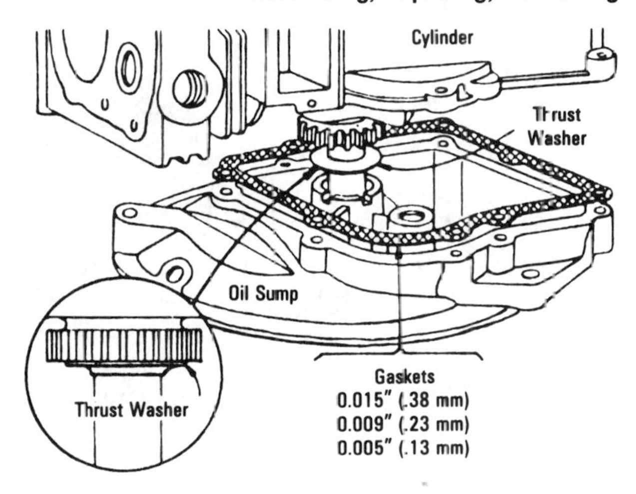

220624 THRUST WASHER for 0.875" (22.23 mm) DIA. CRKSFT.
220708 THRUST WASHER for 1" (25.40 mm) DIA. CRKSFT.
222949 THRUST WASHER for 1.181" (30.00 mm) DIA. CRKSFT.
222951 THRUST WASHER for 1.378" (35.00 mm) DIA. CRKSFT.

Fig. 15–7. Correcting crankshaft end play. *(Courtesy B & S)*

0.005″ (0.13 mm), 0.009″ (0.23 mm), or 0.015″ (0.38 mm) may be added in various combinations to attain the proper end play.

If the end play is more than 0.008″ (0.20 mm) with one 0.015″ (0.38 mm) gasket in place, a thrust washer is available to be placed on the crankshaft power take-off end between the gear and the crankcase cover or sump. Additional gaskets of 0.005″ (0.13 mm) or 0.009″ (0.23 mm) will then have to be added to the 0.015″ (0.38 mm) gasket for proper end play. *Note:* On aluminum models never use less than a 0.015″ (0.38 mm) gasket (Fig. 15–7).

Sump Installation

On models 100900 and 130900, use the spring washer on the cam gear as shown in Fig. 15–8. To protect the oil seal while assembling the crankcase cover, put oil or grease on the sealing edge of the oil seal. Wrap a piece of thin cardboard around the crankshaft so the seal will

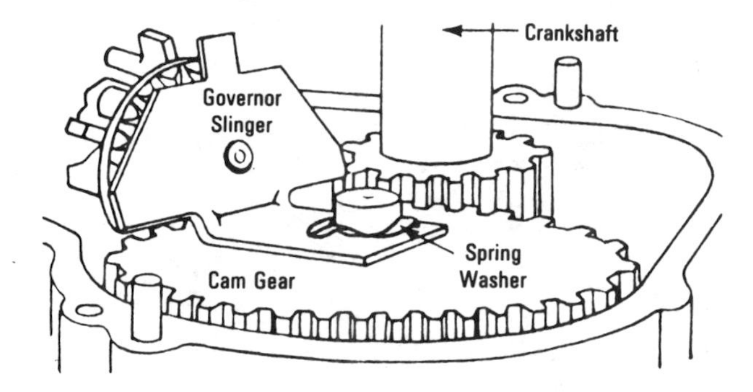

Fig. 15–8. Note the location of the spring washer. *(Courtesy B & S)*

slide easily over the shoulder of the crankshaft. If the sharp edge of the oil seal is cut or bent under, the seal may leak.

Checking End Play—The end play may be checked by assembling a dial indicator on the crankshaft with the pointer against the crankcase. Move the crankshaft in and out. The indicator will show the end play. Fig. 15–9 shows the dial indicator in place. The other method is to assemble a pulley to the crankshaft and measure the end play with a feeler gage (Fig. 15–9). End play should be 0.002″ to 0.008″ (0.05 to 0.20 mm).

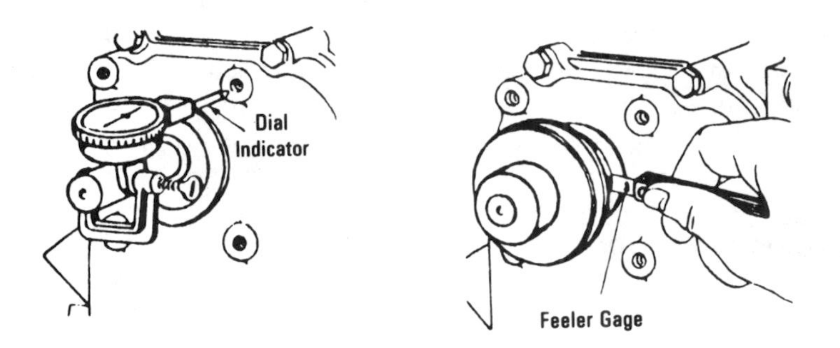

Fig. 15–9. Dial indicator used to check end play in the crankshaft. The feeler gage is also used to check end play. *(Courtesy B & S)*

Installation of Crankshaft and Cam Gear on Cast-Iron Engines

Cast-Iron Engines with Plain Bearings—Assemble the tappets to the cylinder, then insert the cam gear. Push the camshaft into the camshaft hole in the cylinder from the flywheel side through the cam gear. With a blunt punch, press or hammer the camshaft until the end is flush with outside of the cylinder on the power take-off side. Place a small amount of sealer on the camshaft plug, then press or hammer it into the camshaft hole in the cylinder at the flywheel side. Install the crankshaft so the timing mark on the teeth of the cam gear is aligned with the timing mark on the crankshaft.

Cast-Iron Engines with Ball Bearings—Assemble the tappets, then insert the cam gear into the cylinder, pushing the cam gear forward into the recess in front of the cylinder. This is on all models except 300000 and 320000 (12 and 16 hp) in the cast-iron engine. Insert the crankshaft into the cylinder. Turn the camshaft and crankshaft until the timing marks align. Then push the cam gear back until it engages the gear on the crankshaft with the timing marks together. Insert the camshaft (Fig. 15–10). Place a small amount of sealer on the camshaft plug. Then press or hammer it into the camshaft hole in the cylinder at the flywheel side.

Checking End Play with Both Types of Engines—The crankshaft

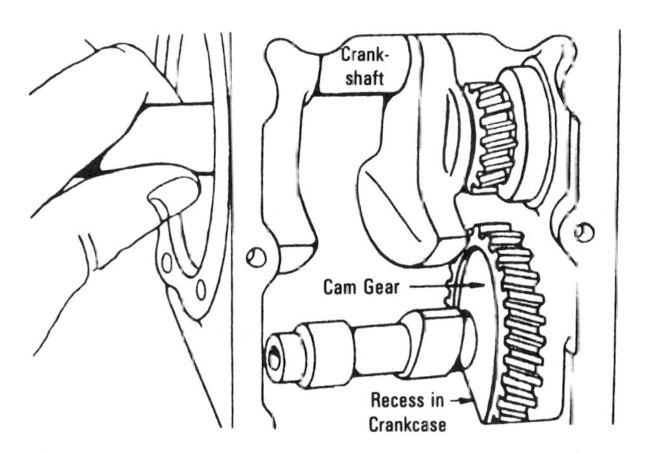

Fig. 15–10. Installing the crankshaft. *(Courtesy B & S)*

end play should be 0.002″ (0.05 mm) to 0.008″ (0.20 mm) with one 0.015″ (0.38 mm) gasket in place. If the end play is less than 0.002″ (0.05 mm), additional gaskets of 0.005″ (0.13 mm) or 0.009″ (0.23 mm) may be added to the 0.015″ (0.38-mm) gasket in various combinations to attain proper end play. If the end play is more than 0.008″ (0.20 mm) with one 0.015″ (0.38-mm) gasket in place, a 0.009″ (0.23 mm) or 0.005″ (0.13 mm) gasket may be used (Fig. 15–11).

If end play is more than 0.008″ (0.20 mm) with one 0.005″ (0.13 mm) gasket in place, a thrust washer is available and is placed on the crankshaft power take-off (PTO) end (Fig. 15–11).

Note: A thrust washer cannot be used on ball-bearing engines.

Checking End Play—On models with a removable base, the end play can be checked with a feeler gage between the crankshaft thrust face and the bearing support on the plain bearing engines (Fig. 15–11). Fig. 15–12 shows how to correct crankshaft end play.

On engine model series 300400 and 320400 (12 and 16 hp), install the breaker plunger and tappets, then insert the cam gear from the power take-off side of the cylinder (Fig. 15–13). Slide the cam gear shaft through the power take-off bearing and into the cam gear (Fig. 15–14). Insert the magneto side gear bearing on the cylinder. Torque the cam gear bearing screws to 85 inch-pounds (1.0 mkp, 9.6 Nm). Install long cam gear shaft bolt (5½″/14 cm) to prevent loss of the shaft (Fig. 15–15).

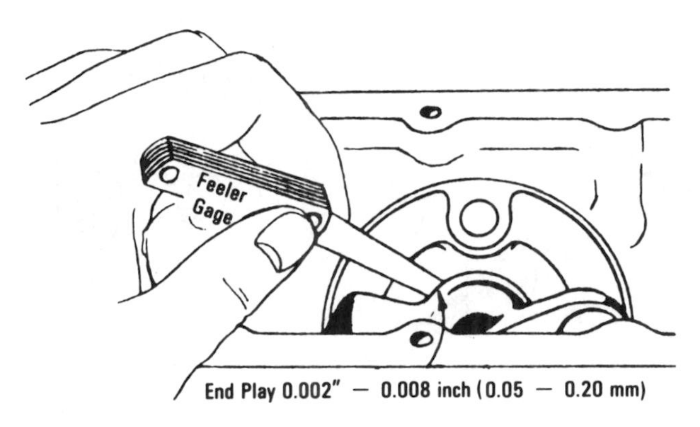

Fig. 15–11. Checking crankshaft end play with a feeler gage. *(Courtesy B & S)*

222949 Thrust Washer for 1.181" (30.00 mm) Dia. Crksft.
222951 Thrust Washer for 1.387" (35.00 mm) Dia. Crksft.

Fig. 15–12. Using a thrust washer. *(Courtesy B & S)*

Checking and Correcting Cam Gear End Play—Cam gear end play tolerance is machined at the factory and normally requires no adjustment unless the magneto side cam gear bearing or cam gear is replaced.

Cam gear play is checked in the same manner as crankshaft end play.

Camshaft end play must be 0.002″ (0.05 mm) to 0.008″ (0.20 mm). If end play is less than 0.002″ (0.05 mm), add service shims to obtain the proper end play. If the end play is more than 0.008″ (0.20 mm), use the service bearing assembly kit, which includes the shims, to obtain the proper end play.

Use chalk or crayon to mark the top of the crankshaft gear tooth,

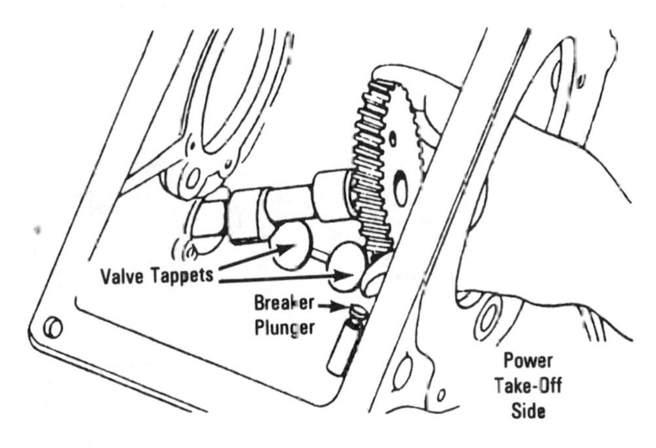

Fig. 15–13. Inserting the cam gear. *(Courtesy B & S)*

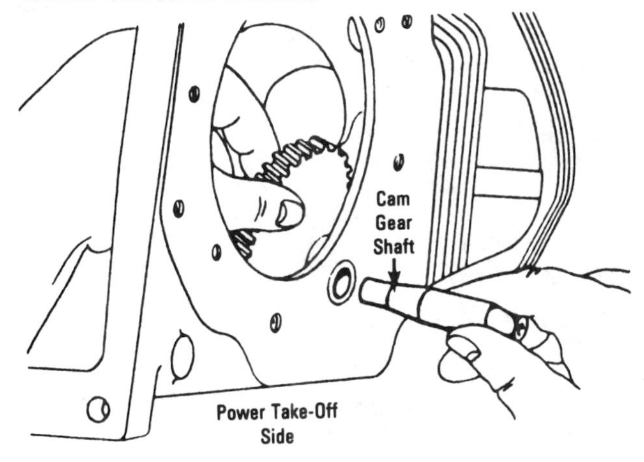

Fig. 15–14. Sliding cam gear shaft. *(Courtesy B & S)*

whose inner end is directly in line with the notch of the timing mark (Fig. 15–16). Align the timing marks on the crankshaft and cam gear. Install the crankshaft. Install the crankshaft carefully to make sure the crankpin is not damaged.

Install the power take-off and magneto side bearing supports. Torque the power take-off support screws to 185 inch-pounds (2.2 mkp, 20.9 Nm). Torque the magneto side support screw to 85 inch-pounds (1.0 mkp, 9.6 Nm) (Fig. 15–17).

Checking and Correcting End Play—Crankshaft end play tolerance is machined at the factory and normally requires no adjustment—that is, unless you replace the bearing supports or crankshaft.

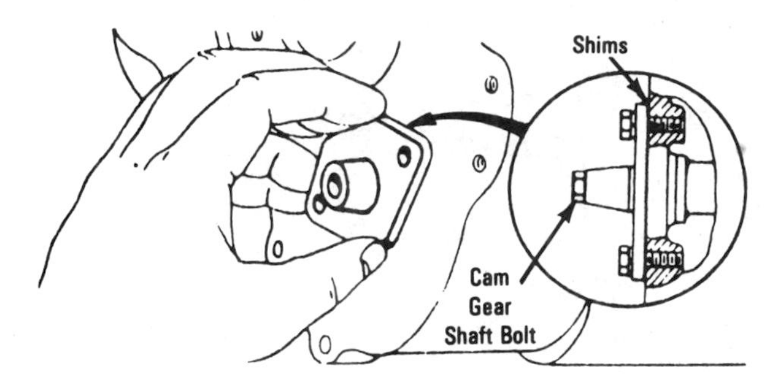

Fig. 15–15. Inserting the cam gear. *(Courtesy B & S)*

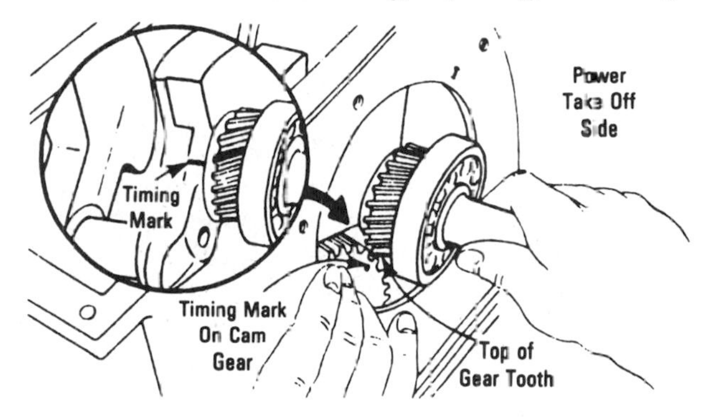

Fig. 15–16. Aligning timing marks. *(Courtesy B & S)*

Crankshaft end play must be 0.002″ to 0.008″ (0.05 to 0.20 mm). If less than 0.002″ (0.05 mm), add service shims to obtain the proper end play. If the end play is more than 0.008 (0.20 mm), use a service bearing support assembly kit, which includes the shims, to obtain the proper end play.

Auxiliary Power Take-Off

This auxiliary power take-off shaft is perpendicular to the crankshaft. It appears on model series 92580, 92980, 94580, 94980, 110980, and 111980 (3.0, 3.5, and 4.0 hp). It rotates at the rate of 1 revolution

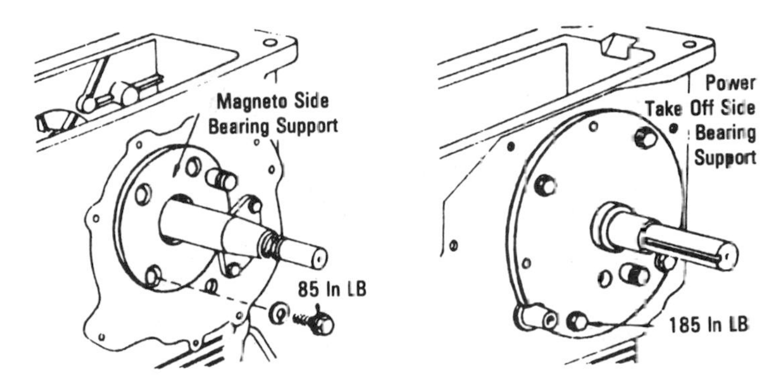

Fig. 15–17. Installing bearing supports. *(Courtesy B & S)*

per every 8½ revolutions of the crankshaft. On these models, the cam gear, worm gear, and oil slinger are factory-assembled and are not available as separate pieces (Fig. 15–18).

To remove the sump, one of the six sump-mounting screws is located under the auxiliary drive cover (Fig. 15–19). Remove the cover and lift out the shaft stop. Now it is time to replace it. Reverse the procedure for removing it and you have it installed properly. The gear had to be slid sideways to expose the head of the sump-mounting screw. A 7⁄16″ (11.11 mm) socket can be used on the screw. When installing the cover (Fig. 15–18), put nonhardening sealant on the cover screws. Fig. 15–20 shows how the power take-off auxiliary shaft works.

Note: If rotation is counterclockwise, the thrust washer is placed next to the worm gear on the camshaft (Fig. 15–20).

Piston, Piston Pin, Connecting Rod, and Rings

The piston pin is a push fit in both piston and connecting rod. On models using a solid piston pin, one end is flat, the other end is recessed. Other models use a hollow pin (Fig. 15–21).

Place a pin lock in the groove at one side of the piston. From the opposite side of the piston insert the piston pin—flat end first with

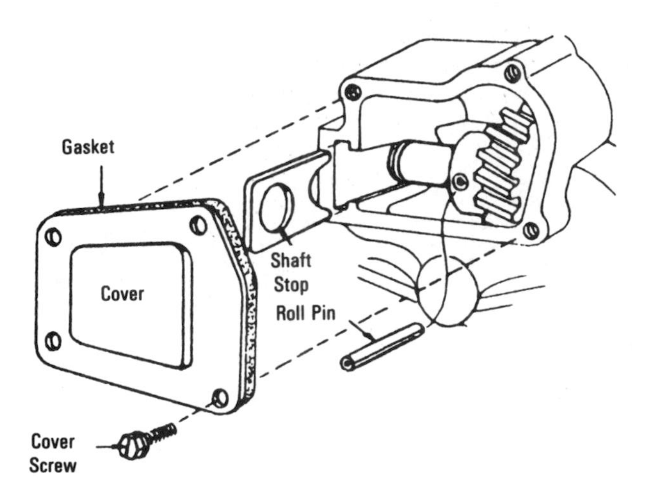

Fig. 15–18. Power take-off (auxiliary) cover. *(Courtesy B & S)*

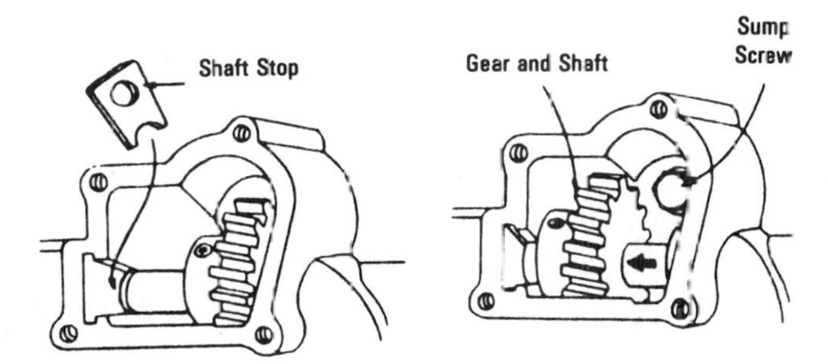

Fig. 15–19. Installing the sump screw. *(Courtesy B & S)*

solid pins, either end with hollow pins—until it stops against the pin lock (Fig. 15–22). Use needle-nose pliers to assemble the pin lock in the recessed end of the piston. Be sure the locks are firmly set in the grooves.

Putting in the Rings—In Fig. 15–22A, the various rings and the proper position of each is shown. Note the center compression ring. The scraper groove should always be down toward the piston skirt

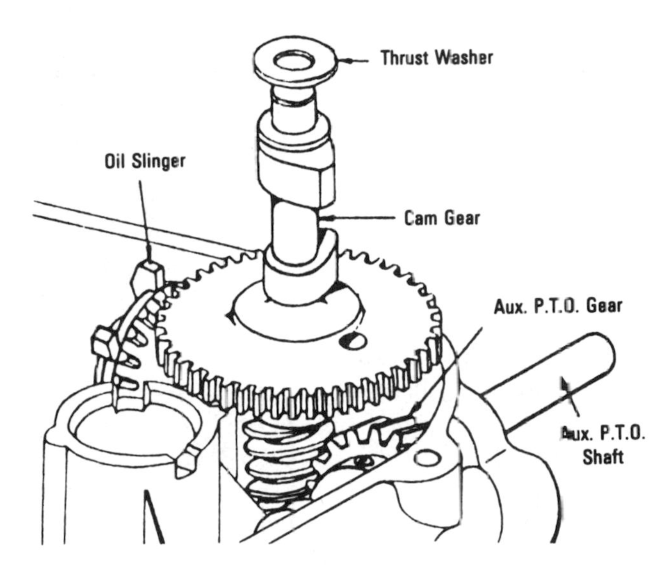

Fig. 15–20. Auxiliary power take-off. *(Courtesy B & S)*

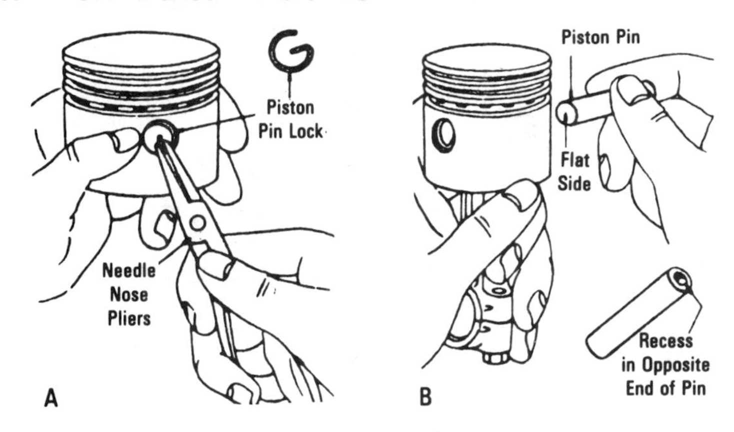

Fig. 15–21. Location of the piston pin and piston pin lock. *(Courtesy B & S)*

(Fig. 15–22B). Be sure the oil return holes are clean and the carbon is removed from all grooves. *Note:* Install the expander under the oil ring when required.

Oil the rings and piston skirt. Then compress the rings with a ring compressor (Fig. 15–23). Fig. 15–23A shows the cast-iron engine's ring compressor and Fig. 15–23B shows the aluminum-engine's ring compressor. Turn the piston and compressor upside down on the bench and push down so the piston head and edge of the compressor band are even while you are tightening the compressor. Draw the compressor up tight to fully compress the rings. Then loosen the compressor very slightly. *Do not attempt to install piston and ring assembly without the ring compressor.*

Note: Pistons used in models 220000 and 250000 (10- and 11-hp) engines have a notch as shown in Fig. 15–24. The notch must face the flywheel side of the cylinder when installed.

Models 300000 and 320000 (12 and 16 hp) have the piston identified with an *F* located next to the piston pin bore. When assembling the piston to the connecting rod, the letter *F* on the piston must be installed so that it appears on the same side as the assembly mark on the connecting rod. An assembly mark on the rod is also used to identify the rod and cap alignment. Install the piston rings as illustrated in Fig. 15–25.

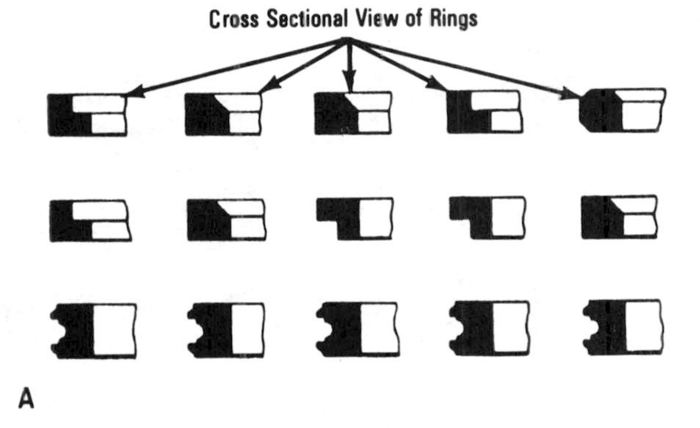

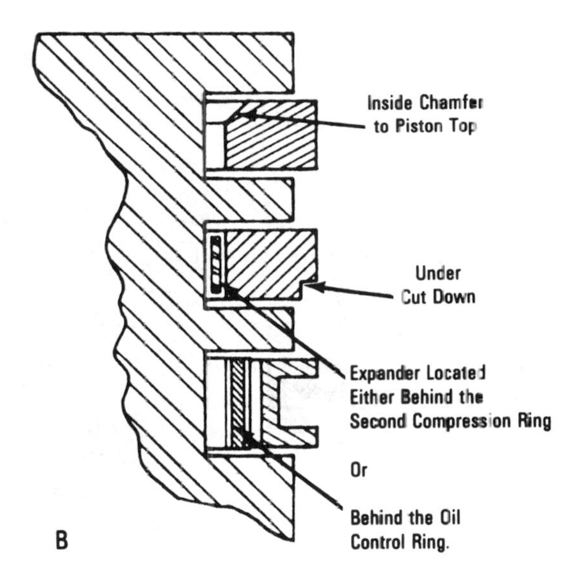

Fig. 15–22. (A) Different rings in cross section. *(Courtesy B & S)* **(B) Placement of rings on a piston.** *(Courtesy Tecumseh)*

Install the piston, connecting rod, and dipper. Piston identification mark *F* and the notch at the top of the piston must be toward the flywheel side. Torque the connecting-rod screw as shown in Table 15–1. Move the connecting rod back and forth on the crankpin to be sure it is free.

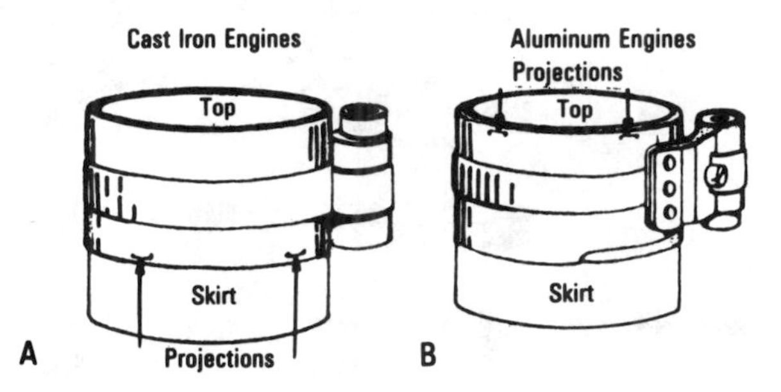

Fig. 15–23. Using the compression rings to install rings. *(Courtesy B & S)* **(A) Cast-iron engine's ring compressor. (B) Aluminum engine's ring compressor.**

On all models except 300000 and 320000, install the piston and rod the same way. Place the connecting-rod and piston assembly (with the rings compressed) into the cylinder bore (Fig. 15–25). Push the piston and the rod down into the cylinder. Oil the crankpin of the crankshaft. Pull the connecting rod against the crankpin and assemble the rod cap so the assembly marks align (Fig. 15–26).

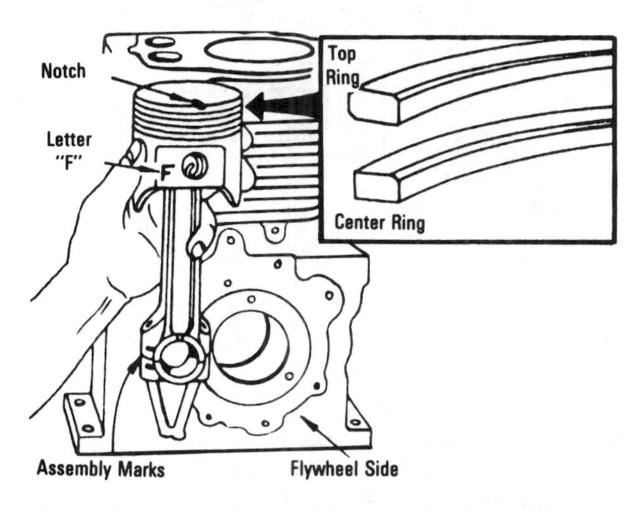

Fig. 15–24. Assembling the piston to the rod. *(Courtesy B & S)*

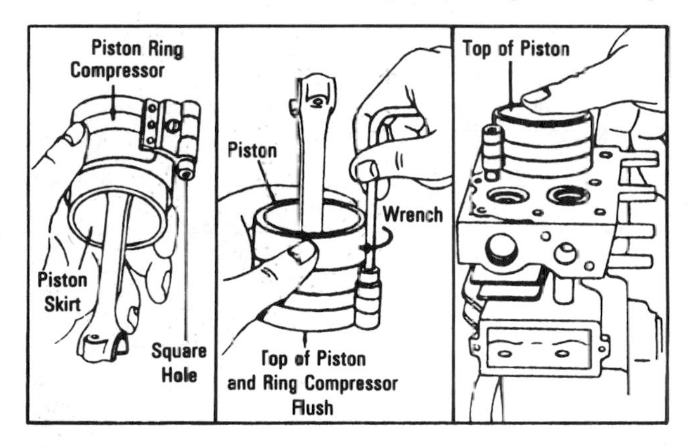

Fig. 15–25. Installing the piston assembly in the cylinder. *(Courtesy B & S)*

Keep in mind that some rods do not have assembly marks as the rod and cap will fit only in *one* position. Use care to make sure they are properly installed. Also notice that some rods may have flat washers under the cap screws. Remove the washers and discard prior to installing the rod.

Assemble the cap screws and screw locks with the oil dipper (if used). Tighten the cap screws to torque shown in Table 15–1. Use a torque wrench as shown in Fig. 15–27A.

Rotate the crankshaft two revolutions to be sure the rod is correctly installed. If the rod strikes the cylinder wall, the connecting rod has been installed wrong or the cam gear is out of time. If the crankshaft operates freely, bend the screw locks against the screw heads as shown in Fig. 15–27B.

Tighten the screws securely. After tightening the rod screws, the rod should be able to move sideways on the crankpin of the shaft. A torque wrench must be used to prevent loose or overtight cap screws, which cause breakage and/or scoring of the rod (Fig. 15–27).

Since torque wrenches come in different units of measurement, all three values are given in Table 15–1 so that you will be able to use any of the three types of wrenches.

Table 15–1. Connecting-Rod Cap Screw Torque

Basic Model Series	Average Torque		
Aluminum Cylinder	Inch-Pounds	Kilo-gram-Meters	Newton-Meters
6B, 60000	100	1.2	11.3
8B, 80000	100	1.2	11.3
82000, 92000, 110000, 111000	100	1.2	11.3
100000, 130000	100	1.2	11.3
140000, 170000, 190000	165	1.9	18.7
250000	185	2.1	21.0
Cast-Iron Cylinder	Inch-Pounds	Kilo-gram-Meters	Newton-Meters
5, 6, N, 8	100	1.2	11.3
9	140	1.6	15.8
14	190	2.2	21.5
19, 190000, 200000	190	2.2	21.5
23, 230000	190	2.2	21.5
240000, 300000, 320000	190	2.2	21.5

(Courtesy B & S)

Oil Dippers and Oil Slingers

Oil Dippers

In the vertical- and horizontal-crankshaft engines, the method of circulating the oil from the sump varies slightly. In the aluminum-alloy the cast-iron engines, the splash system is used (Fig. 15–28). The dipper dips into the oil reservoir in the base of the engine (sump). It has no pump or moving parts. Install the connecting rod and dipper according to the engine model shown in Fig. 15–28.

In the 8- and 10-horsepower models that Tecumseh makes, the dipper is slightly different (Fig. 15–29). The oil passage holes are

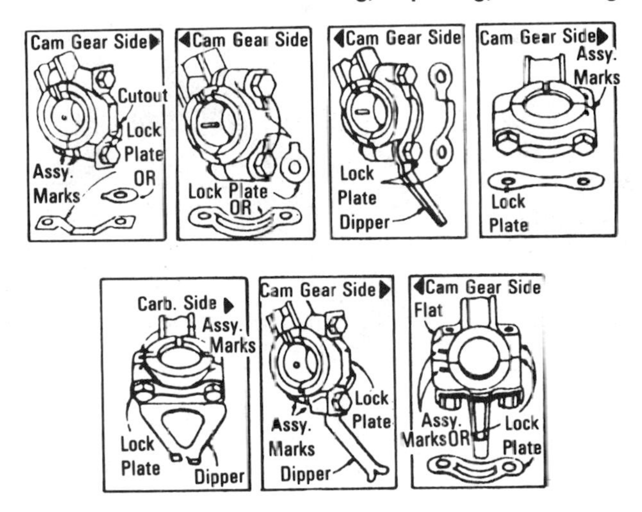

Fig. 15–26. Installation of the connecting rod. *(Courtesy B & S)*

shown also. When reinstalling a used connecting rod, *always* use the torque that will be retained when the connecting rod is replaced.

This connecting rod must be installed with the match mark facing out of the cylinder, toward the power take-off end of the crankshaft.

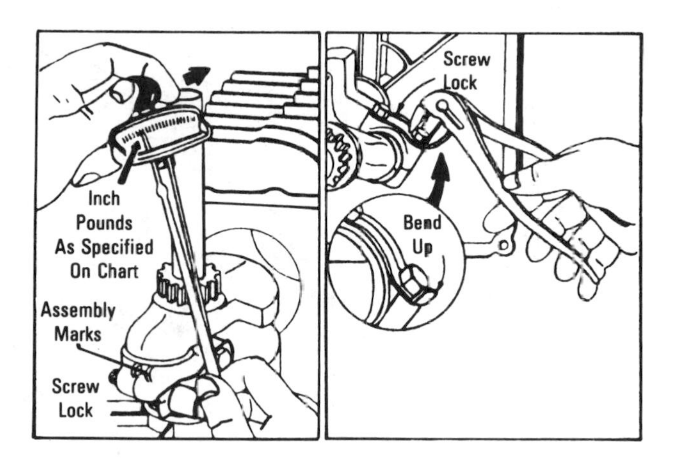

Fig. 15–27. Bending the screw locks and using a torque wrench. *(Courtesy B & S)*

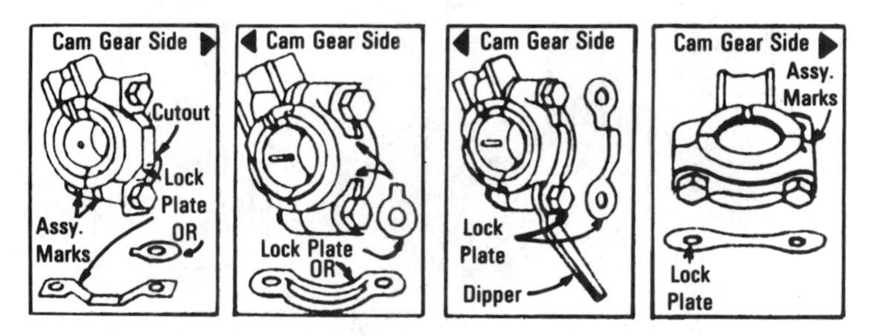

Fig. 15–28. Connecting-rod installation for horizontal-crankshaft engines. *(Courtesy B & S)*

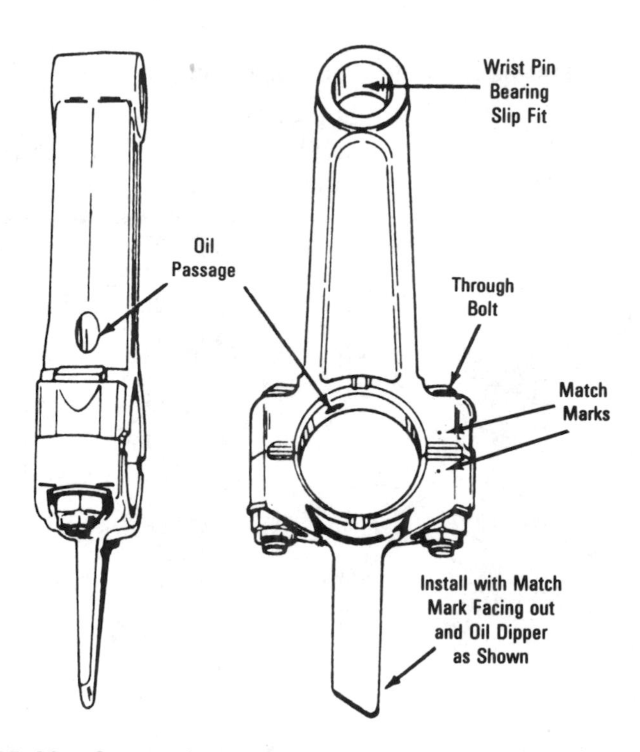

Fig. 15–29. Connecting rods for Tecumseh 8-, 10-, and 12-horsepower engines. *(Courtesy Tecumseh)*

This will ensure the correct positioning of the oil dipper. Install as shown in Fig. 15–30.

The heads of the through bolts must be seated tight against the machined shoulder on the connecting rod. Failure to check this may result in a false torque reading and premature failure.

Before installing, clean the connecting-rod bearing surfaces with a clean cloth. Rods are coated with lead, which will slightly oxidize in storage, and this oxidation must be removed.

Oil Slingers

The oil slinger is driven by the cam gear. Old-style slingers using a die-cast bracket assembly have a steel bushing between the slinger and the bracket. Replace the bracket on which the oil slinger rides if it is worn to a diameter of 0.49″ (12.4 mm) or less. Replace the steel bushing if worn (Fig. 15–30A).

Newer-style oil slingers have a stamped-steel bracket. The unit comes as a complete assembly (Fig. 15–30B). A spring washer is used only on models (100900–130900 (4.0–5.0 h.p.). Inspect the gear teeth, old and new style. Replace if worn (Fig. 15–31).

Chapter 9 goes into detail on how the splasher and the slingers work. The horizontal-crankshaft engines use the dipper As the dipper

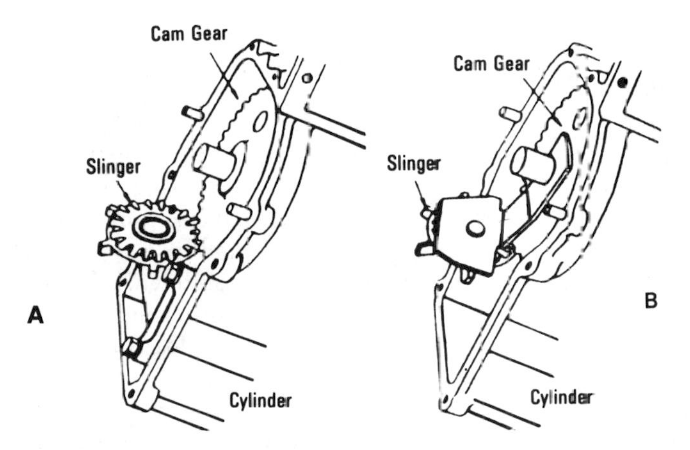

Fig. 15–30. Oil slinger and bracket. *(Courtesy B & S)*

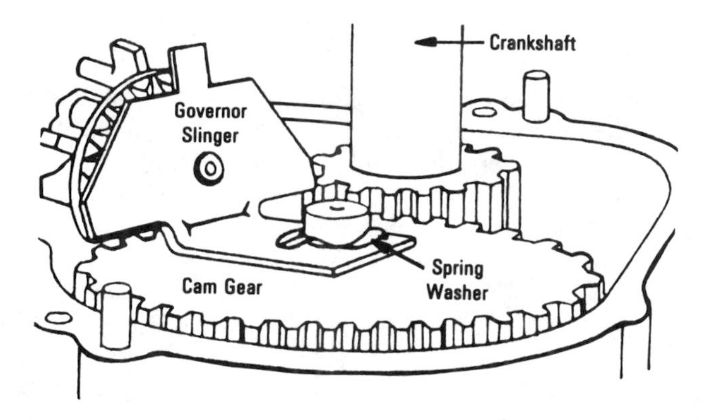

Fig. 15–31. Oil slinger and bracket for a vertical-crankshaft engine. *(Courtesy B & S)*

hits the pool of oil below, it splashes oil inside the engine crankcase. This causes oil to be deposited on the connecting rod and the cylinder walls. In the vertical-crankcase type of engine it would be difficult for a slinger on the bottom of the connecting rod to splash oil. Therefore it is necessary to design a different method of splashing the oil when no oil pump is used. In the vertical-crankshaft types, the oil slinger causes the oil in the sump to be splashed. This will lubricate the crankshaft bearings, the connecting-rod bearings, and the cylinder walls.

Mechanical Governor

There are two types of engine governors, the air-vane type and the mechanical type. The vane type is shown in Fig. 15–32. Note its location. It will be attached to the carburetor once the carburetor is placed on the engine. This type uses the air generated by the flywheel fan blades to place pressure on the vane, thus controlling the linkage and its effect on the carburetor throttle. The governor spring tends to open the throttle. Air pressure against the air vane tends to close the throttle. The engine speed at which these two forces balance is called the *governed speed.* The governed speed can be varied by changing the governor spring tension or by changing the governor spring. Air vane and mechanical governor adjustments will be covered in the appendix.

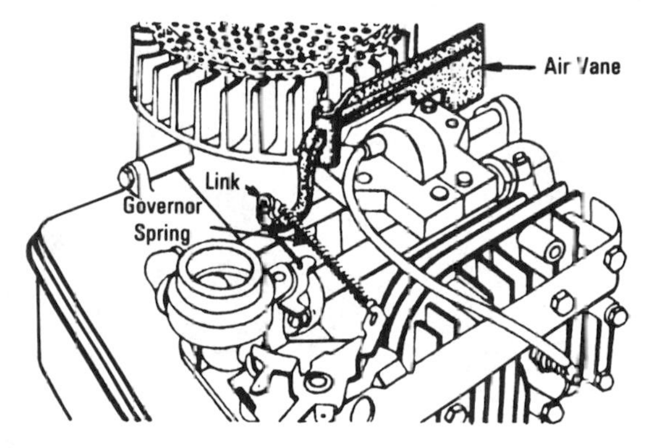

Fig. 15–32. Typical air vane governor. *(Courtesy B & S)*

There are many variations and each model of the various manufacturers' engines shows signs of in-house creativity.

Speed Limits—To comply with specified top governed speed limits, the manufacturers of engines either calibrate the governor springs or provide an adjustable top speed limit. Calibrated springs or an adjustable top speed limit will allow no more than a desired top speed when the engine is operated on a rigid test stand. The top engine speed should be checked with a tachometer when the engine is operated on the completely assembled machine it will drive.

If the test is run on a lawnmower, the mower should be operated on blacktop, concrete, or some hard surface so that a load on the blade and engine is not present.

After a new governor spring is installed, check the engine's top governed speed with an accurate tachometer. Run the engine at half throttle to allow the engine to reach normal operating temperature before measuring the speed with the tachometer.

Reassembly of the Mechanical Governor—Fig. 15–33 shows the governor housing and gear assembly in an exploded view. To assemble the governor crank, bushing, and lever to the housing, push the governor crank into the housing, lever end first. Slip the bushing onto the shaft. Then thread the bushing into the housing and tighten securely. Place the lever on the shaft with the governor crank in the position shown in Fig. 15–34. Place the governor gear on the shaft in the cylinder. Place the gasket on the governor housing and then assemble the

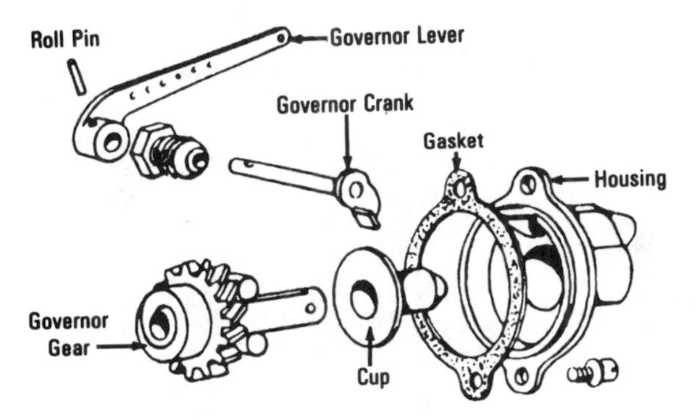

Fig. 15–33. Governor housing and gear assembly. *(Courtesy B & S)*

governor housing to the cylinder and tighten it in place with two mounting screws. This is the mechanical governor found on the cast-iron models N, 6, and 8.

Adjustment—Adjustment of the governor will be left until the carburetor has been placed on the engine and the springs can be hooked into their appropriate holes. There is no adjustment between the governor lever and the governor crank on these models (N, 6, and 8). However, the governor action can be changed by inserting the governor link or spring in different holes of the governor and throttle levers. Fig. 15–35 shows these adjustments, to be made *after* the engine has been reassembled and is ready for starting.

In general, the closer to the pivot end of the lever, the smaller the

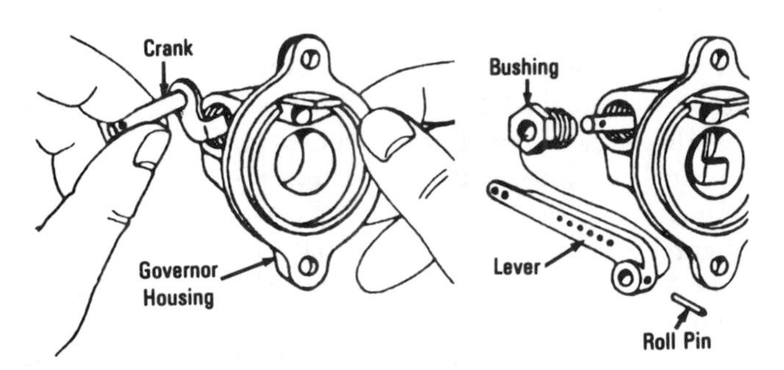

Fig. 15–34. Installing crank and lever for mechanical governor. *(Courtesy B & S)*

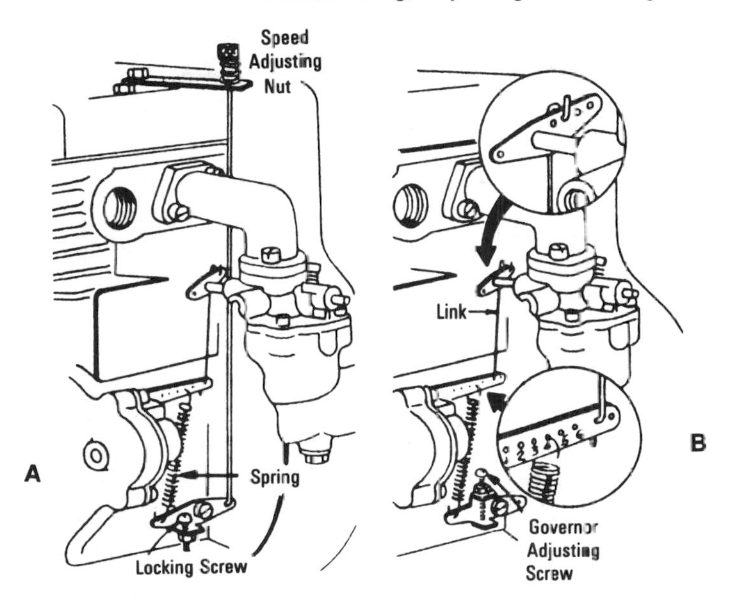

Fig. 15–35. Mechanical governor linkage. *(Courtesy B & S)*

difference between the load and no-load engine speed. The engine will begin to "hunt" if the spring is brought too close to the pivot point. The farther from the pivot end, the less tendency to "hunt," but it causes a greater speed drop with an increasing load. If the governed speed is lowered, the spring can usually be moved closer to the pivot. The standard setting is shown in Fig. 15–35B.

Mechanical Governor on the Aluminum Engine—To reassemble the mechanical governor on models 6B, 8B, 60000, 80000, and 140000, take a look at the exploded view shown in Fig. 15–36. As you can see, this is slightly different arrangement and will need to be assembled with care. Push the governor lever shaft into the crankcase cover, threaded end first. Assemble the small washer on the inner end of the shaft. Then screw the shaft into the governor crank follower by turning the shaft counterclockwise. Tighten securely. Turn the shaft until the follower points down as shown in Fig. 15–37. Place the washer on the outside end of the shaft. Install the roll pin. The leading end of the pin should just go through the shaft so that the pin protrudes from only one side of the shaft.

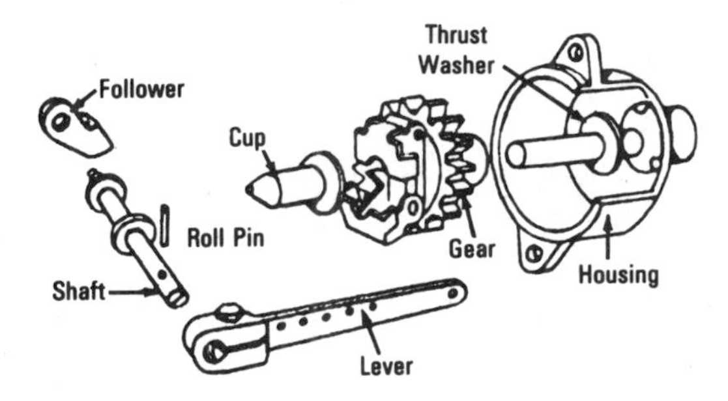

Fig. 15–36. Mechanical governor parts. *(Courtesy B & S)*

Place the thrust washer and then the governor gear on the shaft in the gear housing. Hold the crankcase cover in a vertical (normal) position and then assemble the housing with the gear in such a position that the point of the steel cup on the gear rests against the crank follower. Tighten the housing in place with two mounting screws (Fig. 15–37).

Assemble the governor lever to the lever shaft with the lever pointing downward at about at 30° angle. Adjustment will be made later, when the carburetor linkage is assembled.

Mechanical Governor on the Cast-Iron Engines—Models 9, 14, 19, 190000, 200000, 23, 230000, 240000, 300000, and 320000 are slightly different in their governor arrangement. Assemble the governor gear and cup assembly on the shaft protruding on the inside of the cylinder. Loosely attach the governor lever to the governor shaft. Later

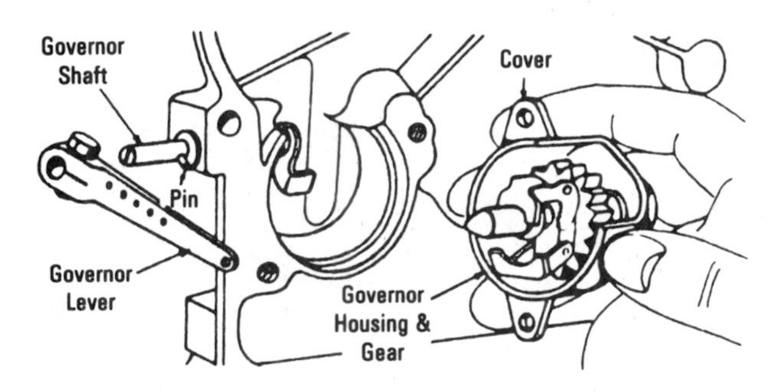

Fig. 15–37. Assembling mechanical governor. *(Courtesy B & S)*

models have a spacer between the governor shaft assembly and the bushing.

After the engine has been assembled, the adjustment can be made by loosening the screw holding the governor lever to the governor shaft. Place the throttle in high-speed position. Hold the throttle in this position and with a screwdriver turn the governor shaft counterclockwise as far as it will go. Tighten the screw holding the governor lever to 35–40 inch-pounds of torque (0.4–0.52 mkp, 4.0–5.0 Nm) (Fig. 15–38). Before cranking the engine, manually move the governor linkage to see if it has any binding.

Mechanical Governor on Aluminum Model 94000—This model illustrates some of the variations that may be encountered in the arrangement of mechanical governors. The mechanical governor used on this model (94000, 3.5 hp) is shown in Fig. 15–39. The governor gear is part of the oil slinger.

To reassemble the governor, insert the governor shaft into the gov-

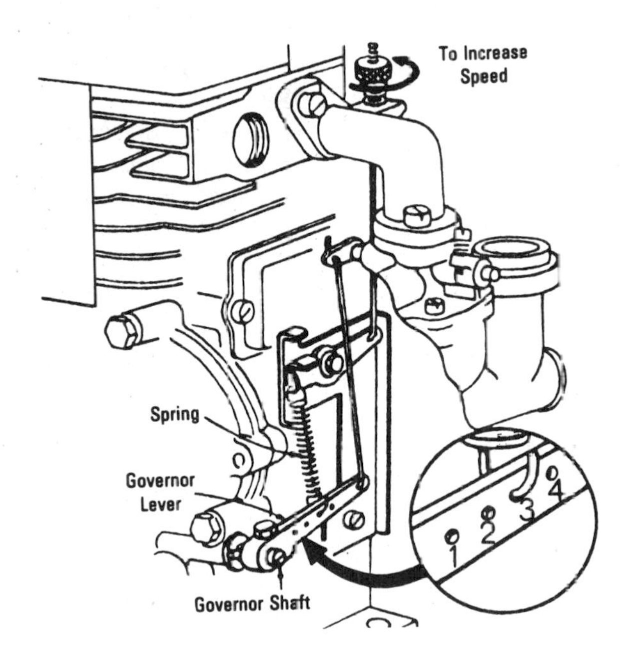

Fig. 15–38. Adjustment of the mechanical governor on the cast-iron engine. *(Courtesy B & S)*

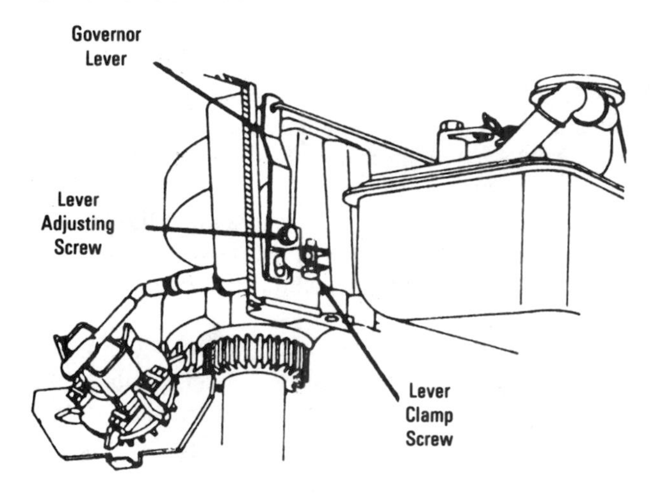

Fig. 15–39. Governor for model series 94000. *(Courtesy B & S)*

ernor bushing from the inside of the cylinder. Then slide the governor lever on the governor shaft and slide the lever down onto the shaft slot. Slide on the lever clamp and start the screw-in adjusting slot on the clamp. Torque the lever screw to 15 inch-pounds of torque (0.17 mkp, 1.7 Nm). Install the oil slinger and governor gear assembly, sump gasket, and oil sump. Place a nonhardening sealant on screw *A* in Fig. 15–40 and tighten all the sump screws.

To adjust the governor, loosen the lever adjusting screw shown in Fig. 15–40. While holding the governor lever and the governor clamp to the left (counterclockwise), tighten the lever adjusting screw to 15 inch-pounds (0.17 mkp, 1.7 Nm).

Install the governor spring once the carburetor has been installed on the engine. Hold the governor spring as shown in Fig. 15–41, with the open end of the small loop down. Hook the large loop in the throttle link loop as shown. Pull the loop toward the throttle lever until the end of the spring loop snaps on. Hook the small loop in the throttle control lever as shown in Fig. 15–41.

On the other models with aluminum cylinders, the horizontal-crankcase models have the governor as shown in Fig. 15–42. Fig. 15–43 shows the vertical-shaft arrangement. Larger aluminum engines are arranged according to Fig. 15–44.

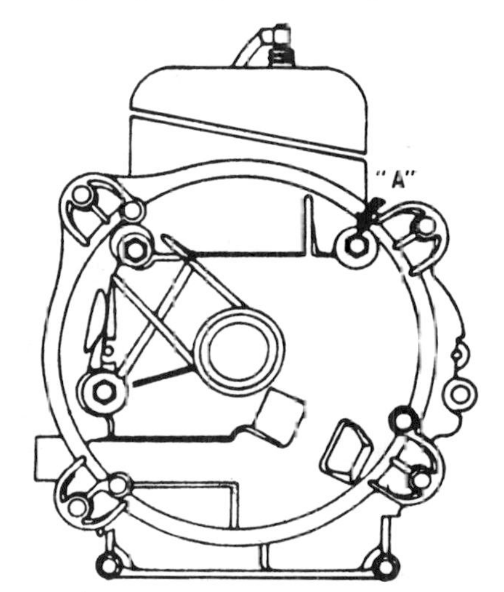

Fig. 15–40. Place nonhardening sealant on screw *A*. *(Courtesy B & S)*

On horizontal-crankshaft models, the governor rides on a short stationary shaft and is retained by the governor shaft, with which it comes in contact after the crankcase cover is secured in place. Press the governor cup against the crankcase cover to seat the retaining ring on the shaft prior to installing the crankcase cover. It is suggested that the assembly of the crankcase cover be made with the crankshaft in the horizontal position. The governor shaft should hang straight down, par-

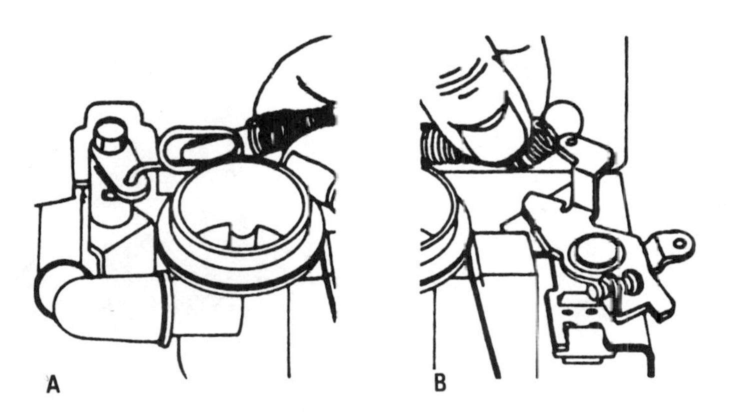

Fig. 15–41. (A) Installing the governor spring. (B) Installation is completed. *(Courtesy B & S)*

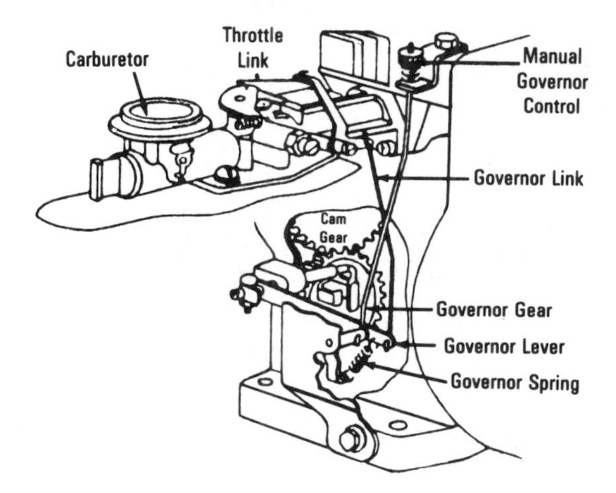

Fig. 15–42. Mechanical governor linkage on a horizontal-shaft engine. *(Courtesy B & S)*

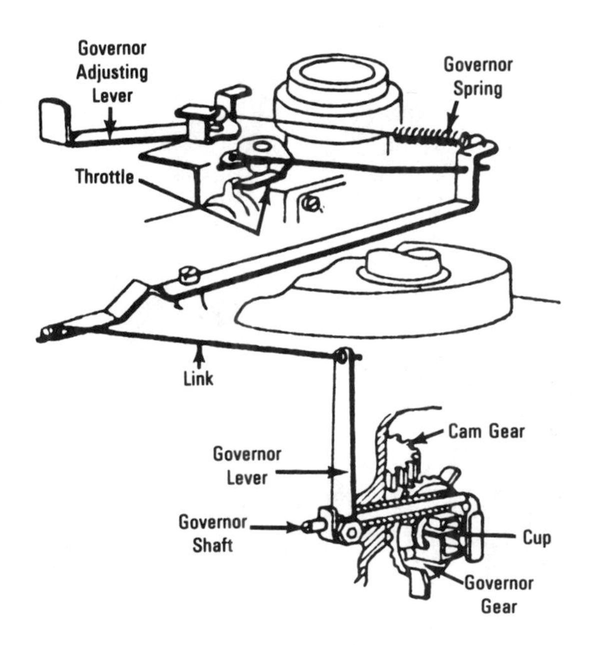

Fig. 15–43. Mechanical governor linkage on a vertical-shaft engine. *(Courtesy B & S)*

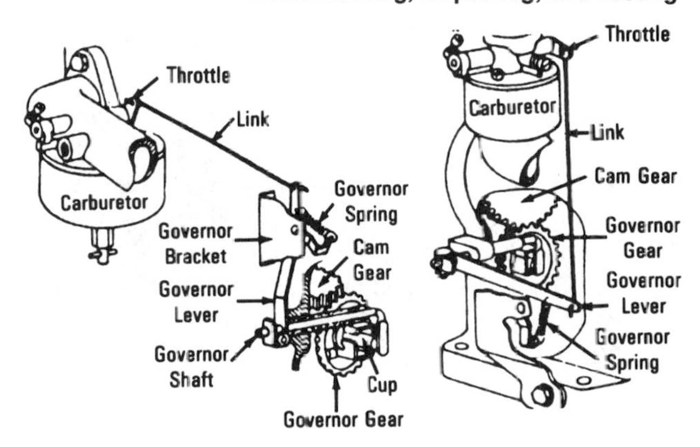

Fig. 15–44. Mechanical governor linkage on large aluminum engines. *(Courtesy B & S)*

allel to the cylinder axis (Fig. 15–45). If the governor shaft is clamped in an angular position, pointing toward the crankcase cover, it is possible for the end of the shaft to be jammed into the inside of the governor assembly, resulting in broken parts when the engine is started. After the crankcase cover and gasket are in place, install the cover screws. Be sure that screw A in Fig. 15–40 has nonhardening sealant on the threads of the screw. Complete the installation of the rest of the gover-

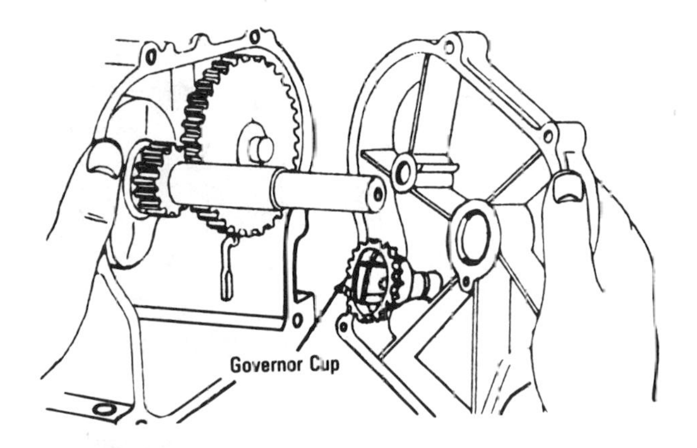

Fig. 15–45. Location of the governor shaft. *(Courtesy B & S)*

nor linkages and carburetor and then adjust the governor shaft and lever.

On vertical-crankshaft models, the governor is part of the oil slinger and is installed as shown in Fig. 15–46. Before installing the sump, be sure that the governor cup is in line with the governor shaft paddle. Install the sump cover and gasket, making sure that screw *A* in Fig. 15–40 has nonhardening sealant on the threads.

Note: On the right-angle auxiliary power take-off models, screw *A* does not need sealant, but the four screws holding the gear sump cover do (Fig. 15–47). Adjustments for the governor linkages can be found in the index.

Installing the Sump or Crankcase Cover

This is the time to install the sump or crankcase cover. These steps have already been shown in a number of illustrations. For instance, Fig. 15–18 shows how the power take-off cover is attached for the auxiliary right-angle drive. Fig. 15–7 shows how the sump is attached and the placement of a thrust washer, if needed. In Fig. 15–45, the horizontal-crankshaft models are shown being assembled.

Be sure to check the end play of the crankshaft before you close up the crankcase or put the cover on. Also check the seals where the crankshaft comes out of the crankcase to make sure they are properly fitted.

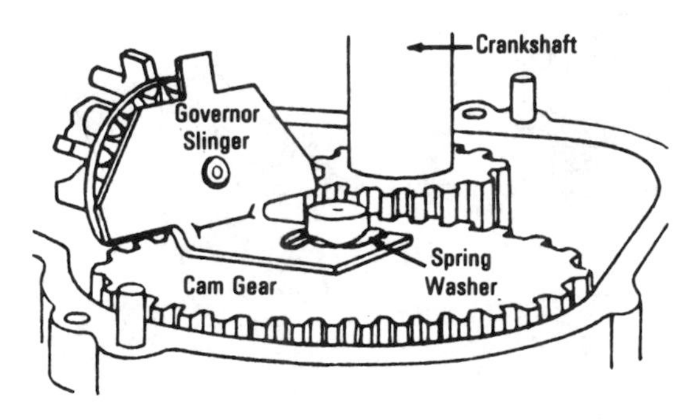

Fig. 15–46. Note the spring on the camshaft after the governor is installed. *(Courtesy B & S)*

Fig. 15–47. Place sealant on the four screws that hold the cover over the auxiliary power take-off section. *(Courtesy B & S)*

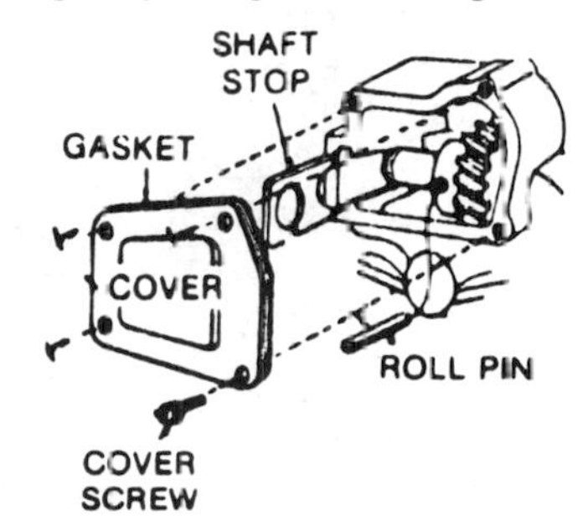

Adjusting Valve Tappet Clearance

Insert the valves in their respective positions in the cylinder. Turn the crankshaft until the exhaust valve is at its highest position Then check the clearance on the intake valve with a feeler gage. Valve tappet clearances are shown in Table 13–1. Grind off the end of the valve stem, if necessary, to obtain the proper clearance. Then turn the crankshaft until the intake valve is at its highest position, and check the exhaust-valve clearance (Fig. 15–48).

Installation of Valves

Some engines use the same spring for the intake and the exhaust side, while others use a heavier spring on the exhaust side. Compare the springs before installing.

If the retainers are held by a pin or collar (Fig. 15–49), place the valve spring and retainer (and cup on model series 9, 14, 19, 20, 23, 24,

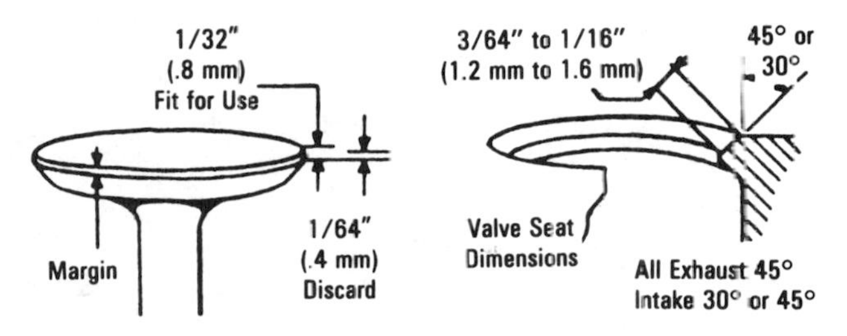

Fig. 15–48. Valve and seat dimensions for Briggs & Stratton engines. *(Courtesy B & S)*

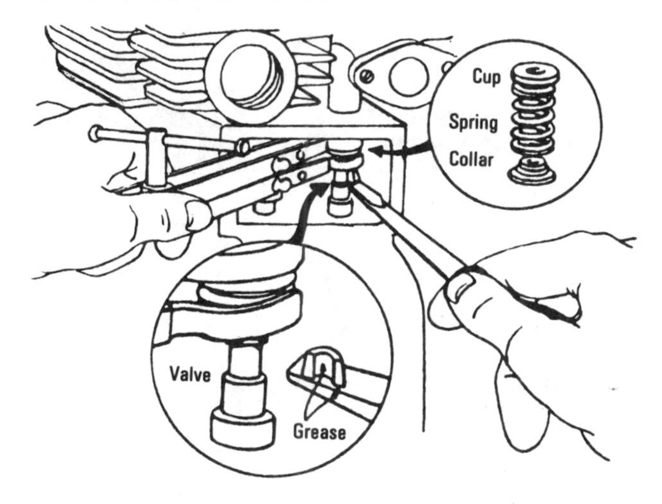

Fig. 15–49. Compressing the spring. *(Courtesy B & S)*

and 32) into the valve spring compressor. Compress the spring until it is solid. Insert the compressed spring and retainer (and cup, when used) into the valve chamber. Then drop the valve into place, pushing the stem through the retainer. Hold the spring up in the chamber, and the valve down, and insert the retainer pin with a pair of needle-nose pliers or place the collars in the valve stem groove. Lower the spring until the retainer fits around the pins or collars, then pull out the spring compressor (Fig. 15–49). Be sure the pin or collars are in place.

If the self-lock retainer (Fig. 15–50C) is used, compress the retainer and the spring with the compressor. The large diameter of the

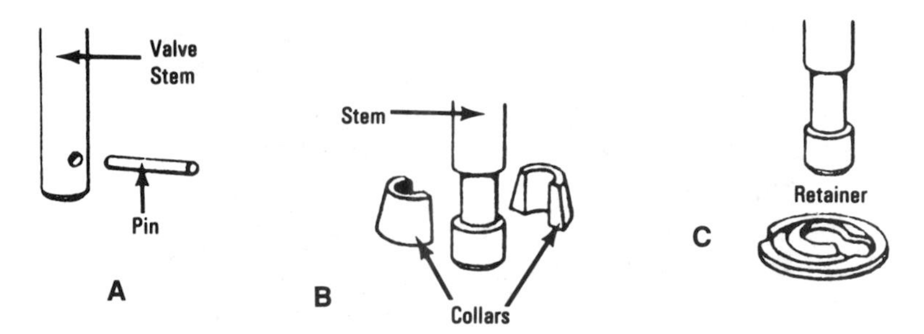

Fig. 15–50. Pin, collars, and retainers for valves. *(Courtesy B & S)*

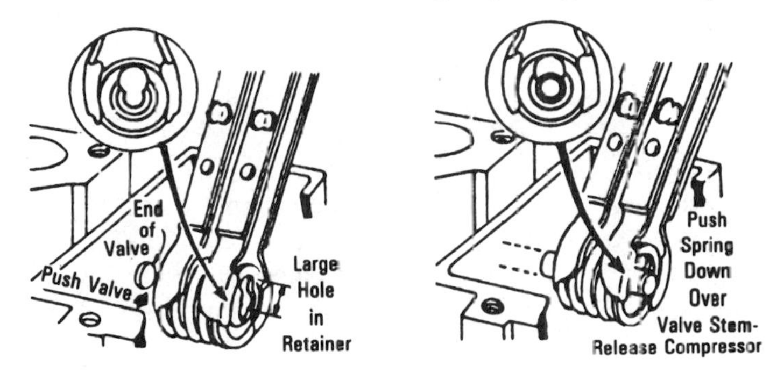

Fig. 15–51. Installing valves. *(Courtesy B & S)*

retainer should be toward the front of the valve chamber (Fig. 15–51). Insert the compressed spring and retainer into the valve chamber. Drop the valve stem through the larger area of the retainer slot and move the compressor so as to center the small area of the valve retainer slot onto the valve stem shoulder. Release the spring tension and remove the compressor.

Ignition System

The next system to merit attention in the sequence of reassembly is the ignition system. Here the breaker points, plunger, armature, coil, and the parts associated with the ignition are taken into consideration. These have been investigated and shown in Chapters 2 and 14. Refer to these chapters for the correct steps in making sure all parts of the system are operational before they are reassembled.

The sequence for checks should be:

1. Coil, armature, and governor blade.
2. Breaker points (internal system).
3. Condenser (internal system).
4. Breaker shaft (Magna-Matic).
5. Primary wire (Magna-Matic).
6. Adjust the armature timing.
7. Adjust and clean breaker points (external type).
8. Breaker point cover replacement.
9. Coil and armature assembly.

10. Adjust rotor timing.
11. Replace the breaker box cover.

Keep in mind that some of these things will vary with the type of engine being reassembled. Here we have taken into consideration the four-cycle engine with internal system and external system of breaker points. The Magna-Matic, being somewhat different, has also been mentioned at the proper reassembly sequence. Know the engine you are working on and make sure you have the ignition system reassembled before you proceed to put the flywheel on the end of the crankshaft.

Flywheel and Starter

Before you place the flywheel on the crankshaft, you need to be sure that all the ignition and starter parts are near at hand. Check your layout in front of you and make sure the points, breaker plunger (if needed), and the clutch and starter are all there and ready to be reassembled.

Examine the exploded views in Fig. 15–1. Look them over for the slight differences in the engines according to model series. These exploded views will allow you to form a mental image of putting the engine back together.

First, you know that the crankshaft has to be in place before the flywheel can be placed on the end. You also know that the ignition parts must be placed on the engine frame before the flywheel can be installed because the flywheel covers the points, the armature, and the coil. The flywheel has to be rotated in order for the breaker points to open and close. You also know that the points have to be adjusted to the proper gap before you install the flywheel. Adjust the points and place the breaker cover over them. Seal the exit point for the wires. Then place the flywheel over the crankshaft end with the Woodruff key in place and move on to the next step.

Keep in mind that some models do not have internally located breaker points. The external types can be mounted *after* the flywheel has been placed on the crankshaft. In all these reassembly instructions you will be looking at a number of types of engines. There are slight differences. You should become aware of the slight differences in procedure as you proceed in gaining experience from working with a number of engines.

Now that the flywheel has been attached to the crankshaft, you can concentrate on replacing the starter or clutch. Recheck the starter motor before mounting it on the engine. The 12-volt gear driver starter shown in Fig. 15–52 uses a helix on the starter motor to drive a pinion gear into engagement with a ring gear attached to the engine flywheel. This cranks the engine. If any sticking occurs or the starter motor does not react properly, inspect the helix and pinion gear for freeness of operation. This sticking must be corrected. Proper operation of the starter depends on the pinion freely moving on the helix.

If the helix is lubricated, it collects dust and dirt. This accumulation can cause trouble as it builds up. If there is some lubrication necessary, spray the helix with a dry silicon and remove the excess before attaching it to the engine.

Fig. 15–53 shows the typical small-engine starter being checked before installation. All parts should have been cleaned by a solvent before reassembly. Varson® or Stanisol may be used. If there is stickiness with the gear and the helix, use a dry silicon spray to lubricate. If this

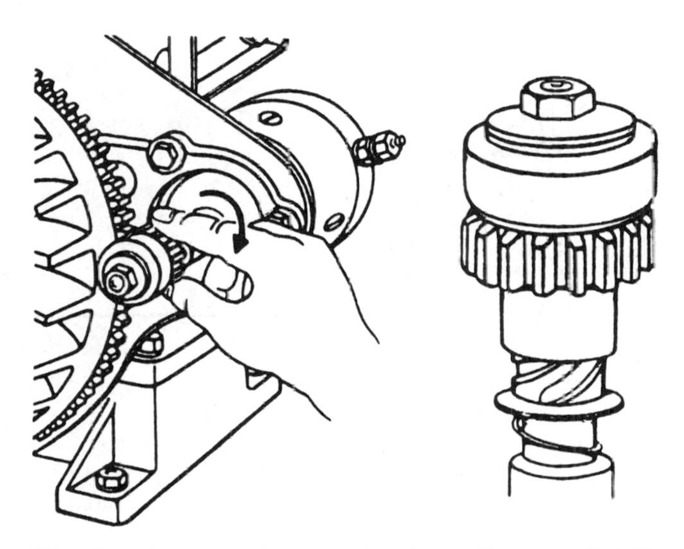

Fig. 15–52. Checking the starter motor drive. *(Courtesy B & S)*

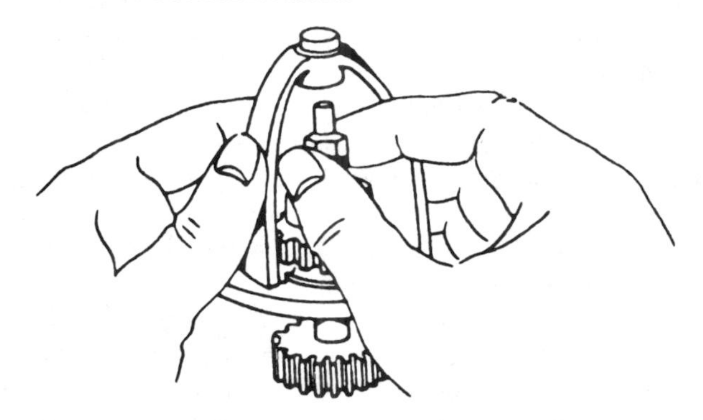

Fig. 15–53. Assembling the starter motor drive. *(Courtesy B & S)*

does not correct the problem, remove it and replace with a new unit. Individual parts of the drive assembly are not available.

Fig. 15–54 shows the 120-volt starter remounted in its proper location on the engine. The 120-volt starter should be Hi-Pot tested before reassembly to the engine. That means you should check between the coils and ground with at least 1000 volts to make sure there are no shorts or breaks in the insulation of the coil. Do not put 1000 volts

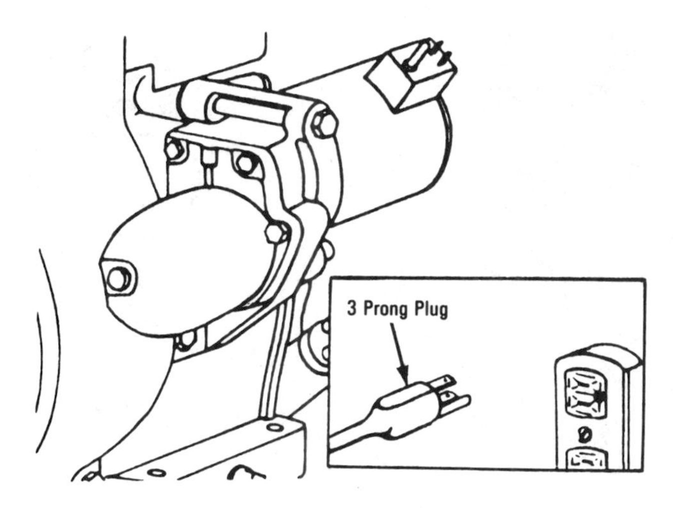

Fig. 15–54. The 120-volt, gear-drive starter. *(Courtesy B & S)*

across the coils. Just touch one coil lead and ground at any time during the testing procedure. Do not become part of the circuit yourself. Use caution.

Adjust the Air Gap to the Flywheel—In order to make sure you have the proper gap, it is best to measure it. This procedure has already been shown in Chapters 5 and 14. However, take a look at Figs. 15–55 and 15–56 to see how the adjustments are made for three different types of armatures. The air gap can be adjusted according to specifications given in Table 14–8.

Check Spark Plug and Install—Clean the spark plug if it has not been done. Clean the gap with a pen knife or wire brush and solvent. Adjust the gap at 0.030″ (0.75 mm) for all models. If the electrodes are burned away or the porcelain is cracked, replace with a new plug (Fig. 15–57). *Do not use abrasive cleaning materials to clean the plug.* The abrasive particles work their way into the engine later and cause damage to the cylinder walls.

The plug should be tightened finger tight, then pulled up snug with a wrench. Snap the high-voltage lead on the spark plug cap.

Breather or Valve Cover

The breather helps create the vacuum needed in the crankcase. The breather is a valve that allows air to flow out the crankcase, but not in. The partial vacuum created by the piston's up-and-down movement

Fig. 15–55. Armature style variations. *(Courtesy B & S)*

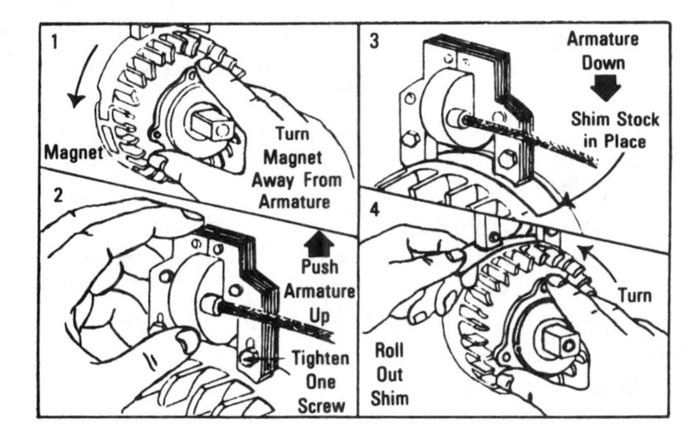

Fig. 15–56. Adjusting the armature air gap. *(Courtesy B & S)*

then allows for normal lubrication to take place without the force of compressed air pushing oil past the rings and through the seals. Use a wire gage to check the space between the fiber disc valve and body. A space of 0.045″ (1.1 mm) is the smallest allowable clearance. In fact, it should be less than that to operate properly. This is a very sensitive valve. Do not push in on the fiber part when checking the clearance. This will distort the valve and cause malfunctioning after you have replaced it (Fig. 15–58).

If the breather is removed for inspection or valve repair, the gasket should be replaced. Tighten the screws securely to prevent oil leakage.

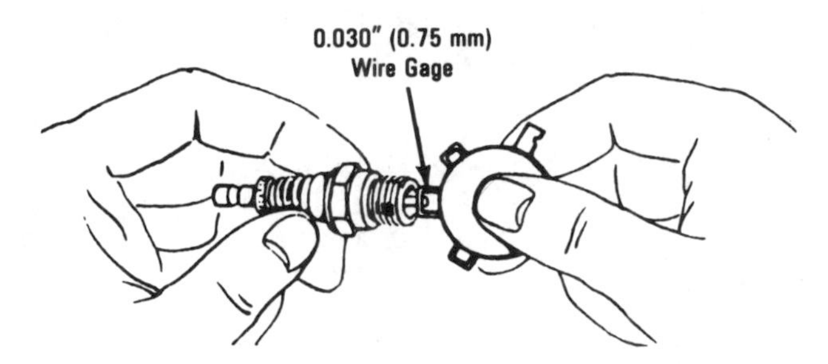

Fig. 15–57. Adjusting the spark plug gap. *(Courtesy B & S)*

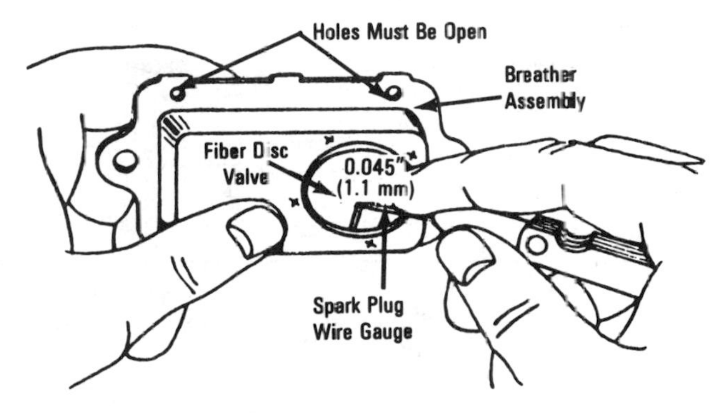

Fig. 15–58. Checking the breather. *(Courtesy B & S)*

Make sure you insert the tubing in the correct place on the valve and the carburetor (Fig. 15–59).

Cylinder Head and Shield

The cylinder head is replaced after a number of jobs have been performed on the cylinder, piston, and valves. Always make sure the ridge and carbon buildup are removed before attempting to remove the piston and ring assembly from the cylinder. This should already have been done previously during the disassembly operation. The cylinder should be deglazed, the valves ground to 45°, and the valve seats to 46° (Fig. 15–60). Lap the valves after grinding for a better seal. Valve guides may be oversized 1/32" (0.794 mm) if worn.

Place the head onto the engine. Torque the head bolts to 200 inch-pounds (230.5 mkp, 22.6 Nm) in 50 inch-pound (57.6 mkp, 5.65 Nm) increments. Tighten in sequence according to Fig. 15–61 or Fig. 15–62. Note that the head bolt marked *1* is 2" (50.8 mm) long. Head bolts *3* and *8* are 1⅜" (34.932 mm) long. All other head bolts are 2¼" (57.15 mm) long. Always use a new head gasket if the cylinder head has been removed.

Later-model engines call for clean surfaces. After cleaning, place the new head gasket on the cylinder and position the head. Slide one Belleville washer (crown toward the bolt head) and one flat washer (sharp edge toward the bolt head) over each bolt as shown in Fig.

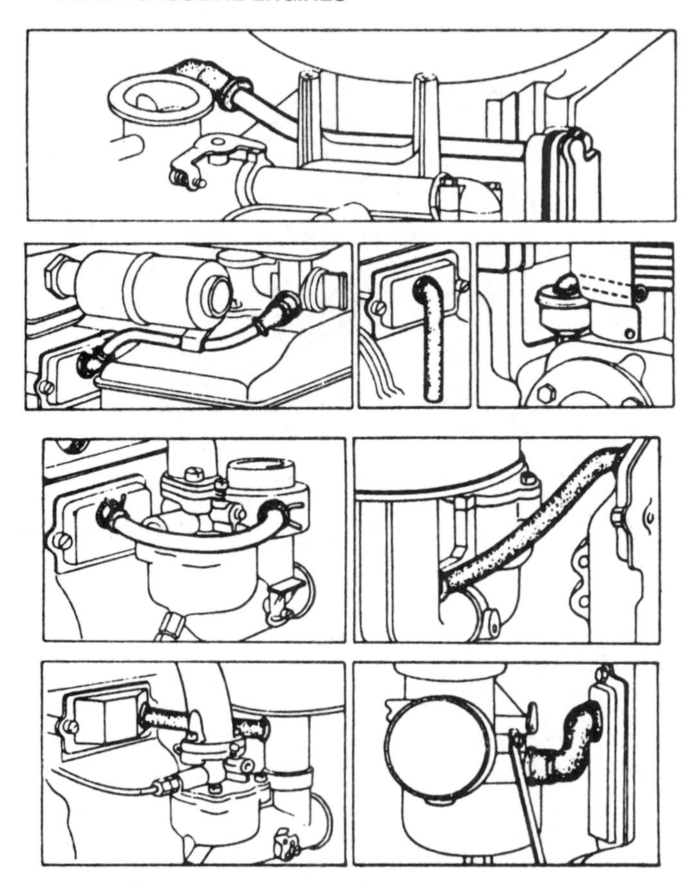

Fig. 15–59. Breather assemblies. *(Courtesy B & S)*

15–63. Insert the bolts in the head with the two shorter bolts in the position shown in Fig. 15–62. The two short bolts for positions *1* and *8* are 2″ (50.8 mm) long; all other bolts are 2¼″ (57.15 mm) long.

Tighten the bolts evenly to 50 inch-pounds (57.6 mkp, 5.65 Nm). Repeat the procedure described previously for the early-model engines. The bolts will have to be retorqued after the engine is run for 15 minutes and allowed to cool down. Retorque in the same sequence shown in Fig. 15–62. If the bolts that hold the shrouding in place are removed, the whole retorquing should be repeated. However, horizontal and vertical engines containing the low-profile head have two sep-

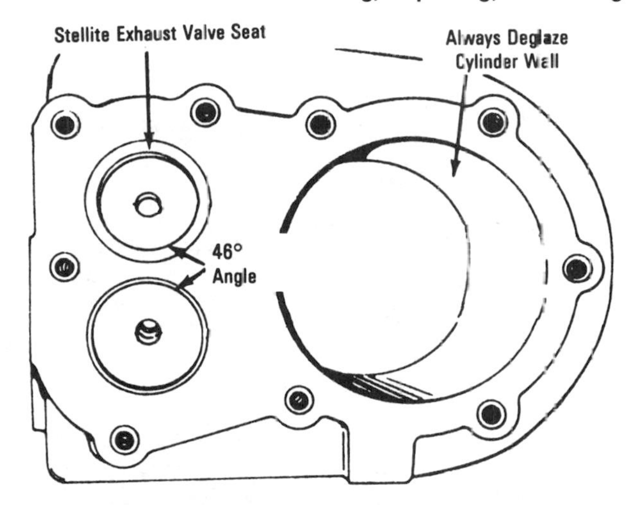

Fig. 15–60. Checking the cylinder before installation of the cylinder head. *(Courtesy Tecumseh)*

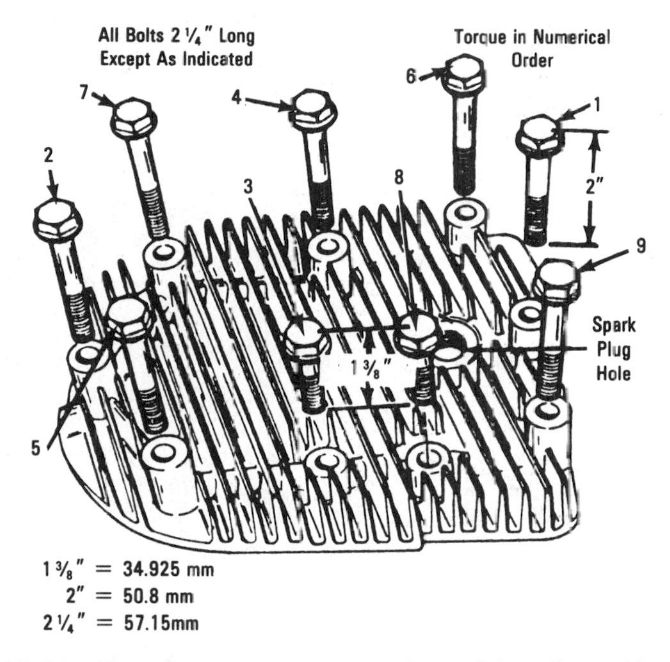

Fig. 15–61. Torquing sequence on early-model engines. *(Courtesy Tecumseh)*

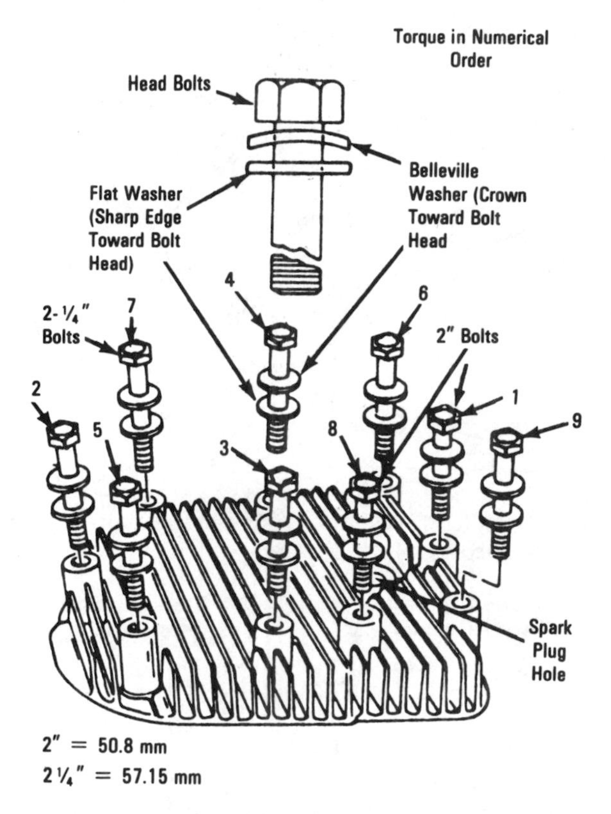

Fig. 15–62. Torquing sequence for later-model engines. *(Courtesy Tecumseh)*

arate bolt mounts that allow the shrouding to be removed without disturbing the cylinder bolts (Fig. 15–63). If the bolts are removed, follow the torquing specifications to reinstall.

Fig. 15–64 is a 16-horsepower, one-cylinder engine with a horizontal crankshaft and power take-off. Note that the shrouding on this engine has air holes that allow air to be taken in to cool the engine. The air cleaner is larger than those seen on the smaller engines.

Other Engine Types

Briggs & Stratton engines have a slightly different arrangement for the cylinder head bolts (Fig. 15–65). Note the position of the differ-

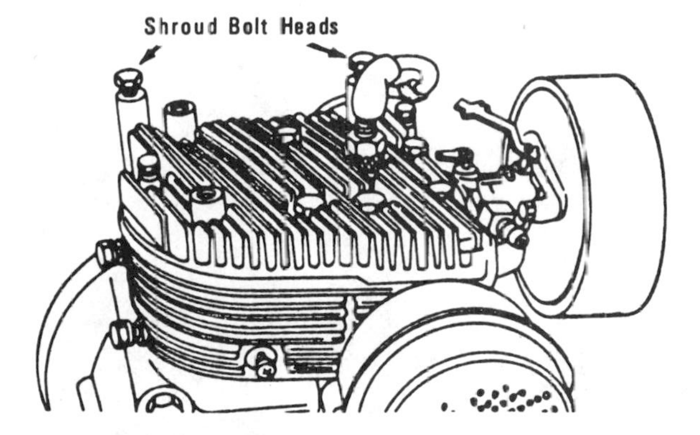

Fig. 15–63. Location of shroud head bolts on a low-profile engine. *(Courtesy Tecumseh)*

ent cylinder head screws so that they are properly reassembled. If a screw is used in the wrong position, it may be too short and not engage enough threads, or it may be too long and bottom on a fin, either breaking the fin or leaving the cylinder head loose.

Assemble the cylinder head with a new head gasket, cylinder head shield, screws, and washers in their proper places. A graphite grease should be used on the aluminum cylinder screws. Do not use a sealer of any kind on the gasket. Tighten the screws down evenly by hand. Use a torque wrench and tighten the head bolts in the sequence shown in Fig. 15–65. Torque to the specifications shown in Table 15–2.

Do not turn one screw down completely before the others. If you do, it may cause a warped cylinder head.

Two-Cycle Engines

The cylinder head is installed easily on some two-cycle engines (Fig. 15–66). This split-crankcase engine uses four cylinder head screws. They are torqued to 90 to 100 inch-pounds (103.7 mkp, 10.17 Nm to 115.3 mkp, 11.3 Nm). Loctite should be applied to the screws before inserting them into the cylinder.

The uniblock two-cycle engine is slightly different (Fig. 15–67). The uniblock engine has a combined crankcase and cylinder design. It provides equal or greater horsepower in a more compact unit. The

Fig. 15–64. An 18-horsepower, L-head, single-cylinder engine. *(Courtesy Tecumseh)*

crankcase compression area has been decreased, allowing for a greater differential in pressure. This means the system will pull the oil-gasoline-air mixture more efficiently. It will compress the mixture tighter, and force it into the combustion chamber with more power. This will purge the exhaust more efficiently. Also, the uniblock engines use the loop scavenge system and the third-port system, which allow greater rpm, hence greater horsepower.

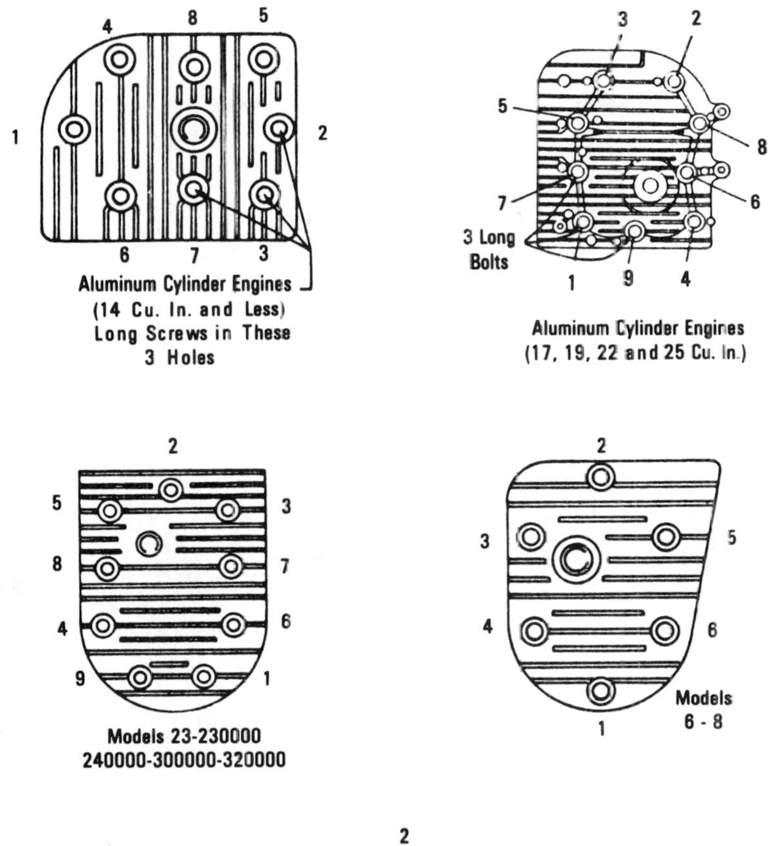

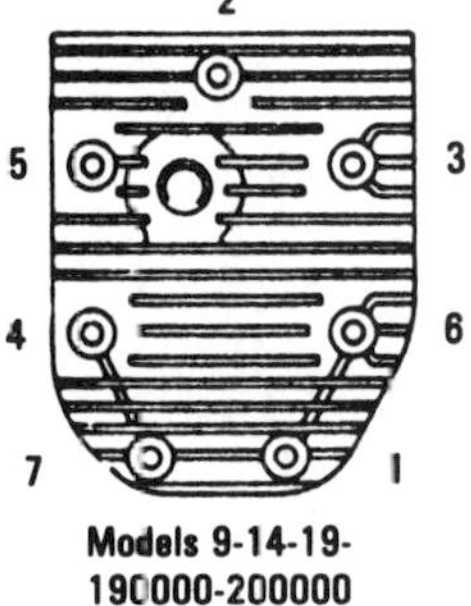

Fig. 15–65. Location of and torquing sequence of the cylinder head bolts on Briggs & Stratton engines. *(Courtesy B & S)*

Table 15–2. Cylinder Head Torque

Basic Model Series	Inch-Pounds	Meter-Kiloponds	Newton-Meters
Aluminum Cylinder			
6B, 60000, 8B, 80000, 82000, 92000, 94000, 110000, 100000, 130000	140	161.3	15.82
140000, 170000, 190000, 220000, 250000	165	190.1	18.65
Cast-Iron Cylinder	**Inch-Pounds**	**Meter-Kiloponds**	**Newton-Meters**
5, 6, N, 8, 9	140	161.3	15.82
14	165	190.1	18.65
19, 190000, 200000, 23, 230000, 240000, 300000, 320000	190	219.0	21.47

(Courtesy B & S)

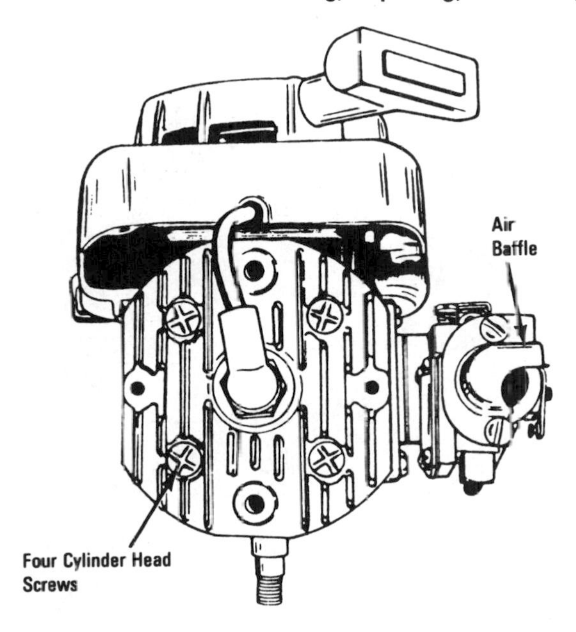

Fig. 15–66. Head bolts on a split-crankcase engine. *(Courtesy Tecumseh)*

The outboard engine head installation on some early 1970 models has a particular identification problem with the cylinder head. The cylinder head may fit to the cylinder in a way that *appears* wrong, but is correct. The head may be like the one shown in Fig. 15–68 with two flats around the circumference, or maybe just one. Important in any case is that the fuel tank bracket mounting hole must have less stock on the boss toward the top of the engine when correctly assembled This is necessary so that the tank bracket positioning slots allow alignment between the tank bracket and the boss mounting holes.

Fig. 15–69 shows that the ridge may have to be removed *before* the piston can be removed. This means that the operation has been done and the piston is back in its proper location in the cylinder. Now, notice that the old gasket is used when installing the piston. Once the piston is in the cylinder, the old gasket is discarded. A new head gasket is called for in the reassembly of the engine. Now, take the cylinder head and attach it, along with a *new* metal head gasket as shown in Fig. 15–70.

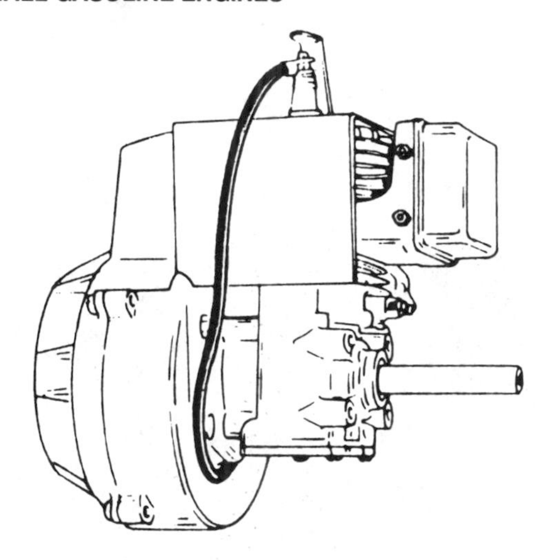

Fig. 15–67. **Uniblock engine.** *(Courtesy Tecumseh)*

Spark Plug and Muffler

Install the spark plug and muffler next. There are different spark plugs available for small engines now. Make sure you have the correct spark plug and it is properly gapped for operation with your engine.

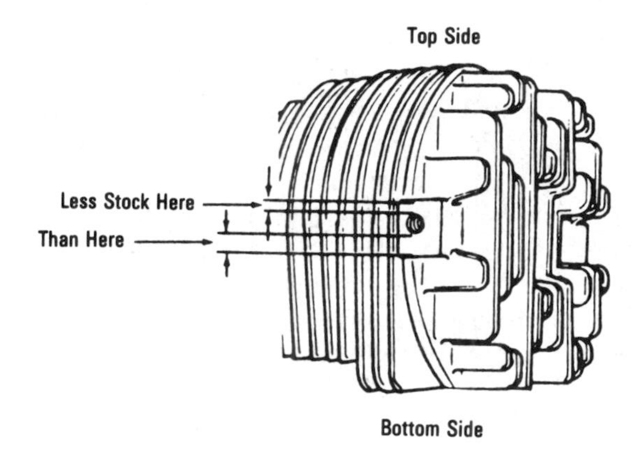

Fig. 15–68. Outboard engine head installation on some 1970 models. *(Courtesy Tecumseh)*

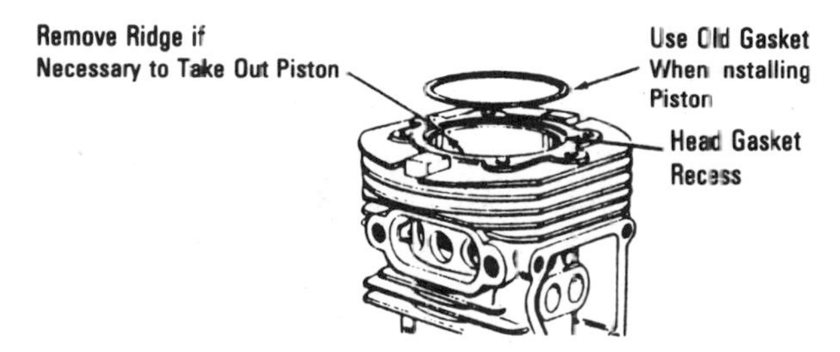

Fig. 15–69. Location of the used gear gasket when installing the piston before putting on the cylinder head. *(Courtesy Tecumseh)*

Refer to Chapter 5 for more details on the spark plug and its installation.

The muffler is a necessary device if you want to be able to retain your hearing. An engine without a muffler can be very loud and hard on the ears. Check with Chapter 8 for more details on the muffler and exhaust system.

Place the intake elbow on the engine and attach the gas tank. Run the hose from the gas tank down to where the carburetor is to be placed.

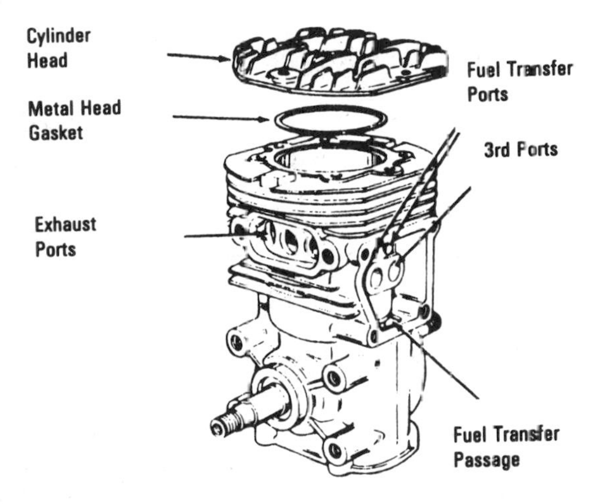

Fig. 15–70. Replacing the cylinder head on a two-cycle engine. *(Courtesy Tecumseh)*

Carburetor and Linkages

The carburetor is mounted next. This device is very different for each engine, so the chapter on carburetion was designed to cover the various types. Take a look at Chapter 7 to obtain the details on the various carburetors and how they are fastened to the engine block. The carburetor linkages and their adjustments will be found in the appendix. There are so many types that it would be somewhat cumbersome to include them here or in the chapter on carburetion. Therefore, you should refer to the appendix for the type of carburetor to be mounted on your particular type of engine.

Make the proper linkages and check the throttle and the governor controls at this time. The engine will be able to run once it is started, if proper adjustments are made.

Check the air vane governor to make sure it is not obstructed and can move freely when the air from the flywheel fan causes it to move.

Check the blower housing. It should be secured to the engine block and free of any obstructions so that the air flow is not impeded.

Check the fuel filter, the tank, and the line.

The air cleaner should be assembled and placed on the carburetor at this time.

Filling the Crankcase with Oil

If you have a four-cycle engine, you will have to fill the crankcase with oil to the specified level. Make sure it is the correct grade of oil for this engine (see Chapter 9).

If you have a two-cycle engine, now is the time to mix the fuel properly, with the correct ratio of oil to gasoline.

Filling the Gas Tank and Starting the Engine

Check all the settings you need to have for the engine you are working with to make sure it will start properly the first time. If not, you will have to make adjustments to the carburetor or go through the troubleshooting procedure to locate the problem.

The gas tank for a four-cycle engine should be filled with fresh gasoline. The fuel tank for the two-cycle should be filled with the correct ratio of gasoline and oil.

Check the spark if the engine does not start the first six times the starter is pulled. Then check the carburetion. You may be flooding the engine. If so, follow the correct procedure for starting the engine while flooded.

Let the engine run for 15 minutes to reach its normal operating temperature. Turn the engine off and then adjust the cylinder head screws once it has cooled enough so you can touch the cylinder head without being burned. Torque the head bolts to their proper torque specifications.

Start the engine again. Adjust the carburetor for the speed recommended for the job to be performed by the engine.

Check out the controls on the remote-control mechanism (Fig. 15–71).

Check to see if the stop wire is working properly and is connected to the carburetor and the engine block. See if the "flick-top" stop switch

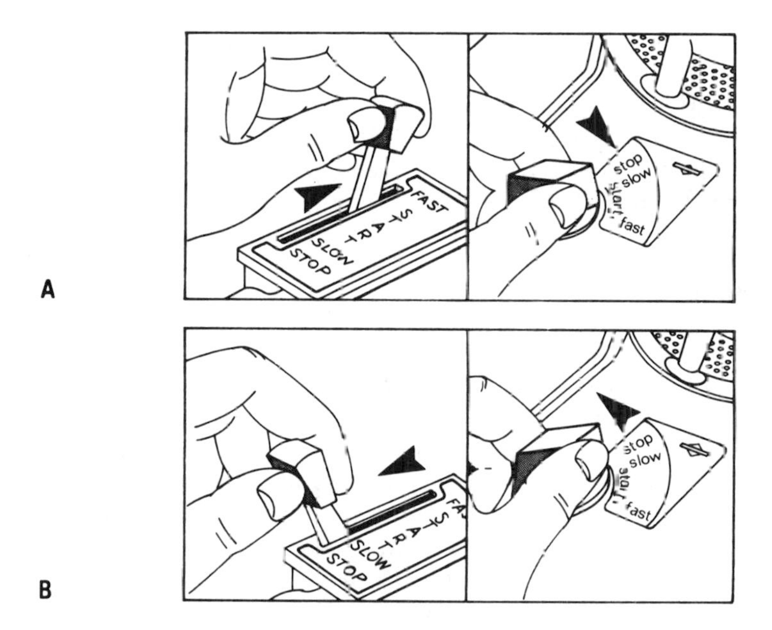

Fig. 15–71. (A) Control to *START* position. (B) Control to *STOP* position. *(Courtesy B & S)*

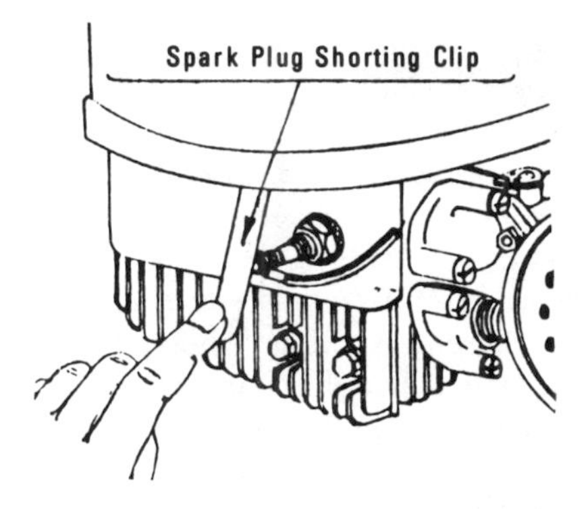

Fig. 15–72. Spark plug shorting clip designed to stop the engine. *(Courtesy Tecumseh)*

is operating properly, if this is the type of switch used for the stop position of the controls (Fig. 15–72).

Fig. 15–73 shows a manufacturer's test stand for new engines. This engine has not yet been painted. Every engine is test-run before it is painted.

If you have tested your engine and it runs properly, now is the time to spray paint it and make it look, as well as run, like new. There is nothing like a good paint job on the outside to make people think the engine has had a complete overhaul and will work like a new one.

Repairing Stripped Threads

In some instances the threads for the dead bolt holes have become stripped or damaged. The first inclination is to throw the block away and start with a new one. This can be very expensive. One of the newer devices designed to correct damaged threads is the Heli-Coil® repair kit. Fig. 15–74 shows the stripped hole and the finished hole with a new stainless-steel thread. There are three steps to making a new thread: (1) Drill the hole to the correct size, given by the instructions on the repair kit; (2) tap the hole that was just drilled; (3) install the new threads from the kit (Fig. 15–74B).

Heli-Coil inserts are precision-formed screw threads of stainless-steel wire having a diamond-shaped cross section (Fig. 15–74C). This

Fig. 15–73. Test stand for checking engines at the factory before they are painted. Every engine is run before it is painted and finished. *(Courtesy Tecumseh)*

diamond-shaped coil is a result of rolling round wire (this gives the wire extremely high strength and hardness) and then coiling it. When installed, the inserts form permanent, conventional threads. Each insert has a tang for installation purposes. It is notched for easy removal so that a through free-running threaded assembly results. The tang is broken off after installation simply by striking it with a piece of rod or, in large sizes and spark plug applications, bending it up and down using long-nosed pliers until it snaps off at the notch.

Inserts are retained in the hole with springlike action (Fig.

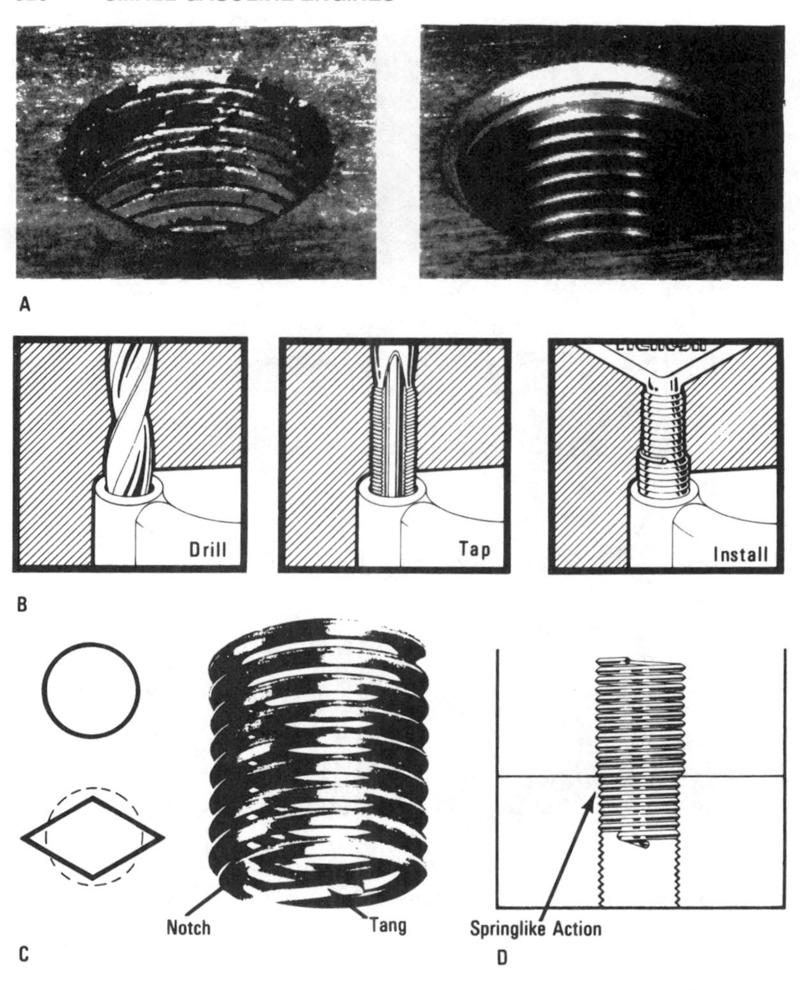

Fig. 15–74. Replacing a stripped threaded hole with a Heli-Coil insert. *(Courtesy Heli-Coil Products)*

15–74D). In the free state they are greater in diameter than the tapped hole into which they are installed. In the assembly operation, the torque applied to the tank reduces the diameter of the leading coil and permits it to enter the hole where the tapped threads are located. When the torque or rotation is stopped, the coils expand with a springlike action, anchoring the insert permanently in place against the tapped hole. No staking, locking, swaging, key, or interference fits are required. Since the insert is made of wire, it automatically adjusts itself to any expansion or contraction of the parent materials.

Chapter Highlights

This chapter has described:

1. How to reassemble an engine.
2. How to inspect an engine and all its individual parts.
3. How to test each part of an engine for proper operation.
4. How to install all parts of an engine properly.
5. How to replace or repair stripped threads on nuts and bolts.

Vocabulary Checklist

actuator
anchoring
auxiliary drive cover
bearing support
cam gear shaft
cast-iron engine
crankpin
Easy-Spin lobe
end play
freeness
helix
interference fit
Loctite
pivot end
protrudes
self-lock retainer
swaging
third port
thrust washer

Review Questions

1. What has to be replaced before the tappets?
2. In the aluminum-alloy engine, what is inserted first, before the crankshaft?
3. What is crankshaft end play?
4. Why do you want to eliminate end play?
5. What is a thrust washer?

6. How do you check for end play?
7. What types of engines *do not* use a thrust washer?
8. Why doesn't the cam gear end play tolerance present a problem?
9. What is a shim? How is the shim used?
10. Where is the auxiliary power take-off located?
11. How fast does the auxiliary power take-off rotate?
12. What is a piston pin?
13. How do you insert the pistons back into the cylinder with new rings?
14. What is a dipper?
15. What is a slinger?
16. What type of engine uses the dipper?
17. What does a governor do?
18. How does the governor work? Mechanical type? Air vane type?
19. On vertical-crankshaft engines the governor is also part of another engine device. What is this device called?
20. How do you clean the spark plug before reassembly?
21. What is the purpose of the breather?
22. How much torque should you apply to a head bolt?
23. What type of grease can be used on screws to secure an aluminum-alloy engine head?
24. What is a "flick-top" switch?

APPENDICES

Appendix I

Millimeter Conversion Chart

mm In.	mm In.	mm In.	mm In.	mm In.	mm In.
0.25=0.0098	11.25=0.4429	22.25= .8760	33.25=1.3091	44.25=1.7421	55.25=2.1752
0.50=0.0197	11.50=0.4528	22.50= .8858	33.50=1.3189	44.50=1.7520	55.50=2.1850
0.75=0.0295	11.75=0.4626	22.75= .8957	33.75=1.3287	44.75=1.7618	55.75=2.1949
1 =0.0394	12 =0.4724	23 = .9055	34 =1.3386	45 =1.7716	56 =2.2047
1.25=0.0492	12.25=0.4823	23.25= .9153	34.25=1.3484	45.25=1.7815	56.25=2.2146
1.50=0.0591	12.50=0.4921	23.50= .9252	34.50=1.3583	45.50=1.7913	56.50=2.2244
1.75=0.0689	12.75=0.5020	23.75= 9350	34.75=1.3681	45.75=1.8012	56.75=2.2342
2 =0.0787	13 =0.5118	24 = 9449	35 =1.3779	46 =1.8110	57 =2.2441
2.25=0.0886	13.25=0.5217	24.25= .9547	35.25=1.3878	46.25=1.8209	57.25=2.2539
2.50=0.0984	13.50=0.5315	24.50= .9646	35.50=1.3976	46.50=1.8307	57.50=2.2638
2.75=0.1083	13.75=0.5413	24.75= .9744	35.75=1.4075	46.75=1.8405	57.75=2.2736
3 =0.1181	14 =0.5512	25 = .9842	36 =1.4173	47 =1.8504	58 =2.2835
3.25=0.1280	14.25=0.5610	25.25= .9941	36.25=1.4272	47.25=1.8602	58.25=2.2933
3.50=0.1378	14.50=0.5709	25.50=1.0039	36.50=1.4370	47.50=1.8701	58.50=2.3031
3.75=0.1476	14.75=0.5807	25.75=1.0138	36.75=1.4468	47.75=1.8799	58.75=2.3130
4 =0.1575	15 =0.5905	26 =1.0236	37 =1.4567	48 =1.8898	59 =2.3228
4.25=0.1673	15.25=0.6004	26.25=1.0335	37.25=1.4665	48.25=1.8996	59.25=2.3327
4.50=0.1772	15.50=0.6102	26.50=1.0433	37.50=1.4764	48.50=1.9094	59.50=2.3425
4.75=0.1870	15.75=0.6201	26.75=1.0531	37.75=1.0531	48.75=1.9193	59.75=2.3524
5 =0.1968	16 =0.6299	27 =1.0630	38 =1.4961	49 =1.9291	60 =2.3622
5.25=0.2067	16.25=0.6398	27.75=1.0728	38.25=1.5059	49.25=1.9390	60.25=2.3720
5.50=0.2165	16.50=0.6496	27.50=1.0827	38.50=1.5157	49.50=1.9488	60.50=2.3819
5.75=0.2264	16.75=0.6594	27.75=1.0925	38.75=1.5256	49.75=1.9587	60.75=2.3917
6 =0.2362	17 =0.6693	28 =1.1024	39 =1.5354	50 =1.9685	61 =2.4016
6.25=0.2461	17.25=0.6791	28.25=1.1122	39.25=1.5453	50.25=1.9783	61.25=2.4114
6.50=0.2559	17.50=0.6890	28.50=1.1220	39.50=1.5551	50.50=1.9882	61.50=2.4213
6.75=0.2657	17.75=0.6988	28.75=1.1319	39.75=1.5650	50.75=1.9980	61.75=2.4311
7 =0.2756	18 =0.7087	29 =1.1417	40 =1.5748	51 =2.0079	62 =2.4409
7.25=0.2854	18.25=0.7185	29.25=1.1516	40.25=1.5846	51.25=2.0177	62.25=2.4508
7.50=0.2953	18.50=0.7283	29.50=1.1614	40.50=1.5945	51.50=2.0276	62.50=2.4606
7.75=0.3051	18.75=0.7382	29.75=1.1713	40.75=1.6043	51.75=2.0374	62.75=2.4705
8 =0.3150	19 =0.7480	30 =1.1811	41 =1.6142	52 =2.0472	63 =2.4803
8.25=0.3248	19.25=0.7579	30.25=1.1909	41.25=1.6240	52.25=2.0571	63.25=2.4901
8.50=0.3346	19.50=0.7677	30.50=1.2008	41.50=1.6339	52.50=2.0669	63.50=2.5000
8.75=0.3445	19.75=0.7776	30.75=1.2106	41.75=1.6437	52.75=2.0768	63.75=2.5098
9 =0.3543	20 =0.7874	31 =1.2205	42 =1.6535	53 =2.0866	64 =2.5197
9.25=0.3642	20.25=0.7972	31.25=1.2303	42.25=1.6634	53.25=2.0965	64.25=2.5295
9.50=0.3740	20.50=0.8071	31.50=1.2402	42.50=1.6732	53.50=2.1063	64.50=2.5394
9.75=0.3839	20.75=0.8169	31.75=1.2500	42.75=1.6831	53.75=2.1161	64.75=2.5492
10 =0.3937	21 =0.8268	32 =1.2598	43 =1.6929	54 =2.1260	65 =2.5590
10.25=0.4035	21.25=0.8366	32.25=1.2697	43.25=1.7028	54.25=2.1358	65.25=2.5689
10.50=0.4134	21.50=0.8465	32.50=1.2795	43.50=1.7126	54.50=2.1457	65.50=2.5787
10.75=0.4232	21.75=0.8563	32.75=1.2894	43.75=1.7224	54.75=2.1555	65.75=2.5886
11 =0.4331	22 =0.8661	33 =1.2992	44 =1.7323	55 =2.1653	66 =2.5984

Millimeter Conversion Chart (Cont'd)

mm In.	mm In.	mm In.	mm In.	mm In.	mm In.
66.25=2.6083	77.25=3.0413	88.25=3.4744	99.25=3.9075	110.25=4.3405	121.25=4.7736
66.50=2.6181	77.50=3.0512	88.50=3.4842	99.50=3.9173	110.50=4.3504	121.50=4.7885
66.75=2.6279	77.75=3.0610	88.75=3.4941	99.75=3.9272	110.75=4.3602	121.75=4.7933
67 =2.6378	78 =3.0709	89 =3.5039	100 =3.9370	111 =4.3701	122 =4.8031
67.25=2.6476	78.25=3.0807	89.25=3.5138	100.25=3.9468	111.25=4.3799	122.25=4.8130
67.50=2.6575	78.50=3.0905	89.50=3.5236	100.50=3.9567	111.50=4.3898	122.50=4.8228
67.75=2.6673	78.75=3.1004	89.75=3.5335	100.75=3.9665	111.75=4.3996	122.75=4.8327
68 =2.6772	79 =3.1102	90 =3.5433	101 =3.9764	112 =4.4094	123 =4.8425
68.25=2.6870	79.25=3.1201	90.25=3.5531	101.25=3.9862	112.25=4.4193	123.25=4.8524
68.50=2.6968	79.50=3.1299	90.50=3.5630	101.50=3.9961	112.50=4.4291	123.50=4.8622
68.75=2.7067	79.75=3.1398	90.75=3.5728	101.75=4.0059	112.75=4.4390	123.75=4.8720
69 =2.7165	80 =3.1496	91 =3.5827	102 =4.0157	113 =4.4488	124 =4.8819
69.25=2.7264	80.25=3.1594	91.25=3.5925	102.25=4.0256	113.25=4.4587	124.25=4.8917
69.50=2.7362	80.50=3.1693	91.50=3.6024	102.50=4.0354	113.50=4.4685	124.50=4.9016
69.75=2.7461	80.75=3.1791	91.75=3.6122	102.75=4.0453	113.75=4.4783	124.75=4.9114
70 =2.7559	81 =3.1890	92 =3.6220	103 =4.0551	114 =4.4882	125 =4.9212
70.25=2.7657	81.25=3.1988	92.25=3.6319	103.25=4.0650	114.25=4.4980	125.25=4.9311
70.50=2.7756	81.50=3.2087	92.50=3.6417	103.50=4.0748	114.50=4.5079	125.50=4.9409
70.75=2.7854	81.75=3.2185	92.75=3.6516	103.75=4.0846	114.75=4.5177	125.75=4.9508
71 =2.7953	82 =3.2283	93 =3.6614	104 =4.0945	115 =4.5275	126 =4.9606
71.25=2.8051	82.25=3.2382	93.25=3.6713	104.25=4.1043	115.25=4.5374	126.25=4.9705
71.50=2.8150	82.50=3.2480	93.50=3.6811	104.50=4.1142	115.50=4.5472	126.50=4.9803
71.75=2.8248	82.75=3.2579	93.75=3.6909	104.75=4.1240	115.75=4.5571	126.75=4.9901
72 =2.8346	83 =3.2677	94 =3.7008	105 =4.1338	116 =4.5669	127 =5.0000
72.25=2.8445	83.25=3.2776	94.25=3.7106	105.25=4.1437	116.25=4.5768	
72.50=2.8543	83.50=3.2874	94.50=3.7205	105.50=4.1535	116.50=4.5866	
72.75=2.8642	83.75=3.2972	94.75=3.7303	105.75=4.1634	116.75=4.5964	
73 =2.8740	84 =3.3071	95 =3.7401	106 =4.1732	117 =4.6063	
73.25=2.8839	84.25=3.3169	95.25=3.7500	106.25=4.1831	117.25=4.6161	
73.50=2.8937	84.50=3.3268	95.50=3.7598	106.50=4.1929	117.50=4.6260	
73.75=2.9035	84.75=3.3366	95.75=3.7697	106.75=4.2027	117.75=4.6358	
74 =2.9134	85 =3.3464	96 =3.7795	107 =4.2126	118 =4.6457	
74.25=2.9232	85.25=3.3563	96.25=3.7894	107.25=4.2224	118.25=4.6555	
74.50=2.9331	85.50=3.3661	96.50=3.7992	107.50=4.2323	118.50=4.6653	
74.75=2.9429	85.75=3.3760	96.75=3.8090	107.75=4.2421	118.75=4.6752	
75 =2.9527	86 =3.3858	97 =3.8189	108 =4.2520	119 =4.6850	
75.25=2.9626	86.25=3.3957	97.25=3.8287	108.25=4.2618	119.25=4.6949	
75.50=2.9724	86.50=3.4055	97.50=3.8386	108.50=4.2716	119.50=4.7047	
75.75=2.9823	86.75=3.4153	97.75=3.8484	108.75=4.2815	119.75=4.7146	
76 =2.9921	87 =3.4252	98 =3.8583	109 =4.2913	120 =4.7244	
76.25=3.0020	87.25=3.4350	98.25=3.8681	109.25=4.3012	120.25=4.7342	
76.50=3.0118	87.50=3.4449	98.50=3.8779	109.50=4.3110	120.50=4.7441	
76.75=3.0216	87.75=3.4547	98.75=3.8878	109.75=4.3209	120.75=4.7539	
77 =3.0315	88 =3.4646	99 =3.8976	110 =4.3307	121 =4.7638	

Appendix 2

Inches to Millimeter Conversion Chart

Fraction	Inches	mm	Fraction	Inches	mm
1/64	0.015625	0.39688	33/64	0.515625	13.09688
[illegible]	0.03125	0.79375	17/32	0.53125	13.49375
3/64	0.046875	1.19063	35/64	0.546875	13.89063
[illegible]	0.0625	1.58750	9/16	0.5625	14.28750
5/64	0.078125	1.98438	37/64	0.578125	14.68438
[illegible]	0.09375	2.38125	19/32	0.59375	15.08125
7/64	0.109375	2.77813	39/64	0.609375	15.47813
[illegible]	0.125	3.17500	5/8	0.625	15.87500
9/64	0.140625	3.57188	41/64	0.640625	16.27188
[illegible]	0.15625	3.97875	21/32	0.65625	16.66875
11/64	0.171875	4.36563	43/64	0.671875	17.06563
[illegible]	0.1875	4.76250	11/16	0.6875	17.46250
13/64	0.203125	5.15938	45/64	0.703125	17.85938
[illegible]	0.21875	5.55625	23/32	0.71875	18.25625
15/64	0.234375	5.95313	47/64	0.734375	18.65313
[illegible]	0.25	6.35000	3/4	0.75	19.05000
17/64	0.265625	6.74688	49/64	0.765625	19.44688
[illegible]	0.28125	7.14375	25/32	0.78125	19.84375
19/64	0.296875	7.54063	51/64	0.796875	20.24063
[illegible]	0.3125	7.93750	13/16	0.8125	20.63750
21/64	0.328125	8.33438	53/64	0.828125	21.03438
11/32	0.34375	8.73125	27/32	0.84375	21.43125
23/64	0.359375	9.12813	55/64	0.859375	21.82804
[illegible]	0.375	9.52500	7/8	0.875	22.22500
25/64	0.390625	9.92188	57/64	0.890625	22.62188
13/32	0.40625	10.31875	29/32	0.90625	23.01875
27/64	0.421875	10.71563	59/64	0.921875	23.41563
[illegible]	0.4375	11.11250	15/16	0.9375	23.81250
29/64	0.453125	11.50938	61/64	0.953125	24.20929
15/32	0.46875	11.90625	31/32	0.96875	24.60625
31/64	0.484375	12.30313	63/64	0.984375	25.00313
[illegible]	0.500	12.70000	1	1.000	25.40000

APPENDIX 3

Governor Linkages and Carburetor Adjustments for Briggs & Stratton Engines

Remote Controls

In general, there are three types of remote controls: governor control, throttle control, choke-a-matic control. Figs. A–1 to A–6 show the operation of these control systems. See the following pages for specific control assemblies and installation hookup by engine model.

Remote Governor Control

The remote governor control regulates the engine speed by changing the governor spring tension, thus allowing the governor to control the carburetor throttle at all times and maintain any desired speed (Fig. A–1).

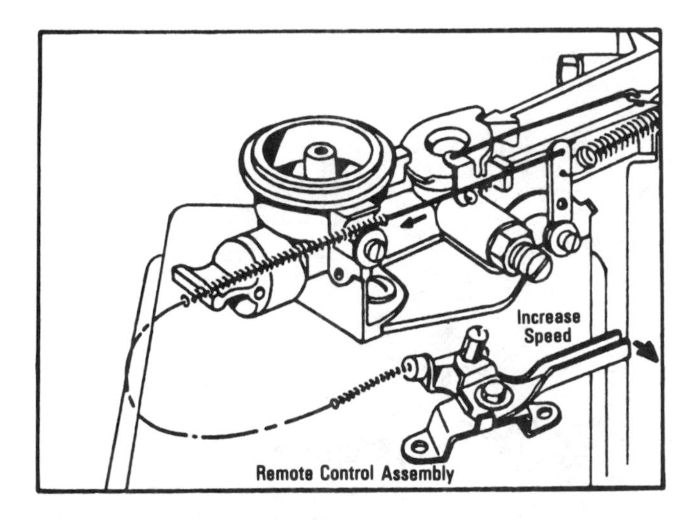

Fig. A–1. Remote Governor Control.

Remote Throttle Control

The remote throttle control is used on an engine having a fixed no-load governed speed setting such as 3600 or 4000 rpm. This control enables an operator to control the speed of an engine, similar to an accelerator used on an automobile. However, when full governed speed is obtained, the governor prevents overspeeding and possible damage to the engine. At any point below the governed speed, the throttle is held in a fixed position and the engine speed will vary with the load (Fig. A–2).

Choke-A-Matic Remote Control

On Choke-A-Matic carburetors, the remote control must be correctly adjusted in order to obtain proper operation of the choke and stop switch.

Note: Remote control system must be mounted on powered equipment in normal operating position before adjustments are made.

Fig. A–4 illustrates typical remote control installations used with Choke-A-Matic carburetors. To adjust, move remote control lever to “Fast” position. Choke actuating lever *A* should just contact choke

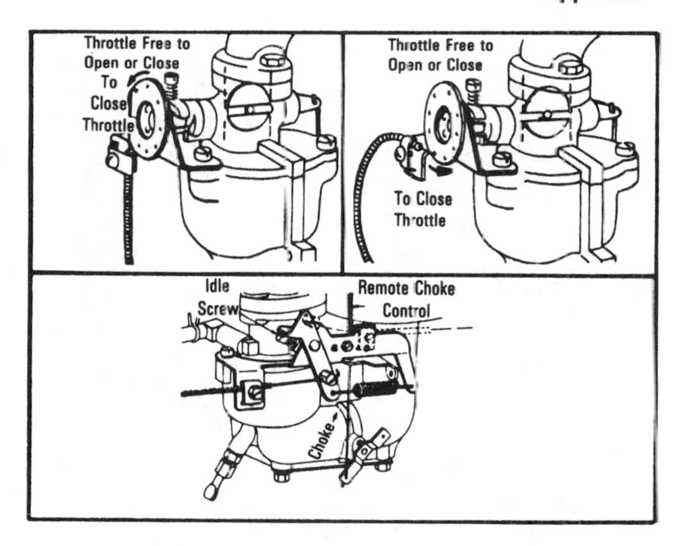

Fig. A–2. Remote Throttle Control.

shaft *B* or link *B* as shown in Fig. A–4. If not, loosen screw *C* slightly and move casing and wire *D* in or out to obtain this condition.

Check operation by moving remote control lever to *START* or *CHOKE* position. Choke valve should be completely closed (Fig. A–5A). Then move remote control level to *STOP* position. Control must contact stop switch blade (Fig. A–5B).

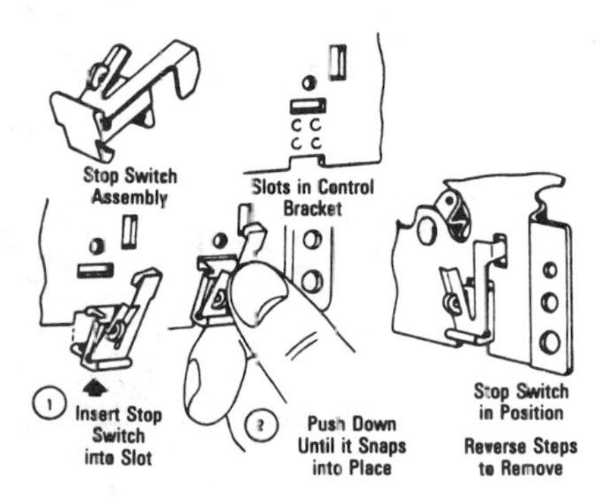

Fig. A–3. Typical Stop Switch Installation.

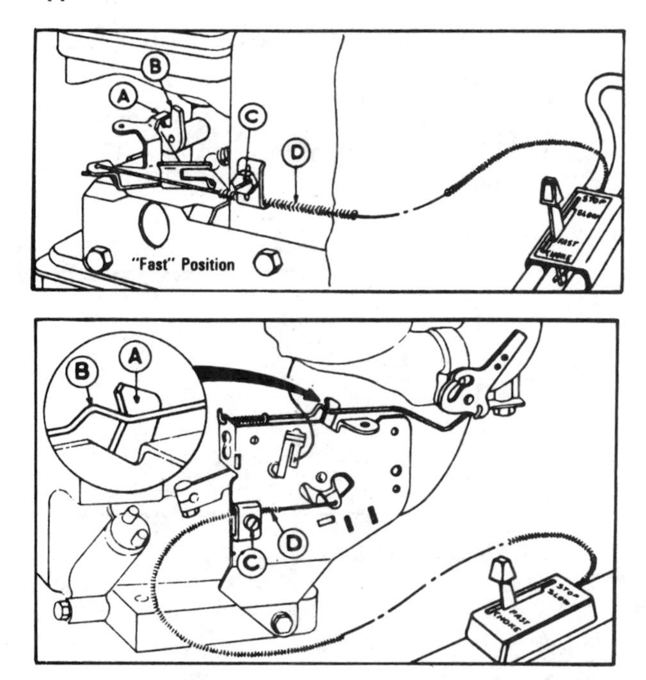

Fig. A–4. Choke-A-Matic Control (Typical).

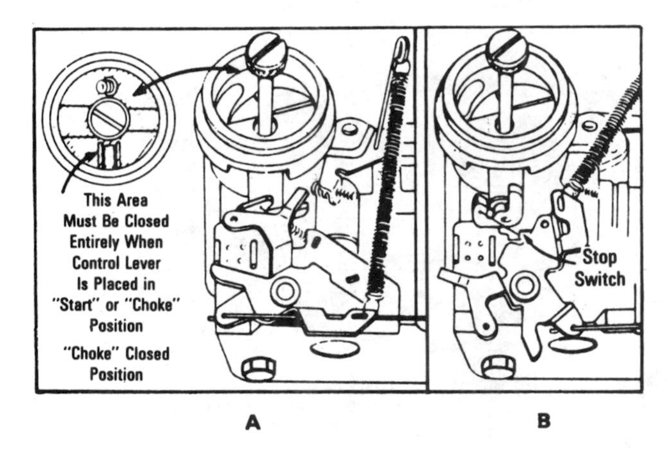

Fig. A–5. Choke and Stop position.

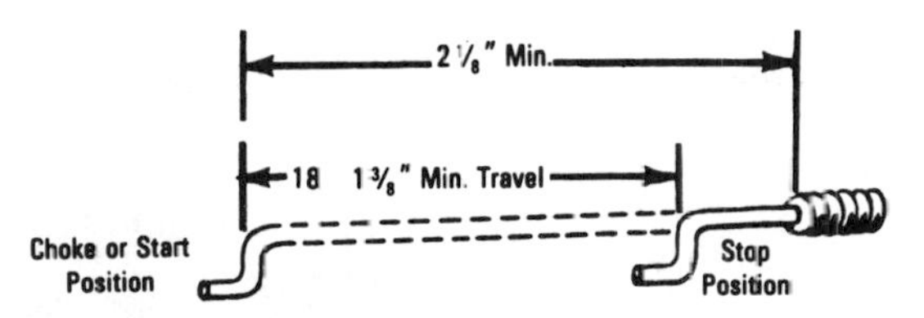

Fig. A–6. Control Wire Travel.

Travel of remote-control wire must be a minimum of 1⅜" (35 mm) in order to achieve full *CHOKE* and *STOP* position (Fig. A–6).

Choke-A-Matic Dial Control Adjustments

Dial controls seldom require adjustment unless blower housing has been removed. To adjust: Place dial control knob in "Start" position. Loosen control wire screw *A;* move lever *C* to full choke position. Allow a ⅛" (3.175 mm) gap between lever and bracket as shown.

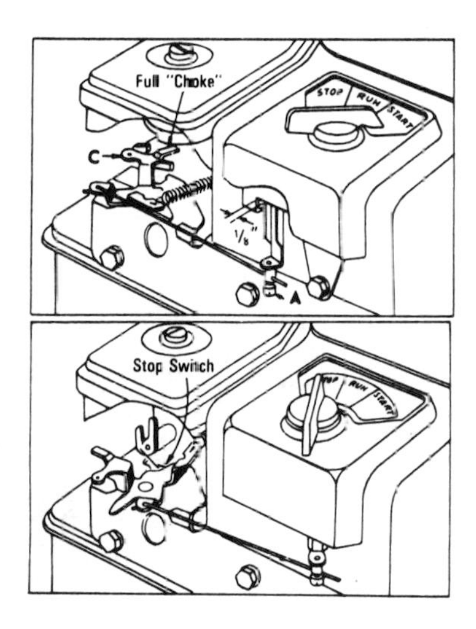

Fig. A–7. Choke-A-Matic Dial Control Adjustments.

MODELS 6, 8, 60000, 80000 Horizontal Crankshaft

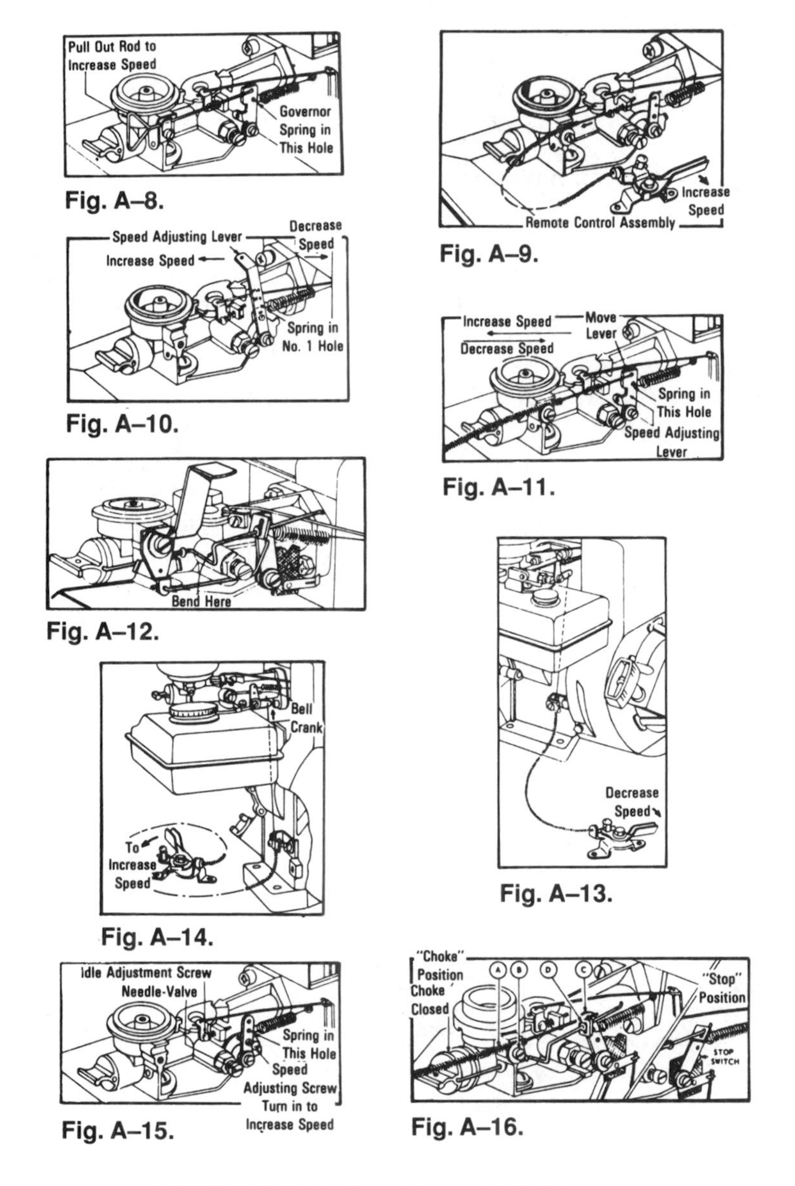

Fig. A–8.

Fig. A–9.

Fig. A–10.

Fig. A–11.

Fig. A–12.

Fig. A–13.

Fig. A–14.

Fig. A–15.

Fig. A–16.

MODELS 6BS, 60100, 61100, 80100, 81100

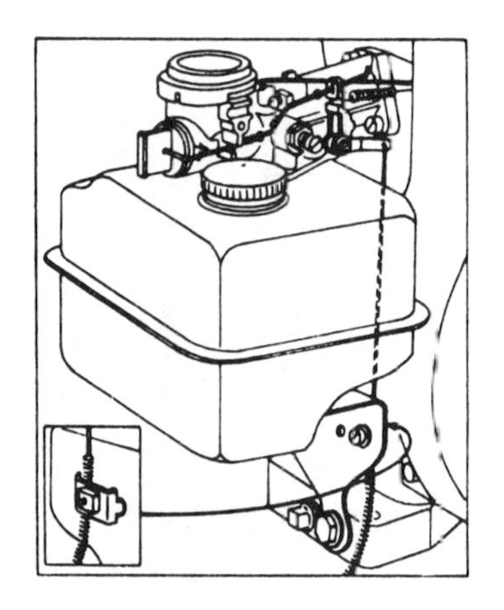

Fig. A–17.

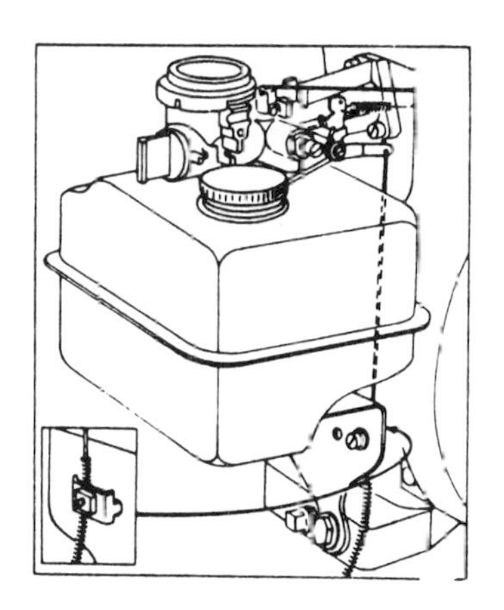

Fig. A–18.

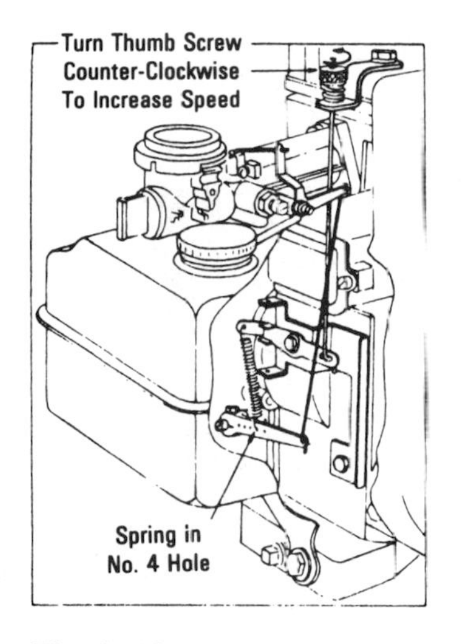

Fig. A–19.

MODELS 60200, 61200, 80200, 81200

MODELS 6, 8, 60000, 80000 Horizontal Crankshaft

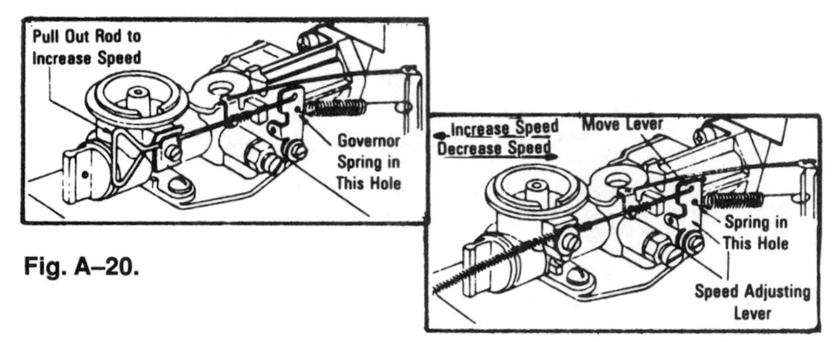

Fig. A–20.

Fig. A–21.

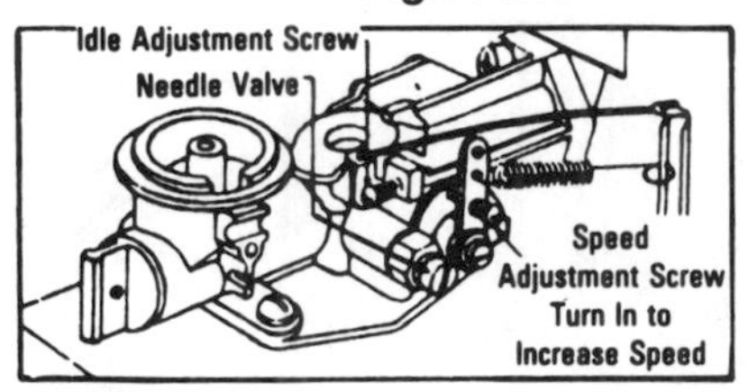

Fig. A–22.

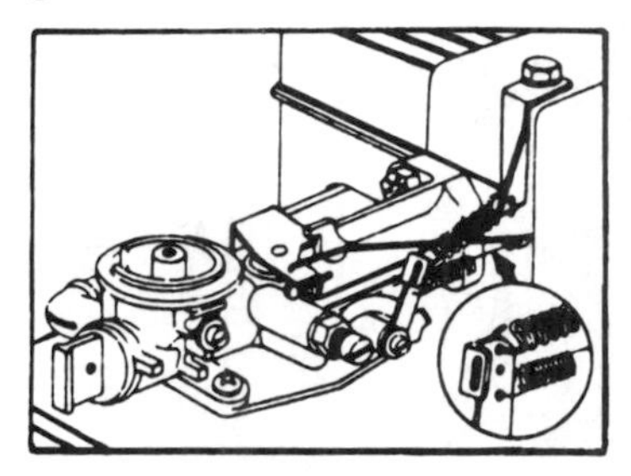

Fig. A–23.

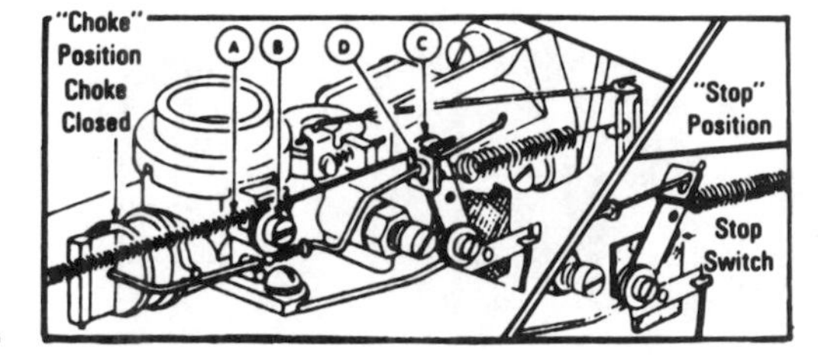

Fig. A–24.

MODELS 60200, 61200, 80200, 81200 (Cont'd.)

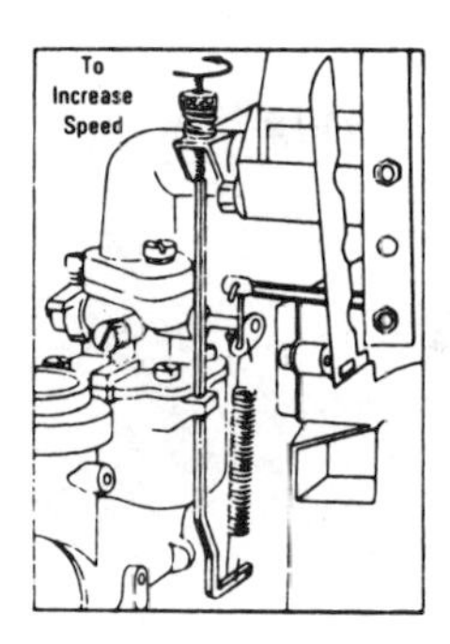

Fig. A–25.

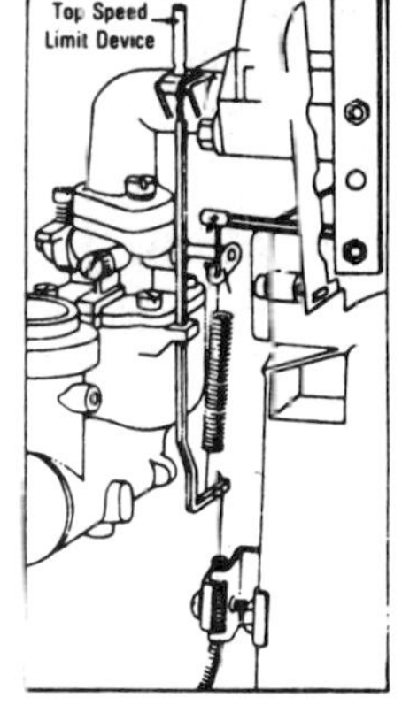

Fig. A–26.

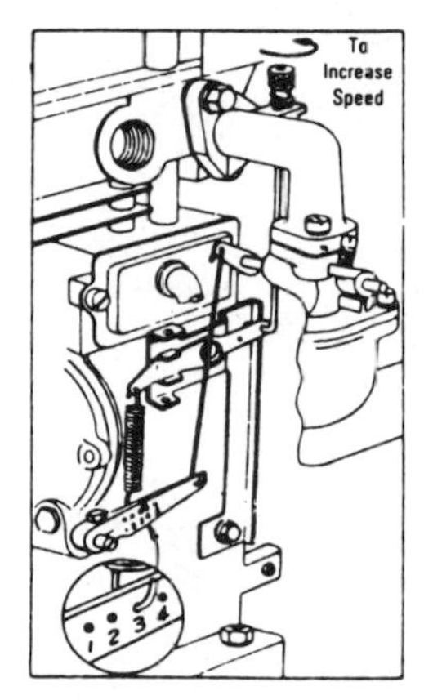

Fig. A–27.

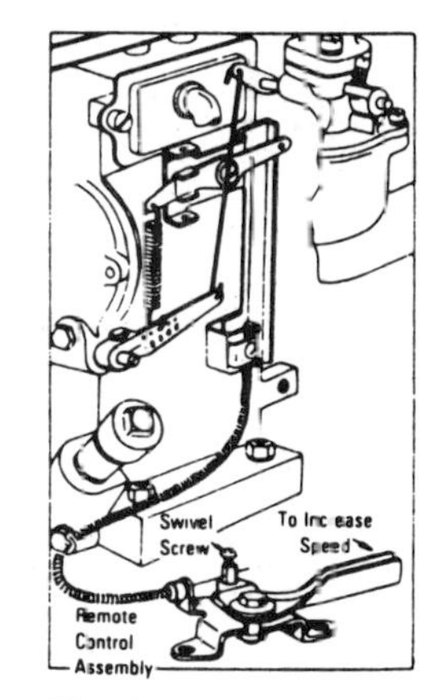

Fig. A–28.

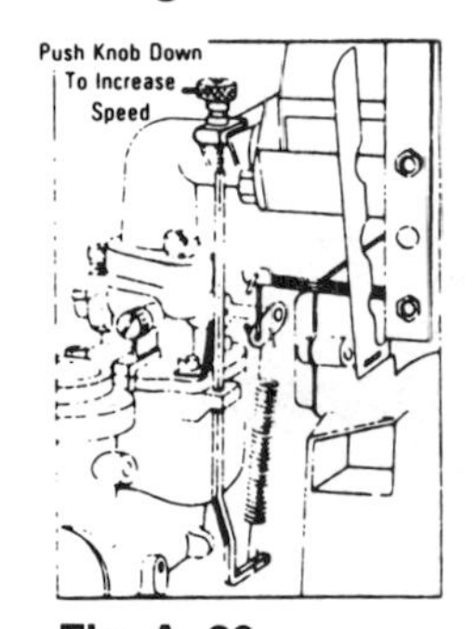

Fig. A–29.

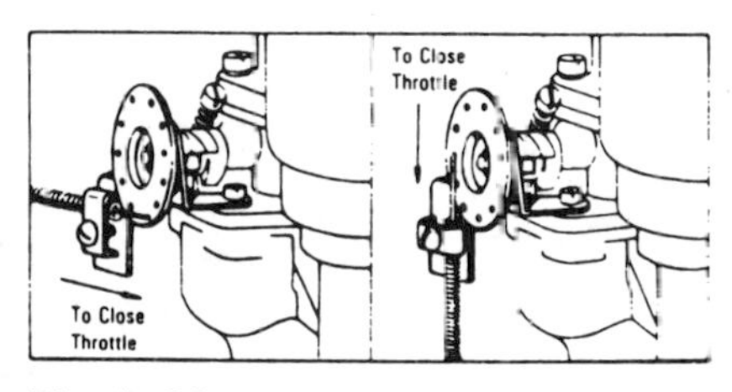

Fig. A–30.

MODELS 6B, 8B, 60300, 60400, 61300, 80300, 80400, 81400

MODELS 6, 8, 60000, 80000 Vertical Shaft

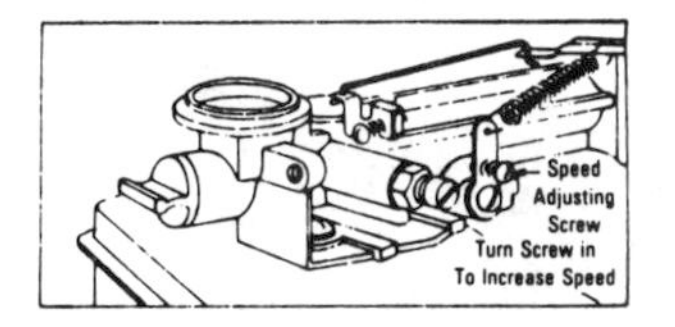

Fig. A–31.

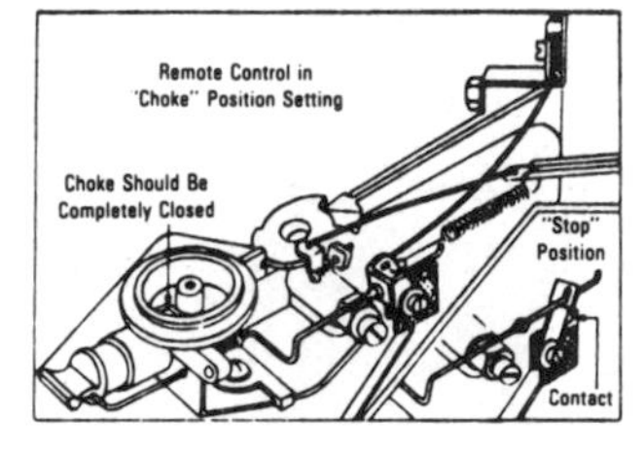

Fig. A–32.

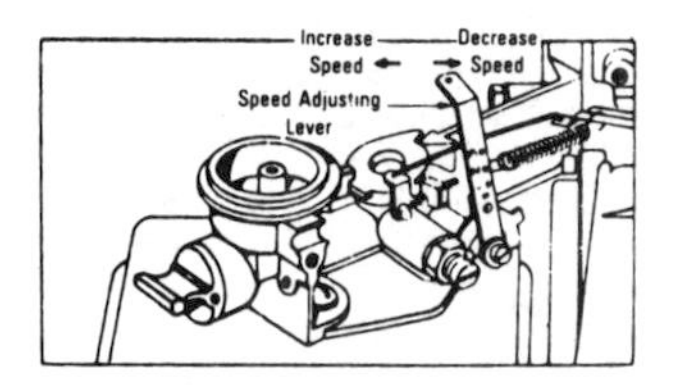

Fig. A–33.

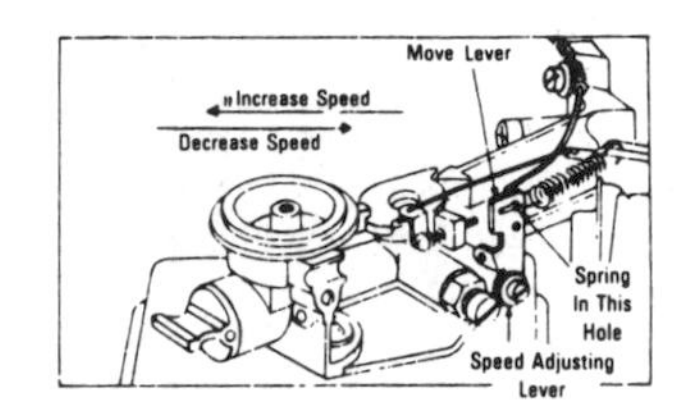

Fig. A–34.

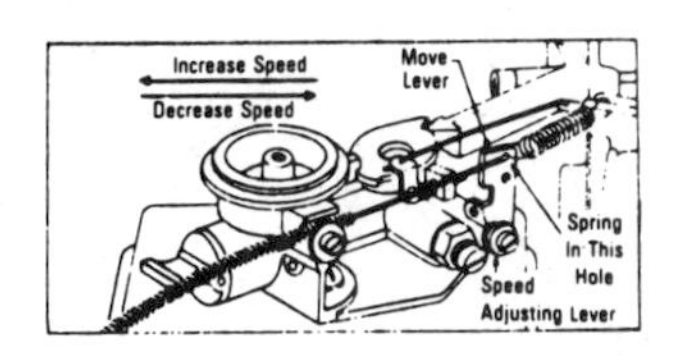

Fig. A–35.

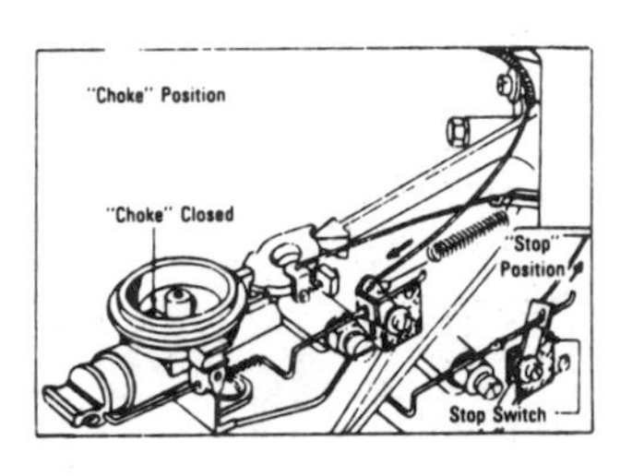

Fig. A–36.

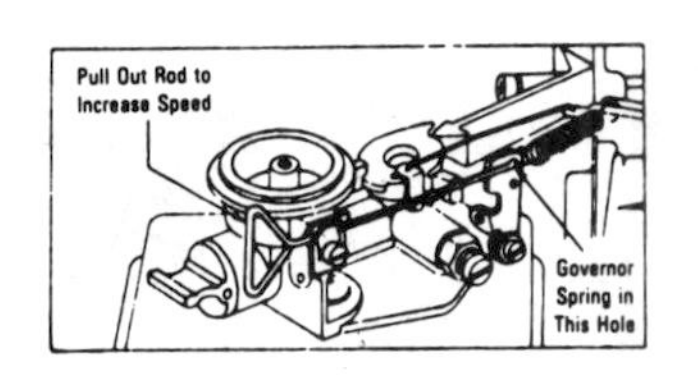

Fig. A–37.

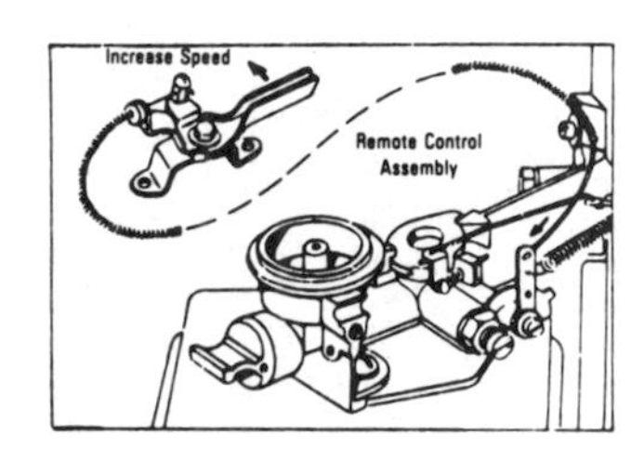

Fig. A–38.

MODELS 6BHS, 60500, 61500, 80500, 81500

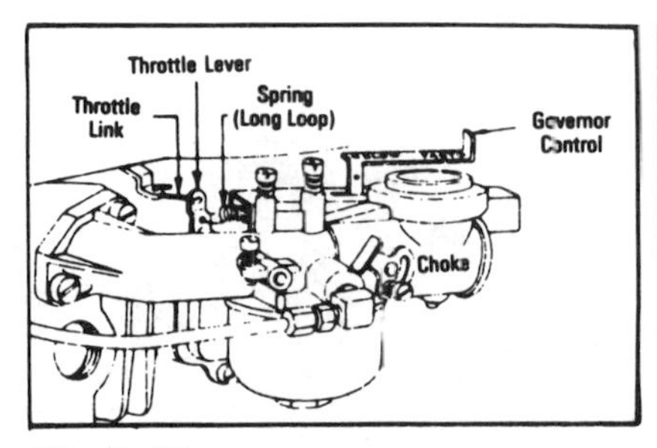

Fig. A–39.

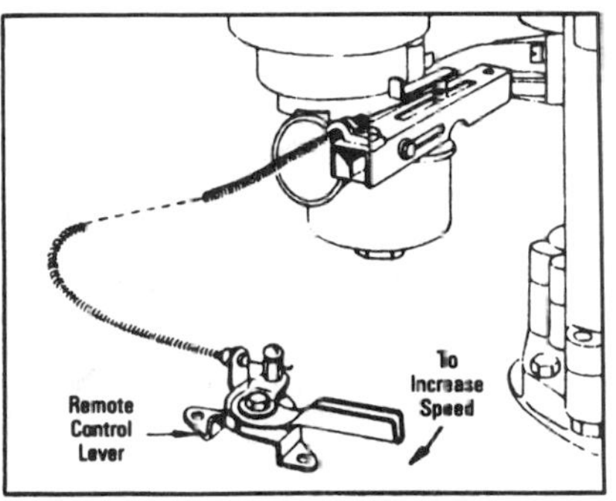

Fig. A–40.

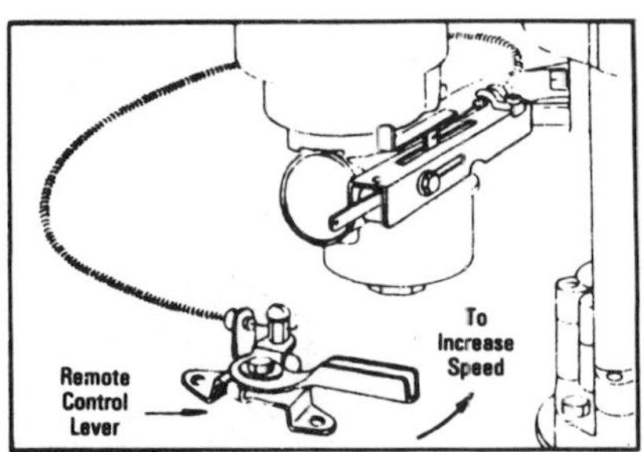

Fig. A–41.

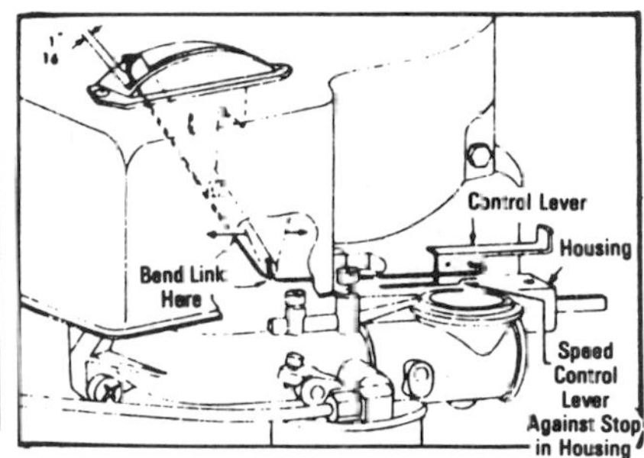

Fig. A–42.

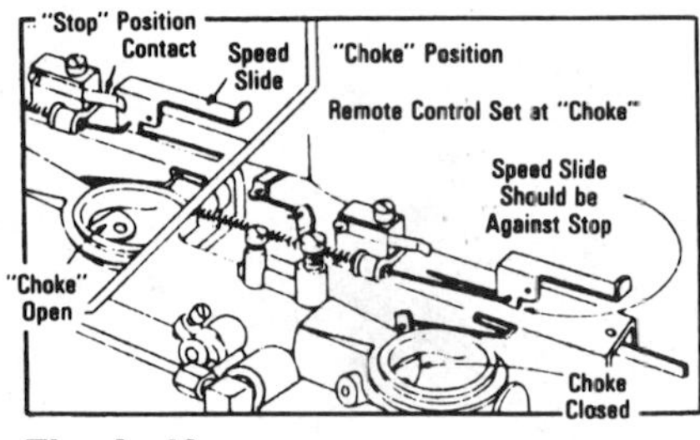

Fig. A–43.

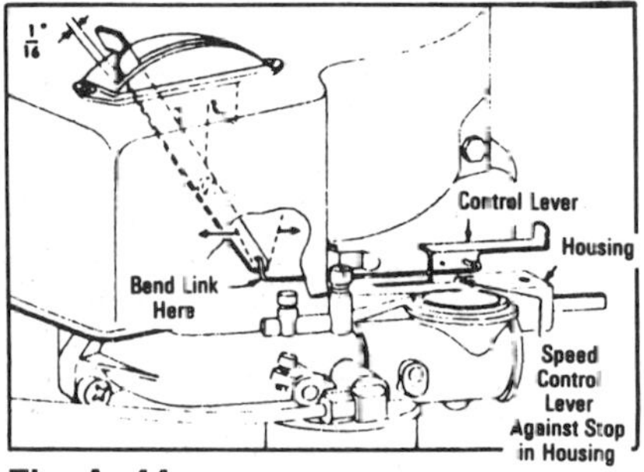

Fig. A–44.

MODELS 6B-H, 8B-HA, 6' 00, 61700, 80700, 81700

MODELS 6, 8, 60000, 80000, 82000, 92000 Vertical Shaft

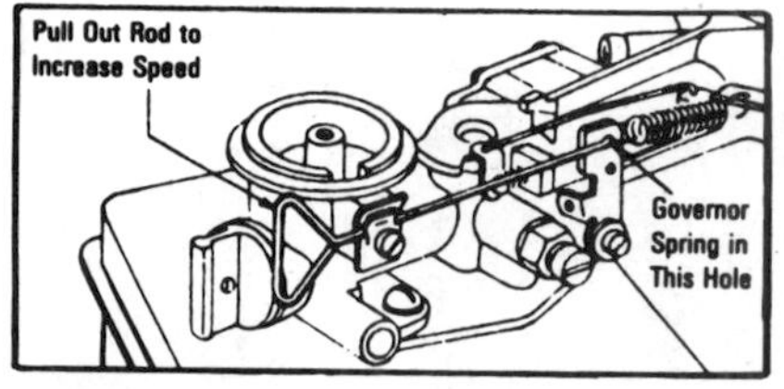

Fig. A–45.

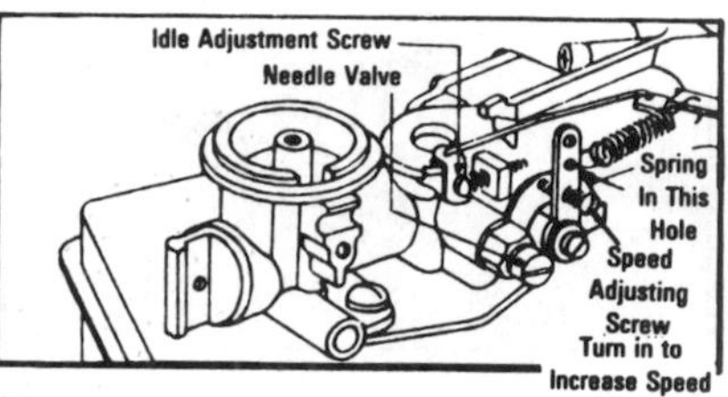

Fig. A–46.

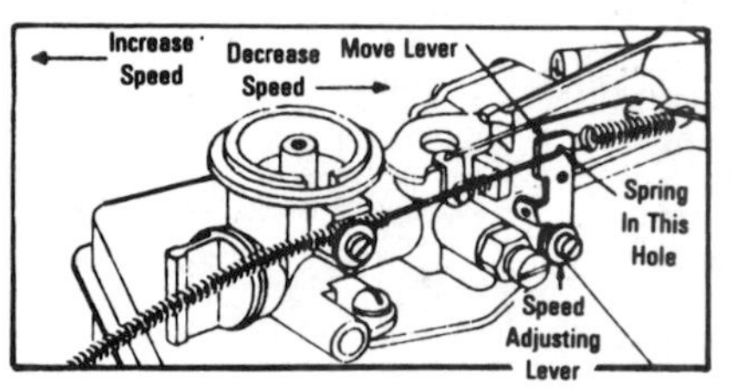

Fig. A–47.

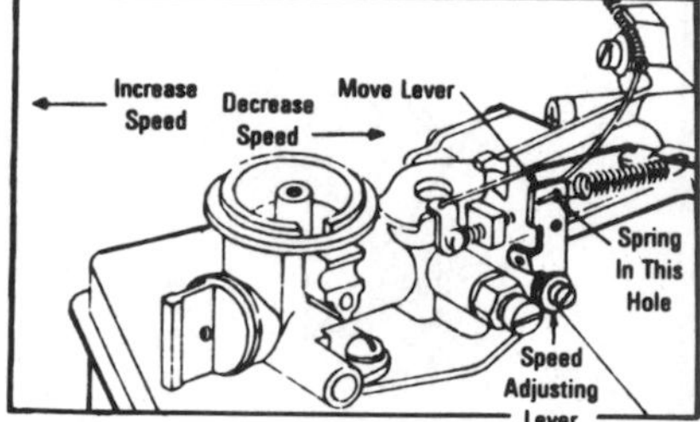

Fig. A–48.

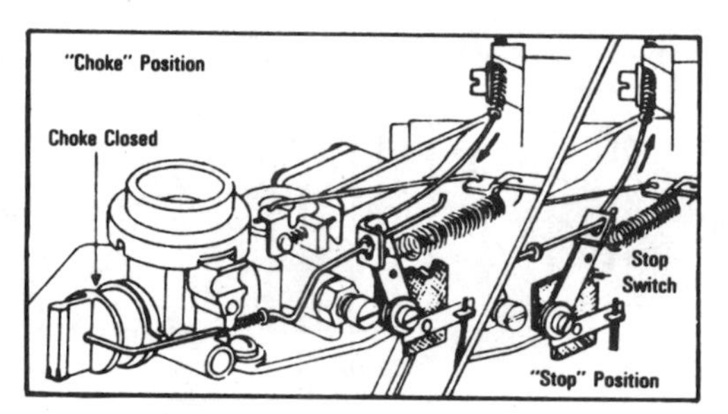

Fig. A–49.

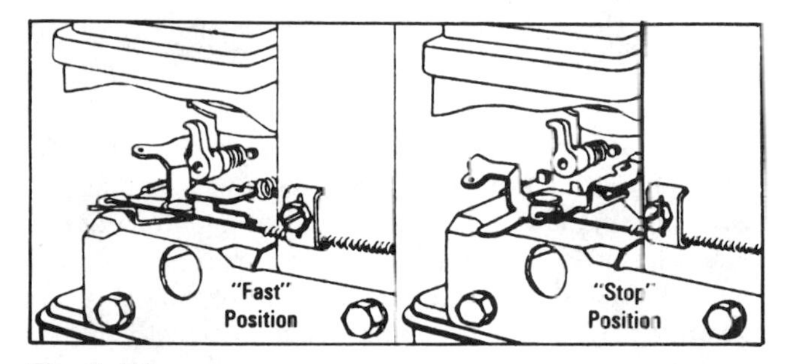

Fig. A–50.

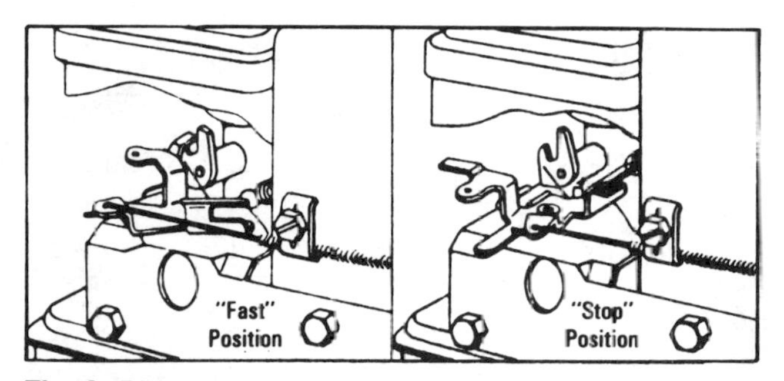

Fig. A–51.

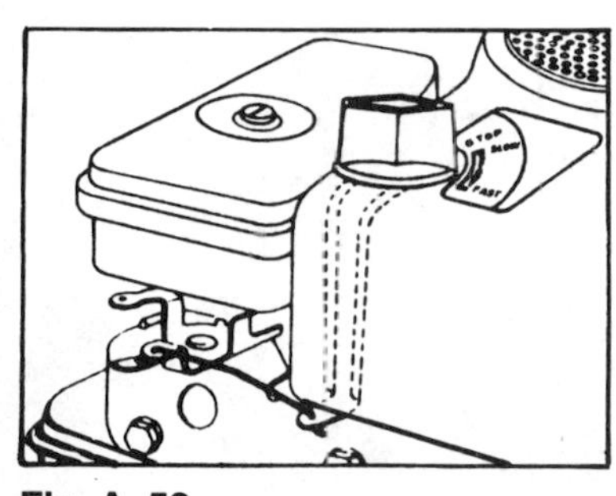

Fig. A–52.

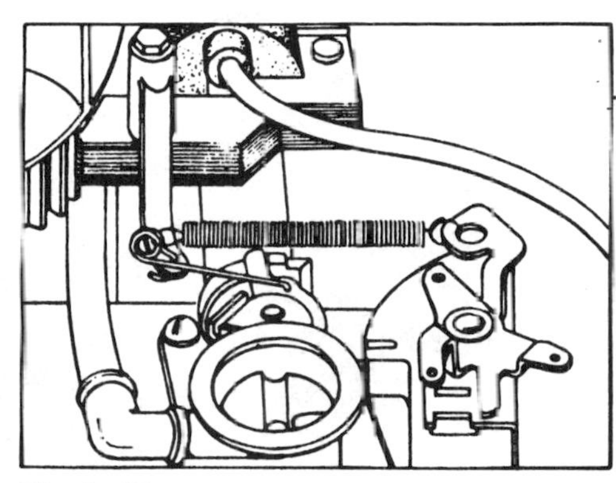

Fig. A–53.

MODELS 82000, 92000

Removal and Installation of Governor Spring on Model Series 92500 and 92900

The governor springs used on engine model series 92500 and 92900 are made with double end loops for a secure attachment and proper governor regulation. Springs with double-end loops are easily removed and installed by following the procedure shown below. Do *not* use a needle-nosed pliers, or the end loops of the governor spring will be deformed. When the governor spring is correctly installed, the spring must be positioned as shown in Fig. A–54.

Correct Position of Spring

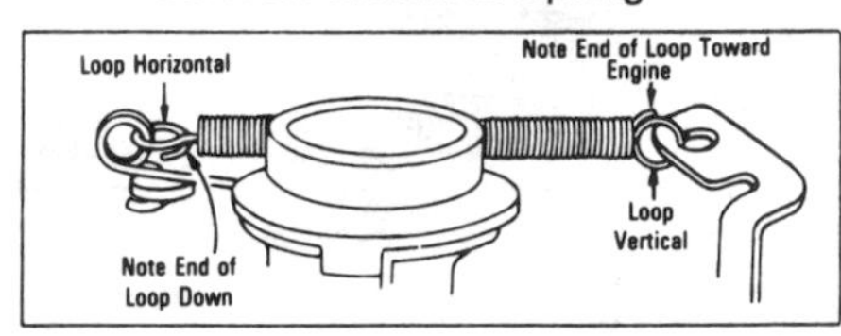

Fig. A–54.

Removing Spring

1. Remove Spring from Control Lever

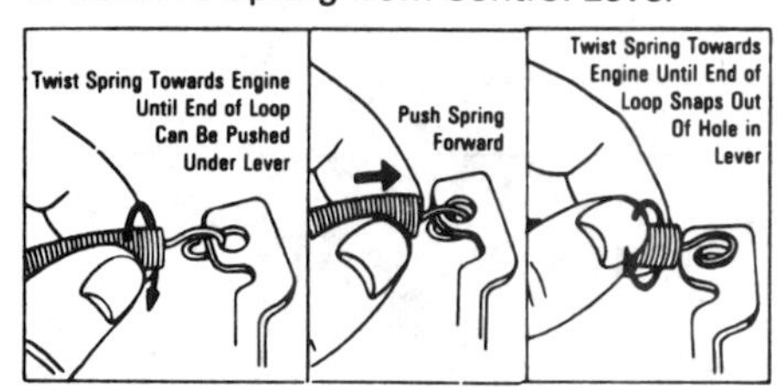

Fig. A–55.

2. Remove Spring from Eyelet in Link

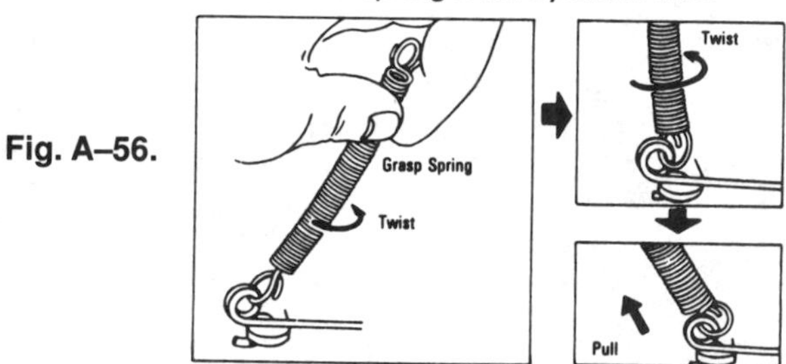

Fig. A–56.

Installing Spring

1. Assemble Spring to Link Eyelet

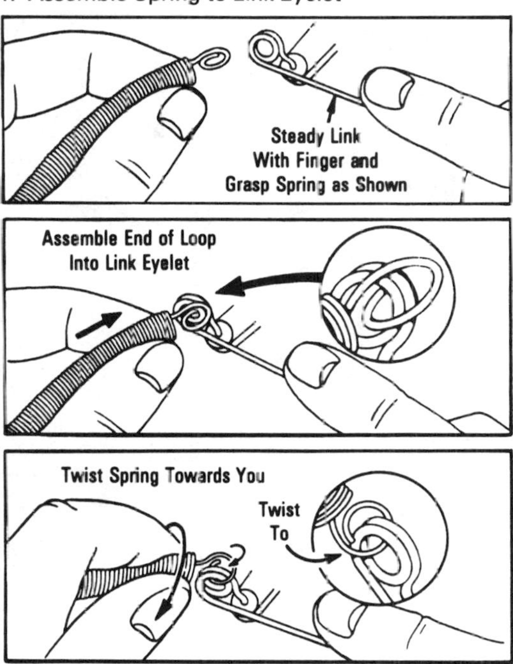

Fig. A–57.

2. Assemble Spring to Control Lever

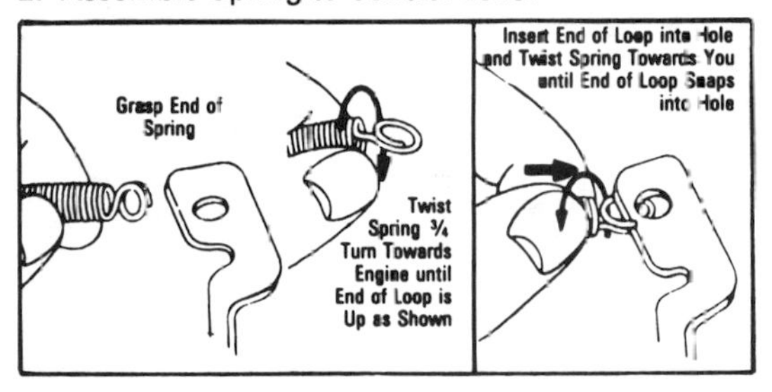

Fig. A–58.

MODELS 100000, 130000

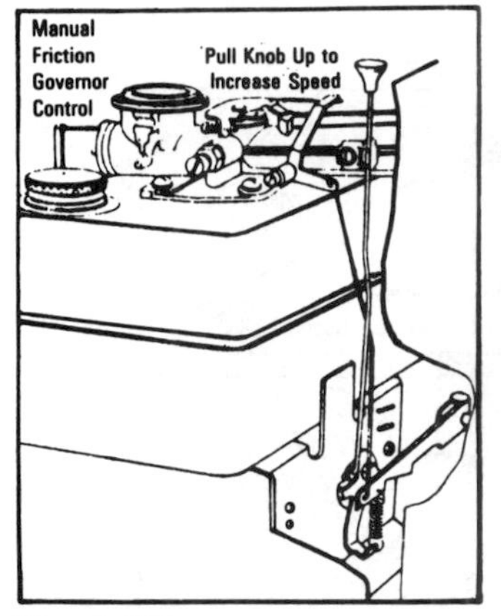

Fig. A–59.

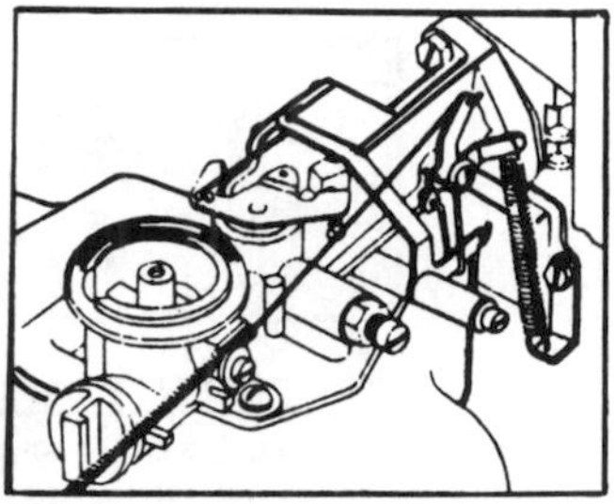

Fig. A–60.

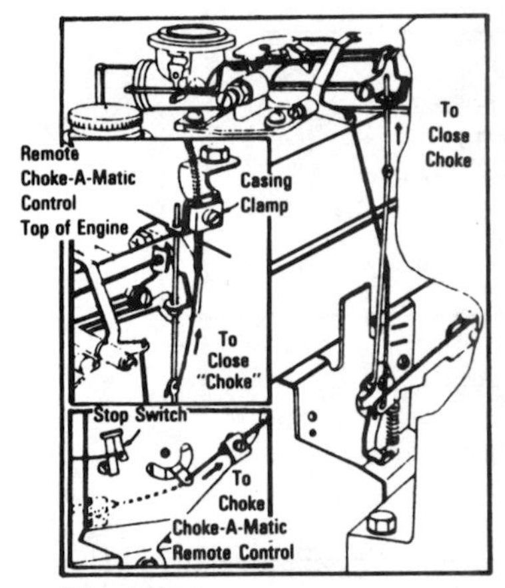

Fig. A–61.

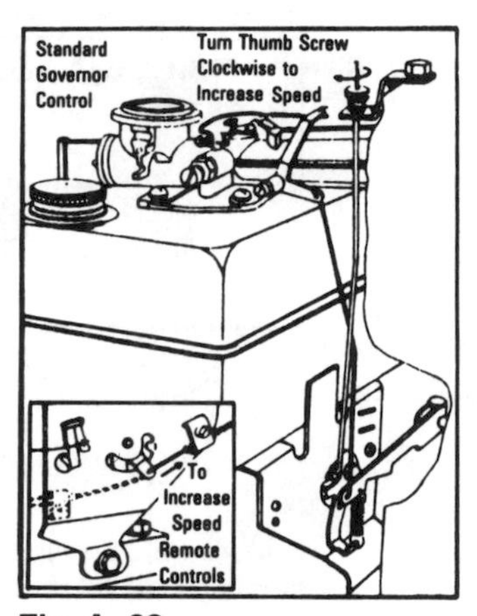

Fig. A–62.

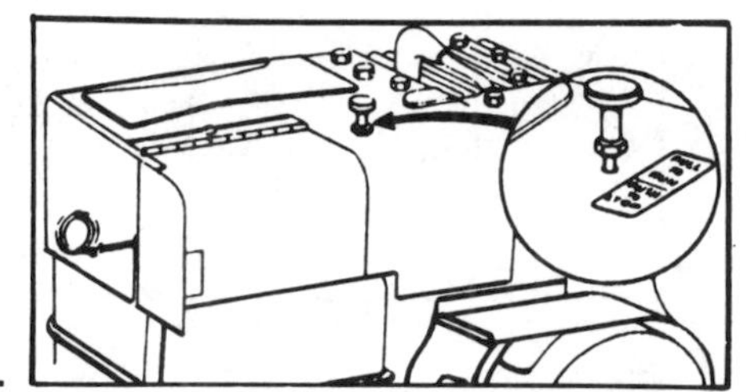

Fig. A–63.

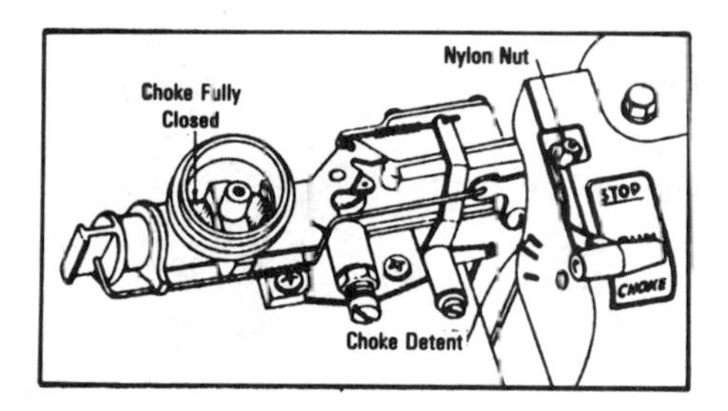

Fig. A–64. Lever-Trol

Place lever in choke detent; if choke is not fully closed, adjust nylon nut (with socket wrench) until choke just closes.

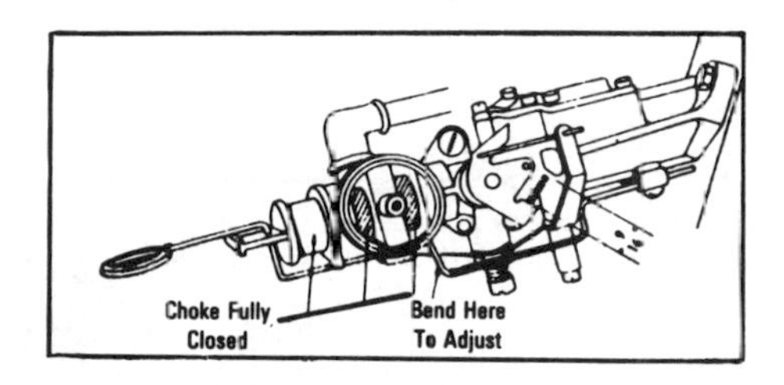

Fig. A–65. Choke Idle Return

Pull lever to choke position. The distance between throttle stamping and throttle screw stop must by $^9/_{16}$" (14.3mm). To adjust, bend lever where shown.

MODELS 100200, 130200

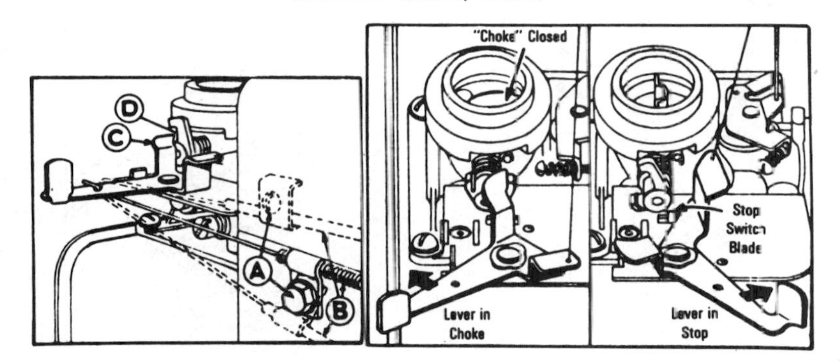

Fig. A–66. **Fig. A–67.**

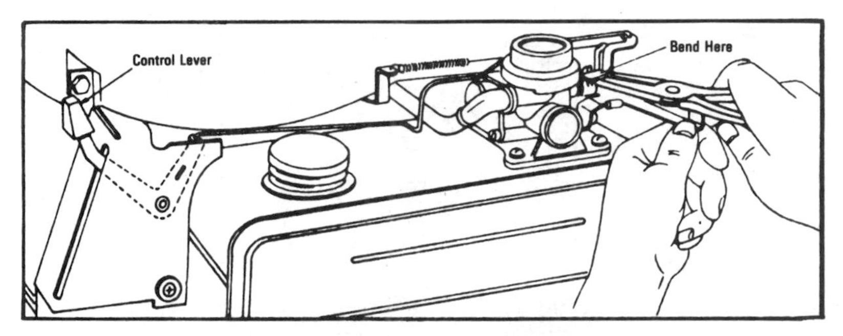

Fig. A–68. Rider Trol.

Place control in choke detent; if choke does not fully close, bend as shown until choke is closed.

MODELS 100900, 130900

MODELS 140000, 170000, 190000

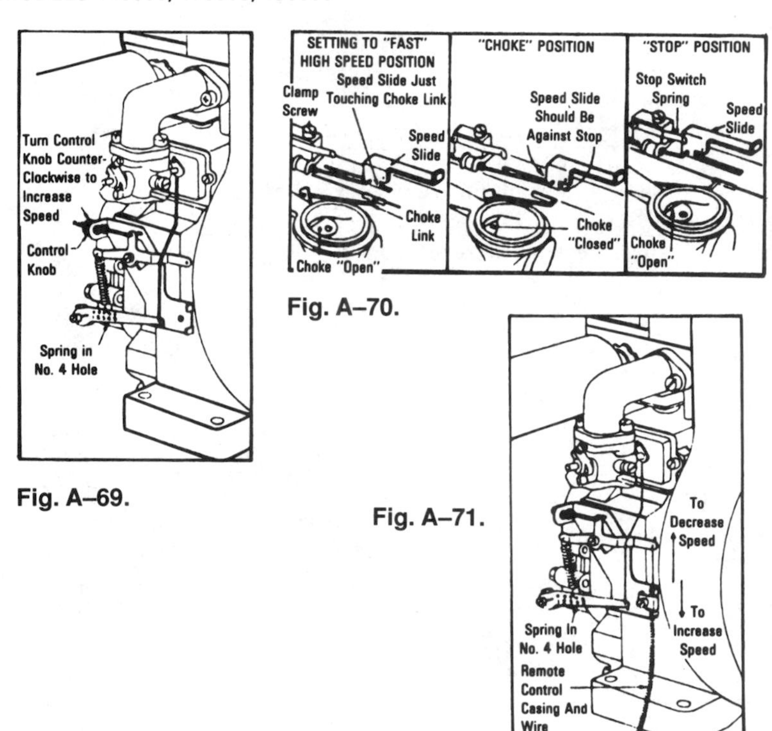

Fig. A–69.

Fig. A–70.

Fig. A–71.

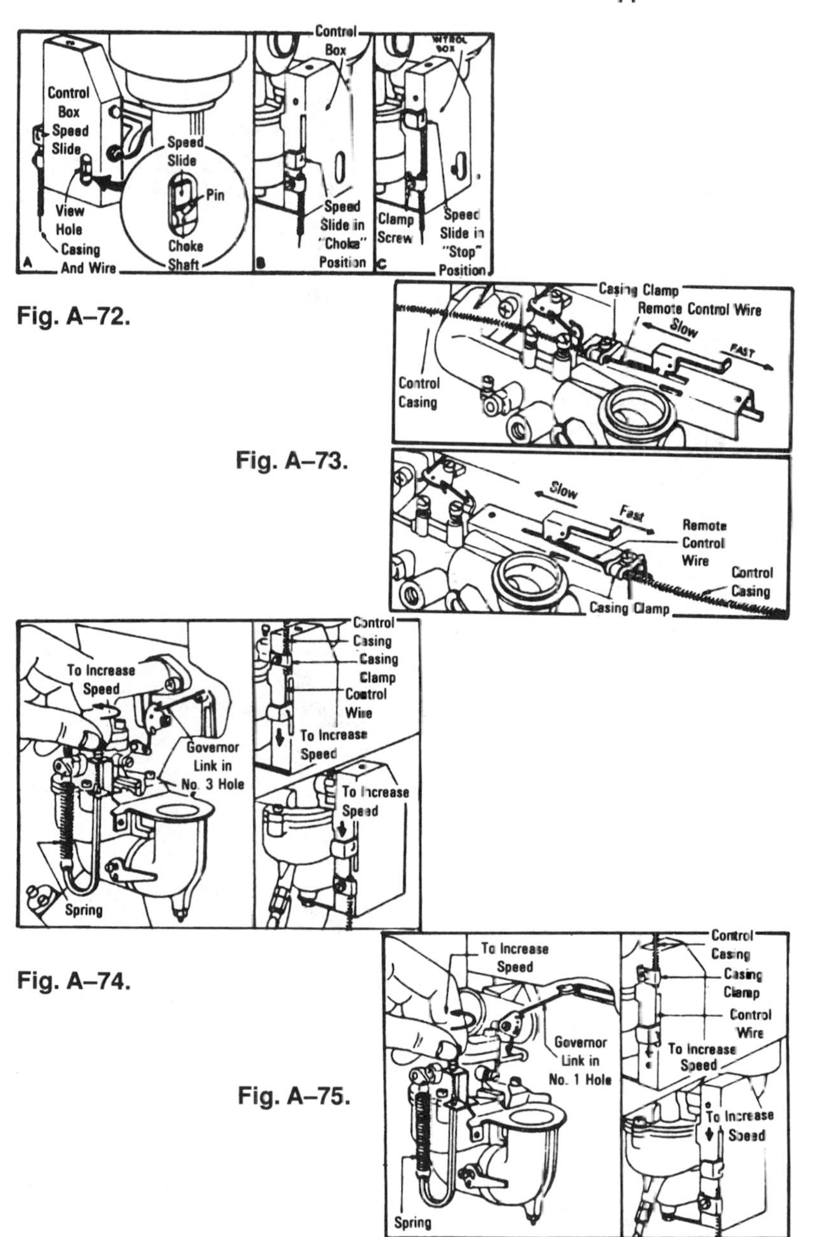

Fig. A–72.

Fig. A–73.

Fig. A–74.

Fig. A–75.

MODELS 140300, 141300, 141700, 142300, 143700

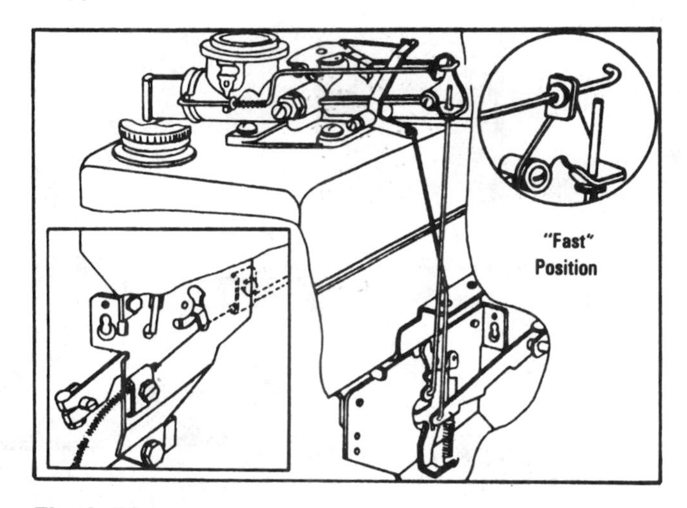

Fig. A–76.

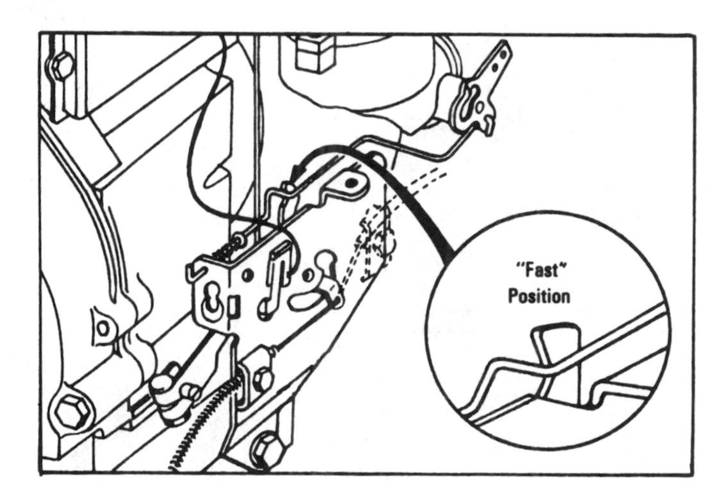

Fig. A–77.

ALUMINUM ENGINE MODELS 144200, 144400, 145200, 145400, 146400, 147400, 190400

MODELS 140000, 170000, 190000

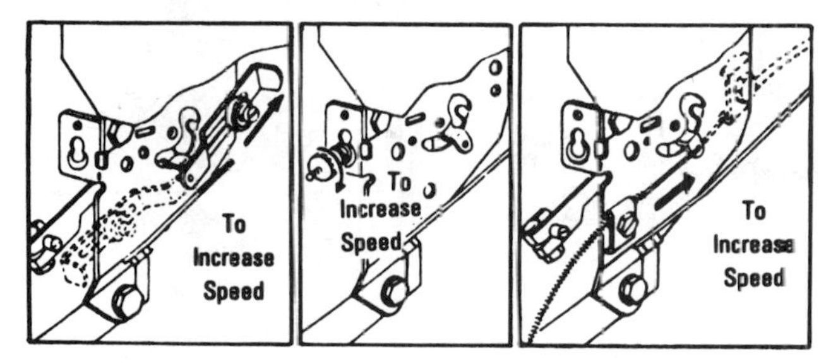

Fig. A–78.

Fig. A–79.

SETTING TOP SPEED-FIG. A-80

TOP SPEED LIMIT SCREW POSITION	NO LOAD TOP SPEED RANGE
None	4000 to 3800 rpm
No. 1 Position	3700 to 3400 rpm
No. 2 Position	3300 to 3000 rpm
No. 3 Position	2900 to 2500 rpm
No. 4 Position	2400 to 1800 rpm

Always set desired no-load top speed at power test by bending end of control lever at the spring anchor.

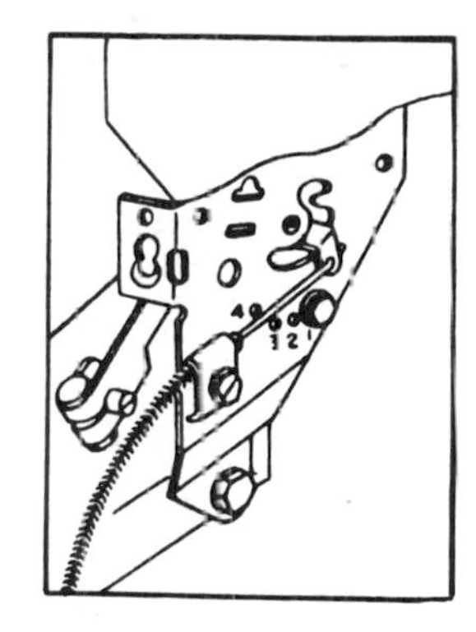

Fig. A–80.

Choke-A-Matic top speed range is 4000 to 3700 rpm with standard spring (top speed limit screw cannot be used).

ALUMINUM ENGINE MODELS 144200, 145200, 145400, 146400, 147400, 190400 (Cont'd)

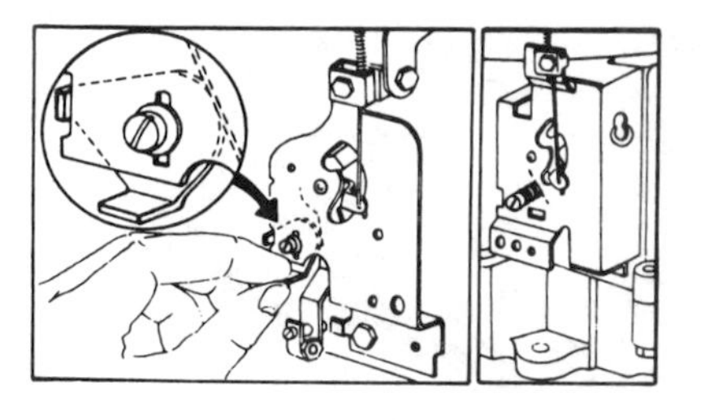

Fig. A–81.

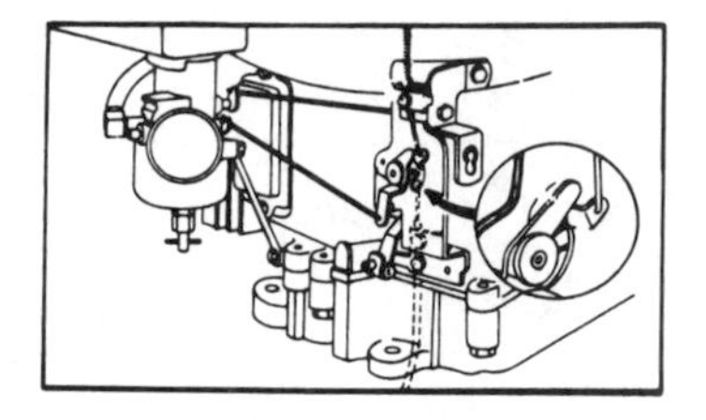

Fig. A–82.

Governed Idle—Fig. A–81

To adjust, first make final carburetor mixture adjustments. Then place remote control in idle position. Hold throttle shaft in closed position with fingers, adjust idle speed screw to 1550 rpm. Release throttle. Set remote control to 1750 rpm. Loosen governed idle stop and place against remote control lever. Tighten governed idle stop.

Adjustable Spring Loaded Screw Type

Follow above procedure, turn screw until it contacts remote control level (Fig. A–81).

Fig. A–83.

ALUMINUM ENGINE MODELS 144700, 145700, 146700, 147700, 170700, 171700, 190700, 191700

MODELS 9, 14, 19, 23, 140000, 170000, 190000, 200000, 230000

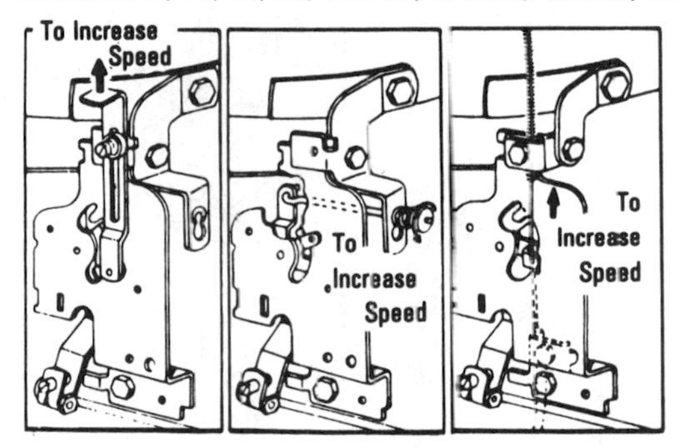

Fig. A–84.

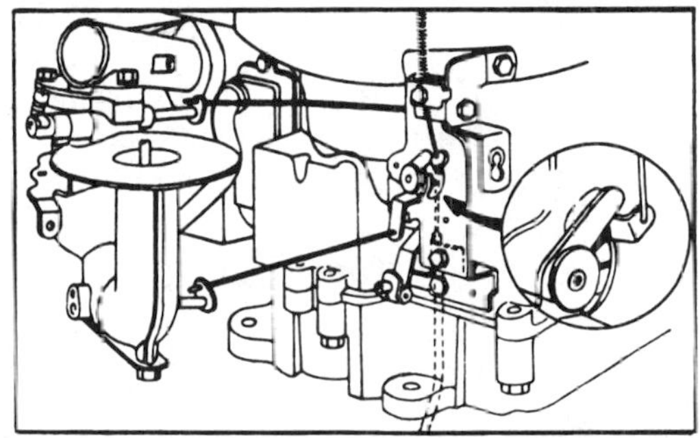

Fig. A–85.

SETTING TOP SPEED-FIG. A-86

TOP SPEED LIMIT SCREW POSITION	NO-LOAD TOP SPEED RANGE
None	4000 to 3500 rpm
No. 1 Position	3400 to 2900 rpm
No. 2 Position	2800 to 2400 rpm

Always set desired no-load top speed at power test by bending end of control lever at the spring anchor.

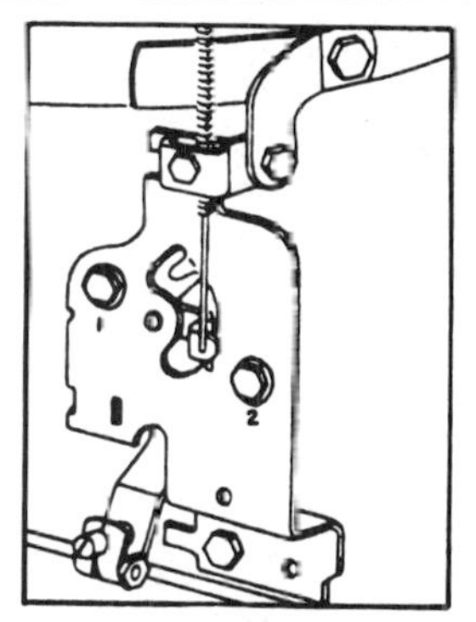

Fig. A–86.

Choke-A-Matic top speed range is 4000 to 3000 rpm with standard spring (top speed limit screw cannot be used).

ALUMINUM ENGINE MODELS 144700, 145700, 146700, 147700, 170700, 171700, 190700, 191700 (Cont'd.)

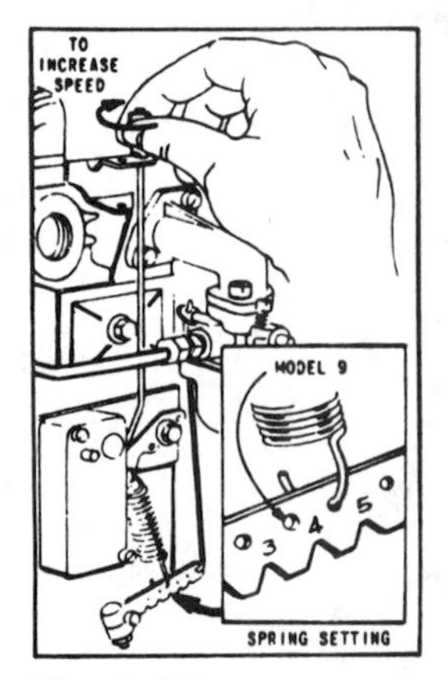

Fig. A–87.

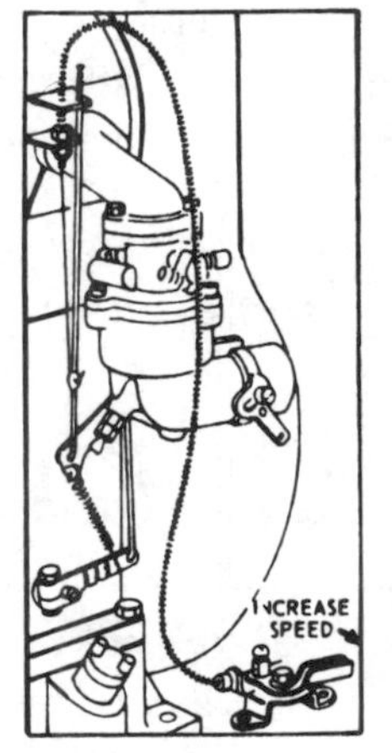

Fig. A–88.

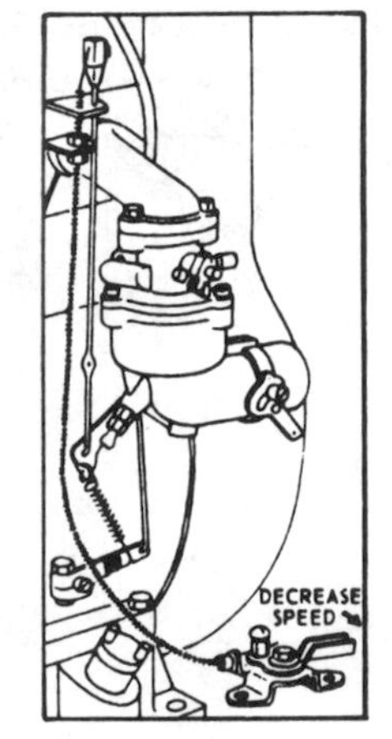

Fig. A–89.

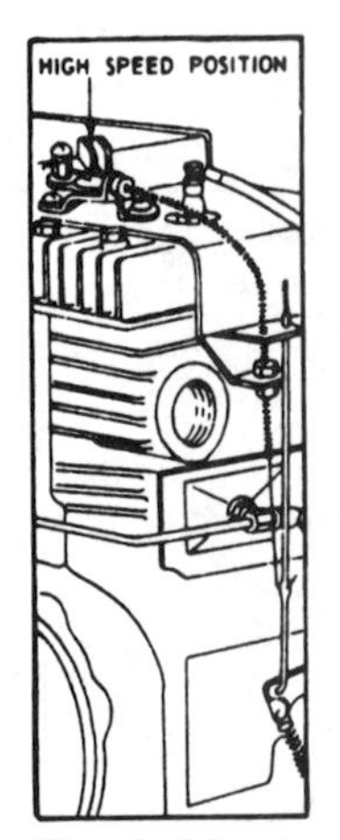

Fig. A–90.

Fig. A–91.

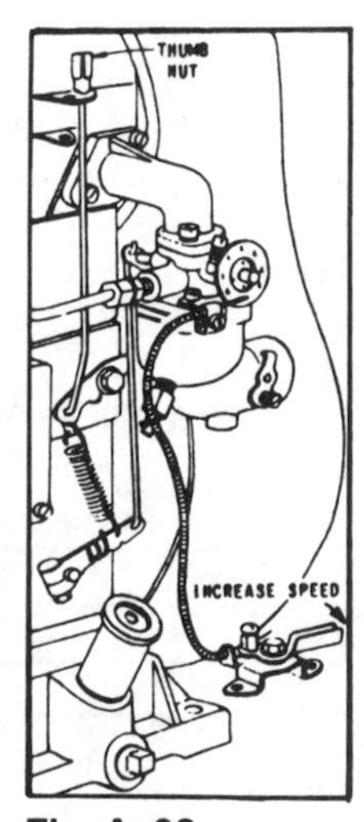

Fig. A–92.

CAST-IRON ENGINE MODELS 9, 14, 19, 23, 190000, 200000, 230000
MODELS 190000, 200000, 230000, 240000, 300000, 320000

Remote Governor Control

Attach remote control casing and wire as shown in Figs. A–93 or A–94. Do not change the position of the small elastic stop nuts. They provide for a governed idle speed and protection against overspeeding.

Thumb Nut Adjustment

Remove thumb nut and upper elastic stop nut. Replace thumb nut and adjust to desired operating speed (Fig. A–95). Do not change the position of the lower elastic stop nut. It provides protection against overspeeding.

Governed Idle

All engines in model series 243400, 300400, 320400 and some model series 23D and 233400 engines use two governor springs as shown in Figs. A–96 and A–97. The shorter spring keeps the engine on governor, even at idle speed. If moderate loads are applied at idle, the engine will not stall.

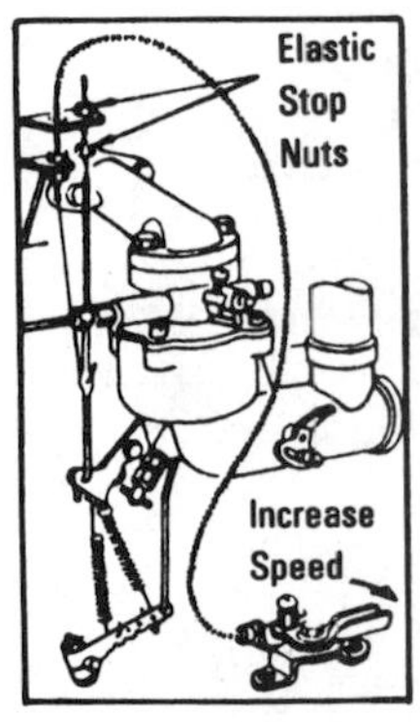

Fig. A–93.

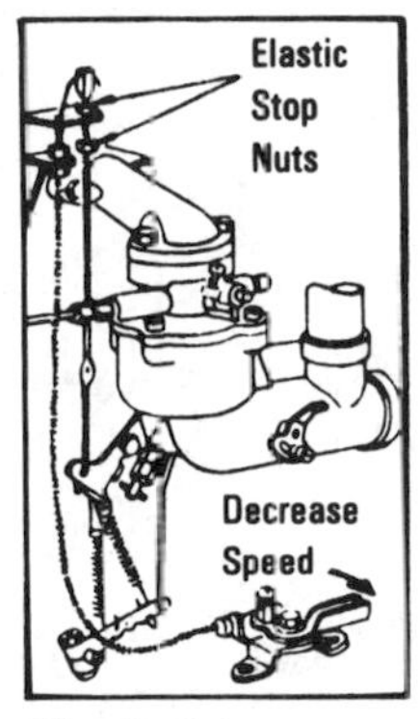

Fig. A–94.

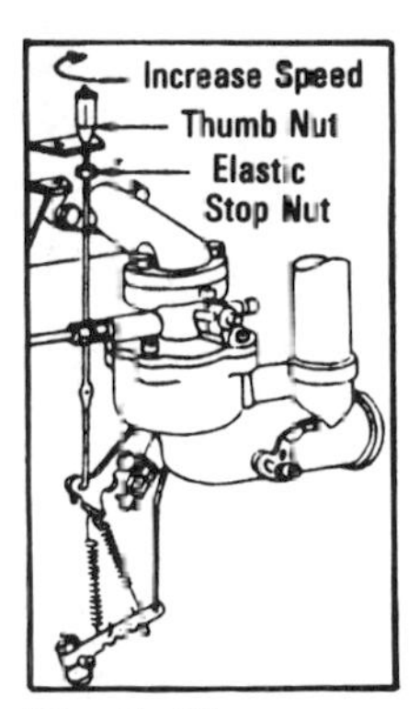

Fig. A–95.

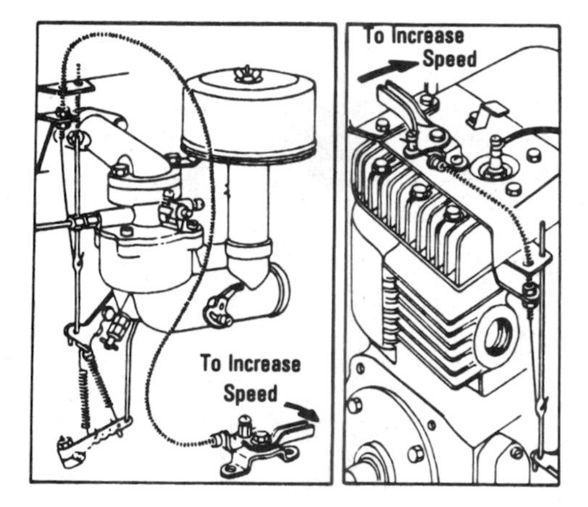

Fig. A–96.

To Adjust for Governed Idle

First make final carburetor mixture adjustments. Then place remote control in idle position. Hold throttle shaft in closed position and adjust idle screw to 1000 rpm. Release the throttle. With remote control in idle position, adjust upper elastic stop nut to 1200 rpm (Fig. A–97).

Fig. A–97.

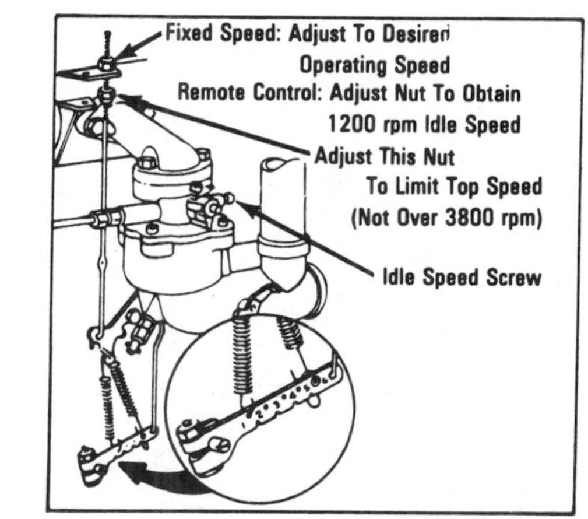

MODELS 190000, 200000, 230000, 240000, 300000, 320000

Appendix 4

Common Specifications for All Popular Engine Models Made by Briggs & Stratton

	Basic Model Series	Idle Speed	Armature: Two Leg Air Gap	Armature: Three Leg Air Gap	Valve Cleaners: Intake	Valve Cleaners: Exhaust	Valve Guide Reject Gage	Torque Specifications: Flywheel Nut Ft. Lbs.	Torque Specifications: Cylinder Head In. Lbs.	Torque Specifications: Conn. Rod In. Lbs.
ALUMINUM	6B, 60000	1750	0.006 0.010	0.012 0.016	0.005 0.007	0.009 0.011	19122	55	140	100
	8B, 80000, 81000, 82000	1750	0.006 0.010	0.012 0.016	0.005 0.007	0.009 0.011	19122	55	140	100
	92000, 94000	1750	0.006 0.010		0.005 0.007	0.009 0.011	19122	55	140	100
	100000	1750	0.010 0.014	0.012 0.016	0.005 0.007	0.009 0.011	19122	60	140	100
	110000	1750	0.006 0.010		0.005 0.007	0.009 0.011	19122	55	140	100
	130000	1750	0.010 0.014		0.005 0.007	0.009 0.011	19122	60	140	100
	140000	1750	0.010 0.014	0.016 0.019	0.005 0.007	0.009 0.011	19151	65	165	165
	170000, 171700•	1750 **	0.010 0.014		0.005 0.007	0.009 0.011	19151	65	165	165
	190000, 191700•	1750 **	0.010 0.014		0.005 0.007	0.009 0.011	19151	65	165	165
	220000, 250000	1750 **	0.010 0.014		0.005 0.007	0.009 0.011	19151	65	165	190
CAST IRON	5, 6, N	1750		0.012 0.016	0.007 0.009	0.014 0.016	19122	55	140	100
	8	1750		0.012 0.016	0.007 0.009	0.014 0.016	19122	55	140	100
	9	1200			0.007 0.009	0.014 0.016	19151	60	140	140
	14	1200			0.007 0.009	0.014 0.016	19151	65	165	190
	19, 190000, 200000•	1200 **	0.010 0.014	0.022 0.026	0.007 0.009	0.014 0.016	19151	115	190	190
	23, 230000	1200 **	0.010 0.014	0.022 0.026	0.007 0.009	0.017 0.019	19151	145	190	190
	243000	1200 **	0.010 0.014		0.007 0.009	0.017 0.019	19151	145	190	190
	300000	1200 **	0.010 0.014		0.007 0.009	0.017 0.019	19151	145	190	190
	320000	1200 **	0.010 0.014		0.007 0.009	0.017 0.019	19151	145	190	190

Common Specifications for All Popular Engine Models Made by Briggs & Stratton (Cont'd)

Crankshaft Reject Size			Main Bearing Reject Gage	Cylinder Bore Std.	Carburetor Type	Initial Carburetor Adjustment All Models Turns Open From Seat	
Mag. Journal	Crankpin	P.T.O. Journal				Needle Valve	Idle Valve
0.8726	0.8697	0.8726	19166	2.375* 2.374	Pulsa-Jet Vacu-Jet	1½	
0.8726	0.9963	0.8726	19166	2.375 2.374	Two Piece Flo-Jet	1½	1
0.8726	0.9963	0.8726	19166	2.5625 2.5615	One Piece Flo-Jet	2½	1½
0.8726	0.9963	0.9976	19166 Mag. 19178 PTO	2.500 2.499	One Piece Flo-Jet (6, 7, 8, 10 & 11 H.P. Vertical Crankshaft)	½	1¼
0.8726	0.9963	0.8726	19166	2.7812 2.7802			
0.8726	0.9963	0.9976	19166 Mag. 19178 PTO	2.5625 2.5615	One Piece Flo-Jet (11 H.P. Horizontal Crankshaft)	1½	1
0.9975	1.090	1.1790	19178	2.750 2.749			
0.9975 1.1790●	1.090	1.1790	19178	3.000 2.999	**Cylinder Resizing**		
0.9975 1.1790●	1.122	1.1790	19178	3.000 2.999	Resize if 0.003 or more wear or 0.0015 out of round on C.I. cylinder engines, 0.0025 out of round on aluminum alloy engines.		
1.3760	1.2470	1.3760		3.4375 3.4365	Resize to 0.010, 0.020 or 0.030 over Standard.		
0.8726	0.7433	0.8726	19166	2.000 1.999	*Model 6B series and early Models 60000 and 61000 series engines have cylinder bore of 2.3125 - 2.3115.		
0.8726	0.7433	0.8726	19166	2.250 2.249			
0.9832	0.8726	0.9832	19117	2.250 2.249	**Ring Gap Reject Sizes**		
1.1790	0.9964	1.1790	19117	2.625 2.624	MODEL	COMP. RINGS	OIL RING
1.1800	.9964 1.1219●	1.1790	19117	3.000 2.999	Alum. Cylinder Models C.I. Cylinder Models	.035" .030"	.045" .035"
1.3769	1.1844	1.3759	19117	3.000 2.999			
Ball	1.3094	Ball	Ball	3.0625 3.0615			
Ball	1.3094	Ball	Ball	3.4375 3.4365	■ With Valve Springs Installed.		
Ball	1.3094	Ball	Ball	3.5625 3.5615	● Synchro-Balance.		

1. Spark plug gap: 0.030 all models.
2. Condenser capacity: 0.18 to 0.24 μF all models.
3. Contact point gap: 0.020 all models.
4. Crankshaft end play: 0.002–0.008 all models.

APPENDIX 5

Governor Linkages and Carburetor Adjustments for HONDA Engines

The HONDA engines also have adjustments that need close attention when servicing or reassemblying after disassembly. The G100, GX120K1, GX160K1, GX240, GX340, GXV270 and GXV340 engines range in horsepower from 2.2 to 13.0.

Carburetor Mechanism (G100)

Main Circuit (Fig. A–98)

When the throttle is opened, enough air is moving through the carburetor air horn to produce an appreciable vacuum in the venturi (1). Since the fuel nozzle is centered in the venturi, atmospheric pressure pushes fuel in the float chamber out through the main nozzle (2) via the main jet (3). As the air flows past the main nozzle and the air jet (4), it meets fuel moving through the air bleed (5). They mix and flow

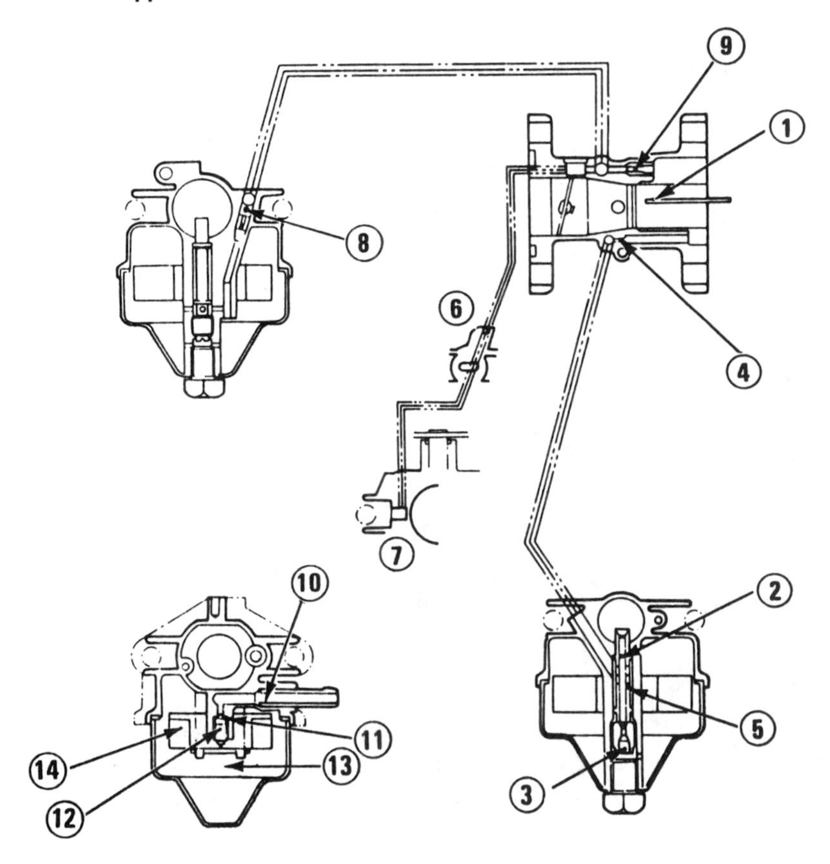

Fig. A–98

past the main nozzle. The rich mixture leans out as it mixes with other air flowing through the air horn to produce the final mixture.

Slow Circuit (Fig. A–98)

When the throttle is opened slightly, a vacuum is produced in the bypass (6) and pilot outlet (7). Under this condition, fuel in the float chamber is pushed out, flowing through the main jet into the passage. The pilot jet (8) meters the fuel as it passes through it.

The metered fuel then meets air that enters via the pilot air jet (9).

Again they mix and flow past the bypass and pilot outlet into the carburetor air horn. The mixture also has a high proportion of fuel. As the mixture is discharged into the air horn, it mixes with other air moving through the air horn, thereby producing the final mixture for slow speed operation.

Float Chamber (Fig. A–98)

The fuel from the fuel tank flows through the fuel passage (10), valve seat (11), and float valve (12), into the float chamber (13). The float (14) then moves up and pushes the float valve into the seat. This shuts the fuel inlet so that no fuel can enter. When the fuel level is lowered, the float moves down, allowing the float valve to move away from the valve seat. In this way a constant fuel level is maintained in the float chamber.

Governor Mechanism (G100)

The engine is equipped with a centrifugal governor which activates the throttle in response to engine speed (Fig. A–99). For example: If the engine is carrying a load and running at rated speed, the speed will drop if the load is increased even slightly. In response to the decreased engine speed, flyweights (2) on the flyweight holders contract by a corresponding amount. Their movement is transmitted through the slider (3) and governor arm shaft (4) to the governor arm (5). The governor arm then moves the rod (6) (in the direction of the arrow) which opens the throttle valve (7). As speed increases it is sensed by the governor, and the rod turns the valve in the opposite direction. This action/reaction sequence soon creates a state of equilibrium which allows the engine to run at nearly the speed as before while carrying greater load. Dumping the load suddenly will cause a rapid increase in speed. But the interaction of the engine-governor-carburetor prevents dangerous racing, and speed soon settles to a constant level.

The maximum speed of the engine is adjusted by turning a 6 mm screw (8) at the control lever (Fig. A–100). The control lever has a number of holes for remote control cable connection as shown.

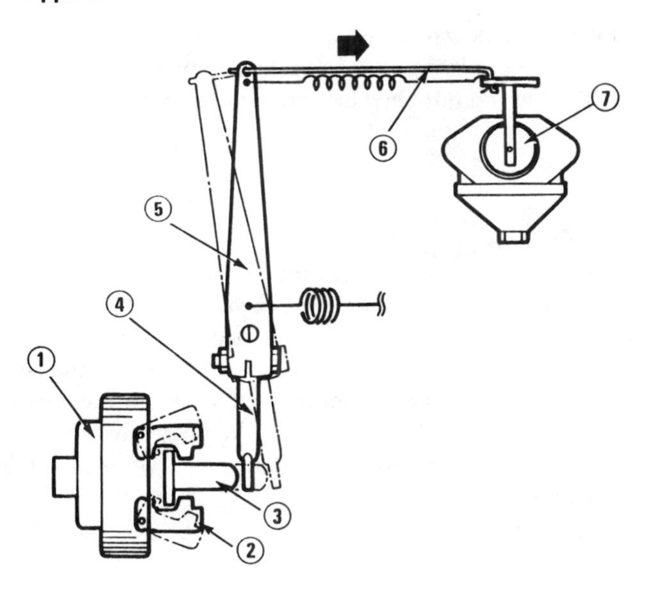

Fig. A–99.

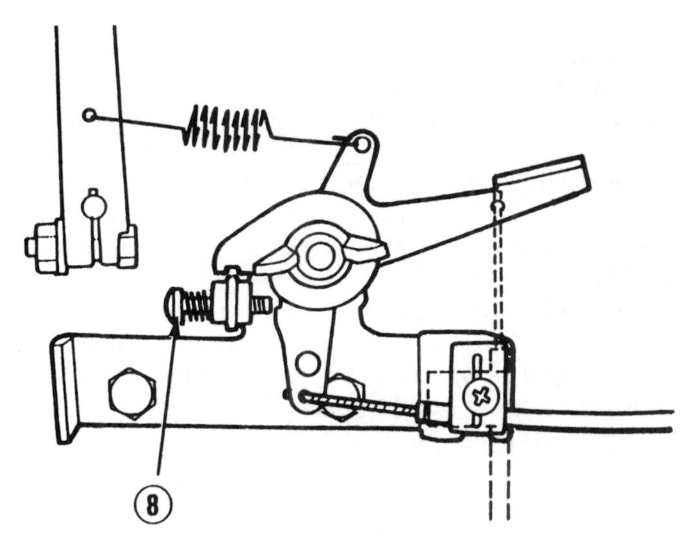

Fig. A–100.

Governor Adjustment (G100)

To adjust the governor, perform the following steps (Fig. A–101):

(1) With the throttle fully open, loosen governor arm attaching bolt and nut.
(2) Turn governor arm shaft clockwise all the way (full-close position) then tighten the bolt and nut.
(3) Start engine. Turn stopper screw in either direction to adjust maximum speed.

Remote Control Choke Kit (GX120K1 and GX160K1)

The metal plate attached to the air cleaner case (A) should be used for cable attachment (Fig. A–102). Stranded cable wire should be used, and cable ends should be used to attach the cable to the choke lever. Commercially available stranded cable may be used.

Remote Control Throttle Kit (GX120K1 and GX160K1)

1) Installation direction (Fig. A–103):
May be installed in two ways, either on the throttle lever side (A) or the cylinder head side (B).

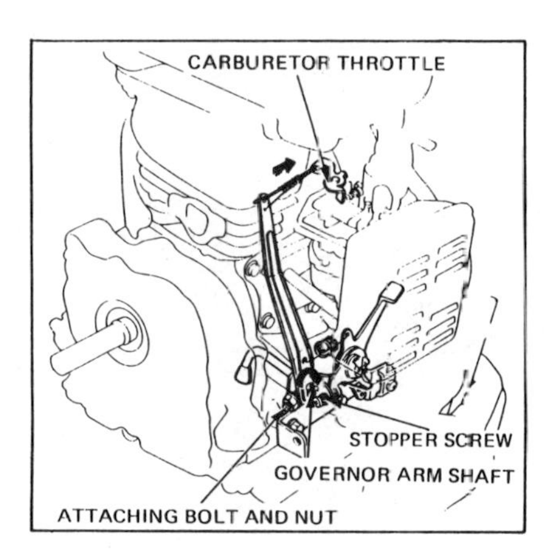

Fig. A–101.

Maximum speed	4,200 rpm

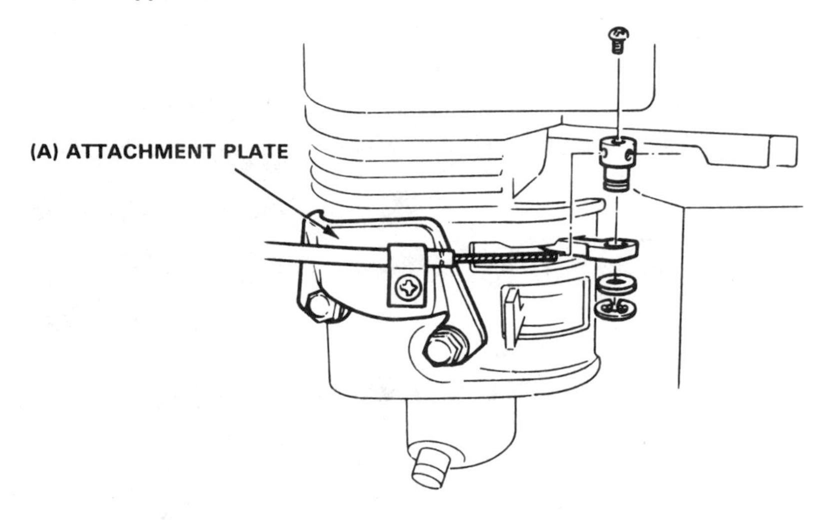

Fig. A–102.

2) Types of cable (Fig. A–103):
Two types can be used, stranded wire (C) and solid wire (E) (commercial products may be used). In the case of stranded wire, a cable end (D) should be used, and in the case of solid wire, attach directly to the hole in the control lever.

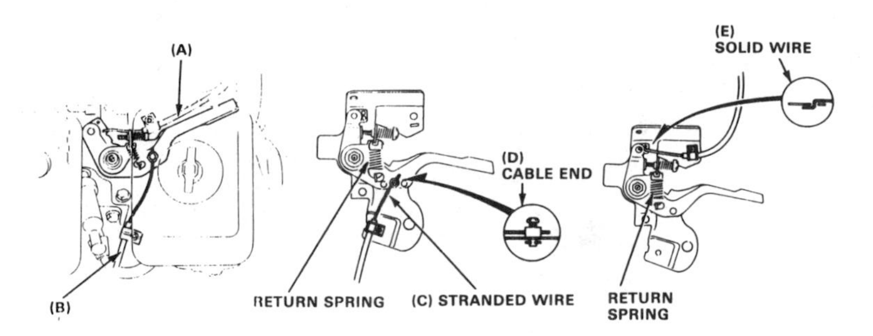

Fig. A–103.

Remote Control Throttle Kit (GX240 and GX340)

A cable for remote throttle control can be attached to the throttle lever in either of the positions shown in Fig. A–104. A solid wire cable core can be formed to hook directly into one of the holes in the throttle lever. Braided wire must be connected to a cable end fitting for attachment to the throttle lever.

Note: It is necessary to loosen the throttle lever friction nut when operating the throttle with a remote control cable.

Remote Control Choke Kit (GX240 and GX340)

Install the choke control bracket between the carburetor and air cleaner (Fig. A–105). Install the remote control cable linkage as shown. Use braided wire cable.

Carburetor (GXV270 and GXV340)

1) Start the engine and allow it to warm up to normal operating temperature.
2) With the engine idling, turn the pilot screw in or out to the setting that produces the highest idle rpm (Fig. A–106). The correct set-

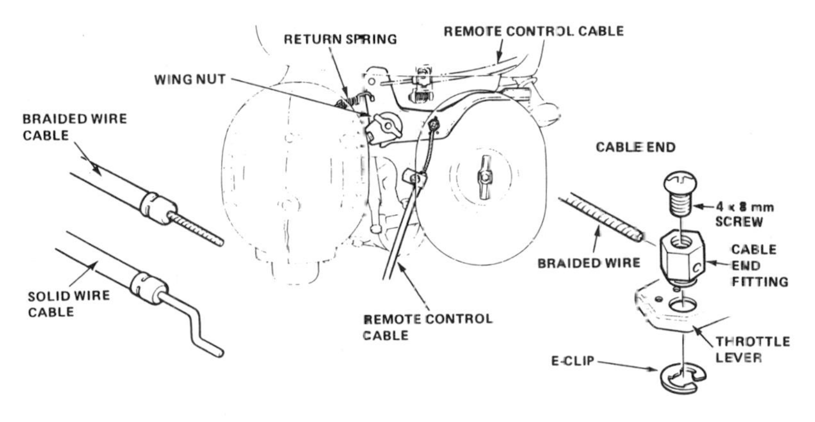

Fig. A–104.

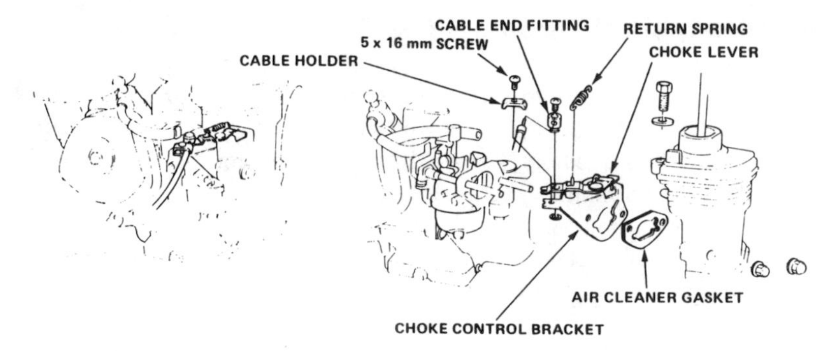

Fig. A–105.

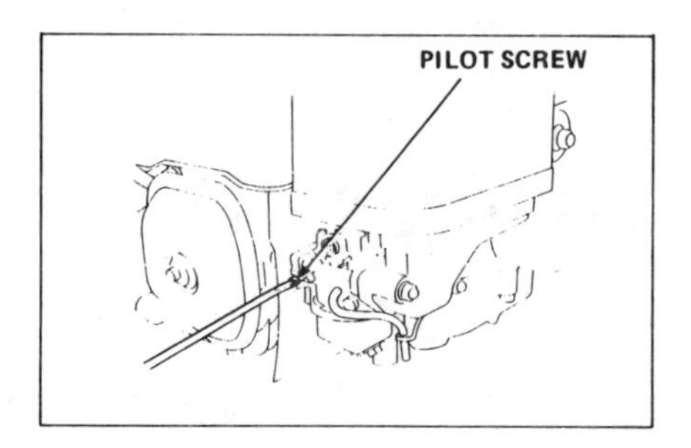

Fig. A–106.

Fig. A–107.

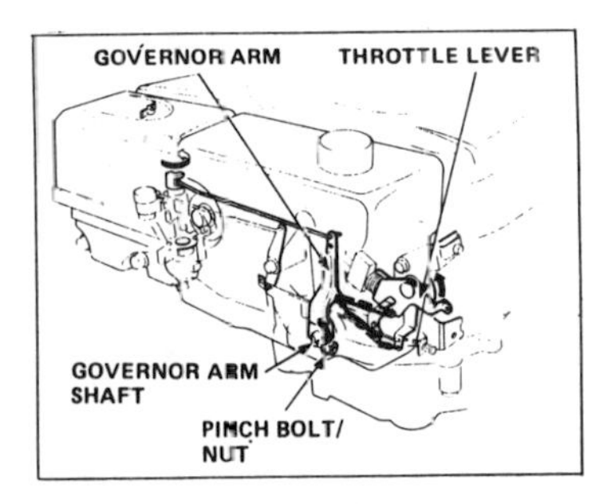

Fig. A–108

ting will usually be obtained at approximately the following number of turns out from the fully closed (lightly seated) position.

Pilot screw opening	GXV 270	2- 1/4 turns out
	GXV 340	2- 1/2 turns out

3) After the pilot screw is correctly adjusted, turn the throttle stop screw to obtain the standard idle speed (Fig. A–107).

Standard idle speed	1,400 ± 150 min^{-1} (rpm)

Governor (GXV270 and GXV340)

1) Loosen the nut on the governor arm pinch bolt, and move the governor arm to fully open the throttle (Fig. A–108).
2) Rotate the governor arm shaft as far as it will go in the same direction the governor arm moved to open the throttle.
3) Start the engine and allow it warm up to normal operating temperature. Move the throttle lever to run the engine at the standard maximum speed, and check the engine speed.

Standard maximum speed	3,400 ± 150 min^{-1} (rpm)

Index

A

B

C

P

R

S

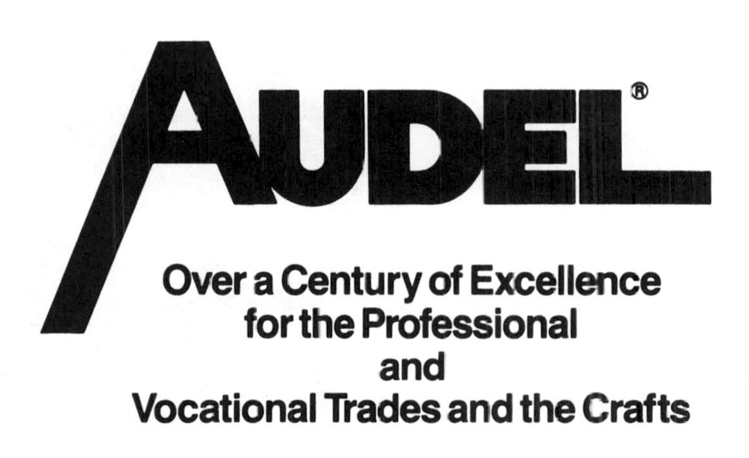
AUDEL®
Over a Century of Excellence
for the Professional
and
Vocational Trades and the Crafts

AUDEL

These fully illustrated, up-to-date guides and manuals mean a better job done for mechanics, engineers, electricians, plumbers, carpenters, and all skilled workers.

CONTENTS

ELECTRICAL

House Wiring (Seventh Edition)

ROLAND E. PALMQUIST;
revised by PAUL ROSENBERG

5 1/2 × 8 1/4 Hardcover 248 pp. 150 Illus.
ISBN: 0-02-594692-7 $22.95

Rules and regulations of the current 1990 National Electrical Code for residential wiring fully explained and illustrated.

Practical Electricity (Fifth Edition)

ROBERT G. MIDDLETON;
revised by L. DONALD MEYERS

5 1/2 × 8 1/4 Hardcover 512 pp. 335 Illus.
ISBN: 0-02-584561-6 $19.95

The fundamentals of electricity for electrical workers, apprentices, and others requiring concise information about electric principles and their practical applications.

Guide to the 1993 National Electrical Code

ROLAND E. PALMQUIST;
revised by PAUL ROSENBERG

5 1/2 × 8 1/4 Paperback 608 pp.
100 line drawings
ISBN: 0-02-077761-2 $25.00

The guide to the most recent revision of the electrical codes—how to read them, understand them, and use them. Here is the most authoritative reference available, making clear the changes in the code and explaining these changes in a way that is easy to understand.

Installation Requirements of the 1993 National Electrical Code

PAUL ROSENBERG

5 1/2 × 8 1/4 Paperback 261 pp.
100 line drawings
ISBN: 0-02-077760-4 $22.00

A handy guide for electricians, contractors, and architects who need a reference on location. Arranging all the pertinent requirements (and only the pertinent requirements) of the 1993 NEC, it has an easy-to-follow format. Concise and updated, it's a perfect working companion for Apprentices, Journeymen, or for Master electricians.

Mathematics for Electricians and Electronics Technicians

REX MILLER

5 1/2 × 8 1/4 Hardcover 312 pp. 115 Illus.
ISBN: 0-8161-1700-4 $14.95

Mathematical concepts, formulas, and problem-solving techniques utilized on-the-job by electricians and those in electronics and related fields.

Fractional-Horsepower Electric Motors

REX MILLER and
MARK RICHARD MILL...

5 1/2 × 8 1/4 Ha...
ISBN: 0-672-2...

NEW EDITION
FOR 1993

The in... ...ntenance re-pa... ...he small-to-moder-ate... ...otors that power home appl... ...d industrial equipment.

Electric Motors (Fifth Edition)

EDWIN P. ANDERSON
and REX MILLER

5 1/2 × 8 1/4 Hardcover 696 pp.
Photos/line art
ISBN: 0-02-501920-1 $35.00

Complete reference guide for electricians, industrial maintenance personnel, and installers. Contains both theoretical and practical descriptions.

Home Appliance Servicing (Fourth Edition)

EDWIN P. ANDERSON;
revised by REX MILLER

5 1/2 × 8 1/4 Hardcover 640 pp. 345 Illus.
ISBN: 0-672-23379-7 $22.50

The essentials of testing, maintaining, and repairing all types of home appliances.

Television Service Manual (Fifth Edition)

ROBERT G. MIDDLETON;
revised by JOSEPH G. BARRILE

5 1/2 × 8 1/4 Hardcover 512 pp. 395 Illus.
ISBN: 0-672-23395-9 $16.95

A guide to all aspects of television transmission and reception, including the operating principles of black and white and color receivers. Step-by-step maintenance and repair procedures.

Electrical Course for Apprentices and Journeymen (Third Edition)

ROLAND E. PALMQUIST

5 1/2 × 8 1/4 Hardcover 478 pp. 290 Illus.
ISBN: 0-02-594550-5 $19.95

This practical course in electricity for those in formal training programs or learning on their own provides a thorough understanding of operational theory and its applications on the job.

Questions and Answers for Electricians Examinations (1993 NEC Rulings Included)

revised by PAUL ROSENBERG

5 1/2 × 8 1/4 Paperback 270 pp.
100 line drawings
ISBN: 0-02-077762-0 $20.00

An Audel classic, considered the most thorough work on the subject in coverage and content. This fully revised edition is based on the 1993 National Electrical Code®, and is written for anyone preparing for the various electricians' examinations—Apprentice, Journeyman, or Master. It provides the license applicant with an understanding of theory as well as of all definitions, specifications, and regulations included in the new NEC.

MACHINE SHOP AND MECHANICAL TRADES

Machinists Library (Fourth Edition, 3 Vols.)

REX MILLER

5 1/2 × 8 1/4 Hardcover 1,352 pp. 1120 Illus.
ISBN: 0-672-23380-0 $52.95

An indispensable three-volume reference set for machinists, tool and die makers, machine operators, metal workers, and those with home workshops. The principles and methods of the entire field are covered in an up-to-date text, photographs, diagrams, and tables.

Volume I: Basic Machine Shop

REX MILLER

5 1/2 × 8 1/4 Hardcover 392 pp. 375 Illus.
ISBN: 0-672-23381-9 $17.95

Volume II: Machine Shop

REX MILLER

5 1/2 × 8 1/4 Hardcover 528 pp. 445 Illus.
ISBN: 0-672-23382-7 $19.95

Volume III: Toolmakers Handy Book

REX MILLER

5 1/2 × 8 1/4 Hardcover 432 pp 300 Illus.
ISBN: 0-672-23383-5 $14.95

Mathematics for Mechanical Technicians and Technologists

JOHN D. BIES

5 1/2 × 8 1/4 Hardcover 342 pp. 190 Illus.
ISBN: 0-02-510620-1 $17.95

The mathematical concepts, formulas, and problem-solving techniques utilized on the job by engineers, technicians, and other workers in industrial and mechanical technology and related fields.

Millwrights and Mechanics Guide (Fourth Edition)

CARL A. NELSON

5 1/2 × 8 1/4 Hardcover 1,040 pp. 880 Illus.
ISBN: 0-02-588591-x $29.95

The most comprehensive and authoritative guide available for millwrights, mechanics, maintenance workers, riggers, shop workers, foremen, inspectors, and superintendents on plant installation, operation, and maintenance.

Welders Guide (Third Edition)

JAMES E. BRUMBAUGH

5 1/2 × 8 1/4 Hardcover 960 pp. 615 Illus.
ISBN: 0-672-23374-6 $23.95

The theory, operation, and maintenance of all welding machines. Covers gas welding equipment, supplies, and process; arc welding equipment, supplies, and process; TIG and MIG welding; and much more.

Welders/Fitters Guide

HARRY L. STEWART

8 1/2 × 11 Paperback 160 pp. 195 Illus.
ISBN: 0-672-23325-8 $7.95

Step-by-step instruction for those training to become welders/fitters who have some knowledge of welding and the ability to read blueprints.

Sheet Metal Work

JOHN D. BIES

5 1/2 × 8 1/4 Hardcover 456 pp. 215 Illus.
ISBN: 0-8161-1706-3 $19.95

An on-the-job guide for workers in the manufacturing and construction industries and for those with home workshops. All facets of sheet metal work detailed and illustrated by drawings, photographs, and tables.

Power Plant Engineers Guide
(Third Edition)

FRANK D. GRAHAM;
revised by CHARLIE BUFFINGTON

5 1/2 × 8 1/4 Hardcover 960 pp. 530 Illus.
ISBN: 0-672-23329-0 $27.50

This all-inclusive, one-volume guide is perfect for engineers, firemen, water tenders, oilers, operators of steam and diesel-power engines, and those applying for engineer's and firemen's licenses.

Mechanical Trades Pocket Manual (Third Edition)

CARL A. NELSON

4 × 6 Paperback 364 pp. 255 Illus.
ISBN: 0-02-588665-7 $14.95

A handbook for workers in the industrial and mechanical trades on methods, tools, equipment, and procedures. Pocket-sized for easy reference and fully illustrated.

PLUMBING

Plumbers and Pipe Fitters Library (Fourth Edition, 3 Vols.)

CHARLES N. McCONNELL

5 1/2 × 8 1/4 Hardcover 952 pp. 560 Illus.
ISBN: 0-02-582914-9 $68.45

This comprehensive three-volume set contains the most up-to-date information available for master plumbers, journeymen, apprentices, engineers, and those in the building trades. A detailed text and clear diagrams, photographs, and charts and tables treat all aspects of the plumbing, heating, and air conditioning trades.

Volume I: Materials, Tools, Roughing-In

CHARLES N. McCONNELL;
revised by TOM PHILBIN

5 1/2 × 8 1/4 Hardcover 304 pp. 240 Illus.
ISBN: 0-02-582911-4 $20.95

Volume II: Welding, Heating, Air Conditioning

CHARLES N. McCONNELL;
revised by TOM PHILBIN

5 1/2 × 8 1/4 Hardcover 384 pp. 220 Illus.
ISBN: 0-02-582912-2 $22.95

Volume III: Water Supply, Drainage, Calculations

CHARLES N. McCONNELL;
revised by TOM PHILBIN

5 1/2 × 8 1/4 Hardcover 264 pp. 100 Illus.
ISBN: 0-02-582913-0 $20.95

The Home Plumbing Handbook
(Fourth Edition)

CHARLES N. McCONNELL

8 1/2 × 11 Paperback 224 pp. 210 Illus.
ISBN: 0-02-079651-X $17.00

This handy, thorough volume, a longtime standard in the field with the professional, has been updated to appeal to the do-it-yourself plumber. Aided by the book's many illustrations and manufacturers' instructions, the home plumber is guided through most basic plumbing procedures. All techniques and products conform to the latest changes in codes and regulations.

The Plumbers Handbook
(Eighth Edition)

JOSEPH P. ALMOND, SR.;
revised by REX MILLER

4 × 6 Paperback 368 pp. 170 Illus.
ISBN: 0-02-501570-2 $19.95

Comprehensive and handy guide for plumbers and pipefitters—fits in the toolbox or pocket. For apprentices, journeymen, or experts.

Questions and Answers for Plumbers' Examinations
(Third Edition)

JULES ORAVETZ;
revised by REX MILLER

5 1/2 × 8 1/4 Paperback 288 pp. 145 Illus.
ISBN: 0-02-593510-0 $14.95

Complete guide to preparation for the plumbers' exams given by local licensing authorities. Includes requirements of the National Bureau of Standards.

HVAC

Air Conditioning: Home and Commercial (Fourth Edition)

EDWIN P. ANDERSON;
revised by REX MILLER

5 1/2 × 8 1/4 Hardcover 528 pp. 180 Illus.
ISBN: 0-02-584885-2 $29.95

A guide to the construction, installation, operation, maintenance, and repair of home, commercial, and industrial air conditioning systems.

Heating, Ventilating, and Air Conditioning Library
(Second Edition, 3 Vols.)

JAMES E. BRUMBAUGH

5 1/2 × 8 1/4 Hardcover 1,840 pp. 1,275 Illus.
ISBN: 0-672-23388-6 $53.85

An authoritative three-volume reference library for those who install, operate, maintain, and repair HVAC equipment commercially, industrially, or at home.

Volume I: Heating Fundamentals, Furnaces, Boilers, Boiler Conversions

JAMES E. BRUMBAUGH

5 1/2 × 8 1/4 Hardcover 656 pp. 405 Illus.
ISBN: 0-672-23389-4 $17.95

Volume II: Oil, Gas and Coal Burners, Controls, Ducts, Piping, Valves

JAMES E. BRUMBAUGH

5 1/2 × 8 1/4 Hardcover 592 pp. 455 Illus.
ISBN: 0-672-23390-8 $17.95

Volume III: Radiant Heating, Water Heaters, Ventilation, Air Conditioning, Heat Pumps, Air Cleaners

JAMES E. BRUMBAUGH

5 1/2 × 8 1/4 Hardcover 592 pp. 415 Illus.
ISBN: 0-672-23391-6 $17.95

Oil Burners (Fifth Edition)

EDWIN M. FIELD

5 1/2 × 8 1/4 Hardcover 360 pp. 170 Illus.
ISBN: 0-02-537745-0 $29.95

An up-to-date sourcebook on the construction, installation, operation, testing, servicing, and repair of all types of oil burners, both industrial and domestic.

Refrigeration: Home and Commercial (Fourth Edition)

EDWIN P. ANDERSON;
revised by REX MILLER

5 1/2 × 8 1/4 Hardcover 768 pp. 285 Illus.
ISBN: 0-02-584875-5 $34.95

A reference for technicians, plant engineers, and the homeowner on the installation, operation, servicing, and repair of everything from single refrigeration units to commercial and industrial systems.

PNEUMATICS AND HYDRAULICS

Hydraulics for Off-the-Road Equipment (Second Edition)

HARRY L. STEWART;
revised by TOM PHILBIN

5 1/2 × 8 1/4 Hardcover 256 pp. 175 Illus.
ISBN: 0-8161-1701-2 $13.95

This complete reference manual on heavy equipment covers hydraulic pumps, accumulators, and motors; force components; hydraulic control components; filters and filtration, lines and fittings, and fluids; hydrostatic transmissions; maintenance; and troubleshooting.

Pneumatics and Hydraulics (Fourth Edition)

HARRY L. STEWART;
revised by TOM STEWART

5 1/2 × 8 1/4 Hardcover 512 pp. 315 Illus.
ISBN: 0-672-23412-2 $19.95

The principles and applications of fluid power. Covers pressure, work, and power; general features of machines; hydraulic and pneumatic symbols; pressure boosters; air compressors and accessories; and much more.

Pumps (Fifth Edition)

HARRY L. STEWART;
revised by REX MILLER

5 1/2 × 8 1/4 Hardcover 552 pp. 360 Illus.
ISBN: 0-02-614725-4 $35.00

The practical guide to operating principles of pumps, controls, and hydraulics. Covers installation and day-to-day service.

CARPENTRY AND CONSTRUCTION

Carpenters and Builders Library (Sixth Edition, 4 Vols.)

JOHN E. BALL;
revised by JOHN LEEKE

5 1/2 × 8 1/4 Hardcover 1,300 pp. 988 Illus.
ISBN: 0-02-506455-4 $89.95

This comprehensive four-volume library has set the professional standard for decades for carpenters, joiners, and woodworkers.

Volume 1: Tools, Steel Square, Joinery

JOHN E. BALL;
revised by JOHN LEEKE

5 1/2 × 8 1/4 Hardcover 377 pp. 340 Illus.
ISBN: 0-02-506451-7 $21.95

Volume 2: Builders Math, Plans, Specifications

JOHN E. BALL;
revised by JOHN LEEKE

5 1/2 × 8 1/4 Hardcover 319 pp. 200 Illus.
ISBN: 0-02-506452-5 $21.95

Volume 3: Layouts, Foundation, Framing

JOHN E. BALL;
revised by JOHN LEEKE

5 1/2 × 8 1/4 Hardcover 269 pp. 204 Illus.
ISBN: 0-02-506453-3 $21.95

Volume 4: Millwork, Power Tools, Painting

JOHN E. BALL;
revised by JOHN LEEKE

5 1/2 × 8 1/4 Hardcover 335 pp. 244 Illus.
ISBN: 0-02-506454-1 $21.95

Complete Building Construction (Second Edition)

JOHN PHELPS;
revised by TOM PHILBIN

5 1/2 × 8 1/4 Hardcover 744 pp. 645 Illus.
ISBN: 0-672-23377-0 $22.50

Constructing a frame or brick building from the footings to the ridge. Whether the building project is a tool shed, garage, or a complete home, this single fully illustrated volume provides all the necessary information.

Complete Roofing Handbook (Second Edition)

JAMES E. BRUMBAUGH
revised by JOHN LEEKE

5 1/2 × 8 1/4 Hardcover 536 pp. 510 Illus.
ISBN: 0-02-517851-2 $30.00

Covers types of roofs; roofing and reroofing; roof and attic insulation and ventilation; skylights and roof openings; dormer construction; roof flashing details; and much more. Contains new information on code requirements, underlaying, and attic ventilation.

Complete Siding Handbook (Second Edition)

JAMES E. BRUMBAUGH
revised by JOHN LEEKE

5 1/2 × 8 1/4 Hardcover 440 pp. 320 Illus.
ISBN: 0-02-517881-4 $30.00

This companion volume to the *Complete Roofing Handbook* has been updated to reflect current emphasis on compliance with building codes. Contains new sections on spunbound olefin, building papers, and insulation materials other than fiberglass.

Masons and Builders Library
(Second Edition, 2 Vols.)

LOUIS M. DEZETTEL;
revised by TOM PHILBIN

5 1/2 × 8 1/4 Hardcover 688 pp. 500 Illus.
ISBN: 0-672-23401-7 $27.95

This two-volume set provides practical instruction in bricklaying and masonry. Covers brick; mortar; tools; bonding; corners, openings, and arches; chimneys and fireplaces; structural clay tile and glass block; brick walls; and much more.

Volume 1: Concrete, Block, Tile, Terrazzo

LOUIS M. DEZETTEL;
revised by TOM PHILBIN

5 1/2 × 8 1/4 Hardcover 304 pp. 190 Illus.
ISBN: 0-672-23402-5 $14.95

Volume 2: Bricklaying, Plastering, Rock Masonry, Clay Tile

LOUIS M. DEZETTEL;
revised by TOM PHILBIN

5 1/2 × 8 1/4 Hardcover 384 pp. 310 Illus.
ISBN: 0-672-23403-3 $14.95

WOODWORKING

Wood Furniture: Finishing, Refinishing, Repairing
(Third Edition)

JAMES E. BRUMBAUGH
revised by JOHN LEEKE

5 1/2 × 8 1/4 Hardcover 384 pp. 190 Illus.
ISBN: 0-02-517871-7 $25.00

A fully illustrated guide to repairing furniture and finishing and refinishing wood surfaces. Covers tools and supplies; types of wood; veneering; inlaying; repairing, restoring and stripping; wood preparation; and much more. Contains a new color insert on stains.

Woodworking and Cabinetmaking

F. RICHARD BOLLER

5 1/2 × 8 1/4 Hardcover 360 pp. 455 Illus.
ISBN: 0-02-512800-0 $18.95

Essential information on all aspects of working with wood. Step-by-step procedures for woodworking projects are accompanied by detailed drawings and photographs.

MAINTENANCE AND REPAIR

Building Maintenance
(Second Edition)

JULES ORAVETZ

5 1/2 × 8 1/4 Paperback 384 pp. 210 Illus.
ISBN: 0-672-23278-2 $11.95

Professional maintenance procedures used in office, educational, and commercial buildings. Covers painting and decorating; plumbing and pipe fitting; concrete and masonry; and much more.

Gardening, Landscaping and Grounds Maintenance
(Third Edition)

JULES ORAVETZ

5 1/2 × 8 1/4 Hardcover 424 pp. 340 Illus.
ISBN: 0-672-23417-3 $15.95

Maintaining lawns and gardens as well as industrial, municipal, and estate grounds.

Home Maintenance and Repair: Walls, Ceilings and Floors

GARY D. BRANSON

8 1/2 × 11 Paperback 80 pp. 80 Illus.
ISBN: 0-672-23281-2 $6.95

The do-it-yourselfer's guide to interior remodeling with professional results.

Painting and Decorating

REX MILLER and GLEN E. BAKER

5 1/2 × 8 1/4 Hardcover 464 pp. 325 Illus.
ISBN: 0-672-23405-x $18.95

A practical guide for painters, decorators, and homeowners to the most up-to-date materials and techniques in the field.

Tree Care (Second Edition)

JOHN M. HALLER

8 1/2 × 11 Paperback 224 pp 305 Illus.
ISBN: 0-02-062870-6 $16.95

The standard in the field. A comprehensive guide for growers, nursery owners, foresters, landscapers, and homeowners to planting, nurturing, and protecting trees.

Upholstering
(Third Edition)

JAMES E. BRUMBAUGH

5 1/2 × 8 1/4 Hardcover 416 pp. 318 Illus.
ISBN: 0-02-517862-8 $25.00

The essentials of upholstering are fully explained and illustrated for the professional, the apprentice, and the hobbyist. Features a new color insert illustrating fabrics, a new chapter on embroidery, and an expanded cleaning section.

AUTOMOTIVE AND ENGINES

Diesel Engine Manual
(Fourth Edition)

PERRY O. BLACK;
revised by WILLIAM E. SCAHILL

5 1/2 × 8 1/4 Hardcover 512 pp. 255 Illus.
ISBN: 0-672-23371-1 $15.95

The principles, design, operation, and maintenance of today's diesel engines. All aspects of typical two- and four-cycle engines are thoroughly explained and illustrated by photographs, line drawings, and charts and tables.

Gas Engine Manual
(Third Edition)

EDWIN P. ANDERSON;
revised by CHARLES G. FACKLAM

5 1/2 × 8 1/4 Hardcover 424 pp. 225 Illus.
ISBN: 0-8161-1707-1 $12.95

How to operate, maintain, and repair gas engines of all types and sizes. All engine parts and step-by-step procedures are illustrated by photographs, diagrams, and troubleshooting charts.

Small Gasoline Engines

REX MILLER and
MARK RICHARD MILLER

5 1/2 × 8 1/4 Hardcover 640
ISBN: 0-672-23414-9

Practical inform... maintain... -cycle en... ...ers, edgers, ...owers, emergency e... ...rs, outboard motors, and ot... ...ment with engines of up to ten hor...power.

Truck Guide Library (3 Vols.)

JAMES E. BRUMBAUGH

5 1/2 × 8 1/4 Hardcover 2,144 pp. 1,715 Illus.
ISBN: 0-672-23392-4 $50.95

This three-volume set provides the most comprehensive, profusely illustrated collection of information available on truck operation and maintenance.

Volume 1: Engines

JAMES E. BRUMBAUGH

5 1/2 × 8 1/4 Hardcover 416 pp. 290 Illus.
ISBN: 0-672-23356-8 $16.95

Volume 2: Engine Auxiliary Systems

JAMES E. BRUMBAUGH

5 1/2 × 8 1/4 Hardcover 704 pp. 520 Illus.
ISBN: 0-672-23357-6 $16.95

Volume 3: Transmissions, Steering, and Brakes

JAMES E. BRUMBAUGH

5 1/2 × 8 1/4 Hardcover 1,024 pp. 905 Illus.
ISBN: 0-672-23406-8 $16.95

DRAFTING

Industrial Drafting

JOHN D. BIES

5 1/2 × 8 1/4 Hardcover 544 pp. Illus.
ISBN: 0-02-510610-4 $24.95

Professional-level introductory guide for practicing drafters, engineers, managers, and technical workers in all industries who use or prepare working drawings.

Answers on Blueprint Reading
(Fourth Edition)

ROLAND PALMQUIST;
revised by THOMAS J. MORRISEY

5 1/2 × 8 1/4 Hardcover 320 pp. 275 Illus.
ISBN: 0-8161-1704-7 $12.95

Understanding blueprints of machines and tools, electrical systems, and architecture. Question and answer format.

HOBBIES

Complete Course in Stained Glass

PEPE MENDEZ

8 1/2 × 11 Paperback 80 pp. 50 Illus.
ISBN: 0-672-23287-1 $8.95

The tools, materials, and techniques of the art of working with stained glass.